# American Educational Research Journal
Volume 57, Number 3—June 2020

947     Effects of a Science of Learning Course on College Students' Learning With a Computer
           *Jeffrey A. Greene, Nikki G. Lobczowski, Rebekah Freed, Brian M. Cartiff, Cynthia Demetriou, and A. T. Panter*

979     Teacher Responses to New Pedagogical Practices: A Praxeological Model for the Study of Teacher-Driven School Development
           *Sławomir Krzychała*

1014    Speaking Volumes: Professional Development Through Book Studies
           *Betty S. Blanton, Amy D. Broemmel, and Amanda Rigell*

1045    Bridging the Gap Between Research and Practice: Predicting What Will Work Locally
           *Kathryn E. Joyce and Nancy Cartwright*

1083    Parent Engagement and Satisfaction in Public Charter and District Schools
           *Zachary W. Oberfield*

1125    Pedagogy and Profit? Efforts to Develop and Sell Digital Courseware Products for Higher Education
           *Matthew D. Regele*

1159    "Dear Future President of the United States": Analyzing Youth Civic Writing Within the 2016 Letters to the Next President Project
           *Antero Garcia, Amber Maria Levinson, and Emma Carene Gargroetzi*

1203    Fostering Democratic and Social-Emotional Learning in Action Civics Programming: Factors That Shape Students' Learning From *Project Soapbox*
           *Molly W. Andolina and Hilary G. Conklin*

1241    Dynamics of Reflective Assessment and Knowledge Building for Academically Low-Achieving Students
           *Yuqin Yang, Jan van Aalst, and Carol K. K. Chan*

1290    Differences at the Extremes? Gender, National Contexts, and Math Performance in Latin America
           *Ran Liu, Andrea Alvarado-Urbina, and Emily Hannum*

1323    The Moderating Effect of Neighborhood Poverty on Preschool Effectiveness: Evidence From the Tennessee Voluntary Prekindergarten Experiment
           *Francis A. Pearman II*

1358    Giving Community College Students Choice: The Impact of Self-Placement in Math Courses
           *Holly Kosiewicz and Federick Ngo*

1392    Will Mentoring a Student Teacher Harm My Evaluation Scores? Effects of Serving as a Cooperating Teacher on Evaluation Metrics
           *Matthew Ronfeldt, Emanuele Bardelli, Stacey L. Brockman, and Hannah Mullman*

*All articles were accepted under the coeditorship of Mark Berends, Francesca López, Julianne C. Turner, and Suzanne Wilson.*

The *American Educational Research Journal* (*AERJ*) is the flagship journal of the American Educational Research Association, featuring articles that advance the empirical, theoretical, and methodological understanding of education and learning. It publishes original peer-reviewed analyses that span the field of education research across all subfields and disciplines and all levels of analysis. It also encourages submissions across all levels of education throughout the life span and all forms of learning. *AERJ* welcomes submissions of the highest quality, reflecting a wide range of perspectives, topics, contexts, and methods, including interdisciplinary and multidisciplinary work.

*American Educational Research Journal* (ISSN 0002-8312) (J589) is published bimonthly—in February, April, June, August, October, and December—on behalf of the American Educational Research Association, 1430 K Street NW, Washington, DC 20005, by SAGE Publishing, 2455 Teller Road, Thousand Oaks, CA 91320. Periodicals postage paid at Thousand Oaks, California, and at additional mailing offices. POSTMASTER: Send address changes to AERA Membership Department, 1430 K Street NW, Washington, DC 20005.

Copyright © 2020 by American Educational Research Association. All rights reserved. No portion of the contents may be reproduced in any form without written permission from the publisher.

**Manuscript Submission:** Authors should submit manuscripts electronically via SAGE Track: http://mc.manuscriptcentral.com/aerj. For more information regarding submission guidelines, please see the Manuscript Submission section of the *American Educational Research Journal*'s website: http://aerj.aera.net.

**Member Information:** American Educational Research Association (AERA) member inquiries, member renewal requests, changes of address, and membership subscription inquiries should be addressed to the AERA Membership Department, 1430 K Street NW, Washington, DC 20005; fax (202) 238-3250; phone (202) 238-3200; e-mail: members@aera.net; website: http://www.aera.net. AERA annual membership dues are $215 (Regular Members), $215 (Affiliate Members), $165 (International Affiliates), $65 (Graduate Students), and $40 (Student Affiliates). **Claims:** Claims for undelivered copies must be made no later than six months following month of publication. Beyond six months and at the request of the American Educational Research Association, the publisher will supply replacement issues when losses have been sustained in transit and when the reserve stock permits.

**Subscription Information:** All nonmember subscription inquiries, orders, back-issue requests, claims, and renewals should be addressed to SAGE Publishing, 2455 Teller Road, Thousand Oaks, CA 91320; telephone (800) 818-SAGE (7243) and (805) 499-0721; fax (805) 375-1700; e-mail journals@sagepub.com; http://journals.sagepub.com. **Subscription Price:** Institutions $1,284; Individuals $85. For all customers outside the Americas, please visit http://www.sagepub.co.uk/customerCare.nav for information. **Claims (nonmembers):** Claims for undelivered copies must be made no later than six months following month of publication. The publisher will supply replacement issues when losses have been sustained in transit and when the reserve stock will permit.

**Copyright Permission:** To request permission for republishing, reproducing, or distributing material from this journal, please visit the desired article on the SAGE Journals website (journals.sagepub.com) and click "Permissions." For additional information, please see www.sagepub.com/journals-permissions.

**Advertising and Reprints:** Current advertising rates and specifications may be obtained by contacting the advertising coordinator in the Thousand Oaks office at (805) 410-7763 or by sending an e-mail to advertising@sagepub.com. To order reprints, please e-mail reprint@sagepub.com. Acceptance of advertising in this journal in no way implies endorsement of the advertised product or service by SAGE, the journal's affiliated society(ies), or the journal editor(s). No endorsement is intended or implied. SAGE reserves the right to reject any advertising it deems as inappropriate for this journal.

**Change of Address:** Six weeks' advance notice must be given when notifying of change of address. Please send the old address label along with the new address to the AERA Membership Department (address above) to ensure proper identification. Please specify name of journal.

---

*Printed on acid-free paper*

# Effects of a Science of Learning Course on College Students' Learning With a Computer

Jeffrey A. Greene
Nikki G. Lobczowski
Rebekah Freed
Brian M. Cartiff
University of North Carolina at Chapel Hill
Cynthia Demetriou
University of Arizona
A. T. Panter
University of North Carolina at Chapel Hill

*First-year courses have been used to bolster college student success, but empirical evidence on their efficacy is mixed. We investigated whether a first-year science of learning course, focused on self-regulated learning, would benefit first-generation college students. We randomly assigned students to a treatment condition involving enrollment in the course, a comparison condition in which students had access to online course materials only, or a control condition. From this larger study, we recruited 43 students to participate in a laboratory task involving learning about the circulatory system with a computer. We found that treatment and comparison students experienced greater changes in conceptual knowledge than the control group, and we found differences in the enactment of monitoring and strategy use across conditions.*

KEYWORDS: computers, first-generation college students, science of learning, self-regulated learning

On average, individuals with a postsecondary degree earn more than their peers who lack one (Kena et al., 2015). Unfortunately, institutions of higher education in the United States continue to struggle to retain and graduate their students (Redford & Hoyer, 2017). Retention and graduation rates are particularly low for first-generation college students (FGCS; Lohfink & Paulsen, 2005; Redford & Hoyer, 2017), typically defined in the United States as those students whose parents have no postsecondary experience. Research has shown that FGCS, who on average comprise 33% of 4-year and community college populations, are less likely to be academically prepared for college and graduate at much lower rates than their peers whose parents have some postsecondary experience, despite FCGS' equivalent desire to succeed (Cahalan & Perna, 2015; X. Chen, 2005; Lauff & Ingels, 2013; Redford & Hoyer, 2017). This disparity between desire and result has led educators and researchers to explore multiple ways of supporting FGCS success.

College and university educators have implemented first-year courses in an attempt to bolster students' resilience, motivation, and knowledge, with presumed subsequent effects on retention and academic achievement (Hofer & Yu, 2003; Strayhorn, 2013; Young & Hopp, 2014). A recent meta-analysis of research on these courses revealed that, on average, they had very small effects on GPA ($\delta$ = .02) and retention ($\delta$ = .11) despite their

---

JEFFREY A. GREENE is a professor of educational psychology and learning sciences in the School of Education at the University of North Carolina at Chapel Hill, 101A Peabody Hall, CB#3500, University of North Carolina at Chapel Hill, Chapel Hill, NC 27599; e-mail: *jagreene@email.unc.edu*. His research interests include effective means of promoting learner achievement and digital literacy, particularly via helping students enact self-regulated learning and epistemic cognition. His work spans domains, including online learning, science, and history.

NIKKI G. LOBCZOWSKI is a postdoctoral fellow in the Human-Computer Interaction Institute at Carnegie Mellon University. She completed her PhD in education at the University of North Carolina at Chapel Hill in the Learning Sciences and Psychology Studies Program. Her research interests include collaboration, social regulation of learning of cognition, motivation, emotions, and behaviors; designing technological interventions for K–12 STEM classrooms; and design-based research.

REBEKAH FREED is a doctoral student in the Learning Sciences and Psychological Studies Program in the School of Education at the University of North Carolina at Chapel Hill. She holds a graduate degree in educational psychology and taught university courses for many years. Her research interests include how motivation and goal pursuit enhance self-regulated learning and conceptual understanding of social science.

BRIAN M. CARTIFF is a doctoral student in the Learning Sciences and Psychological Studies Program in the School of Education at the University of North Carolina at Chapel Hill. He has degrees in chemistry and education and was a high school science teacher for more than 20 years. His research interests include epistemic cognition, critical thinking, and nature of science understanding, as well as how these factors influence public understanding of science.

CYNTHIA DEMETRIOU is the Associate Vice Provost for Student Success and Retention Innovation at the University of Arizona where she is also a faculty affiliate in the Center for the Study of Higher Education and the Department of Educational Policy Studies.

A. T. PANTER is the senior associate dean for undergraduate education for the College of Arts and Sciences and a professor of psychology and neuroscience in the L. L. Thurstone Psychometric Laboratory at University of North Carolina at Chapel Hill. She is past president of the American Psychological Association's Division on Quantitative and Qualitative Psychology. She develops instruments, research designs, program evaluations, and data-analytic strategies for applied research questions in higher education, personality, and health.

often-high financial and human resources costs (Permzadian & Credé, 2016). That meta-analysis revealed that courses focused on particular academic topics (e.g., history, science) were less effective than those focused on adjustment and orientation to higher education (e.g., review of campus resources and policies, instruction on study skills, time management, and learning strategies). The higher efficacy of courses focused on helping students "learn how to learn" (Hofer & Yu, 2003, p. 30) is not surprising when viewed through the lens of theory and research on self-regulated learning (SRL; Greene, 2018; Zimmerman, 2013). Self-regulated learners have indeed learned how to learn: They are able to pursue valued academic goals via planning, monitoring, controlling, and evaluating their cognition, motivation, behavior, and emotions, such as by using effective strategies and knowing how to self-motivate. The knowledge, skills, and dispositions to enact SRL are not intuitive or innate but can be acquired (Bjork, Dunlosky, & Kornell, 2013) and are predictive of academic performance (Dent & Koenka, 2016; Dignath & Büttner, 2008). Early research on SRL focused on its effects on achievement outcomes, but more recently researchers and educators have also become interested in SRL as a process (Azevedo, 2005; Greene et al., 2015; Schunk & Greene, 2018; Winne & Perry, 2000) and how students' enactment of SRL processes can be influenced (Dignath & Büttner, 2008; Graham, Harris, MacArthur, & Santangelo, 2018). Research findings suggest that the effects of first-year courses on retention and achievement may be significantly mediated by the degree to which they enhance students' ability to self-regulate their learning (Paris & Paris, 2001; Zimmerman, 2013); therefore, SRL may be a particularly important predictor of FGCS success.

In this study, which is part of a larger project on the short-term and long-term outcomes of a first-year science of learning course for FGCS (e.g., Hofer & Yu, 2003), we investigated whether this course had effects on a very specific, but common, aspect of the college student experience: learning with a computer-based learning environment (CBLE; Azevedo, 2005). We randomly assigned students to a treatment condition (i.e., enrollment in a face-to-face science of learning course), a comparison condition (i.e., access to readings from the course but no enrollment in the actual course), and a control condition (i.e., business as usual with no access to course or materials). Then, at the end of the semester, we recruited participants to engage in a laboratory study in which they were asked to think aloud (Ericsson & Simon, 1993; Greene, Deekens, Copeland, & Yu, 2018) as they used a computer to learn about the circulatory system. The think-aloud protocol (TAP) data allowed us to capture the kinds of SRL processing participants enacted, and we used pretest and posttest measures to assess changes in declarative and conceptual knowledge as a result of using the computer. We hypothesized that treatment participants would show greater changes in knowledge scores from pretest to posttest than from comparison or control participants. Also, we investigated how SRL processing differed across these groups and how this processing predicted changes in knowledge over the course of the task. These data afforded a window into both product (i.e., changes in knowledge) and process (i.e., TAP) effects of the science of learning course and shed light on how such courses change the ways in which FGCS enact the knowledge, skills, and dispositions necessary for success in higher education (Azevedo, 2005).

## Literature Review

### First-Generation College Students

Although researchers have defined FGCS status in a variety of ways (Demetriou, 2014), it is generally understood in the United States that FGCS are college students whose parents did not attend college or postsecondary education (Demetriou, Meece, Eaker-Rich, & Powell, 2017). Researchers and educators have endeavored to understand how to support FGCS and increase the likelihood of their college success (Perna & Thomas, 2006). Supporting FGCS benefits them during college as well as after they graduate (Garriott, Hudyma, Keene, & Santiago, 2015). Understanding the characteristics, challenges, and assets unique to FGCS is essential to determining how to support them effectively.

In terms of background, FGCS are more likely than non-FGCS, known as continuing-generation college students (CGCS), to be older, married, part-time students; live off campus; receive financial aid; and work full time while enrolled in school (Demetriou et al., 2017; Pike & Kuh, 2005). FGCS more often come from diverse sociocultural and socioeconomic backgrounds compared with CGCS (Demetriou, 2014). FGCS are more likely to be from an underrepresented ethnic background and to speak a home language other than English (Bui, 2002; Redford & Hoyer, 2017). FGCS often find differences between the sociocultural norms and values of their homes and their K-12 context, as well as their higher education context (Stephens, Markus, Fryberg, Johnson, & Covarrubias, 2015). These differences make it more difficult for FGCS to acquire college readiness knowledge and skills (e.g., SRL) during their K-12 education (Gamez-Vargas & Oliva, 2013; Stephens et al., 2015).

Studies suggest that FGCS are less academically prepared for college than CGCS (Choy, 2001; Redford & Hoyer, 2017). FGCS report feeling less prepared for college and fear failing in college more than CGCS (Pike & Kuh, 2005). They are more likely to have lower scores on the SAT, take fewer college credits, receive lower grades, need more remedial assistance, and withdraw from or repeat courses they attempt, as compared with CGCS (X. Chen, 2005). However, FGCS do not differ from CGCS in terms of their desire to succeed in college (Lauff & Ingels, 2013). Also, FGCS report knowing less about college social environments than their CGCS peers, which can lead to stress (Pike & Kuh, 2005).

Whereas many FGCS experience challenges in their transition to college, they are also likely to possess unique strengths. For example, in one study, FGCS of color from a low socioeconomic status demonstrated more resilience in the face of challenges than CGCS (Morales, 2014). As FGCS come from a more diverse socioeconomic and sociocultural background, they also demonstrate greater diversity in their goals (e.g., more group and community-related goals) that assist in completing the different types of coursework in college (Stephens et al., 2015). Depending on the context the students found themselves in, identifying as FGCS was also a source of strength for them in school (McKinley & Brayboy, 2004). In conclusion, FGCS who get into college certainly have the potential to succeed, but they are less likely to persist to graduation, on average, for a variety of reasons. It is important to provide environmental supports that help them succeed in college, so that they can persist to reach their educational goals. One support that universities can provide for FGCS is first-year courses.

**First-Year Courses**

Administrators and educators at many institutions of higher education have turned to first-year courses in their attempt to improve the academic performance and retention of their students. Retention benefits these institutions in terms of their academic mission of promoting student success, but also financially; replacing students who drop out costs more money than retaining them, and students doing better academically need fewer support services (Permzadian & Credé, 2016). Unfortunately, the empirical evidence regarding the efficacy of first-year courses, in terms of students' retention and academic performance, is mixed (Permzadian & Credé, 2016).

*Four Main Types of First-Year Courses*

Based on the literature, there are four main types of first-year courses: transition, academic themed, discipline themed, and remedial themed (Barefoot, 1992; Porter & Swing, 2006). Most institutions offer only one type of course, whereas some institutions provide a mixed format, in which students can choose from different types of seminar courses (Porter & Swing, 2006). Instructors leading first-year transition courses intend to help students acclimate to college life. Topics covered in these courses typically include study skills, connecting with faculty and staff, orientation to the campus and college life, and personal wellness (Rogerson & Poock, 2013). Findings on the effectiveness of this type of seminar are varied. Barton and Donahue (2009) found that enrollment in a first-year seminar transition course led to increases in GPA (Cohen's $d$ = 0.19 as calculated based on reported $t$ and $df$), but not retention, as compared with students who chose to take a shortened orientation course (i.e., either 1 week during the summer or a one-credit course during the fall semester). Cambridge-Williams, Winsler, Kitsantas, and Bernard (2013), on the other hand, found that students enrolled in transition courses had higher long-term academic retention (i.e., 4-year retention rate for treatment students was 75.0% compared with 59.9% for control), and higher graduation rates (i.e., 7-year graduation rate for treatment students was 68.7% compared with 55.9% for control), than students not enrolled in the courses. They also found that students self-reported higher short-term self-efficacy and SRL skills. Additionally, Miller and Lesik (2014) found evidence of higher retention for students in transition courses compared with students not enrolled in these courses, but only for retention during the students' first year (i.e., transition course students' odds of dropout were 40% compared with those of peers who did not participate in transition course) and graduation in their fourth year (i.e., transition course students were 1.9 times more likely to graduate).

A variety of instructors, including graduate students and full-time faculty from many disciplines, teach academic-themed first-year seminars. For these seminars, the instructors usually focus on an academic topic of their choice but are required to integrate common themes into their lessons (e.g., how to be reflective learners; Zerr & Bjerke, 2016). Research on this type of seminar is limited, but Zerr and Bjerke (2016) notably compared these academic-themed first-year courses with first-year transition courses and found no differences in student retention or GPA, but they did find higher student engagement. They also found that students rated these seminars negatively due to the amount of work required.

Discipline-themed first-year seminars offer students an introduction to a specific major or academic area (e.g., introduction to computer science), or preprofessional skills (e.g., public speaking or leadership; Black, Terry, & Buhler, 2016). Black et al. (2016) studied a variety of discipline-themed first-year courses and found that students enrolled in the specialized (e.g., introduction to computer science) and business-themed courses had statistically significantly higher retention rates from the fall to the spring semester than students enrolled in the generalized (e.g., elementary group dynamics) and English courses, as well as any courses tailored to transfer students.

Remedial-themed first-year courses focus on study skills and adjustment to college (e.g., time management, test preparation, and career planning). The main difference between remedial-themed and other types of seminars that also teach skills and adjustment techniques (e.g., transition seminars) is the type of students enrolled in the course. Namely, institution officials usually reserve remedial-themed first-year courses for at-risk, transfer, or struggling students (Forster, Swallow, Fodor, & Fousler, 1999). Researchers have reported that students enrolled in remedial courses exhibit higher grades than would be predicted based on their previous performance (Cone & Owens, 1991). These students also had higher retention rates (i.e., an increase in retention rate from 7% to 53%, Cone, 1991; Forster et al., 1999) and gains in study skills (Forster et al., 1999) compared with other at-risk cohorts that were not offered the courses.

*Comparison of the Different Types of First-Year Seminars*

Although many researchers have looked at these various types of courses separately, some have looked across the different types to determine which is the most effective. In a meta-analysis focused on first-year courses at the collegiate level, Hattie, Biggs, and Purdie (1996) investigated the effectiveness of teaching students learning or study skills, which were a central component to several different types of courses. They found that the overall effect of these courses at the university level (i.e., Cohen's $d$ = 0.27) was below the average effect of other typical interventions in education. Despite more positive attitudes as a result of enrollment in these courses, students did not demonstrate substantive performance gains. The authors concluded that study skills training was relatively ineffective.

In another meta-analysis, Permzadian and Credé (2016) used categorizations from Barefoot (1992) to code seminars as extended orientation, academic, or hybrid. Unfortunately, they excluded seminars focused on study skills because Hattie et al. (1996) already had discussed the overall effectiveness of these types of courses. Overall, they found that first-year courses had almost no effect on first-year GPA ($\delta$ = .02) and only a small effect on 1-year retention rates ($\delta$ = .11). However, they found variance in efficacy. Using moderator analyses, they found the characteristics that had the largest effects on GPA were hybrid courses (i.e., a combination of extended orientation and academic content, $\delta$ = .11), implementation at a 2-year institution ($\delta$ = .22), studies published in a peer-reviewed publication ($\delta$ = .09), and randomized ($\delta$ = .40) or ex post facto with matching research designs ($\delta$ = .11). For 1-year retention, the characteristics that produced the largest effects were extended orientation ($\delta$ = .12), stand-alone courses (i.e., compared with those connected to other courses as part of a learning community, $\delta$ = .12), courses taught by faculty or staff (i.e., compared with classes taught by graduate students, $\delta$ = .10), and courses that targeted all first-year students ($\delta$ = .12). Finally, they hypothesized that the low overall effect sizes may have been due to a lack of randomized experiments on these courses.

*A Fifth Type of First-Year Seminar*

Missing from these meta-analyses, however, is a fifth type of first-year seminar: learning to learn (i.e., science of learning). Even though these courses share some topics with transition courses (e.g., test-taking strategies, metacognitive skills), they are entrenched in educational and cognitive psychology research, including work on learning and memory (Hofer & Yu, 2003). These courses are typically inclusive of all students (i.e., not just remedial students). Although the literature on this type of course is limited, there are promising findings. For example, Tuckman and Kennedy (2011) found that students enrolled in learning to learn courses had higher GPAs ($\gamma = .11$), retention rates, and graduation rates (i.e., expected odds of graduation for students in the course were 1.69 times greater than control) than those first-year students not enrolled in the course. Hofer and Yu (2003) found that students in these courses reported lower test anxiety by the end of the semester and "significant increases in . . . memorization, elaboration, organization, deep processing, planning, and metacognition" (p. 31, Cohen's $d$ ranged from 0.62 to 1.22). Finally, Wingate (2007) suggested that most study skills seminars do not support students holistically (i.e., being able to apply these skills to all courses) and that their focus should shift to learning how to learn (i.e., SRL), thereby changing students' "perceptions, learning habits, and epistemological beliefs" (p. 395).

**Self-Regulated Learning**

SRL refers to how "learners systematically activate and sustain their cognitions, motivations, behaviors, and affects" (Schunk & Greene, 2018, p. 1) toward attaining their personal learning goals. Quite a few distinct models of SRL have been proposed, but most of them have common assumptions: (1) learners are active participants in the construction of the knowledge; (2) learners have the abilities to observe and alter different aspects of their learning, as necessary; (3) learners enact the abilities in (2) based on their goals, criteria, or standards; and (4) that SRL processing serves as a mediator in the relationship between personal or contextual characteristics and learning outcomes (Pintrich, 2000). Most SRL models (e.g., Winne & Hadwin, 2008; Zimmerman, 2013) also include phases of SRL in which learners enact processes preceding learning (e.g., self-motivating, planning), during learning (e.g., monitoring progress against standards and toward goal achievement, changing plans or strategies as needed), and then after learning (e.g., reflecting, evaluating; Greene, 2018).

Research in SRL has developed greatly over the past 40 years (Winne, 2018), including investigations into the various kinds and aspects of cognition, metacognition, motivation, and strategy use that are critical for learning (Vandevelde, Van Keer, Schellings, & Van Hout-Wolters, 2015). These studies have been conducted with populations ranging from graduate students (Mullen, 2011; Whipp & Chiarelli, 2004), to undergraduate students (Kauffman, 2004; Moos & Azevedo, 2008), high school students (Winters & Azevedo, 2005), middle school students (Cleary & Kitsantas, 2017; Eom & Reiser, 2004; Kramarski & Mizrachi, 2006), and, more recently, elementary populations (Neitzel & Connor, 2017; Vandevelde et al., 2015), with the vast majority of these investigations revealing high correlations between effective SRL and learning performance. Also, research has shown strong relations between SRL and students' knowledge gains when using CBLEs to acquire

## Effects of a Science of Learning Course

conceptual understanding via text-based and multimodal information sources (Azevedo, 2005; Dent & Koenka, 2016; Greene et al., 2015; Winters, Greene, & Costich, 2008). However, much of this research has been correlational in nature, precluding claims of causality.

The vast majority of empirical research has been focused on the predictive validity of various SRL processes, either individually (e.g., practice testing; Adesope, Trevisan, & Sundararajan, 2017) or aggregated in some manner (Cleary & Kitsantas, 2017; Deekens, Greene, & Lobczowski, 2018). Less has been done to investigate the relationships between various SRL processes using objective measures, such as observation and performance (Ben-Eliyahu & Bernacki, 2015), though existent findings show that learners who accurately evaluate and calibrate their comprehension (Alexander, 2013) have greater tendencies to recognize ineffective learning strategies and switch to new ones (Binbasaran Tüysüzoğlu & Greene, 2015; Dunlosky & Thiede, 2013). Likewise, Deekens et al. (2018) found more frequent enactment of monitoring processes predicted more frequent use of deep-level strategies (i.e., strategies that foster elaboration and recall), which in turn predicted learning performance, above and beyond the effect of prior knowledge. Clearly, there is a need for further research on the role of conditional, contingent, and adaptive SRL processing (e.g., planning, monitoring, enacting of strategies, and evaluating progress and learning; Ben-Eliyahu & Bernacki, 2015) in learning how to help students enact those processes efficiently and effectively, and how changes in SRL processing relate to academic outcomes (Schunk & Greene, 2018).

Intervention research has revealed that SRL can be taught within courses and that students do seem to benefit from such instruction (e.g., P. Chen, Chavez, Ong, & Gunderson, 2017; Zepeda, Richey, Ronevich, & Nokes-Malach, 2015). However, there have not been systematic investigations of whether SRL training effects transfer to other courses or context. If one of the purported goals of SRL research is to help students truly self-regulate, then instruction on SRL should transfer beyond the context in which it was learned (Alexander, Dinsmore, Parkinson, & Winters, 2011). This is one of the goals of first-year learning to learn courses (Hofer & Yu, 2003).

### Measuring Self-Regulated Learning

One limiting factor in the research on how to bolster SRL has been the challenge of effective and efficient measurement of SRL processing. In the past, researchers have measured SRL predominantly by self-report surveys and other instruments administered outside the actual context in which the SRL occurred (i.e., offline measures; Winne & Perry, 2000) such as structured interviews and teacher ratings. These types of measures have their advantages, including the ease with which they can be administered in large-scale testing (Schellings & Van Hout-Wolters, 2011) and their general feasibility, but their asynchronous administration requires the assumption that learners, or their teachers, can accurately recall relevant cognitive, metacognitive, motivational, behaviors, and affective processing. Furthermore, self-report measures present a closed set of response options, generated by researchers, that may not represent all the ways in which learners engage in SRL or present those ways in language that learners can understand. There is evidence that scores on self-report measures may not correspond to actual learner behavior (Veenman, 2007, 2011a, 2011b; Winne & Perry, 2000). In addition, self-report instruments, such as the ubiquitous Motivated Strategies for Learning Questionnaire (Pintrich, Smith, Garcia, & McKeachie, 1991), are too coarse grained (i.e., at a level that does not pick up on contextual details; Karabenick & Zusho, 2015), which has impelled investigators to use online (i.e., capturing SRL as it occurs) methodologies, including trace data (Perry & Winne, 2006), eye tracking (Scheiter, Schubert, & Schüler, 2018; Trevors, Feyzi-Behnagh, Azevedo, & Bouchet, 2016), and verbal report protocols (Azevedo & Cromley, 2004; Greene & Azevedo 2009; Greene, Deekens, et al., 2018).

Ericsson and Simon (1980, 1993) established that TAPs could help researchers gain insight into thinking without being too disruptive. These protocols involve learners verbalizing, but not explaining, all of their thoughts while working on a learning task (e.g., "I didn't understand what I just read, I'm going to read it again"). These data can reveal SRL processing (e.g., judgment of learning). Veenman, Elshout, and Groen (1993) found that TAPs might slow learners' performance in a learning activity, but they do not seem to alter regulatory processing. Subsequent research (e.g., Bannert & Mangelkamp, 2008) has supported these conclusions.

Perhaps more important, many researchers have successfully employed TAPs to reveal relations between online SRL processing and learning outcomes (e.g., Azevedo, Moos, Greene, Winters, & Cromley, 2008; Greene & Azevedo, 2007; Vandevelde et al., 2015). TAPs provide data that can capture SRL's dynamic, conditional, contingent, and contextual aspects (Greene, Deekens, et al., 2018; Ben-Eliyahu & Bernacki, 2015) that retrospective measures such as self-report instruments cannot elicit (Greene et al., 2015). TAPs' open-ended nature also provides richer data than self-report instruments because participants are able to verbalize freely, instead of being restricted to Likert-type items found in surveys (Greene, Robertson, & Costa, 2011). Such data allow students to share what they perceive they are actually doing and thinking, as opposed to being forced to choose among only those activities listed on a self-report measure.

**Purpose of This Study**

Research on academic performance suggests that the majority of college students and, in particular, many FGCS, would benefit from additional instruction in SRL (Greene, 2018; Zimmerman, 2013). First-year seminars can provide an extended, focused opportunity to directly instruct SRL and provide opportunities for practice with support and feedback. In this study, we randomly assigned FGCS to a treatment science of learning course with a strong focus on SRL, a comparison condition in which students had access to course materials but no actual instruction or support, or a business-as-usual control condition. Then, at the end of the semester, we recruited these college students to participate in a laboratory study of their ability to enact SRL while using a computer to learn. We chose this method of data collection to gather multimodal data on how FGCS students transferred their SRL skills to new learning contexts. We had one hypothesis and two research questions. First, we hypothesized that treatment participants would outperform participants in the other groups in terms of their acquisition of declarative and conceptual knowledge as a result of learning with the computer. Our research questions were "In what ways do participants in each group enact SRL processing differently?" and "How does SRL processing relate to conceptual knowledge at posttest?" We expected that participants who took the science of learning course would more frequently enact SRL processing, including more frequent monitoring and use of effective strategies, than students in the comparison or control conditions. We did not have expectations about differences in SRL processing between comparison and control conditions.

## Method

### Participants

During spring 2016, fall 2016, and spring 2017 semesters, we recruited FGCS at a highly selective university in the southeastern United States via email to participate in a study on the effects of a science of learning course on FGCS success. The students who agreed to participate in the study were randomly assigned into the experimental course condition to take the science of learning class, a comparison condition that had access to course materials online but did not enroll in the actual course, or a control condition where they did not take the course or receive access to any course materials. From this larger study including 137 students, we recruited 43 students to participate in the laboratory study described here. Of those 43 students who agreed to participate in the laboratory study, 20 participants were from the science of learning course condition, 11 participants were from the comparison condition, and 12 students were from the control condition. Recruitment for the laboratory portion of our study was a challenge, despite high participant incentives, requiring us to extend data collection across multiple semesters and sections of the science of learning course. Participants' ages ranged from 18 to 35 years, with a mean age of 20.72 years ($SD$ = 3.14). There were 11 male and 32 female participants. Average reported GPA was 3.08 ($SD$ = 0.46), with a range of 1.90 to 3.89. The most common majors were psychology (13), biology (8), and exercise and sport science (6).

**Procedures**

This science of learning course was designed as a learning to learn class (e.g., Hofer & Yu, 2003). The course was part of a U.S. Department of Education (2016) "First in the World" grant, called The Finish Line Project (FLP; https://www2.ed.gov/programs/fitw/index.html). The FLP focused on piloting well-researched programs and supports that increase FGCS success in college and progress to degree. As a part of the FLP grant, this study was advertised to FGCS via email. It was explained to students that if they agreed to participate in this study, they would be randomly placed into one of the three conditions, and that, as they participated in the study, they could be eligible to participate in a secondary research study, which would involve separate compensation. To ensure equitable benefits for all participants, those FGCS randomly assigned to comparison or control conditions were told that they would have priority if they wanted to take the course the following semester. For this laboratory study, students could only participate in their first semester in the larger study (e.g., a student who was in the comparison condition in spring 2016 and did not participate but then was placed in the treatment class in fall 2016 would be ineligible to participate in this study, because they had access to course materials for two semesters).

Toward the end of the semester in which they participated, students were invited to take part in a second laboratory portion of the research study. For this laboratory part of our research, we advertised for participants in each condition differently. In the treatment condition, we advertised by emailing the course students twice, and we advertised in class once. We advertised in the comparison and control conditions by emailing participants twice. We offered monetary compensation, in the form of a gift card redeemable at multiple businesses (e.g., Apple, Target) to incentivize participation in the laboratory study.

*The Science of Learning Course*

The science of learning course met twice a week for 1 hour and 15 minutes each. It was 3 credit hours, and there were between 13 and 20 students over multiple sections of the course and three semesters. In this course, students were expected to gain an understanding of the conceptual and empirical foundations of the science of learning, and they were also asked to apply this understanding to coursework, exams, and their own education. The goals for this course were for the students to be able to critically evaluate learning and education claims in the scholarly literature and media as well as be knowledgeable on how student motivation, deep learning strategies (Dinsmore, 2017), and self-regulation (Zimmerman, 2013) relate to academic success. Topics discussed in the course included academic SRL, internal and external factors of motivation, growth mind-set (i.e., implicit theories of intelligence; Dweck & Master, 2008), the information processing system, knowledge and expertise, intelligence, goal setting, emotions and cognition, technology and learning, sleep, exercise, and deep learning strategies, including spaced versus massed practice and elaboration. Specifically, we provided students with direct instruction on SRL and strategies and how they could use them to succeed in college. We used Zimmerman's (2013) model of SRL as a foundational aspect of the course, in terms of both providing students with an understanding of how to learn more effectively and also as an organizer of other topics in the course such as the roles of monitoring and strategy use in learning. In addition, we gave them opportunities in class to practice the various aspects of SRL during authentic learning activities and provided them with feedback.

We followed numerous procedures to ensure consistency in the course over the three semesters. First, we had the same two instructors teach the course. In the first semester, both instructors taught together (i.e., one instructor of record, one teaching assistant). For the next semester, the two instructors each taught a class separately. In the last semester, one of the instructors taught the course. Next, we created common assessments to use in each course across all three semesters, including weekly quizzes, five homework paper assignments, a midterm exam, and a final exam. Students completed the homework paper assignments independently, reflecting on how the material applied to their own personal and academic experiences (e.g., reflecting on their own SRL skills and identifying strengths and areas for improvement). Finally, the lesson plans were also similar across semesters. These lessons consisted of short lectures, class discussions, partner activities, small group discussions, videos and other media, and formative assessment questions using PowerPoint. Overall, our efforts to ensure consistency across semesters added to the quality and rigor of our research design.

*Laboratory Study Procedures*

To conduct our laboratory study of how students used a CBLE to acquire conceptual understanding, we used a procedure similar to the one used by Azevedo and colleagues (Azevedo, Gutherie, & Seibert, 2004; Azevedo, Johnson, & Burkett, 2015). Each learning session was conducted in a room with one participant and one researcher. First, the participants were informed of the length of time for the learning session and that they were able to opt out at any time without penalty. No participants chose to opt out. On agreeing to continue, the participant read and signed the laboratory study consent form that was approved by the university's institutional review board. Then, the participants had 15 minutes to complete the pretest. Participants were given the instructions to complete each page in the order provided, without skipping ahead or moving backward. The participants received no instructional material during the pretest and they were not told that the posttest would be identical to the pretest.

Next, on completion of the pretest, the participants received a tour of the relevant CBLE pages on heart, blood, and the circulatory system, within Microsoft Encarta (Microsoft Corporation, 2007). Then, participants were taught how to think aloud (Ericsson & Simon, 1993; Greene, Deekens, et al., 2018), including instructions to verbalize all thoughts and reading while interacting with the CBLE. Participants were able to practice thinking aloud within the CBLE using text irrelevant to the learning task, and researchers provided feedback on their verbalizations. Participants were told that they should verbalize but not explain their thinking, per Ericsson and Simon's (1993) protocol.

At this point, the participants were presented with the learning task in written form, which asked them to use the CBLE to learn about the circulatory system. The researcher read this task aloud to them, which included the following in bold: "Make sure you learn about the different parts and their purpose, how they work both individually and together, and how they support the human body." The learning task, which also reminded the participants to think aloud while learning, was posted next to the computer. Participants were given 30 minutes to learn in the CBLE without interruption. Participants were allowed to take notes, but were not required to do so, and informed that they would not be able to use them for the posttest.

The researcher audio- and videotaped the learning session, and we used screen capture software to record the computer screen. If the participants were silent for more than 2 seconds, the researcher prompted them to verbalize their thoughts by saying, "Keep talking, please." Such prompts occurred rarely and sporadically. Additionally, verbal time prompts were given to the participant at 20, 10, and 2 minutes left. After the 30 minutes, all recording devices were turned off, the CBLE was closed, and any notes taken by the participants were collected.

Last, participants were told that they had 15 minutes to complete the written posttest. They were not given access to the learning materials or their notes. On completion of the posttest, participants completed a demographic questionnaire. After the demographic questionnaire was completed, the researcher debriefed each participant about the study and asked each one of them to refrain from sharing the specific details of their participation in the study with other potential participants. In total, participant time in the laboratory study did not exceed 90 minutes.

## Measures

### Knowledge Measures

We collected pretest and posttest knowledge measures based on the work of Azevedo et al. (2004). These included matching (i.e., connecting

14 terms with their appropriate definition) and labeling (i.e., correctly identifying various parts of the heart) tasks, as well as a conceptual essay. For our declarative measures, two trained graduate students scored the matching and labeling portions of each participant's pretest and posttest for accuracy against an answer key. To score the conceptual knowledge measures, our mental model rubric included 12 possible scores that progressed from no understanding to complete understanding of the circulatory system. The 12 scores were (1) no understanding, (2) basic global concept, (3) basic global concept with purpose, (4) basic single-loop model, (5) single-loop model with purpose, (6) advanced single-loop model, (7) single-loop model with lungs, (8) advanced single-loop model with lungs, (9) double-loop concept, (10) basic double-loop model, (11) detailed double-loop model, and (12) advanced double-loop model. The 12 scores can be clustered into three categories (i.e., low, intermediate, and high conceptual understanding) based on understanding of the circulatory system as involving lungs (i.e., intermediate) and consisting of a double loop (i.e., high). Inter-rater reliability was calculated. The overall inter-rater reliability between scorers was 77.34%.

## Think-Aloud Protocol Coding

We used TAPs to capture students' SRL. We transcribed these TAPs and coded them using a codebook and scheme from Azevedo et al. (2008), adapted for this study (see Online Supplementary File A in the online version of the journal). Within our coding scheme, 31 microlevel self-regulatory processes (e.g., taking notes, monitoring use of strategies) could be coded based on transcript content. Segments consisted of a word or a group of words that corresponded to one of the microlevel SRL processes. For example, if a participant stated, "I don't understand that," the statement would be coded as a Judgment of Understanding, with a negative valence (JOU−). If a segment was not codable or multiple microlevel codes could have reasonably been assigned to it, we coded the segment as No Code (NC) and ignored it during the analyses. Each microlevel code also fell within a macrolevel code category, such as planning or strategy use (Greene & Azevedo, 2009). We also aggregated certain microlevel codes into macrolevel deep-level strategy use and surface-level strategy use variables (Dinsmore, 2017; see Online Supplementary File B in the online version of the journal). These macrolevel variables were developed based on evidence in the extant research stating which high-utility strategies facilitate deep-level learning or surface-level learning of the material (Dunlosky, Rawson, Marsh, Nathan, & Willingham, 2013) as well as the research conducted by Deekens et al. (2018). We chose two microlevel SRL codes, Feeling of Recognition, with a negative valence (FOR−), and JOU− to investigate based on Binbasaran Tüysüzoğlu and Greene (2015) and Ben-Eliyahu and Bernacki's (2015) work on the contingent nature of SRL. This work suggests that learners who are more accurate in assessing their

*Greene et al.*

understanding are also more likely to enact metacognitive control and enact effective SRL strategy use (Zimmerman, 2013).

*Demographic Questionnaire*

The demographic questionnaire included questions about participants' gender, age, and academic major. It asked participants to list the biology courses they had previous taken and designate whether or not the circulatory system was covered in each course listed. Last, the questionnaire asked participants to list relevant work experience related to health or medicine.

**Coding Process for TAP Data**

As mentioned previously, each participant transcription was coded twice, independently, by two members of the research team. After coding the transcripts independently, the researchers met to resolve any discrepancies in their coding. The first author was consulted if the two research team members could not come to a consensus on a given code. This type of coding is sometimes referred to as a "two-pass" coding. It is based on work conducted by Chi (1997) and is similar to procedures used by Herrenkohl and Cornelius (2013). Using a two-pass coding takes into account the complexity of coding and ensures that at least two researchers agree on a code for a given segment. Using this procedure ensures greater objectivity via consensus than methods where a single researcher codes a subset of the transcriptions, which is typical in studies that report inter-rater reliability. Additionally, previous research has shown that independent coders can achieve an acceptable level of reliability using this coding scheme (e.g., Greene & Azevedo, 2009).

It has been argued that reliability in coding methods may not be as important as predictive validity (Hammer & Berland, 2014). We assert that statistical calculations of inter-rater agreement are not particularly appropriate for this study. This is because two research team members coded each segment of every transcription independently, reconciled their coding, resolved disagreements, and came to a consensus through a process of consultation. Indeed, measures of reliability are more appropriately used in studies where raters code some subset of the data without verification from another researcher. This was not the procedure for this study. In fact, calculations of SRL TAP data coding agreements are rarely performed. This is due to the complexity of coding, as well as because researchers prioritize accuracy over inter-rater reliability. Though it is rare, when they are calculated, measures of SRL TAP coding agreements tend to be conducted on a very small subset of data (Bannert, Reimann, & Sonnenberg, 2014). In our case, our "two-pass" method, which ensures that two coders agree on every code assigned to every transcription, reflects our emphasis on predictive validity over inter-rater reliability. We privilege this, as demonstrated by our methods, despite its increase on resource demands.

Furthermore, our coding scheme has been used in numerous studies (e.g., Greene, Copeland, Deekens, & Yu, 2018; Greene, Costa, Robertson, Pan, & Deekens, 2010; Greene et al., 2015), with findings aligned with SRL theory, showing support for the predictive validity of the coding. In addition, other research teams have employed coding schemes that were derived from Azevedo and Cromley (2004), as we have here. These numerous studies published employing these coding schemes are more evidence that our coding scheme can be reliably implemented by other research teams (e.g., Greene et al., 2015; Johnson, Azevedo, & D'Mello, 2011; Moos, 2013). An additional source of evidence that these procedures can produce research and inferences with strong validity and reliability has been the recent proliferation of research with TAPs that do not only derive from this work but also utilize similar coding schemes and inter-rater procedures (e.g., De Backer, Van Keer, & Valcke, 2014; Dinsmore & Alexander, 2016).

**Data Analysis**

To analyze the knowledge measure data, first we established that pretest and posttest scores across semesters were not statistically significantly different from each other with an analysis of variance (all $p > .05$). Given these findings, we collapsed the data across semesters and conducted separate mixed analysis of covariance (ANCOVA) analyses (i.e., 2 × 3, pretest-posttest knowledge scores, and three conditions) for each set of knowledge measures: the two declarative knowledge measures (i.e., matching and labeling) and the conceptual knowledge measure (i.e., essay mental model), with students' number of previously-taken relevant courses as a covariate to account for prior experience in the academic domain. To analyze the SRL TAP data, we used measured variable path analysis to investigate how prior knowledge (i.e., pretest scores and number of previous courses taken) predicted monitoring (i.e., FOR− and JOU−) and how those monitoring codes predicted deep-level and surface-level learning strategies and subsequent posttest performance, per Deekens et al. (2018). Also, we investigated whether the frequency of monitoring and strategy use differed by condition.

We treated the knowledge measure data as continuous, normally distributed variables. For the TAP data, we summed the frequency of each participant's codes and then accumulated the total of each microlevel code into corresponding macrolevel aggregated codes (i.e., deep-level and surface-level strategy use). We treated our SRL TAP data as counts in all measured variable path models (Greene, Costa, & Dellinger, 2011).

## Table 1
### Descriptive Statistics for Pretest and Posttests

|  | M (SD) | Range | Skewness (SE) | Kurtosis (SE) |
|---|---|---|---|---|
| Previous courses | 1.65 (1.41) | 0–6 | 1.03 (0.36) | 1.15 (0.71) |
| Pretest matching | 8.12 (3.85) | 1–13 | −0.15 (0.36) | −1.23 (0.71) |
| Pretest labeling | 2.30 (3.11) | 0–12 | 1.51 (0.36) | 1.66 (0.71) |
| Pretest essay | 6.79 (3.14) | 1–12 | 0.11 (0.36) | −0.55 (0.71) |
| Posttest matching | 11.42 (2.39) | 5–13 | −1.41 (0.36) | 0.92 (0.71) |
| Posttest labeling | 7.56 (3.10) | 0–14 | −0.76 (0.36) | 0.25 (0.71) |
| Posttest essay | 9.02 (2.92) | 1–12 | −0.68 (0.36) | −0.10 (0.71) |
| Feeling of recognition minus | 2.07 (2.53) | 0–13 | 2.43 (0.36) | 8.20 (0.71) |
| Judgment of understanding minus | 1.56 (2.14) | 0–8 | 1.50 (0.36) | 1.55 (0.71) |
| Deep-level strategy use | 9.51 (8.75) | 0–42 | 1.74 (0.36) | 3.94 (0.71) |
| Surface-level strategy use | 18.30 (11.05) | 1–49 | 1.06 (0.36) | 1.11 (0.71) |

*Note.* M = mean; SD = standard deviation; SE = standard error.

## Results

### Descriptive Statistics, Correlation Matrix, and Semester Analyses

Through descriptive statistics we found that, on average, the participants' mean declarative (i.e., matching and labeling) and conceptual (i.e., essay) knowledge scores increased from pretest to posttest (see Table 1). During this task, participants verbalized a lack of recognition more often than a lack of understanding and more frequently enacted surface-level strategies than deep-level strategies. From a correlation matrix of the variables (see Table 2), we found strong positive, statistically significant relationships among many of the knowledge measures. Notably, there were correlations between the pretest labeling scores and the number of previous courses taken with similar content, the frequency of negative judgments of understanding and the use of surface-level strategies, the frequency of deep-level strategy use and essay scores, the frequency of negative feelings of recognition and judgments of understanding, and the frequency of deep-level strategy use and negative feelings of recognition. We did not, however, find statistically significant relationships between the frequency of surface-level strategy use and scores on any of the knowledge measures. On the other hand, the frequency of deep-level strategy use was positively, statistically significantly correlated with our posttest conceptual knowledge measure (i.e., essay). Finally, the number of previous relevant courses taken did not correlate with two of our three pretest knowledge measures, suggesting that it might contribute useful additional information beyond our knowledge measures; therefore, we included it as a covariate in our analyses.

Table 2
Correlation Matrix

|   |   | 1 | 2 | 3 | 4 | 5 | 6 | 7 | 8 | 9 | 10 | 11 |
|---|---|---|---|---|---|---|---|---|---|---|----|----|
| 1 | Previous courses | — | 0.24 | .46** | .19 | .26 | .33* | −.22 | −.11 | −.23 | .03 | .05 |
| 2 | Pretest matching |  | — | .59** | .61** | .68** | .61** | .30 | −.13 | −.29 | .05 | −.17 |
| 3 | Pretest labeling |  |  | — | .54** | .38* | .61** | .25 | −.17 | −.25 | .20 | .01 |
| 4 | Pretest essay |  |  |  | — | .51** | .62** | .36* | −.02 | −.25 | .08 | .01 |
| 5 | Posttest matching |  |  |  |  | — | .73** | .22 | −.001 | −.46** | .08 | −.24 |
| 6 | Posttest labeling |  |  |  |  |  | — | .34* | −.06 | −.40** | .25 | −.01 |
| 7 | Posttest essay |  |  |  |  |  |  | — | .09 | −.09 | .31* | −.24 |
| 8 | FOR− |  |  |  |  |  |  |  | — | .33* | .52** | .04 |
| 9 | JOU− |  |  |  |  |  |  |  |  | — | .21 | .40** |
| 10 | Deep-level strategy use |  |  |  |  |  |  |  |  |  | — | .25 |
| 11 | Surface-level strategy use |  |  |  |  |  |  |  |  |  |  | — |

*Note.* FOR = Feeling of Recognition; JOU = Judgment of Understanding. Correlations involving variables with count distributions (i.e., FOR−, JOU−) should be interpreted with caution given that they violate the assumption of a normal distribution.
*$p < .05$. **$p < .01$.

## Knowledge Score Analysis

To investigate our hypothesis, we conducted three repeated-measures ANCOVA analyses (i.e., one for each knowledge measure: matching, labeling, and essay), with knowledge score as the within-subjects factor (i.e., pretest to posttest), condition as a between-subjects factor, and number of relevant previous courses as a covariate. For each analysis, Box's test of equality of covariance matrices was statistically nonsignificant; thus, we found no evidence of any violation of this assumption. In each analysis, there was a statistically significant (all $ps < .001$) and practically significant (all partial $\eta^2 = .384-.649$) within-subjects main effect, indicating that, on average, participants in each condition increased their declarative (i.e., matching and labeling) and conceptual (i.e., essay) knowledge over the course of the task. Previous courses did not have a statistically significant main effect in any ANCOVA, nor was the interaction of this covariate and the within-subjects factor statistically significant except for the conceptual knowledge analysis ($p = .006$, partial $\eta^2 = .177$). Likewise, there were no statistically significant interactions between the within-subjects factor and condition in the declarative knowledge ANCOVAs, but there was a statistically ($p = .040$) and practically (partial $\eta^2 = .152$) significant interaction for the conceptual knowledge ANCOVA.

Examination of adjusted means (see Figure 1) showed that, on average, participants in the control condition possessed unexpectedly high conceptual knowledge of the learning task at pretest, but they did not gain much conceptual knowledge as a result of learning with the computer. Notably, this was not a ceiling effect, as control participants' mean score on the essay posttest was 8.89, well below the maximum possible score of 12 on the measure. On the other hand, both treatment and comparison participants showed strong changes in knowledge from pretest to posttest, with treatment participants' slope greater than that of comparison participants. These differences in conceptual knowledge acquisition begged the question of why or how participants learned during the task, which was investigated using our SRL TAP process data.

## Process Data Analysis

Given the lack of condition differences on the declarative knowledge measures, for the SRL processing analysis we focused solely on the conceptual knowledge measures. Our research question regarding SRL processing was "In what ways do participants in each group enact SRL processing differently and how does SRL processing relate to conceptual knowledge at posttest?" To test relations among knowledge and SRL process variables, we posited a path model similar to the one tested in Deekens et al. (2018). In this model, prior knowledge (i.e., essay pretest) and previous relevant coursework predicted the frequency of monitoring enacted during

*Effects of a Science of Learning Course*

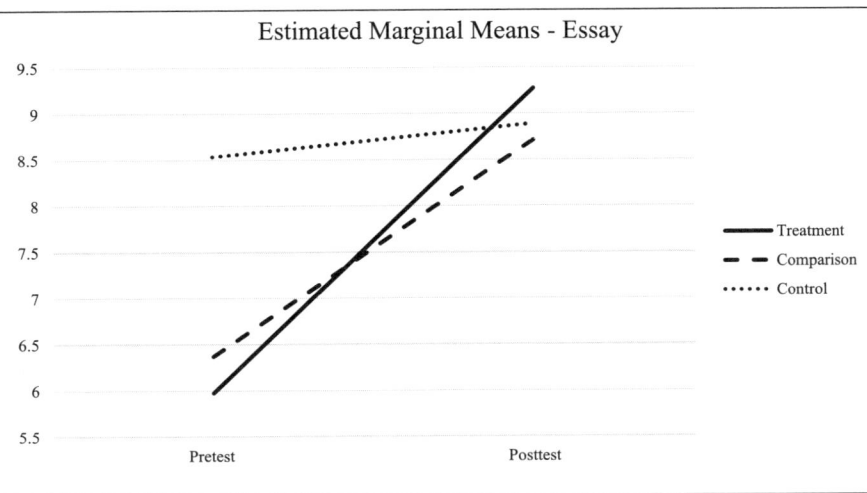

*Figure 1.* **Estimated marginal means for essay scores at pretest and posttest, by condition.**

learning (i.e., FOR− and JOU−), which in turn predicted the frequency of different kinds of strategies enacted (i.e., deep- and surface-level strategy use; Dinsmore, 2017), which then predicted performance on the knowledge essay posttest. The knowledge essay posttest was regressed on the knowledge pretest and previous coursework, as well. The condition to which participants were randomly assigned was posited to predict all monitoring, strategy use, and posttest variables, but in the final model, only those paths that were statistically significant were retained (see Figure 2). Finally, we tested paths from the essay pretest and previous coursework to all monitoring and strategy use variables and retained only those that were statistically significant. Based on examination of distributions and fit indices, we determined that the FOR− variable was best modeled as following a zero-inflated Poisson distribution and that the JOU-, deep-level strategy, and surface-level strategy variables were best modeled using a negative-binomial distribution (Greene, Costa, & Dellinger, 2011).

Our final model was estimated normally within Mplus 7.2 and had 31 free parameters, with a log-likelihood of −721.11. Neither data-model fit indices nor chi-square tests of data-model fit are available for path models with endogenous variables modeled with count distributions (i.e., zero-inflated Poisson, negative-binomial). The path coefficients illustrated in Figure 2 should be interpreted as the relationship between the criterion and the predictor above and beyond relationships between that criterion and other predictors.

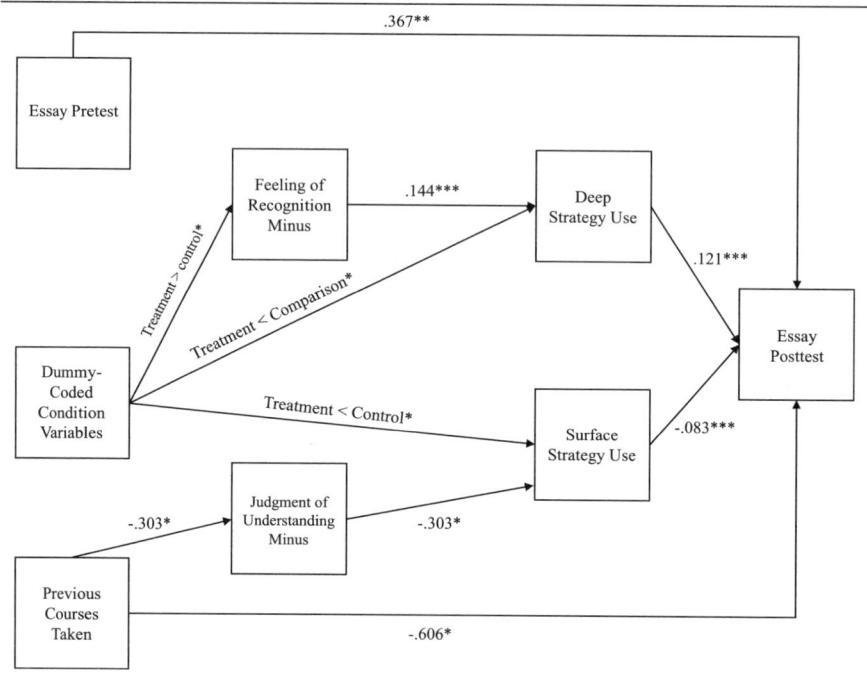

*Figure 2.* **Path analysis involving think-aloud protocol self-regulated learning (SRL) process data.**
*Note.* Statistically nonsignificant paths are not shown. "Dummy-coded condition variables" represent two dummy-coded condition variables. Paths representing variables regressed on dummy-coded condition variables illustrate the condition comparison that was statistically significant; all other comparisons were not statistically significant.
\*$p < .05$. \*\*$p < .01$. \*\*\*$p < .001$.

As expected, pretest and posttest essay scores were positively related. More frequent use of deep-level strategies was positively related to posttest essay scores, whereas more frequent use of surface-level strategies was negatively related, again as would be expected given past research (e.g., Deekens et al., 2018; Dinsmore, 2017). Verbalizations coded as FOR− (e.g., "I do not think I have seen this before") were positively related to the frequency of deep-level strategy use, whereas verbalizations of JOU− (e.g., "I do not understand what I am reading") were negatively related to frequency of surface-level strategy use. As expected, participants who took more courses related to the biology content were less likely to verbalize that they did not understand something, but our measure of prior conceptual knowledge (i.e., essay pretest) was not statistically significantly related to any SRL monitoring or strategy use variables. The number of previous courses taken was negatively related to essay posttest scores, which was unexpected; this represents the relationship between such courses and posttest performance after accounting for prior knowledge. Finally, there were three relationships of note involving our conditions. Treatment participants more often verbalized FOR− than control participants, suggesting a difference in the use of monitoring between these groups. Treatment participants less often verbalized using deep-level strategies than the comparison participants, which was unexpected. Also, treatment participants less frequently reported using surface-level strategies than the control participants. Overall, these SRL process findings support the literature in terms of the posited relations among prior knowledge, monitoring, strategy use, and performance (Deekens et al., 2018; Dinsmore, 2017; Greene, 2018; Moos & Azevedo, 2008) and provide support for some but not all of our expected effects of our treatment on SRL processing.

## Discussion

In the United States, low retention and graduation rates, particularly among FGCS, have led university administrators and researchers to turn to first-year courses in an attempt to improve FGCS' skills and increase their likelihood of success (Redford & Hoyer, 2017; Young & Hopp, 2014). Unfortunately, the efficacy of these seminars remains unclear (Permzadian & Credé, 2016). Existing empirical evidence favors first-year courses focused on teaching students the science of learning, in particular SRL knowledge, skills, and dispositions (Hofer & Yu, 2003; Zimmerman, 2013). Therefore, in this study, we randomly assigned FGCS to a science of learning course, a comparison condition with access to the materials from the course but no instruction, or a control business-as-usual condition. Then, we recruited 43 of these FGCS to participate in a laboratory study of how they used a computer to learn about the circulatory system. We found that participants, on average, increased both their declarative and conceptual knowledge over the course of the learning task. However, treatment and comparison condition participants experienced greater increases in conceptual knowledge scores from pretest to posttest than control students, with the treatment participants achieving the largest increase. We had hypothesized that the treatment condition participants would outperform the other groups; therefore, the comparison groups' performance was unexpected.

Our research question involved how SRL processing related to performance and what differences existed in that processing across conditions. The frequency of deep-level strategy use positively predicted conceptual knowledge performance, whereas more frequent use of surface-level strategies was negatively related to such performance, as found in other research (Deekens et al., 2018; Dinsmore, 2017). Participants who more frequently verbalized a feeling of not recognizing something also more frequently enacted deep-level strategies, suggesting that participants were using monitoring information to control learning and calibrate their learning strategies to their needs (Alexander, 2013; Zimmerman, 2013). The more often a participant verbalized a lack of understanding, the less often they stated that they were using surface-level strategies, which further supports the theoretical connection between monitoring and control (Ben-Eliyahu & Bernacki, 2015). These findings, which were common across conditions, support models of SRL in that they show the sequential and conditional nature of monitoring, strategy use, and learning (Ben-Eliyahu & Bernacki, 2015; Binbasaran Tüysüzoğlu & Greene, 2015; Zimmerman, 2013).

Our science of learning course included a focus on using deep- as opposed to surface-level strategies (Dinsmore, 2017), and we found treatment participants did this via the path through monitoring, supporting the role of SRL as a mediator of the relationship between individual differences such as prior knowledge and learning performance (Zimmerman, 2013). This suggests that the course was effective in helping FGCS acquire monitoring and strategy use knowledge and skills, two phenomena thought to be important across many contexts (Alexander et al., 2011; Greene, 2018). On the other hand, comparison participants more frequently verbalized using deep-level strategies than treatment participants. These analyses shed light on why treatment and comparison participants acquired more conceptual knowledge over the course of the learning task than their control condition peers: They more frequently enacted the kinds of monitoring and strategy use associated with learning (Dinsmore, 2017; Dunlosky et al., 2013).

To our knowledge, this is the first study to utilize random assignment as well as both product and process data (Azevedo, 2005) to investigate the effects of a first-year science of learning seminar on FGCS. As such, it makes a significant contribution to the literature regarding how to improve the efficacy of first-year courses by focusing them on SRL and strategy use (Permzadian & Credé, 2016). Such seminars provide FGCS with the knowledge, skills, and dispositions necessary for success in college, and in the case of this study, those necessary for enacting what has become a key component of college success: learning with a computer (Greene, Copeland, et al., 2018). Such transfer, from the classroom to the laboratory and computer context, is promising and deserves further research (Alexander et al., 2011). Researchers should also compare learning to learn courses with other first-year initiatives, with populations including both FGCS as well as other students (e.g., general, at-risk, and transfer students). Additionally, researchers can extend our study to capture the effect of courses like ours on learning within the course itself by analyzing in-vivo classroom academic outcome or trace data students generate when interacting with technology (e.g., learning management systems; Bernacki, 2018).

## Limitations

There are a few limitations regarding the design of our study. First, we had issues with recruitment, despite generous incentives. This required us to incorporate data collection from multiple semesters, which despite no systematic differences in course instruction of student pretest performance in our laboratory study, may have introduced some confounding effects. Our difficulties with recruitment also resulted in a less than optimal sample size and concerns about selection effects. Our small sample size limited us to being able to only investigate a subset of the possible relations among SRL and learning. Finally, the nature of the laboratory setting in which we studied the participants' learning processes limits the external validity of our findings, thus restricting our ability to generalize to different learning environments.

## Future Directions for Practice

As institutions of higher education expand their study of distal outcomes (e.g., retention, graduation) to include proximal ones such as student satisfaction and lifelong success, there is a growing need for rigorous research on not only what works but also why, as required by the What Works in Education Clearinghouse and other education organizations (Honig, 2009; Kinzie & Kuh, 2016). Our findings, based on a randomized control trial, support the development of first-year courses focused on the science of learning with an emphasis on explicit instruction of SRL as well as frequent opportunities for practice and feedback. In particular, the first specific suggestion for developers of first-year courses, derived from our findings, is that SRL should be taught as a set of related processes, rather than separate components, with ample opportunities for practicing monitoring and control in authentic contexts so that students learn how calibration informs effective learning strategy use (Paris & Paris, 2001; Weinstein & Acee, 2013). Second, many students reported a lack of knowledge regarding the research on effective strategy use (e.g., Dunlosky et al., 2013); therefore, explicit instruction and practice using deep strategies is also likely a key component of a successful learning to learn course (Zepeda et al., 2015). Finally, we suggest that course developers include frequent opportunities for students to reflect on their academic work and how SRL could be used to strengthen it. Course evaluation and anecdotal feedback indicated that students found such activities very helpful. All these suggestions should be informed by other research on first-year seminars, such as Permzadian and Credé's (2016) findings that the most efficacious courses combined extended orientation and academic content; were not connected to other courses as a part of a learning community; were taught by faculty or staff, not graduate students; and were targeted for all first-year students. Finally, as Wingate (2007) suggested, instructors should explicitly endorse students' generalization of deep learning strategies beyond a single course to promote true self-regulation of learning throughout their college career and life.

## Conclusion

In sum, we found evidence that a science of learning course can help FGCS acquire conceptual knowledge using a computer via more effective SRL processing. This evidence of transfer from the classroom to the laboratory task is compelling and makes a strong case for continued development and refinement of first-year seminar learning to learn courses as a way to bolster academic performance and retention. More research is needed to determine whether such first-year seminars lead to transfer of SRL knowledge, skills, and dispositions to other college courses and outcomes. Our study provides a model for doing such work utilizing online measures of SRL and suggests the need for future research in authentic contexts.

### Notes

Supplemental material is available for this article in the online version of the journal.
This research is part of the Finish Line Project (P116F140018; Panter, PI; Demetriou, Executive Director), which is funded by the U.S. Department of Education's "First in the World" grant program. The opinions expressed are those of the authors and do not represent the views of the U.S. Department of Education.

### References

Adesope, O. O., Trevisan, D. A., & Sundararajan, N. (2017). Rethinking the use of tests: A meta-analysis of practice testing. *Review of Educational Research, 87*, 659–701.

Alexander, P. A. (2013). Calibration: What is it and why it matters? An introduction to the special issue on calibrating calibration. *Learning and Instruction, 24*, 1–3. doi:10.1016/j.learninstruc.2012.10.003

Alexander, P. A., Dinsmore, D. L., Parkinson, M. M., & Winters, F. I. (2011). Self-regulated learning in academic domains. In B. Zimmerman & D. Schunk (Eds.), *Handbook of self-regulation of learning and performance* (pp. 393–407). New York, NY: Routledge.

Azevedo, R. (2005). Using hypermedia as a metacognitive tool for enhancing student learning? The role of self-regulated learning. *Educational Psychologist, 40*, 199–209.

Azevedo, R., & Cromley, J. G. (2004). Does training on self-regulated learning facilitate students' learning with hypermedia? *Journal of Educational Psychology, 96*, 523–535.

Azevedo, R., Guthrie, J. T., & Seibert, D. (2004). The role of self-regulated learning in fostering students' conceptual understanding of complex systems with hypermedia. *Journal of Educational Computing Research, 30*, 87–111.

Azevedo, R., Johnson, A., & Burkett, C. (2015, July). *Does training of cognitive and metacognitive regulatory processes enhance learning and deployment of processes with hypermedia?* Paper presented at the Annual Meeting of the Cognitive Science Society, Pasadena, CA.

Azevedo, R., Moos, D. C., Greene, J. A., Winters, F. I., & Cromley, J. C. (2008). Why is externally regulated learning more effective than self-regulated learning with hypermedia? *Educational Technology Research & Development, 56*, 45–72.

Bannert, M., & Mengelkamp, C. (2008). Assessment of metacognitive skills by means of instruction to think aloud and reflect when prompted: Does the verbalization method affect learning? *Metacognition and Learning, 3*, 39–58.

Bannert, M., Reimann, P., & Sonnenberg, C. (2014). Process mining techniques for analysing patterns and strategies in students' self-regulated learning. *Metacognition and Learning, 9*, 161–185.

Barefoot, B. O. (1992). *Helping first-year college students climb the academic ladder: Report of a national survey of freshman seminar programming in American higher education* (Doctoral dissertation, College of William and Mary). Retrieved from https://scholarworks.wm.edu/cgi/viewcontent.cgi?article=1790&context=etd

Barton, A., & Donahue, C. (2009). Multiple assessments of a first-year seminar pilot. *JGE: The Journal of General Education, 58*, 259–278.

Ben-Eliyahu, A., & Bernacki, M. L. (2015). Addressing complexities in self-regulated learning: A focus on contextual factors, contingencies, and dynamic relations. *Metacognition and Learning, 10*, 1–13.

Bernacki, M. L. (2018). Examining the cyclical, loosely sequenced, and contingent features of self-regulated learning: Trace data and their analysis. In D. H. Schunk & J. A. Greene (Eds.), *Handbook of self-regulation of learning and performance* (2nd ed., pp. 370–387). New York, NY: Routledge.

Binbasaran Tüysüzoğlu, B., & Greene, J. A. (2015). An investigation of the role of contingent metacognitive behavior in self-regulated learning. *Metacognition & Learning, 10*, 77–98.

Bjork, R. A., Dunlosky, J., & Kornell, N. (2013). Self-regulated learning: Beliefs, techniques, and illusions. *Annual Review of Psychology, 64*, 417–444.

Black, A., Terry, N., & Buhler, T. (2016). The impact of specialized courses on student retention as part of the freshman experience. *Allied Academies International Conference: Academy of Educational Leadership, 20*, 85–92.

Bui, K. V. T. (2002). First-generation college students at a four-year university: Background characteristics, reasons for pursuing higher education, and first-year experience. *College Student Journal, 36*, 3–11.

Cahalan, M., & Perna, L. (2015). *Indicators of higher education equity in the United States: 45 year trend report*. Washington, DC: The Pell Institute.

Cambridge-Williams, T., Winsler, A., Kitsantas, A., & Bernard, E. (2013). University 100 orientation courses and living-learning communities boost academic retention and graduation via enhanced self-efficacy and self-regulated learning. *Journal of College Student Retention: Research, Theory & Practice, 15*, 243–268.

Chen, P., Chavez, O., Ong, D. C., & Gunderson, B. (2017). Strategic resource use for learning: A self-administered intervention that guides self-reflection on effective resource use enhances academic performance. *Psychological Science, 18*, 774–785.

Chen, X. (2005). *First generation students in postsecondary education: A look at their college transcripts* (NCES 2005-171). Washington, DC: U.S. Department of Education, National Center for Education Statistics.

Chi, M. T. H. (1997). Quantifying qualitative analyses of verbal data: A practical guide. *Journal of the Learning Sciences, 6*, 271–315.

Choy, S. (2001). *Students whose parents did not go to college: Postsecondary access, persistence, and attainment* (NCES 2001-126). Washington, DC: U.S. Department of Education, National Center for Education Statistics.

Cleary, T. J., & Kitsantas, A. (2017). Motivation and self-regulated learning influences on middle school mathematics achievement. *School Psychology Review, 46*, 88–107.

Cone, A. L. (1991). Sophomore academic retention associated with a freshman study skills and college adjustment course. *Psychological Reports, 69*, 312–314.

Cone, A. L., & Owens, S. K. (1991). Academic and locus of control enhancement in a freshman study skills and college adjustment course. *Psychological Reports, 68*(3 Suppl.), 1211–1217.

De Backer, L., Van Keer, H., & Valcke, M. (2014). Promoting university students' metacognitive regulation through peer learning: The potential of reciprocal peer tutoring. *Higher Education, 70*, 469–486.

Deekens, V. M., Greene, J. A., & Lobczowski, N. G. (2018). Monitoring and depth of strategy use in computer-based learning environments for science and history. *British Journal of Educational Psychology, 88*, 63–79.

Demetriou, C. (2014). *Reflections at the finish line: The activities, roles, and relationships of college success for first-generation college students.* (Doctoral dissertation). Retrieved from https://cdr.lib.unc.edu/concern/dissertations/cj82k8251

Demetriou, C., Meece, J., Eaker-Rich, D., & Powell, C. (2017). The activities, roles, and relationships of successful first-generation college students. *The Journal of College Student Development, 58*, 19–36.

Dent, A. L., & Koenka, A. C. (2016). The relation between self-regulated learning and academic achievement across childhood and adolescence: A meta-analysis. *Educational Psychology Review, 28*, 425–474.

Dignath, C., & Büttner, G. (2008). Components of fostering self-regulated learning among students: A meta-analysis on intervention studies at primary and secondary school level. *Metacognition and Learning, 3*, 231–264.

Dinsmore, D. L. (2017). *Strategic processing in education.* New York, NY: Routledge.

Dinsmore, D. L., & Alexander, P. A. (2016). A multidimensional investigation of deep-level and surface-level processing. *Journal of Experimental Education, 84*, 213–244.

Dunlosky, J., Rawson, K. A., Marsh, E. J., Nathan, M. J., & Willingham, D. T. (2013). Improving students' learning with effective learning techniques: Promising directions from cognitive and educational psychology. *Psychological Science in the Public Interest, 14*, 4–58.

Dunlosky, J., & Thiede, K. W. (2013). Four cornerstones of calibration research: Why understanding students' judgments can improve their achievement. *Learning and Instruction, 24*, 58–61.

Dweck, C. S., & Master, A. (2008). Self-theories motivate self-regulated learning. In D. Schunk & B. Zimmerman (Eds.), *Motivation and self-regulated learning: Theory, research, and applications* (pp. 31–51). Mahwah, NJ: Lawrence Erlbaum.

Eom, W., & Reiser, R. A. (2000). The effects of self-regulation and instructional control on performance and motivation in computer-based instruction. *International Journal of Instructional Media, 27*, 247–260.

Ericsson, K. A., & Simon, H. A. (1980). Verbal reports as data. *Psychological Review, 87*, 215–251.

Ericsson, K. A., & Simon, H. A. (1993). *Protocol analysis: Verbal reports as data* (Rev. ed.). Cambridge: MIT Press.

Forster, B., Swallow, C., Fodor, J. H., & Fousler, J. E. (1999). Effects of a college study skills course on at-risk first-year students. *NASPA Journal, 36*, 120–132.

Gamez-Vargas, J., & Oliva, M. (2013). Adult guidance for college: Rethinking educational practice to foster socially-just college success for all. *Journal of College Admission, 221*, 60–68.

Garriott, P. O., Hudyma, A., Keene, C., & Santiago, D. (2015). Social cognitive predictors of first- and non-first-generation college students' academic & life satisfaction. *Journal of Counseling Psychology, 62*, 253–263.

Graham, S., Harris, K. R., MacArthur, C., & Santangelo, T. (2018). Self-regulation and writing. In D. H. Schunk & J. A. Greene (Eds.), *Handbook of self-regulation of learning and performance* (2nd ed., pp. 138–152). New York, NY: Routledge.

Greene, J. A. (2018). *Self-regulation in education*. New York, NY: Routledge.

Greene, J. A., & Azevedo, R. (2007). Adolescents' use of self-regulatory processes and their relation to qualitative mental model shifts while using hypermedia. *Journal of Educational Computing Research, 36*, 125–148.

Greene, J. A., & Azevedo, R. (2009). A macro-level analysis of SRL processes and their relations to the acquisition of sophisticated mental models. *Contemporary Educational Psychology, 34*, 18–29.

Greene, J. A., Bolick, C. M., Jackson, W. P., Caprino, A. M., Oswald, C., & McVea, M. (2015). Domain-specificity of self-regulated learning processing in science and history. *Contemporary Educational Psychology, 42*, 111–128.

Greene, J. A., Copeland, D. Z., Deekens, V. M., & Yu, S. (2018). Beyond knowledge: Examining digital literacy's role in the acquisition of understanding in science. *Computers & Education, 117*, 141–159.

Greene, J. A., Costa, L.-J., & Dellinger, K. (2011). Analysis of self-regulated learning processing using statistical models for count data. *Metacognition & Learning, 6*, 275–301.

Greene, J. A., Costa, L.-J., Robertson, J., Pan, Y., & Deekens, V. (2010). Exploring relations among college students' prior knowledge, implicit theories of intelligence, and self-regulated learning in a hypermedia environment. *Computers & Education, 55*, 1027–1043.

Greene, J. A., Deekens, V. M., Copeland, D. Z., & Yu, S. (2018). Capturing and modeling self-regulated learning using think-aloud protocols. In D. H. Schunk & J. A. Greene (Eds.), *Handbook of self-regulation of learning and performance* (2nd ed., pp. 323–337). New York, NY: Routledge.

Greene, J. A., Robertson, J., & Costa, L.-J. C. (2011). Assessing self-regulated learning using think-aloud protocol methods. In B. J. Zimmerman & D. Schunk (Eds.), *Handbook of self-regulation of learning and performance* (pp. 313–328). New York, NY: Routledge.

Hammer, D., & Berland, L. K. (2014). Confusing claims for data: A critique of common practices for presenting qualitative research on learning. *Journal of the Learning Sciences, 23*, 37–46.

Hattie, J., Biggs, J., & Purdie, N. (1996). Effects of learning skills interventions on student learning: A meta-analysis. *Review of Educational Research, 66*, 99–136.

Herrenkohl, L. R., & Cornelius, L. (2013). Investigating elementary students' scientific and historical argumentation. *Journal of the Learning Sciences, 22*, 413–461.

Hofer, B. K., & Yu, S. L. (2003). Teaching self-regulated learning through a "learning to learn" course. *Teaching of Psychology, 30*, 30–33.

Honig, M. I. (2009). What works in defining "what works" in educational improvement: Lessons from education policy implementation research, directions for future research. In G. Sykes, B. Schneider, & D. N Plank (Eds.), *Handbook of education policy research* (pp. 333–347). New York, NY: Routledge.

Johnson, A., Azevedo, R., & D'Mello, S. (2011). The temporal and dynamic nature of self-regulatory processes during independent and externally assisted hypermedia learning. *Cognition and Instruction, 29*, 471–504.

Karabenick, S. A., & Zusho, A. (2015). Examining approaches to research on self-regulated learning: Conceptual and methodological considerations. *Metacognition and Learning, 10*, 151–163.

Kauffman, D. F. (2004). Self-regulated learning in web-based environments: Instructional tools designed to facilitate cognitive strategy use, metacognitive processing, and motivational beliefs. *Journal of Educational Computing Research, 30,* 139–161.

Kena, G., Musu-Gillette, L., Robinson, J., Wang, X., Rathbun, A., Zhang, J., Wilkinson-Flicker, S., Barmer, A., & Dunlop Velez, E. (2015). *The condition of education 2015* (NCES 2015-144). Washington, DC: U.S. Department of Education, National Center for Education Statistics.

Kinzie, J., & Kuh, G. (2016, November). *Report prepared for the Lumina Foundation: Review of student success frameworks to mobilize higher education.* Bloomington: Indiana University, Center for Postsecondary Research.

Kramarski, B., & Mizrachi, N. (2006). Online discussion and self-regulated learning: Effects of instructional methods on mathematical literacy. *Journal of Educational Research, 99,* 218–230.

Lauff, E., & Ingels, S. J. (2013). *Education longitudinal study of 2002 (ELS: 2002): A first look at 2002 high school sophomores 10 years later* (NCES 2014-363). Washington, DC: U.S. Department of Education, National Center for Education Statistics.

Lohfink, M. M., & Paulsen, M. M. (2005). Comparing the determinants of persistence for first-generation and continuing-generation students. *Journal of College Student Development 46,* 409–428.

McKinley, B., & Brayboy, J. (2004). Hiding in the ivy: American Indian students and visibility in elite educational settings. *Harvard Educational Review, 74,* 125–152.

Microsoft Corporation. (2007). Encarta premium. [Computer software]. Redmond, WA: Microsoft.

Miller, J. W., & Lesik, S. S. (2014). College persistence over time and participation in a first-year seminar. *Journal of College Student Retention: Research, Theory & Practice, 16,* 373–390.

Moos, D. C. (2013). Examining hypermedia learning: The role of cognitive load and self-regulated learning. *Journal of Educational Multimedia and Hypermedia, 22,* 39–61.

Moos, D. C., & Azevedo, R. (2008). Self-regulated learning with hypermedia: The role of prior domain knowledge. *Contemporary Educational Psychology, 33,* 270–298.

Morales, E. E. (2014). Learning from success: How original research on academic resilience informs what college faculty can do to increase the retention of low socioeconomic status students. *International Journal of Higher Education, 3,* 92–102.

Mullen, C. A. (2011). Facilitating self-regulated learning using mentoring approaches with doctoral students. In B. J. Zimmerman, D. H. Schunk, B. J. Zimmerman, & D. H. Schunk (Eds.), *Handbook of self-regulation of learning and performance* (pp. 137–152). New York, NY: Routledge/Taylor & Francis Group.

Neitzel, C., & Connor, L. (2017). Messages from the milieu: Classroom instruction and context influences on elementary school students' self-regulated learning behaviors. *Journal of Research in Childhood Education, 31,* 548–560.

Paris, S. G., & Paris, A. H. (2001). Classroom applications of research on self-regulated learning. *Educational Psychologist, 36,* 89–101.

Permzadian, V., & Credé, M. (2016). Do first-year seminars improve college grades and retention? A quantitative review of their overall effectiveness and an examination of moderators of effectiveness. *Review of Educational Research, 86,* 277–316.

Perna, L. W., & Thomas, S. L. (2006). *Commissioned report for the national symposium on postsecondary student success: Spearheading a dialog on student-success, a framework for reducing the college success gap and promoting success for all.* Washington, DC: National Postsecondary Education Cooperative.

Perry, N. E., & Winne, P. H. (2006). Learning from learning kits: gStudy traces of students' self-regulated engagements with computerized content. *Educational Psychology Review, 18,* 211–228.

Pike, G. R., & Kuh, G. D. (2005). First- and second-generation college students: A comparison of their engagement and intellectual development. *The Journal of Higher Education, 76,* 276–300.

Pintrich, P. R. (2000). The role of goal orientation in self-regulated learning. In M. Boekaerts, P. Pintrich, & M. Zeidner (Eds.), *Handbook of self-regulation* (pp. 451–502). San Diego, CA: Academic Press.

Pintrich, P. R., Smith, D. A. F., Garcia, T., & McKeachie, W. J. (1991). *A manual for the use of the Motivated Strategies for Learning Questionnaire (MSLQ).* Ann Arbor: University of Michigan, National Center for Research to Improve Postsecondary Teaching and Learning.

Porter, S. R., & Swing, R. L. (2006). Understanding how first-year seminars affect persistence. *Research in Higher Education, 47,* 89–109.

Redford, J., & Hoyer, K. M. (2017). *First-generation and continuing-generation college students: A comparison of high school and postsecondary experiences.* Washington, DC: Institute of Education Sciences, National Center for Education Statistics.

Rogerson, C. L., & Poock, M. C. (2013). Differences in populating first year seminars and the impact on retention and course effectiveness. *Journal of College Student Retention: Research, Theory & Practice, 15,* 157–172.

Scheiter, K., Schubert, C., & Schüler, A. (2018). Self-regulated learning from illustrated text: Eye movement modelling to support use and regulation of cognitive processes during learning from multimedia. *British Journal of Educational Psychology, 88,* 80–94.

Schellings, G., & Van Hout-Wolters, B. H. (2011). Measuring strategy use with self-report instruments: Theoretical and empirical considerations. *Metacognition and Learning, 6,* 83–90.

Schunk, D. H., & Greene, J. A. (2018). Historical, contemporary, and future perspectives on self-regulated learning and performance. In D. H. Schunk & J. A. Greene (Eds.), *Handbook of self-regulation of learning and performance* (2nd ed., pp. 1–15). New York, NY: Routledge.

Stephens, N. M., Markus, H. R., Fryberg, S. A., Johnson, C. S., & Covarrubias, R. (2015). Unseen disadvantage: How American universities' focus on independence undermines the academic performance of first generation college students. *Journal of Personality and Social Psychology, 21,* 420–429.

Strayhorn, T. L. (2013). Academic achievement: A higher education perspective. In J. Hattie & E. M. Anderman (Eds.), *International guide to student achievement* (pp. 16–18). New York, NY: Routledge.

Trevors, G., Feyzi-Behnagh, R., Azevedo, R., & Bouchet, F. (2016). Self-regulated learning processes vary as a function of epistemic beliefs and contexts: Mixed method evidence from eye tracking and concurrent and retrospective reports. *Learning and Instruction, 42,* 31–46.

Tuckman, B. W., & Kennedy, G. J. (2011). Teaching learning strategies to increase success of first-term college students. *Journal of Experimental Education, 79,* 478–504.

U. S. Department of Education. (2016). *First in the world program.* Retrieved from https://www2.ed.gov/programs/fitw/index.html

Vandevelde, S., Van Keer, H., Schellings, G., & Van Hout-Wolters, B. (2015). Using think-aloud protocol analysis to gain in-depth insights into upper primary school children's self-regulated learning. *Learning and Individual Differences, 43,* 11–30.

Veenman, M. V. J. (2007). The assessment and instruction of self-regulation in computer-based environments: A discussion. *Metacognition and Learning, 2*(2–3), 177–183.

Veenman, M. V. J. (2011a). Alternative assessment of strategy use with self-report instruments: A discussion. *Metacognition and Learning, 6*, 205–211.

Veenman, M. V. J. (2011b). Learning to self-monitor and self-regulate. In R. Mayer & P. Alexander (Eds.), *Handbook of research on learning and instruction* (pp. 197–218). New York, NY: Routledge.

Veenman, M. V. J., Elshout, J. J., & Groen, M. G. M. (1993). Thinking aloud: Does it affect regulatory processes in learning? *Tijdschrift voor Onderwijsresearch, 18*, 322–330.

Weinstein, C. E., & Acee, T. W. (2013). Helping college students become more strategic and self-regulated learners. In H. Bembenutty, T. J. Cleary, & A. Kitsantas (Eds.), *Applications of self-regulated learning across diverse disciplines* (pp. 197–236). Charlotte, NC: Information Age Press.

Whipp, J. L., & Chiarelli, S. (2004). Self-regulation in a web-based course: A case study. *Educational Technology Research & Development, 52*(4), 5–22.

Wingate, U. (2007). A framework for transition: Supporting "learning to learn" in higher education. *Higher Education Quarterly, 61*, 391–405.

Winne, P. H. (2018). Cognition and metacognition within self-regulated learning. In D. H. Schunk & J. A. Greene (Eds.), *Handbook of self-regulation of learning and performance* (2nd ed., pp. 36–48). New York, NY: Routledge.

Winne, P. H., & Hadwin, A. F. (2008). The weave of motivation and self-regulated learning. In D. Schunk & B. Zimmerman (Eds.), *Motivation and self-regulated learning: Theory, research, and applications* (pp. 297–314). Mahwah, NJ: Lawrence Erlbaum.

Winne, P. H., & Perry, N. E. (2000). Measuring self-regulated learning. In M. Boekaerts, P. R. Pintrich, & M. Zeidner (Eds.), *Handbook of self-regulation* (pp. 531–566). San Diego, CA: Academic Press.

Winters, F. I., & Azevedo, R. (2005). High-school students' regulation of learning during computer-based science inquiry. *Journal of Educational Computing Research, 33*, 189–217.

Winters, F. I., Greene, J. A., & Costich, C. M. (2008). Self-regulation of learning within computer-based learning environments: A critical analysis. *Educational Psychology Review, 20*, 429–444.

Young, D. G., & Hopp, J. M. (2014). *2012–2013 National survey of first-year seminars: Exploring high-impact practices in the first college year* (Research Report No. 4). Columbia: University of South Carolina, National Resource Center for the First-Year Experience and Students in Transition.

Zepeda, C. D., Richey, J. E., Ronevich, P., & Nokes-Malach, T. J. (2015). Direct instruction of metacognition benefits adolescent science learning, transfer, and motivation: An in vivo study. *Journal of Educational Psychology, 107*, 954–970.

Zerr, R. J., & Bjerke, E. (2016). Using multiple sources of data to gauge outcome differences between academic-themed and transition-themed first-year seminars. *Journal of College Student Retention: Research, Theory & Practice, 18*, 68–82.

Zimmerman, B. J. (2013). From cognitive modeling to self-regulation: A social cognitive career path. *Educational Psychologist, 48*, 135–147.

Manuscript received February 28, 2018
Final revision received June 17, 2019
Accepted June 27, 2019

# Teacher Responses to New Pedagogical Practices: A Praxeological Model for the Study of Teacher-Driven School Development

### Sławomir Krzychała
*University of Lower Silesia*

*This article focuses on the teacher community as an agent of school development, and in the context of teacher engagement in new educational practices, it discusses how school change can be analyzed as a process of creating and transforming professional knowledge (orientation pattern). The qualitative research was conducted in 2015–2016 at 12 schools participating in an innovative tutoring program in Wrocław (Poland). A total of 12 group discussions and 52 individual interviews were interpreted using Mannheim's documentary method. As a result, a typology of the four forms of new professional orientation patterns—niche, instrumental, apparent, and synergic activities—was elaborated, and in a case study, they were applied as a theoretical model to the sociogenetic analysis of the school development process.*

Keywords: pragmatic knowledge, professional learning community, school development, school tutoring, teachers

## Introduction

Professional learning communities have been thoroughly examined in the past two decades not only in the context of teacher professional development but also as a significant space in which to develop and test new pedagogical practices (Stoll, Bolam, McMahon, Wallace, & Thomas, 2006; Vangrieken, Meredith, Packer, & Kyndt, 2017) and to empower teachers as agents of school development (Watson, 2014). These expectations are strengthened by the belief that the teachers' cooperation is predominantly supportive, has pro-developmental potential, and focuses attention and commitment on shared values, visions, and goals (Gallagher, Griffin,

---

SŁAWOMIR KRZYCHAŁA is an associate professor in the Faculty of Education, University of Lower Silesia, ul. Strzegomska 55, 53-611 Wrocław, Poland; e-mail: *slawomir.krzychala@dsw.edu.pl*. His research focuses on the learning environment changes in educational institutions, the sociocultural processes of creating practical knowledge, and the teachers' professional development.

Ciuffetelli Parker, Kitchen, & Figg, 2011; Hord & Sommers, 2008; Parker, Patton, & Tannehill, 2012). In the examination of the teachers' learning communities, a significance is assigned to reflexivity, reflexive practices (Hofer, 2017; Pultorak, 2010; Schön, 1983), and the collective building of professional knowledge (Parker et al., 2012; Popp & Goldman, 2016). However, the meta-studies show that in the studies of professional communities, higher attention is paid to maintaining and improving cooperation within the teaching team than to the process of constructing new professional knowledge (Camburn & Han, 2017; Vangrieken, Dochy, Raes, & Kyndt, 2015; Vangrieken et al., 2017). In the context of teacher engagement in initiating and extending new educational practices at school, this article discusses how the scope and sustainability of teacher-driven school development can be examined by reconstructing the process of creating and transforming professional knowledge.

Systematic reviews reveal that studies on school improvement embedded in the teachers' participation are dominated by the examination of short-term activities in situationally limited contexts of team meetings or classroom lessons (Gallagher et al., 2011; Parker et al., 2012; Tam, 2015; Vangrieken, Dochy, Raes, & Kyndt, 2013, 2015). The explanation for this state of the research may be partly based on the idea that teachers themselves prefer cooperation that is oriented toward lesson preparation, teaching effectiveness, and problem-solving issues that arise on an ongoing basis rather than toward new didactic frameworks and the critical review of a school's organizational culture (Vangrieken et al., 2015; Zhang & Yin, 2017). Wells and Feun (2007), as well as Camburn and Han (2017), argue that deeper collaboration going beyond sharing teaching scenarios and discussing particular lessons and student results appears to be challenging. Along similar lines, Vangrieken et al. (2017) find that the research on teacher communities as a context for professional learning is dominated by analyses of the teacher's reporting about situationally limited cooperation, and they postulate that studying the sociocultural aspects of development and the impact of different formations of the teachers' involvement could fill a gap in the current empirical studies.

The studies on institutional change from the cultural-historical (Beatty & Feldman, 2012; Engeström, Kajamaa, Lahtinen, & Sannino, 2015) and sociogenetic perspectives (Amling & Vogd, 2017; Bohnsack, Pfaff, & Weller, 2010) indicate key methodological issues that may extend more critical and differential insight into the process of introducing new practices in the teachers' communities:

- The sociocultural and institutional changes can be examined as a process of creating and reproducing pragmatic knowledge immersed in the flow of practice at two levels of knowledge: knowledge *about* action and knowledge *in* action (Schön, 1983).

- The sociogenetic (Bohnsack, 2017a) and cultural-historical (Engeström, 2015) research perspectives accentuate observed or reported episodes not as a separate unit of analysis but as interconnected stages within the process of change.
- A study on the process of change should remain sensitive to nonlinear and not always progressive development. The experience of crisis, contradictions, and helplessness can potentially represent not only a stage of failure but also an opportunity to question an established practice and a germ of critical developmental prospects (Beatty & Feldman, 2012; Rajala, Kumpulainen, Rainio, Hilppö, & Lipponen, 2016).
- A reconstruction of the development of new practices should cover the whole spectrum of experience, not only the episodes of collaborative engagement based on shared visions, values, and goals. The process of creating new pragmatic knowledge involves a wide range of both individual and collective experiences (Vogd & Amling, 2017), which not only complement one another but also relate to contradictions and asynchrony (Desimone & Garet, 2015; Skott & Møller, 2017).

This article contributes to the research conducted from a sociocultural perspective, on teacher communities as agents of school development. I propose a theoretical data-based model for interpreting the teachers' responses to new pedagogical practices as a sociohistorical/sociogenetic process of transforming the teacher's professional knowledge embedded in the flow of school activity. As a primary analytical formula, I apply the praxeological approach of Karl Mannheim's (1952a, 1952b) sociology of knowledge (Bohnsack, 2017a, 2017b). In this perspective, the knowledge—including the teachers' professional knowledge—is empirically reconstructed as a pragmatic *orientation pattern* that reflects the manner (*modus operandi*) in which this knowledge "is leading or orientating practical action" (Bohnsack, 2017a, p. 200). Almost analogously, as in Schön's (1983) proposal, two levels of orientation patterns are distinguished: *orientation schemes* and *orientation frameworks* (Bohnsack, 2017a). Orientation schemes include the teachers' knowledge *about* action mediated in linguistic and symbolic communication means, which are used to comment on and coordinate activities. The orientation framework exceeds the teachers' declarations and reporting on practice and includes the habitual and atheoretic knowledge-*in*-action resources, mediated directly *in* the practice itself (Bohnsack, 2017a; Krzychała, 2019).

In the praxeological perspective, I focus on a new practice as a turning point in the professional learning process. The teachers—as a collective and multi-individual agent of change—launch change as a chain of practice transformations: The initiated alteration is to be considered in the context of existing pedagogical practice and, above all, as an opening of a further development sequence. The way teachers respond to the new practice reflects the understanding and the pragmatic meaning of the undertaken improvements. The central research question concerns how the teachers' knowledge *about* and *in* action is transformed by the involvement in the new pedagogical practices and how the teacher experience designates the course of school development.

An exploration of the change process in pedagogical practice was carried out as a study of a school tutoring program implemented in 29 public schools in Wrocław, Poland, from 2008 to 2016. The institutional and strategic objective of the program was to implement an individualized pedagogical counseling/advising model supplementing the work with students in classroom settings (Drozd & Zembrzuska, 2013; Krzychała, 2018). The teachers became tutors and were individually chosen by the students. The tutor-tutee cooperation lasted for 3 years of the student's school attendance and included regular monthly meetings (tutorials). The tutoring as a new school practice was planned as a bottom-up initiative, and except for the concept of tutoring and the general outline of the tutorials presented in the initial training, the teachers did not receive a ready-made scheme for tutoring. None of the public schools in Poland had previously conducted such activities. During the first years of the program, the teacher-tutors were expected to develop a detailed organizational model suited to a specific school. This task naturally prompted the teachers to individual experimentation with the new professional role. They shared their know-how and engaged in joint coordination of school development. This process of change took 5 to 7 years and included the improvement of both individual tutoring skills and institutional arrangements.

At the turn of 2015 and 2016, teachers from 12 junior high schools participated in registered semistructured group discussions in every school, and then these same four to six teachers were invited to individual interviews. The analysis of this multidimensional process of change embraced complex and long-term processes of pedagogical transformation beyond the initial phase of implementation of the new practice by a small group of pioneers and enthusiasts. In this way, the analysis took into account different variants of practice transformation, problems related to the growing scale of change, the contradictions and failures experiences, and the relationships between the individual and collective learning processes of the teacher-tutors. The analysis of narratives by using the documentary method was the basis for the elaboration of a theoretical model that showed four forms of new pedagogical orientation patterns, such as *niche, instrumental, apparent,* or *synergic* change at school. This general model—as it will be presented in the case study of one of Wrocław's schools—can then be used for the sociogenetic reconstruction of the process of both individual and collective professional development.

## Theoretical Framework: Pragmatic Orientation Pattern

In the inquiry into the process of constructing the teachers' pragmatic knowledge development in the course of responding to the new pedagogical practice, the analytical distinction between the two levels of knowledge, that is, the level of knowledge *about* action and that of knowledge *in* action

(Schön, 1983), is of crucial relevance. This differentiation can be thoroughly demonstrated within the praxeological sociology of knowledge developed by Karl Mannheim in his papers from 1921 to 1924 (Mannheim, 1964, 1980; also Bohnsack, 2017a, 2017b).

Mannheim (1952a) describes two layers of knowledge immersed in practice: (1) an immanent (*explicit*) meaning/understanding and (2) a documentary (*implicit*) meaning/understanding:

- An immanent understanding is intentionally assigned by subjects to the activity in which they participate and which they shape. It includes a goal, intention, and motive of action. Mannheim also defines this meaning as *communicative* knowledge. The observers or researchers can perceive it to the extent that participants in social practices share their experiences with linguistic and symbolic means at the *commonsense* level (Bohnsack, 2017a).
- Documentary understanding connects the present activity with broader sociocultural contexts of experience, such as personal and professional biographies, and informal and institutional spaces of involvement. Mannheim also identifies this meaning as *conjunctive* knowledge. Observers or researchers can perceive it only as an integral part of an activity, as an atheoretical outline incorporated into the style and proficiency of the practice, as the *experience sense* mediated directly in the flow of engagement (Krzychała, 2019). It is reflected independently of the subject's intention and declarations, in the modus operandi of the performance (Bohnsack, 2017a, 2017b).

In the analysis of narratives, communicative knowledge can first be reconstructed, as it encompasses the description of the organization, methods, both formal and informal rules, and justifications employed in practice. This level of professional knowledge can be defined as an orientation scheme (Bohnsack, 2017a).

However, not all activities entirely overlap with the orientation scheme. In addition to the goals, intention, and motive underlying the described activities, a practical sense can be detected. This dimension is referred to as an orientation framework, which can be considered the product of the experiential processes and knowledge experienced in the modus operandi of this practice (Bohnsack, 2017b, p. 104). In other words, the manner and proficiency in which subjects carry out tasks reveal visible, but not necessarily explicitly formulated, sense and understanding of the activity.

A certain cautiousness is recommended when interpreting activities in an institution by using only the orientation scheme (Amling & Vogd, 2017; Bohnsack, 2014). Even in strongly hierarchical organizations, "lived hierarchies" function parallel to formal ones (Mensching, 2008). Similarly, in the research on schools as organizations, significant attention should be paid to the generation of meanings beyond the orientation scheme (Asbrand & Martens, 2018; Welling, Breiter, & Schulz, 2015). An analysis of orientation frameworks reveals many contradictions in the practice that the participants sometimes seem not to recognize or to justify on the level of the orientation scheme (Vogd & Amling, 2017). Professionals note some such inconsistencies, justifying them when faced with given conditions, by the need for flexibility and as a way to reconcile contradictory obligations or as personal or group variants of the interpretation of rules. Such habits and practices are perceived "as if" they were compliant with the rules (Ortmann, 2004).

The analytical reconstruction of the orientation framework poses a particular challenge for researchers. This level of professional knowledge can be understood only "in the totality of experience" (Mannheim, 1980, p. 218) through the integral reconstruction of both (1) the practice in which the modus operandi of the activity is performed and activated and (2) the practice in which the *experience sense* is developed as well as the sociocultural history of the constructing orientation framework. Mannheim (1952a) defined this methodological strategy as a *genetic* explanation, and it became a fundamental element of the documentary reconstruction of pragmatic knowledge. As long as individual biography and development are embedded in multidimensional references to sociocultural spaces of experience, such as belonging to a particular generation, social and economic stratification, or cultural and political determinants, the sociogenetic approach combines an individual and collective perspective. Analyzing the category of the generation as a collective space of experience, Mannheim (1952b) shows that collective relationships do not necessarily follow from direct interactions but rather from an anchoring in identical or similar sociohistorical and cultural structures.

Despite the analytical distinction, the orientation scheme and the orientation framework are immersed in the same practice and form an integral orientation pattern, regardless of whether they are in a relationship of homologation, supplementation, or contradistinction (Bohnsack, 2014). In this sociogenetic perspective, the teachers' pragmatic response to the new pedagogical practice can be reconstructed as a process of constructing and reconstructing a new orientation pattern on the interconnected levels of the orientation scheme (procedures, categories, instructions, declarations) and the orientation framework (atheoretical knowledge, professional habitus, and a sense of practice). In the presented research, the interpretation aims to consider the processual reconstruction of school praxis changes caused by including new praxis (tutoring) into the school system and by confronting the challenges that result from the perceived contradictions between the school curriculum and the tutoring program.

## Study

### School Tutoring Program

The Wrocław school tutoring program—intended as a bottom-up initiative for changing public schools in Wrocław—was introduced to 29 junior high schools (gymnasia) in the period from 2008 to 2016 (Drozd & Zembrzuska, 2013; Krzychała, 2018). The program introduced a new pedagogical practice of individualized tutor-tutee meetings (tutorials), and this seemingly simple intervention involved teachers in the broader cycle of school development.

School tutoring can be considered as a form of personalized education. The tutoring method was developed at the Universities of Oxford and Cambridge and was based on an individualized tutor-tutee relationship and peer-tutoring in small groups (Ashwin, 2005; Moore, 1968; Palfreyman, 2008). In afterschool learning and tutoring centers, the tutoring method was also adapted to the needs of elementary and secondary programs of general education (K–12) to enhance school achievements and literacy (Dawson, 2010; Ehri, Dreyer, Flugman, & Gross, 2007; Smith, Cobb, Farran, Cordray, & Munter, 2013). In a different form, the Wrocław school tutoring program was developed as an integral part of the schools' curriculum (Drozd & Zembrzuska, 2013; Krzychała, 2018). Each student chose a teacher at the school as a tutor, with whom they met at least once a month at a 30- to 40-minute meeting; the meetings were usually conducted on an individual basis and sometimes in a small group of several tutees cooperating with the teacher-tutor. The tutoring involved advising students not only in relation to their school achievements but also concerning their well-being at school, social and learning skills, and independent out-of-school activities in the areas of art, sports, and volunteering. The strategic goal of the program was to empower each student, not only those with special needs or exceptional talents but also those "invisible" students with "average" achievement who pass almost unnoticed through subsequent stages of education. In the first school year, the tutors assisted students in understanding their talents and learning styles. In the second year, the tutors encouraged students to set their interests, and they accompanied the students in achieving their personal learning goals. In the third year, the objectives were extended to the choice of further educational and vocational career.

In 2008, based on the general guidelines prepared by consultants from the Tutors' Collegium established in Wrocław for the project, the first three Wrocław schools began the tutoring program (a total of 14 teacher-tutors in 6 classes/students groups; Budzyński, Traczyński, & Czekierda, 2009). The teacher-tutors participated in initial training during which they became familiar with the general idea and the goals of tutorials. However, they did not receive a ready-made model for organizing tutoring at school. From the beginning, it was assumed that in each school, the teachers would autonomously develop individualized tutorial care models corresponding to the resources and needs of a particular school. Counseling based on an individual tutor-tutee relationship and a free choice of goals was a new challenge for teachers and differed from the school's dominant modus operandi of the classroom-based and subject-oriented work. The new pedagogical practice opened up space for the development of new orientation patterns in the teacher community.

In 2014, as many as 29 schools (among them 21 junior high schools) and 442 teacher-tutors participated in the school tutoring program. The tutorial program was run until 2016, when the Polish Parliament passed an act reforming the Polish school system of general education and replaced the three-level model K–6, 7–9, and 10–12 with two levels K–8, 9–12 (European Commission/EACEA/Eurydice, 2018, p. 23). As a result, the junior high schools were successively closed. This systemic reform finally stifled the Wrocław tutoring program. However, in the study, which was conducted at the turn of 2015 and 2016—almost at the last moment before the termination of the junior high schools—12 group and 52 individual interviews were recorded. The narratives document the process of tutoring development both at the level of the organizational change created by the tutors' teams as well as at the level of becoming a tutor represented by the actions of the individual teachers. The variety of organizational forms worked out and the diversity of the individual and collective experiences allowed us to develop a theoretical model of four types of teacher responses to the new pedagogical practice.

**Participants and Research Design**

Teachers from 12 public junior high schools were invited to participate in the research (10 with tutoring, 2 control schools). The study was financed and conducted independently of the school tutoring program. As the sample was composed contrastively, the schools differed in size, student achievements, and the scope of implementation of the tutoring program (Table 1).

The research materials were collected in three stages. First, together with the teachers, the research team prepared ethnographic descriptions of the school, including information on the socioeconomic status of the students and the organization of teaching, tutoring, and other school activities. Next, in each school, a 60-minute unstructured teacher group discussion on tutoring experiences was recorded. Afterward, 52 individual in-depth interviews (at least four at each school) were conducted with teachers who had previously participated in the group discussion: Teachers with differing experiences of teaching and tutoring were invited. The interviews were transcribed,[1] and the institutional and the interviewees' personal information was anonymized.[2]

The triangulation of individual interviews and group discussions proved to be particularly inspiring for the analysis. The narratives of the teachers in the interviews and group discussions were not identical. The comparison of differences allowed for the reconstruction of the relationships between (1) the individual experiences of tutor-tutee relations and becoming a tutor and (2) the collective processes of introducing into the teacher-tutors'

## Table 1
### For the Schools Participating in the Study, Basic Information and the Scope of the Implementation of the School Tutoring Program.

| School Code | Number of Years of Tutoring (Upto the 2015/2016 School Year) | Total Number of Students (Percentage Covered by Tutoring)[a] | Total Number of Teachers (Percentage of Tutors)[a] | Average Result in Final External Exam[b] (%) |
|---|---|---|---|---|
| C | 2 | 410 (25) | 60 (15) | 107 |
| S | 3 | 110 (25) | 15 (100) | 92 |
| A | 5 | 110 (80) | 20 (50) | 92 |
| K | 5 | 150 (85) | 30 (55) | 87 |
| B | 6 | 120 (30) | 15 (70) | 125 |
| E | 6 | 320 (50) | 35 (40) | 96 |
| G | 6 | 520 (30) | 60 (45) | 117 |
| H | 6 | 790 (90) | 85 (60) | 111 |
| F | 7 | 100 (95) | 20 (50) | 88 |
| L | 7 | 350 (100) | 30 (90) | 103 |
| P | Control school | 330 (—) | 35 (—) | 104 |
| R | Control school | 460 (—) | 40 (—) | 115 |
| Mean | 5.4[c] | 312 (61) | 34 (57) | 104 |

*Note.* The examination scores in the literature, mathematics, and natural sciences have been converted so that the national distribution approximates a normal distribution, the mean corresponds to 100%, and the standard deviation corresponds to 15%.
[a]Rounded figures.
[b]Results of 2015 were taken from the Educational Value-Added database (ewd.edu.pl/wskazniki/gimnazjum).
[c]Excluding control schools.

community organizational and cultural changes in the pedagogical practice of the school and learning. Thus, having a sociogenetic character, the fundamental research question is how tutoring—as a new pedagogical praxis—is developed by the teacher-tutors; that is, in what practices are the professional orientation patterns transformed and by what teacher experiences are the sociocultural paths of school development marked?

### The Documentary Method

The material was subsequently analyzed in compliance with the principle of Mannheim's documentary method (1952a, 1980), which has been currently developed by R. Bohnsack (2014, 2017a, 2017b) and other researchers (Amling & Vogd, 2017; Asbrand & Martens, 2018; Bohnsack et al., 2010; Krzychała, 2019; Loos, Nohl, Przyborski, & Schäffer, 2013).

The documentary method includes the *formulating* (also *formulative*) of an interpretation, then the *reflecting* (also *reflective*) of the interpretation of the empirical material, and, finally, the generating of theory as a multidimensional *typology* (Bohnsack, 2014, 2017a). The five primary data processing steps in the documentary interpretation are as follows (Krzychała, 2019):

- *Formulating analysis I* comprises the thematic division and identification of the structure of text passages constituting the basic units for subsequent analysis.
- *Formulating analysis II* aims to determine the species and forms of expression used by the narrators in particular passages—for example, the use of description, telling, argumentation, metaphors, and gestures. The formulating of the interpretation allows for the description of the orientation schemes to be expressed at the level of the communicative knowledge.
- *Reflecting analysis I* focuses on capturing the meanings that extend beyond the literal content of speech and reveals orientation frameworks evident to narrators and that are anchored in atheoretical knowledge. Analogously as in the conversation analysis (Rapley, 2007, p. 72), researchers take into account the sequence of passages, the order in which the topics emerge, and references to the statements of the previous speaker or the topic mentioned above. The first statement is treated as a *proposal*, followed by a *reaction*. However, in addition, in the documentary analysis (Bohnsack, 2017a, p. 211), the researchers also take into account the third turn of a reaction to the reaction (called also a *conclusion*), in which the author of the proposal can address whether and how the proposal was understood in a reaction. The primary utterance chain in the reflecting analysis I contains a *proposal-reaction-conclusion*, and the attention is focused on the conclusion in which the researchers can observe the confirmation of habitual understanding or the need to draw a difference or misunderstanding of the reaction if it was anchored in a different experience sense.
- *Reflecting analysis II* introduces to the analysis a comparison with other nonadjacent passages from discussions or with statements recorded in other interviews (comparison inside the case and between cases). In particular, the researchers identify different ways of presenting similar topics and tasks depending on the various biographical, social, and institutional conditions in which the orientation frameworks are created and reproduced. In the presented study, the comparison takes into account differences in the individual experience of teachers within a specific school and related not only to different factors—for example, differences in work experience, position in school, and the tasks undertaken—but also to the biographical and sociocultural background. It also considers the differences between teacher teams in different schools; due to the diversity of the organizational culture or the community of students attending school, these teacher teams are confronted with various challenges in their space of experience.

In formulating the interpretation, the collected narratives are read as a *description* of activity at the school. In reflecting the interpretation, the narratives are also seen as a *document* of the teacher's understanding reflected in the way in which the themes emerge and from which perspective they are addressed. In the pilot research, I participated in two peer meetings of teacher-tutors from several schools; in these peer meetings, the participants spontaneously shared their experiences and discussed current problems to be solved. The comparative analysis showed that the same dynamics were documented in the group discussions recorded during the study. These discussions were also moderated spontaneously by the teachers themselves and only partially by the interviewers. In Mannheim's (1952a) perspective, we can treat group discussions as a *document* of collectively shared or differed experiences to the extent that the teachers' orientation patterns are *reflective* of the content and dynamics of the discussed activity.

The interpretations resulting from the formulating and reflecting analyses become the basis for a theoretical generalization:

- The creation of a *typology* as a theoretical model aims to define based on reconstructed sociogenetic differences in the history of their emergence in practice and their inclusion in new contexts of practice, multidimensional *types* of orientation patterns. The researchers generate a typology in accordance with the principle of theoretical saturation until the introduction to the analysis of subsequent cases does not change the key types but only increases the set of situational examples and allows for the understanding of nuances in the empirical variants of the orientation patterns.

As a result, four key types of teachers' responses to the new pedagogical practice were identified: *synergic, niche, instrumental*, and *apparent* activities. These types were reconstructed based on the analysis of the processes of the creation of the orientation patterns. In line with the sociogenetic strategy, the exploratory significance of the typology does not reduce to the assignment of individual schools to ideal types but functions to facilitate the understanding of the processes of school development spearheaded by the teachers introducing a new pedagogical practice into the school activity system. This analytical potential of typology is demonstrated by applying it to a detailed sociogenetic reconstruction of a teacher-driven change process in one of the Wrocław schools.

## Findings

### New Orientation Patterns in the Teacher Activity

In the idea of school tutoring, the implicit challenges to overcoming contradictory expectations include (1) a classroom system that focuses on school achievements and meeting the curriculum requirements and (2) a tutoring program that presupposes individualized cooperation is adjusted to the potential and interests of the students and is conducted within an informal relationship between the tutor and the tutee. Implementing the individualized model of tutoring in the "traditional" model of a school

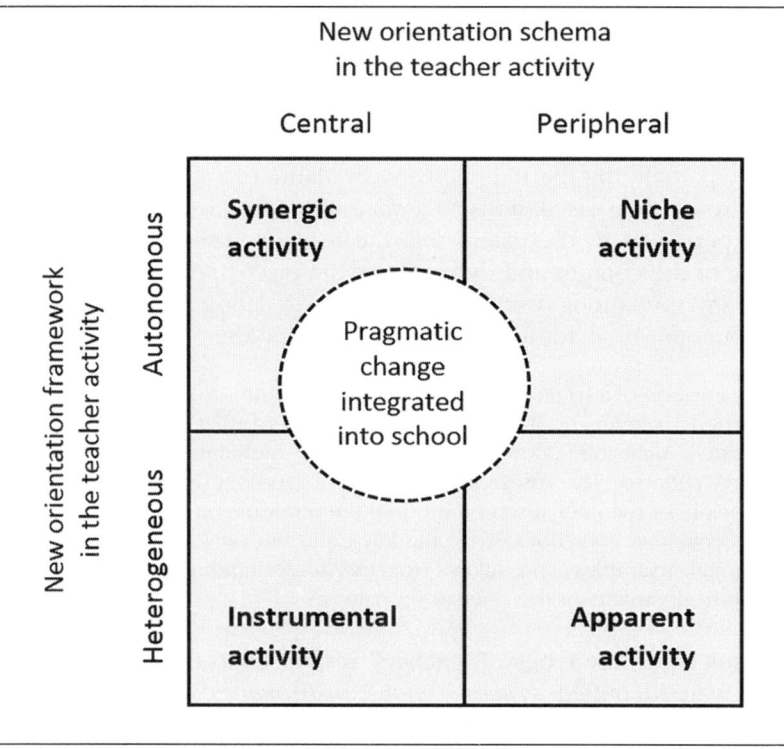

*Figure 1.* **Typology of integrating the new orientation pattern in the teacher activity.**
*Source.* Author's elaboration based on Krzychała (2018, p. 333).

requires the teachers' learning community to effectuate a change in the school system and adapt it to the new pedagogical tasks.

Based on the research data, four change variations were identified that specify the new orientation pattern's inclusion—both as a tutoring *scheme* and as a *framework*—into the school practice (Figure 1). Schools in which the tutoring orientation scheme has been integrated into the school program and has become a key and almost indispensable element characterizing a proficient style of work at school (*central change*) can be distinguished. In contrast, there are schools in which tutoring is not perceived as a strategic element of the school work organization but rather as one of many additional activities, which, at best, makes the school's offerings more attractive (*peripheral change*).

The development of the orientation framework results from a different understanding of the meaning and significance of tutoring experiences. The teachers have a different sense of the nature and extent of the changes that the tutoring practice contributes to school development. On one hand, the tutoring experience generates a new quality and a new pedagogical perspective in thinking about the student, the teacher, their relationship, commitment, and even about learning and teaching. It provides the teacher's community with a critical perspective for the self-evaluation of activities at school, not only tutoring activities (*autonomous change*). On the other hand, tutoring is admittedly perceived as a new method of organizing individual activities with the student; however, it does not result in a new quality of work at school. It remains subordinate to the already dominant goals and strategies of work in a specific school (*heteronomous change*). In such cases, tutoring does not become a critical impulse to think about school in a different way, and it does not provide a new voice in the discussion and reflection on the school performance.

The analytical distinction between the changes in the orientation scheme (program and organization) and the orientation framework (style and sense) underlays a model that describes four ways of incorporating the new orientation in the school practice (Figure 1). Correspondingly, tutoring constituted as an innovative pedagogical practice can be categorized in four forms: *synergic, niche, instrumental,* and *apparent* activities. In the retail analysis, particular attention is focused on schools with a developed and tested synergistic tutoring model. However, this was not observed in a pure and isolated form. In everyday school activities, synergistic tutoring coexists with elements of other forms of tutoring activity. The pragmatic form of synergic activity, observed in schools E, F, G, H, K, and L (Table 1), was identified as a pragmatic change integrated into the school. The four theoretical types can be treated as ideal types, which—although derived from historical and empirical observations—do not finally describe social phenomena but can be used as heuristic propositions and analytical categories to understand specific forms of experience (Weber, 1978). Following this analytical strategy, the typology of integrating the new pedagogical practice into the school activity system is described, and then—by using the example of school H—the way in which a specific teacher team develops through a dynamic and complex process a new pragmatic orientation pattern of synergistic tutoring is demonstrated.

*Niche Activity*

Niche activity describes tutoring that was not yet fully integrated into the school curriculum. In the *peripheral* form, the tutoring program only involves some students and classes. In the initial phase of introducing the tutoring program in most schools, niche tutoring was observed when a small group of highly motivated teachers became involved in tutorials with students. With full conviction, they realized the scenarios and methods learned during the tutors' initial training. At the same time, the experience of tutoring relations is so qualitatively different from the experience associated with previous activities at school that it begins to be perceived by teachers as an experience that has *autonomously* changed their thinking about the essence of the educational impact. The pioneers of change and informal leaders of change begin to emerge. A characteristic feature of niche tutoring is the simultaneous learning of tutoring methods, trying them out, and searching for original modifications. This search results from the need to gain experience, on one hand, and from the lack of verified organization, on the other hand. Teachers remember this time as especially inspiring for their individual development and as a period of intensive cooperation in their tutors' team. Even after a long period of time had passed, the story of the beginnings of tutoring was also told by the teachers, who did not originally directly participate in it and joined the team of tutors after a few years. From the sociogenetic perspective, referring to this phase, often by contrasting it with the tutoring model developed later, helps understand the direction of further development.

Niche tutoring, as an initial stage, constitutes a transitional form (schools E, H, G, F, K, L, and S; Table 1). In one of the schools (E), niche tutoring took a permanent form. Initially, it was performed by a large team of teachers and comprised approximately half of the students. After the school management changed, it ceased to be considered a strategic activity. Teachers did not receive further financial or organizational support. They had to organize time on their own to meet the students in tutorials after the lesson, and over time, this activity was limited to a small group of enthusiasts.

*Apparent Activity*

An apparent activity describes situations in which tutoring is officially run as a program; however, teachers, despite their organizational and personal effort, do not experience the benefit of tutoring activities. From the *heteronomous* perspective, tutoring is seen as an additional burden. Two general forms of apparent tutoring are documented. The first is related to the marginalization of tutoring from the very beginning and constitutes a permanent form of apparent activities (schools A and B). The second results from the incompatibility between work methods and the students' needs and abilities. In this form, the teachers later notice that tutoring is not effective (school H). In this case, it becomes a transitional form since it initiates an intense discussion on the sense of further engagement. It also poses a new challenge for teachers to develop their new model of tutoring at school.

Even after the tutoring program had been running for several years (school A for 5 years and school B for 6 years), the permanent form of apparent tutoring was identified. Interestingly, whereas only one in three junior high school students participated in the tutoring activities at school B (30%), at school A, as many as 80% of the students were officially involved in tutoring activities with a designated tutor. However, in a high-performing school A and a discipline-oriented school B, supporting the students' individualized objectives seemed contradictory to the unquestionable school routine. In both the schools, the tutoring program was maintained because it could be presented in the school's advertising as a concern for the individual talents or particular needs of the students. The tutoring—as can be observed in the narrative of the teachers at school A—turned out to be a reproductive model of the apparent activity:

> Excerpt GRD-A 366-381[1, 2]
> Af1: *At the beginning, when we were starting all this work as tutors, I remember we ran after some of them all over the place (.) to make an appointment.*
> ⌊ Af3: *We even had to put together the timetable.*
> Af1: *Yes. We used to set up the timetable so that it suited them; we would catch the children (y) before they left school (y), right? Remember? You were making arrangements with me. None of us is doing it anymore, are we?*
> Agr: *No.*
> Af1: *To force the child to do so, (.) to organize such meetings and-. They know that we are at their disposal. If they want to talk to us, they, we are (.) ready to talk, but (.) we don't force them, and we also don't catch them*
> *(hm) so that the schedule is honored, you know, because we have to meet with them a few times per week or per month(.). We do not (m) chase after them as it has been up to now. At least, I am not doing it anymore.*

In the quoted excerpt, the sociogenesis of the apparent change is documented in the experience of both a potential tutoring leader (Af1) and the teachers' team. Teacher Af1 had high hopes for tutoring; however, her enthusiasm dwindled. Additionally, this change in attitude did not result only from being overloaded with too many tasks or difficulties in reconciling tutoring with other school responsibilities. The tutor does not see any qualitative difference when working with tutoring. Tutors experience chaos in tutorial activities, including chasing after children, catching them, and the impossibility of implementing the timetable they devised. In their opinion, this "chase" was children oriented "so that it suited them"; however, there was no building of relationships or meeting with the children. This personal experience of Af1 is reinforced by the group experience, as evidenced by the linguistic change from "me" to "us" and confirmation by the group that no one is "catching" children anymore. This withdrawal occurred as the teachers felt they were at the students' disposal; this resulted in the students having a sense of freedom and left them with the responsibility for taking the initiative in tutoring activities. From a theoretical point of view, these experiences are consistent with tutoring activities; however, the experience of the teachers at school A is quite different from that of teachers in a niche or synergic tutoring situation. The illusion of the teachers from school A that they have "really" worked hard in using the tutoring method and that tutoring "really" does not work is authentic because the teachers have no other point of reference. The abandonment of "chasing after them" seems to be a rational action from their collective point of view to avoid further confrontation with failure. At the same time, it is a trap that protects them from the aspect of the tutoring experience that might call their convictions into question. In this situation, apparent tutoring occurs on a minimal scale as an additional *peripheral* activity. It is evaluated from the perspective of the dominant *heteronomous* activity strategies operating at the school and on the same level as the other activities in the school. Thus, in the long run, it turns out to be superfluous.

## Instrumental Activity

Tutoring as an instrumental activity is introduced as one of the appreciated methods, and it is perceived as an effective and efficient way of motivating students. Most often, its value results from the warming up of the teacher's image and the appreciation of the student's strengths, which facilitates the interventional and disciplining work of the class tutor. Over time, the teachers have found that the students can still be allowed a wide margin of freedom during tutorials, while working systematically to solve problems that are in line with school objectives. The teacher-tutors report that tutoring helps them "manage their classes" to even out differences among students and to truly "model" them:

> **Excerpt GRD-L 202-209**
> Lf6: *So, I also think that the tutor's role is that of a guardian. And this is what I like in tutoring that(.) that well, children have (m) this closer contact with the teacher, but well, that simply, well (.) two teachers manage a class of thirty and not just one teacher, right? And, for example, then (y) one worked as a head of the class, and that is the one (.) without these proper relations, and then that tutor did his job. Basically, (.) it was not exactly clear what it was all about, right? I didn't get it at that time.*

In tutoring, it is crucial to focus on the students' strengths and support them in their autonomy. In instrumental tutoring, this orientation has been overshadowed by the need to respond to educational deficits and difficulties and to strive to unify the group of students. In the course of the tutorials they run, the teacher-tutors notice the differences in how the students work; however, such differences are not interpreted from the personalistic and dialogical perspective. They are used to develop the students' scheme of modeling consistent with the mission of the school and the educational program. The main driving force behind instrumental tutoring is, as the teacher Lf6 puts it, "closer contact with the teacher." The purpose of the intervention is masked by a caring face and techniques of subtle disciplining (Foucault, 2001). In instrumental tutoring, the care for a person has a hidden program of a *good guardian* who knows what is suitable for the tutees. Tutoring becomes a technique of controlling the freedom and subjectivity of the student. "These are techniques which allow individuals to work on themselves to become different kinds of subjects. Such technologies, however, could also bind us to categories in which individuals are subjected to social control" (Wong, 2013, p. 6).

Instrumental tutoring as a transitional form could be observed in schools E, F, and H. Tutors, who worked with students who were in the group at risk of social exclusion, were looking for effective methods. When they became convinced that tutoring could serve as a potentially useful method, they became involved in a direct relationship with the students whom they had previously avoided due to fear of manipulation and abuse of trust by the students. However, the tutoring meetings changed the tutors' attitudes toward the students in the course of school tutoring development:

**Excerpt GRD-E 1494–1497**
Ef8: *For me, tutoring gives the school sort of a human face, so that I am a human being, and I think it also gives a lot to kids, and makes them think I am not a teacher, but a human being, and he is not just a student but also a human being. I think that is it.*

At first, the significant effect of the tutoring was an image of a "human face" of the school. Over time, however, a critical discussion developed among the teachers who began to see increasingly that the student "is not just a student but also a human being." Experiencing the contradiction between the idea of a personalized pedagogy and the form of tutoring that had been developed thus far was no longer accepted, and the tutors have begun to learn new patterns of the teacher-student relationship.

In one school (L), instrumental tutoring remained in a permanent form and coexisted with activities that can be identified as synergic tutoring. The teachers in this school combined tutoring with tests, diagnosis, and devising work schedules with the students, all of which were subsequently included in the student's personal records. The cooperation with the tutees within the framework of tutoring has been strongly formalized; however, the individual teachers within this bureaucratic framework have developed space for open and deeper relationships with the pupils. Although the group discussion did not include a critical reflection on this tutoring practice, during the individual interviews, two teachers began to review this form of tutoring critically. For the time being, concealing instrumental tutoring by emphasizing its effectiveness allows the group of tutors in school L to maintain the illusion of an individualized educational process.

*Krzychała*

*Synergic Activity*

Synergic activity refers to tutoring that is adopted as one of the central elements of the school activity. Tutoring inspires and influences the development of the entire school curriculum, and in a school, educational performance stimulates and promotes further development of tutoring beyond the boundaries set by the first tutoring experiences and methods acquired during training. Synergic tutoring can be observed in schools with a more extended period—that is, from 5 to 7 years—for developing a new orientation pattern. In the research, these schools are marked with the symbols E, F, G, H, K, and L.

In synergic tutoring, even when they experience difficulties and a chronic lack of time for all tasks, the teacher-tutors' attention is no longer focused on organizing work and combining it with other responsibilities. The essential point of reference is the orientation framework—that is, the individual and team experience gained in tutoring with students and in cooperation with other tutors. Tutoring is seen as an *autonomous* understanding and for pedagogical activity, a *central* concept that covers the entire learning period students spend at school. The tutors recognize the potential of the students and the students' diverse needs, including extracurricular ones beyond school achievements. The tutors are much more comfortable in dealing with difficult situations such as conversations with hyperactive or taciturn students who adopt a passive or confrontational attitude in their relationship with the tutor. The tutors adhere less strictly to the tutorial scenarios and methods learned from their initial training:

> **Excerpt GRD-F 348-361**
> Ff2: *Later on. For example, (yy) (.) just after such integration classes and these,, and now I want to do exactly this (yy) in my class in the second semester, even though it is the first grade.*
> *Still, I would like to set such goals now, that is, agree with the children on the goals to achieve. Normally, I would do it in the second grade, and most often, in the third grade, when we have got to know each other better. But, for example, when I am looking at my children, it seems to me that I should do it right now. Especially because they themselves are saying: "At the end of the year I would like to, for example, have an average of four point-something, right?" And that's with, that's how I did it. Actually, these were the children's decisions, and they signed next to that note to confirm. And at the end of the year, we checked whether they made it or not or in which areas they were successful and in which ones, not.*

The teacher Ff2 does not perceive her current work with the tutees as a simple repetition of the good practices elaborated in the initial phase of a tutoring project. She can modify her approach flexibly in response to her interactions with the students. While planning tutorial activities, the teachers fuse and mediate several perspectives: (1) the recognition of the tutees' perspectives and expectations ("especially because they themselves are saying"), (2) the teacher's own professional overview of the students' needs and development potential ("when I am looking at my children"), (3) the timing of the intervention in the broader process of tutoring and the cycle of the students development at school ("I should do it right now"), and (3) the flexible application of tutorial methods ("Normally, I would do it in the second grade", "And at the end of the year, we checked whether they made it or not").

In reality, however, the examined cases do not represent pure synergic tutoring. Thus, they can be considered to be pragmatic forms of synergic tutoring. The pragmatic form of synergic change integrated into schools has been developed and tested in at least one full 3-year cycle of work with the students (from the first grade to the end of junior high school attendance). In the subsequent cycle, the teachers refer not only to the ideas of tutoring but also to their own experiences. They do not focus on the current organization of meetings. They perceive the tutor-educated relationship as a long-term process of counseling and broadening the scope of cooperation. The pragmatic form of synergic tutoring emerges as a result of compromises and the arrangement and expanding of new forms. The pragmatic form of tutoring moves more or less away from the theoretical school tutoring model, with which teachers become acquainted during the initial training and which is described in the tutoring program. Understanding the essence of activity in the schools with synergic tutoring requires not only observing the activity itself but also analyzing the ways of developing a new orientation pattern.

### School in the Process of Change: A Case Study

The process of creating a new orientation pattern can be presented by using the sociogenetic reconstruction of the changes experienced by the teachers implementing tutoring at schools. Initially, the school tutoring program (Budzyński et al., 2009) assumed that the implementation of the new orientation pattern would be carried out in two model stages (Figure 2). In the first stage (Figure 2; Pathway ①), selected teachers were supposed to introduce tutoring ideas in a small number of classes and test organizational solutions adequate to the needs of the school. In the second stage (Figure 2; Pathway ②), additional teachers and students would be involved in the tutoring until this pragmatic change became an integral part of the school's system of activity. No contradictions in the implementation of the program were foreseen.

However, such a model change process was not observed in any of the 12 schools participating in the study. The paths for developing a pragmatic model of school tutoring not only differed from the model expectations but also differed from each other. One universal process of change in teacher

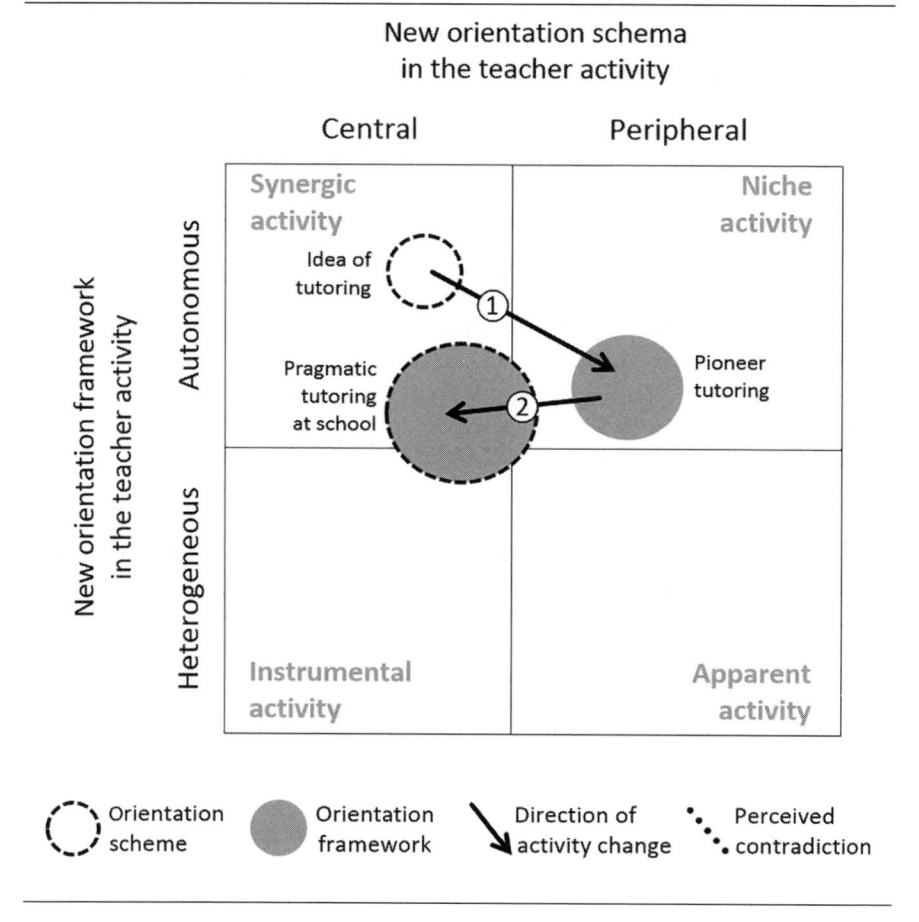

*Figure 2.* **The expected model process of school tutoring implementation.**
*Source.* Author's elaboration.

activity cannot be identified. Nevertheless, even if the process differs significantly from the model assumptions, the proposed model can be applied to reconstruct the process of change in a particular school, such as in school H. It is the largest school participating in the study (Table 1). The tutoring program was enacted for 7 years in this school. The process of change in school H is presented graphically in Figure 3. At first, the process seems complicated. This is because it, in fact, is. The teachers did not describe the process in an orderly manner or as a step-by-step process with subsequent stages. During the interviews, all the elements of social history appeared simultaneously in the narrative. Nevertheless, enabling the understanding of the pragmatic form of the orientation pattern developed in this school, the

**New orientation schema
in the teacher activity**

|  | Central | Peripheral |
|---|---|---|
| Autonomous | **Synergic activity**<br>Idea of tutoring<br>Pragmatic tutoring at school H | **Niche activity**<br>Anticipation of 'privatized' tutoring<br>Spontaneous tutoring |
| Heterogeneous | **Instrumental activity**<br>Use of tutoring to correct behavior | **Apparent activity**<br>Redundant tutoring |

New orientation framework in the teacher activity

Legend:
- Orientation scheme
- Orientation framework
- Direction of activity change
- Perceived contradiction

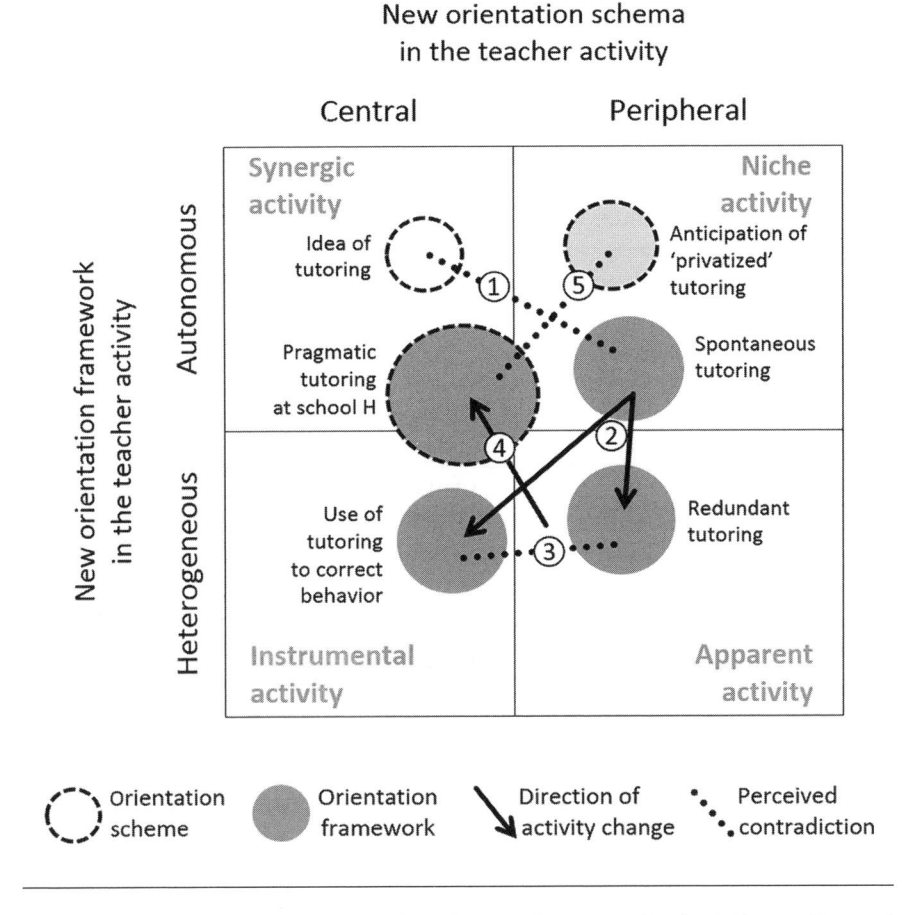

*Figure 3.* **The process of changing the teacher-tutors' orientation pattern at school H.**
*Source.* Author's elaboration.

five stages of change experienced in this teacher team were identified. The purpose of sociogenetic reconstruction is neither to chronologically reconstruct particular stages nor to recognize how tutoring was perceived in the past. The interpretation focuses on the pragmatic knowledge that teachers document here and now in their narratives. The references to individual and collective experiences, including those of the past or those to which the interviewees have already distanced themselves or barely remembered, are narrated—according to the perspective documented in the interviews—from the present point of the experience sense. The sociogenic perspective is reflected in the founding stories about the beginnings of tutoring at the school and about turning points, which were reported in interviews also by teachers who did not directly participate in these events but currently identify themselves with them. The primary question in the analysis is not how precisely the specific stages looked in the past (which could be an adequate task for a longitudinal study) but how their meaning is reflected in the present point of view. Figure 3 should, therefore, be read as the overall representation of a current pragmatic orientation pattern with its retrospective and anticipative links. In interviews, however, this *total* and *simultaneous* character of the experience sense is translated into a *sequential* narrative, (Bohnsack, 2014), and only the reflecting analyses I and II reveal the connection between the temporally distant components of the experience.

## Becoming a Tutor and a Tutoring School: A Sociogenetic Reconstruction

The first stage of change in school H involved the experience of conflict between the tutoring based on a goal-setting strategy concept, which the teachers knew from the training, and the tutoring practice undertaken by six teacher-tutors in the first two classes (Figure 3: Pathway ①). "The beginnings were tragic. The children did not know what it was all about" (Hf2, GRD-H 232–233). The tutors did not see any opportunity to adapt the method to the students' needs. The students were not able to set their own goals that could be discussed in the tutorials. However, the tutees took the initiative because they saw a new area of activity for themselves that was different from school education: "They became very much involved" (Hf2, GRD-H 235). They proposed their own areas of activity. When remembering that time, the tutors emphasized that they were learning to tutor from their students. The tutors and tutees formed an informal learning community. From the present perspective, the dichotomy between the adopted model knowledge about action (orientation scheme) and the spontaneously shaped understanding *in* practice (orientation framework) is highlighted. The tutor Hf2 remembers how the students thanked her after 3 years for their spontaneous meetings, which were quite different from what the tutors had expected:

> **Excerpt GRD-H 248-254**
> Hf2: *They brought me such a nice bouquet of red roses and thanked me for the fact that they could show me their card tricks, which they did not show in other classes because they did not have satisfactory results. And that they could go with me to the park. That on Saturday, I was not ashamed to have ice cream with them in the mall. So, these are such <u>trivial things</u> for us, and to them, they simply must have seemed very, very important.*

At that time, through interactions with the students, the first tutors developed a pragmatic orientation framework in which it was essential to leave space for the students' free activity, accept their interests, and build relaxed tutor-tutee relationships. The tutorials were held only partially at school. However, the teachers considered the change to be vital because the students felt much better at school. They intended to extend the activity to all classes.

In the second stage, after 2 years, other tutors and the next classes joined the program (Figure 3: Pathway ②). The experience of spontaneous tutoring gained by their colleagues provided the point of reference for the next tutors. The new tutors tried to repeat the pattern of spontaneous tutoring. For some of the tutors, this turned out to be an opportunity for an informal and subtle influence on the students. Several tutors set a goal for themselves of making at-risk students take responsibility for their learning and maintaining safety at school. They shaped tutoring according to these goals:

> **Excerpt GRD-H 206-209**
> Hf3: *From the beginning, at all times, such modeling. Someone from the outside might call it something negative, but it seems to me that modeling in this manner should come first, and only afterward, teaching openness to the world.*

Another group of tutors began working with students who had good learning outcomes. Tutoring proved unnecessary when offered to students who seemed not to need it:

> **Excerpt GRD-H 263-265**
> Hf2: *They do not show any behavioral problems, but neither do these children see such kind of need (.) to come and talk, to establish any different bond.*

As the school tutoring program expanded, various bottom-up initiatives of the niche, instrumental and apparent tutoring models were developed. The tutors met regularly and discussed their diverse experiences. Over time, during these conversations, critical discussions began to occur about the sense of conducting the instrumental and apparent tutoring, which visibly differed from the working style of the classes that initiated niche tutoring. Additionally, organizational problems became more apparent, as the teachers had to combine spontaneous tutoring with the numerous tasks they performed at school.

These discussions initiated the third stage of change (Figure 3; Pathway ③). Tutors Hf1 and Hf2, who were considered to be the group leaders in the tutor team, played a decisive role. Both tutors began tutorials in new classes; however, they shared the opinion that these students "did not need tutoring." The students did not see the need to enter into a new tutor relationship exceeding the established relationship with the teacher in the lesson. The perceived contradiction represented by the different reception of tutoring by the students became a call for rethinking the sense of tutoring at the school and devising a new model. The tutors critically discussed their experience and developed a new understanding of the tutor's role, which they described in interviews as becoming a "significant adult." In the group discussion, the tutors reproduced their thoughts on why they had developed a new pedagogical point of view and what they wanted to keep in school tutoring:

> **Excerpt GRD-H 968–985**
> Hf2: *What gives me satisfaction in tutoring is the conversation with young people. Of course, when they open up, when they start talking about their problems and, and, and they are waiting for some advice, for what you will say. It is often the case, even I have noticed that, that they perceive me <u>differently</u> than a parent. That I am for them, as if, such a (y), they do realize I am older, so as if I were <u>their</u> parent, but I am also a partner for conversation, who is not necessarily the one that will tell them this, this, this, and that is bad.*
> Hf3: *The greatest satisfaction for me is listening because they just like speaking. And I just listen, (yy) yes, I just listen. At some point, when it is necessary, of course, I interrupt with a question. But I like most to listen. When they,, I see that the <u>more</u> I listen and the more they say, the less tense and more relaxed they become, as simple as that.*
> Hm8: *When a child comes with a problem, it means that he or she trusts us. That is (.) as a proof of the greatest trust for me, i.e., so they don't go to, let's say, their parent, but they come to me. Well, problems can be different, different kinds, right?*

This reflection began the fourth stage of change, which the teachers place at the third and fourth years of the tutoring project at the school (Figure 3; Pathway ④). On one hand, they refer to their rich and varied experiences (orientation framework); on the other hand, they developed a new tutoring discourse defined as a meeting of an adolescent and a significant adult (orientation scheme). In the interviews, regarding the tutoring program at the school, the tutors spoke about a "second" and "new" start, which addresses the experienced contradiction resulting from attempts to continue the "first" model of tutoring. The tutors introduced regular 2-hour tutorials into the school timetable. This made it possible to organize regular meetings in a large school and to freely choose a tutor not only from their class teachers but also from among all the teachers in the school. They initiated tripartite tutor-tutee-parent meetings, where together they discussed the student's strengths and plans. The parents were initially skeptical about coming to these meetings "because, also based on their own experience from the school years, being called to school usually meant that (.) there was a problem" (Hf2, GRD-H 796–797). These meetings facilitated the development of a new model of goal-oriented work that was also applicable for talented students.

The school has already trained more teachers than the number of tutors needed. Therefore, not all tutors run tutorial meetings; however, they are familiar with the principles and goals of the tutoring method. The teachers as a whole group, not only the tutors themselves, have reorganized the activities already undertaken at the school, giving them new meaning in the context of the tutoring organization. City trips and slumber parties organized at the beginning of the school year have ceased to be just a way of spending leisure time. They have gained meaning as a place for teachers and students to meet in extracurricular situations, thereby allowing the tutees to choose their tutor more consciously. The launching of a self-service kitchen—although not directly related to tutoring—has facilitated the transfer and—as the teachers call it—the dispersal of the tutorial style to other meetings in the school. In interviews, the tutors emphasized that this was their original idea for school tutoring, which they had been testing over the past 3 years.

The sharing and the affirmation of the tutorial orientation pattern are documented in the way teachers talk about initiating tutor-tutees meetings at school:

> Excerpt GRD-H 408-410
> Hf5: *Yes, a, (y)- so everything is such that very often, it is . . .*
> ⌊ Hf1: *(yay)*
> . . . *such contact during a break, please, because I have to tell you something. I say, well, well, let's us go to the classroom, or (.) (y) I say, maybe, there is <u>some occasion</u>, maybe we'll meet with five people or something else, isn't it? We just have a kitchen at school, and this is a very good place to do everything. Because you can talk <u>by the way . . .</u>*
> ⌊ Hf1: *Yes, yes.*
> . . . <u>*while*</u> *making cookies, teas, or other such things, so that is the right moment . . .*
> ⌊ Hf3: *Sometimes, you need to make a reservation.*
> . . . *for tutoring, and this (.) arrangement is very nice, because it is very, very, very, so to speak, it facilitates the whole process.*
> ⌊ Hf1: *The kitchen is on the ground floor, yes.*
> ⌊ Hf2: *And children like it.*
> ⌊ Hf3: *They like it very much @.@*

This excerpt was initiated by the moderator's question: "How do you organize these tutorial meetings?" (Mf1, GRD-H 380). The first answers were quite casual, including responses containing phrases such as "it is, so that much depends on what my students expect" (Hf1), "very different, because" (Hf5), and "however, it depends on" (Hf3, GRD-H 381–389). The teachers did not give a simple answer, although it would probably not have been difficult for them to describe an organizational tutorial scheme. For them, however, a different perspective is important: the presentation of the principle that characterizes the developed tutoring formula. The moderator's question was transformed into a *proposal*, and the above fragment is a *reaction* in which the proposed perspective of understanding the topic of the tutorial organization is confirmed and specified. Tutor Hf5 describes the principle by recalling examples of situations in which a tutor-tutee contact is initiated. The organization of tutorials is a response to the students' initiative and a flexible use of opportunities by the teacher; therefore, the organization of the tutorials "depends on" other factors. The tutor emphasizes the importance of "some occasions", acting "by the way," and "while." The way other teachers interfere in her statements indicates that this is not just an isolated perspective. They complement and reinforce the statements of their colleague. This orientation pattern is shared by the whole team. The specific meaning of the "it depends" is further confirmed in the conversation's *conclusion*, in which the teachers ensure that the students prefer when the meetings are initiated in this way. Further on, the teachers develop this theme and make it clear why students like these meetings: "This is also an option that kids like very much and you can also, talk about things because different things are going on and they can talk to each other" (Hf1 in GRD-H 408–410). The timing and relevance of the tutorial meetings do not result from the implementation of the plan (although this one was recently prepared) but is conditional on the readiness of the tutor and the tutee to negotiate how the meeting will be conducted; this enables the students to enter into a dialogue regarding tutoring and to engage in peer tutoring. The fact that the tutors reconcile their professional perspective with the perspective of the tutees is one of the distinguishing features of synergistic tutoring.

However, a motif of a new crisis appears in these interviews. In 2016, the city council withdrew funding from the tutoring program and undertook preparations to close the junior high schools. The program was terminated. The teachers know that they will not be able to continue this form of tutoring. In the interviews, the teachers consistently mentioned the issue of the new fifth stage that they have entered and the challenge of how to maintain the tutoring experience in the new schools.

> **Excerpt IDI-Hf8 648-654**
>
> Hf8: *This year, we have already had the information that there will be no tutoring because there is no money. A colleague, who used to work with us said she liked her tutoring very much. And now she will still come up to join her tutored class during my lessons, and she will continue to help somehow because she likes it so much. That is why my impression is that this tutoring, it will survive. It will be carried out in a slightly different formula.*

This fifth stage was anticipated by the teachers (Figure 3; Pathway ⑤). If it happens soon, everyday practice will verify the speculations mentioned above. The teacher Hf8 and her colleague are planning how they can continue their tutors' cooperation on the official completion of the tutoring project. They assume that their engagement can "be carried out in a slightly different formula." This "different formula" already includes activities undertaken on one's own. For the time being, "privatized" tutoring appears to be an imagined solution. This challenge was not yet addressed in the group discussion. The problem came to light in individual interviews, not spontaneously but in response to an additional question about the expected future of tutoring. These images are still vague. The germ of the anticipated change is already addressed as the fear of whether it will be possible to keep the tutoring experience outside of the team and outside the school that has already integrated a new pedagogical orientation pattern.

*Embedding Synergic Change at School: A Comparative Analysis*

In the analyzed 12 schools, especially in the course of Reflective Analysis I of group discussions and Reflective Analysis II, in which individual interviews of the teachers participating in the discussion were included in the comparative analysis, different sociogenetic paths for creating patterns of pedagogical orientation can be reconstructed. In this way, in the presented case of school H, the second stage, in which the individual attempts to reproduce the orientation scheme worked out by the pioneers played a decisive role, was reconstructed in more detail. Guidance from the training and the pioneers' suggestions/tips proved to be insufficient and inadequate preparation for work in the more diverse classes. As the first to begin the tutoring project, the teachers designated two classes in which they had already seen the need to improve relations with the students. In the following years, the tutoring included classes in which the students—as the teachers expressed it—"did not need tutoring" in the form that had already been developed.

In Reflective Analysis II, the strategy of comparing various schools was applied to identify different approaches in which similar issues were addressed. This comparison verifies, on one hand, the orientation patterns developed in the course of Reflective Analysis I, and on the other hand, it allows for the further specification of the sociogenetic determinants of the development of the teachers' orientation patterns. Introducing the case of school F into the comparative analysis highlights the specificity of school H. School F is the smallest school in the research, and it is attended by students at risk of social exclusion and with low educational achievement (Table 1). In school F, only two primary stages of change in the school activity were reconstructed (Krzychała, 2019): one within a niche activity, the second as a transition to a synergistic activity that spread a tutorial orientation pattern across the whole school. The form of instrumental or apparent tutoring did not appear. Even before the beginning of the tutorial program, the teachers formed a harmonious team and took many initiatives to develop an individual approach to the students with numerous problems in learning and social relationships. The teachers assessed the idea of school tutoring as directly addressing their pedagogical needs. This process of change is very similar to the model assumption (Figure 2) but differs from the first stage, which is marked by an experience of contradiction. The teachers began to carry out tutorial meetings with accuracy. However, they were soon confronted with the problem of regularity and punctuality as well as the setting of personal goals, which turned out to be a challenge for the students. After a series of informal discussions, the tutor-teachers began to turn tutorial meetings with small groups of students into social projects, such as preparation for homework, training for sports street competitions, or a project to renovate a neglected courtyard in a housing estate. This change in the orientation pattern is documented in one of the teacher's statements:

> **Excerpt IDI-Ff4 165–169**
> Ff4: *I gave up on some high-flying, big goals for only small tasks; (.) small, if they do not come out, (.) there is no punishing finger; yes, we are only trying, we are looking further. It doesn't go out with every child, but (.) in general, if there is a good relationship, you can go further.*

The teachers saw the effects of work based on building "small tasks." Teacher Ff4 expressed it: "I have engaged them <u>in action</u>. In action, not in chatting, but in action" (GRD-F 396–397). At the same time, the teachers expressed concern as to whether this was still tutoring, as their actions differed significantly from the model of tutorial meetings learned at the training. In principle, each teacher conducted activities with groups of students in a completely different way, such as sports activities, psychotherapeutic workshops, social projects, or lessons with individually defined learning objectives. Over time, such a remodeled tutorial program became an integral part of the lessons and additional projects.

A comparative analysis among the schools that developed a synergic form of tutoring (H and F as well as E, G, K, and L) revealed that depending on the social structure of the students, there were different forms of tutorial work. In working with students who achieve good learning outcomes, the tutoring has retained the form of individualized meetings focused on the setting and achievement of objectives not related to the school curriculum. However, in work with average or low achievers, the initial formulas for regular meetings was replaced by work in small groups, with a focus on identifying deficits and needs and supporting social and learning competence—as in school F and partly in school H. In all schools with a pragmatic form of synergistic tutoring, although in different variants, a change in the teaching in classes was observed.

The teacher-tutors transformed the modus operandi of the tutoring—the orientation of dialogical and personalized cooperation with the tutees—into other areas of school activity. This change occurred at the level of the orientation framework, not the orientation scheme. The teachers did not design new strategies for teaching; however, they did notice a qualitative change in their classroom work. The tutors describe this influence as "scattered tutoring" as "acting without naming this tutoring" (IDI-Ef9 1500). This change in the way in which lessons are taught is to some extent similar to the changes that were observed in other studies on improving dialogue practices in the classroom (Snell & Lefstein, 2017) and on the influence of the teacher's knowledge about students in the teaching process (Hill & Chin, 2018; Sadler, Sonnert, Coyle, Cook-Smith, & Miller, 2013). Snell and Lefstein (2017) examined the impact of a professional development program ("Towards Dialogue") designed to promote the teachers' interactional awareness and sensitivity. They observed the classroom interactional patterns' change in the wake of that intervention; however, the change could still be considered as a niche change in teaching. The effect of differentiated interactions was particularly noticeable in relation to students who were perceived as academically able; in relation to other learners, with whom interactions were more sporadic and less challenging, the change was only marginal. A different effect was noted in the Wrocław tutoring program. The teacher-tutors considered students from a broader perspective than merely the context of the subject being taught, and they set tasks that also involved the weaker students. The difference can be attributed to the orientation on individualized understanding and support of the student in the tutoring, while in the project Towards Dialogue, the focus was on the organization of the lesson. This explanation can also be supported by Hill and Chin's (2018) study. The authors prove that teachers with more knowledge of the students engaged in more remediation of student misconceptions and more often used student thinking in instruction, while teachers with less knowledge about the students limit interventions to regroup students and reteach content.

## Discussion

In this study, a theoretical model for reconstructing the process of transformation of professional orientation patterns was designed based on the inquiry into the school tutoring program expanding the class- and lesson-based curriculum. Analytically identified both as an orientation scheme and as an orientation framework, the praxeological concept of orientation patterns enables an analysis of the school development not only at the level of the description of institutional solutions but also as tacit agreements and a habitual understanding. In a long-term change in educational practice, elements of *synergic, niche, instrumental,* and *apparent* activities can swap, intertwine, or coexist as transitional or permanent forms of incorporating new pedagogical practices in schools.

The study of the school tutoring program covered 12 schools, including 8 schools developing tutoring from 5 to 7 years. In six schools, various pragmatic forms of integrating synergistic tutoring into the school's system were reconstructed. In these schools, the pedagogical innovation does not cover a selected group of students but becomes an integral part of the school's curriculum. The analysis showed that although the tutoring process lasted up to 7 years, the process of changing schools was still open. However, due to the dissolving of junior high schools in the results of the school system reform initiated in Poland in 2016 (European Commission/EACEA/Eurydice, 2018), it was not possible to further investigate these teams of teachers. The study was designed in such a way that the object of analysis was the teachers' experiences. It took into account neither the interviews with students nor the direct observation of the tutorial meetings or classroom teaching. Further study on the sociogenetic process of teacher knowledge transformation could benefit by merging data from different sources to reconstruct the professional learning community experience and school improvement from multidimensional perspectives.

School development may not be linear and homogeneous; furthermore, based on specific experiential spaces in a school, it can vary in numerous forms. Presumably, if the research could be extended to other schools, we would be able to obtain an even more diverse picture of the possible trajectories of changes. The proposed general model of theoretical typology—ideal types in Webber's sense—is designed as an analytical tool that can help in the reconstruction of particular cases. Moreover, it allows for the interpretation of critical and contradictory moments in the process of emerging new orientation patterns. From this perspective, *apparent* and *instrumental* activities can also be an essential element of the experience, as they inspire critical reflection and open new challenges for school development. For those who support teams of teachers, both at the level of school administration and at the level of counseling and supervising, this observation sets an essential objective to be attentive to the conflicts and challenges that have been experienced, while at the same time not closing the space for experience by arbitrarily accelerating decisions and setting preestablished goals. Sharing ideas and goals—as in the case of apparent change—can only turn out to be an illusion of change.

Concerning far-reaching changes, the teacher community can constitute itself by learning how to overcome contradictions within the school practice. Engeström (2015) argues that the creation of new forms of activity can just be initiated when solving problems and challenges turn out to be impossible on the basis of orientation patterns shared by the community. Significant changes in practice are opened up through confrontation with double-bind situations that are often experienced as "personal crises," "breaking away," "turning points," or "moments of revelation" (Engeström, 2015, p. 122). This conclusion is in line with the reconstruction of the school development process, which led to pragmatic forms of synergistic tutoring (schools E, F, G, H, K, and L). Initially, the first phase of school changing was marked in part by the experience of contradiction between the idea and practice of tutoring and in part by the subsequent experience of contradiction between the niche tutoring and extending tutoring when successive teacher-tutors and student-tutees were included. The practice becomes a space for expansive learning if it comprises "a social, societally essential dilemma that cannot be resolved through separate individual actions alone—but in which joint cooperative actions can push a historically new form of activity into emergence" (Engeström, 2015, p. 131). These "joint cooperative actions" seem particularly crucial when after the enthusiastic *niche* change stage, the new practice cannot simply be multiplied in the institution.

The strategy of triangulation of group and individual interviews applied in the study indicates the potential for further research on the relationships between the individual and collective experiences of teachers. In the analyzed interviews, we observed, as in the cases of school H and F, the desynchronization of individual and collective changes. Among other individual characteristics (e.g., related to the subject taught and the profile of pedagogical preparation as well as personal and educational biography, social experience of exclusion or advancement), differences in the teachers' careers and in their range of levels of experience and proficiency can vary significantly in the way in which they influence how teachers within the same school incorporate new orientation patterns.

Further studies should take into account the differences in the experience of change encountered by teachers who begin working at a school and include the new orientation pattern in their professional development from the very beginning. On the other hand, further analysis may involve experienced teachers whose professional development is connected with the already acquired and the new orientation patterns' integration and coordination based on a critical and pragmatic revision of both. In the Wrocław schools, leaders of the change emerged from both groups of teachers (Krzychała, 2018). From this perspective, the professional praxis community can be examined from the perspective of a multi-individual experience environment in which educational change creates a knot of activity (Engeström, 2007) that focuses on different points of view and a sense of practice.

The proposed model of interpreting the teachers' responses to the new educational practice may be used in a further study on educational change, regardless of whether this change is situated in the area of individualized teaching/tutoring or democratic, multicultural, and inclusive education. The continuation of research on institutional changes from the praxeological perspective in diverse cultural and educational contexts, taking into account the long-term and multidimensional process of transforming professional knowledge, has the potential not only to verify the proposed model but also to spark new methodological and practical inspirations in advancing the research on the professional learning communities.

## ORCID iD

Sławomir Krzychała https://orcid.org/0000-0002-1339-4657

## Notes

I thank Agnieszka Zembrzuska, PhD, for coordinating the field study at 12 schools in Wrocław and PhD students from the University of Lower Silesia for their assistance in moderating 64 interviews. This study was financed by the National Science Centre in Kraków, Poland (Project No. 2014/15/B/HS6/03116) and conducted in the period August 2015 to March 2018.

[1] The transcript symbols are based on the simplified transcription version proposed by Bohnsack (2014). The basic characters used in the article are as follows: (.) marks a short pause, suspended utterance; the underlined text indicates words that are spoken loudly or accentuated; ⌊ begins an utterance that overlaps with other utterances; and (hm) indicates incomprehensible fragments, in which the words represent the most likely text of the utterance.

[2] Names and proper names were replaced with symbols (e.g., Af1, Af2—female teachers from school A; Bm1—a male teacher from school B; Mf1—the interview moderator; Bgr—an utterance by several people at the same time; GRD-B—group discussion at school B; and IDI-Bf1—an individual interview with Bf1).

## References

Amling, S., & Vogd, W. (Eds.). (2017). *Dokumentarische Organisationsforschung: Perspektiven der praxeologischen Wissenssoziologie* [Documentary organizational research: Perspectives of the praxeological knowledge sociology]. Opladen, Germany: Verlag Barbara Budrich.

Asbrand, B., & Martens, M. (2018). *Dokumentarische Unterrichtsforschung* [Documentary teaching research]. Wiesbaden, Germany: Springer Fachmedien Wiesbaden.

Ashwin, P. (2005). Variation in students' experiences of the "Oxford Tutorial." *Higher Education, 50*, 631–644.

Beatty, I. D., & Feldman, A. (2012). Viewing teacher transformation through the lens of cultural-historical activity theory (CHAT). *Education as Change, 16*, 283–300.

Bohnsack, R. (2014). Documentary method. In U. Flick (Ed.), *The SAGE handbook of qualitative data analysis* (pp. 217–233). London, England: Sage. doi:10.4135/9781446282243.n15

Bohnsack, R. (2017a). Praxeological sociology of knowledge and documentary method: Karl Mannheim's framing of empirical research. In D. Kettler & V. Meja (Eds.), *The Anthem companion to Karl Mannheim* (pp. 199–220). London, England: Anthem Press.

Bohnsack, R. (2017b). *Praxeologische Wissenssoziologie* [Praxeological knowledge sociology]. Opladen, Germany: Verlag Barbara Budrich.

Bohnsack, R., Pfaff, N., & Weller, W. (Eds.). (2010). *Qualitative analysis and documentary method in international educational research*. Opladen, Germany: Verlag Barbara Budrich.

Budzyński, M., Traczyński, J., & Czekierda, P. (2009). *Tutor we wrocławskim gimnazjum: Program innowacji pedagogicznej* [A Tutor at the Wrocław gymnasium: Pedagogical innovation program]. Unpublished Internal document of the Tutors' College, Wrocław, Poland.

Camburn, E. M., & Han, S. W. (2017). Teachers' professional learning experiences and their engagement in reflective practice: A replication study. *School Effectiveness and School Improvement, 28*, 527–554.

Dawson, W. (2010). Private tutoring and mass schooling in East Asia: Reflections of inequality in Japan, South Korea, and Cambodia. *Asia Pacific Education Review, 11*, 14–24.

Desimone, L. M., & Garet, M. S. (2015). Best practices in teachers' professional development in the United States. *Psychology, Society and Education*, 7, 252–263.
Drozd, E., & Zembrzuska, A. (2013). School tutoring as a concept and a support method in student's development. *Forum Oświatowe*, 2, 167–175.
Ehri, L. C., Dreyer, L. G., Flugman, B., & Gross, A. (2007). Reading rescue: An effective tutoring intervention model for language-minority students who are struggling readers in first grade. *American Educational Research Journal*, 44, 414–448.
Engeström, Y. (2007). *From teams to knots. Learning in doing: Social, cognitive and computational perspectives*. Cambridge, England: Cambridge University Press.
Engeström, Y. (2015). *Learning by expanding: An activity-theoretical approach to developmental research*. Cambridge, England: Cambridge University Press.
Engeström, Y., Kajamaa, A., Lahtinen, P., & Sannino, A. (2015). Toward a grammar of collaboration. *Mind, Culture, and Activity*, 22, 92–111.
European Commission/EACEA/Eurydice. (2018). *The structure of the European education systems 2018/19: Schematic diagrams*. Luxembourg: Publications Office of the European Union. Retrieved from https://eacea.ec.europa.eu/national-policies/eurydice/content/structure-european-education-systems-201819-schematic-diagrams_en
Foucault, M. (2001). Truth and power. In M. Foucault (Ed.), *Essential works of Foucault (1954–1984): Vol. 3. Power* (pp. 111–133). London, England: New Press.
Gallagher, T., Griffin, S., Ciuffetelli Parker, D., Kitchen, J., & Figg, C. (2011). Establishing and sustaining teacher educator professional development in a self-study community of practice: Pre-tenure teacher educators developing professionally. *Teaching and Teacher Education*, 27, 880–890.
Hill, H. C., & Chin, M. (2018). Connections between teachers' knowledge of students, instruction, and achievement outcomes. *American Educational Research Journal*, 55, 1076–1112.
Hofer, B. K. (2017). Shaping the epistemology of teacher practice through reflection and reflexivity. *Educational Psychologist*, 52, 299–306.
Hord, S. M., & Sommers, W. A. (2008). *Leading professional learning communities: Voices from research and practice*. Thousand Oaks, CA: Corwin Press.
Krzychała, S. (2018). *Nauczyciel-tutor: Prakseologiczna rekonstrukcja tutoringu szkolnego* [Teacher-tutor: A praxeological reconstruction of the school tutoring]. Kraków, Poland: Oficyna Wydawnicza "Impuls".
Krzychała, S. (2019). Professional praxis community in a dialogical perspective: Towards the application of Bakhtinian categories in the documentary method. *Forum Qualitative Sozialforschung*, 20, Article 17. doi:10.17169/FQS-20.1.3073
Loos, P., Nohl, A.-M., Przyborski, A., & Schäffer, V. B. (Eds.). (2013). *Dokumentarische Methode: Grundlagen—Entwicklungen—Anwendungen* [The documentary method: Basics-development-applications]. Opladen, Germany: Verlag Barbara Budrich.
Mannheim, K. (1952a). On the interpretation of "Weltanschauung." In P. Keckskemeti (Ed.), *Essays on the sociology of knowledge* (pp. 33–83). London, England: Routledge & Kegan Paul.
Mannheim, K. (1952b). The problem of generations. In P. Keckskemeti (Ed.), *Essays on the sociology of knowledge* (pp. 276–320). London, England: Routledge & Kegan Paul.
Mannheim, K. (1964). Beiträge zur Theorie der Weltanschauungs-Interpretation [1921–1922; On the Interpretation of the Worldview (translated title)]. In *Wissenssoziologie* (pp. 91–154). Berlin, Germany: Hermann Luchterhand Verlag.

Mannheim, K. (1980). *Strukturen des Denkens [1922–1924; Structures of thinking* (translated title)]. Frankfurt am Main, Germay: Suhrkamp.
Mensching, A. (2008). *Gelebte Hierarchien: Mikropolitische Arrangements und organisationskulturelle Praktiken am Beispiel der Polizei* [Living hierarchies: Micro-political arrangements and organizational-cultural practices on the example of the police]. Wiesbaden, Germany: VS Verlag für Sozialwissenschaften.
Moore, W. G. (1968). *The tutorial system and its future*. Oxford, England: Pergamon Press. Retrieved from http://worldcatlibraries.org/wcpa/oclc/189181
Ortmann, G. (2004). *Als ob: Fiktionen und Organisationen* [As if: Fictions and organizations]. Wiesbaden, Germany: VS Verlag für Sozialwissenschaften.
Palfreyman, D. (2008). *The Oxford Tutorial: "Thanks, you taught me how to think"* (2nd ed.). Oxford, England: OxCHEPS.
Parker, M., Patton, K., & Tannehill, D. (2012). Mapping the landscape of communities of practice as professional development in Irish physical education. *Irish Educational Studies, 31*, 311–327.
Popp, J. S., & Goldman, S. R. (2016). Knowledge building in teacher professional learning communities: Focus of meeting matters. *Teaching and Teacher Education, 59*, 347–359.
Pultorak, E. G. (Ed.). (2010). *The purposes, practices, and professionalism of teacher reflectivity*. Lanham, MD: Rowman & Littlefield.
Rajala, A., Kumpulainen, K., Rainio, A. P., Hilppö, J., & Lipponen, L. (2016). Dealing with the contradiction of agency and control during dialogic teaching. *Learning, Culture and Social Interaction, 10*, 17–26.
Rapley, T. (2007). *Doing conversation, discourse and document analysis*. London, England: Sage.
Sadler, P. M., Sonnert, G., Coyle, H. P., Cook-Smith, N., & Miller, J. L. (2013). The influence of teachers' knowledge on student learning in middle school physical science classrooms. *American Educational Research Journal, 50*, 1020–1049.
Schön, D. A. (1983). *The reflective practitioner: How professionals think in action*. New York, NY: Basic Books.
Skott, C. K., & Møller, H. (2017). The individual teacher in lesson study collaboration. *International Journal for Lesson and Learning Studies, 6*, 216–232.
Smith, T. M., Cobb, P., Farran, D. C., Cordray, D. S., & Munter, C. (2013). Evaluating math recovery: Assessing the causal impact of a diagnostic tutoring program on student achievement. *American Educational Research Journal, 50*, 397–428.
Snell, J., & Lefstein, A. (2017). "Low ability," participation, and identity in dialogic pedagogy. *American Educational Research Journal, 55*, 40–78.
Stoll, L., Bolam, R., McMahon, A., Wallace, M., & Thomas, S. (2006). Professional Learning Communities: A Review of the Literature. *Journal of Educational Change, 7*, 221–258.
Tam, A. C. F. (2015). The role of a professional learning community in teacher change: A perspective from beliefs and practices. *Teachers and Teaching, 21*, 22–43.
Vangrieken, K., Dochy, F., Raes, E., & Kyndt, E. (2013). Team entitativity and teacher teams in schools: Towards a typology. *Frontline Learning Research, 2*, 86–98.
Vangrieken, K., Dochy, F., Raes, E., & Kyndt, E. (2015). Teacher collaboration: A systematic review. *Educational Research Review, 15*, 17–40.
Vangrieken, K., Meredith, C., Packer, T., & Kyndt, E. (2017). Teacher communities as a context for professional development: A systematic review. *Teaching and Teacher Education, 61*, 47–59.
Vogd, W., & Amling, S. (2017). Einleitung: Ausgangspunkte und Herausforderungen einer dokumentarischen Organisationsforschung [Introduction: Origins and

challenges of the documentary organizational research]. In S. Amling & W. Vogd (Eds.), *Dokumentarische Organisationsforschung: Perspektiven der praxeologischen Wissenssoziologie* (pp. 9–42). Opladen, Germany: Verlag Barbara Budrich.

Watson, C. (2014). Effective professional learning communities? The possibilities for teachers as agents of change in schools. *British Educational Research Journal, 40*, 18–29.

Weber, M. (1978). *Economy and society: An outline of interpretive sociology.* Berkeley: University of California Press.

Welling, S., Breiter, A., & Schulz, A. H. (2015). *Mediatisierte Organisationswelten in Schulen* [Mediatized organizational worlds at schools]. Wiesbaden, Germany: Springer Fachmedien Wiesbaden.

Wells, C., & Feun, L. (2007). Implementation of learning community principles: A study of six high schools. *NASSP Bulletin, 91*, 141–160.

Wong, J. (2013). Self and others: The work of "Care" in Foucault's Care of the Self. *Philosophy Faculty Publications, 6*. Retrieved from http://scholars.wlu.ca/phil_faculty/6

Zhang, Z., & Yin, H. (2017). Effects of professional learning community and collective teacher efficacy on teacher involvement and support as well as student motivation and learning strategies. In R. Maclean (Ed.), *Education in the Asia-Pacific Region: Vol. 38. Life in schools and classrooms: Past, present and future* (pp. 433–452). Singapore: Springer.

Manuscript received May 23, 2018<br>
Final revision received June 12, 2019<br>
Accepted July 1, 2019

# Speaking Volumes: Professional Development Through Book Studies

Betty S. Blanton
*Knox County Schools*
Amy D. Broemmel
Amanda Rigell
*University of Tennessee*

*This research describes a professional book study experience and offers insight into its use in supporting professional development. Framed in situated learning theory, this qualitative case study examined the perceptions of 12 educators who voluntarily participated in multiple professional book studies over 4 years. Two major themes were found in the data. The Process Theme encompassed what occurred within the professional book studies and participants' perceptions of the studies. The Outcomes Theme provided insight into how participants changed instructional practices, academic thinking, and personal beliefs. The book studies provided components of effective professional development and principles of adult learning. Participants believed that the book study groups provided professional development that met their needs in more powerful ways than traditional professional development.*

---

BETTY S. BLANTON, PhD, is an instructional coach at Powell Middle School in Knoxville, TN. Her varied background includes extensive work in both literacy and special education, and her work focuses primarily on meaningful ways to support the professional development of practicing teachers. Her work with adult learners has led her to align with constructivist learning theory.

AMY D. BROEMMEL is an associate professor and Director of Graduate Studies in the Department of Theory and Practice in Teacher Education at the University of Tennessee, Knoxville, A221 Bailey Education Complex, Knoxville, TN 37996; e-mail: *broemmel@utk.edu*. Her research is primarily qualitative, focusing on teacher development in both preservice and in-service settings.

AMANDA RIGELL is a doctoral student and graduate teaching assistant in Literacy Education at the University of Tennessee, Knoxville. Her research interests include adaptive teaching and creative insubordination, reading motivation and engagement, and the reading-writing connection.

KEYWORDS: book clubs, in-service teachers, knowledge communities, professional development, situated learning theory

During the past two decades, teachers have increasingly been placed at the forefront of a purported crisis in U.S. education (Cizek, 1999; Lemann, 2010). In the current atmosphere of political reform in public education, wherein teacher expertise is often identified as the single most important factor in student improvement (Gulamhussein, 2013), professional development activities have taken on an importance and scrutiny heretofore unseen in American education. It is our perspective that school districts rarely provide teachers with high-quality professional development (PD) opportunities that could lead to long-term transformation in instructional practices.

Mitchell (2013) described professional development as "the process whereby an individual acquires or enhances the skills, knowledge, and/or attitude for improved practice" (p. 390). Vu (2019) suggested that professional learning should be collaborative, dialogic, reflective, and job-embedded. In order to ultimately increase student achievement, Hirsh (2009) recommended that effective PD activities should be comprehensive, sustained over time, and provide intensive support. Furthermore, PD activities require their designers to plan around critical factors, including (1) consideration of the needs and perspectives of adult learners (Gregson & Sturko, 2007), and (2) incorporation of the elements of successful PD (Nevills, 2003; Porter, Garet, Desimone, Yoon, & Birman, 2000).

## Professional Development for Teachers as Adult Learners

Teachers are adult learners first. They bring a different set of needs to learning than children do (Trotter, 2006). They require that educational opportunities consider both their life experience and learning needs (Avalos, 2011; Hunzicker, 2011; Trotter, 2006). Adult learners compare new ideas with their current beliefs, expectations, and understandings of how the world works (Day, Lovato, Tull, & Ross-Gordon, 2011). In addition, adult learners prefer to participate in andragogical learning approaches (Chan, 2010), those in which the adult learner has a major role in planning the direction and goal of the learning experience (Day et al., 2011).

Lawler and King (2000) recommended six adult learning principles when implementing any learning activity for educators:

- create a climate of respect,
- encourage active participation,
- build on participant experience,
- employ collaborative inquiry,
- learn for action, and
- empower the participants.

*Blanton et al.*

Incorporating these adult learning principles increases the chance for buy-in and change (Gravani, 2012). Gravani's (2012) study of educators found that unless participants are allowed to direct their own learning and it is immediately applicable, they will not be committed to implementing what was introduced. According to Hunzicker (2011), "When professional development is supportive, job-embedded, instructionally focused, collaborative, and ongoing, teachers are more likely to consider it relevant and authentic, which is more likely to result in teacher learning and improved teaching practice" (p. 178). Because adult learners' needs are often not considered in PD planning, instructional change does not occur (Gravani, 2012; Le Fevre, 2014). A more finely grained analysis of teacher professional development in the United States underscores the lack of consideration for adult learners' needs in typical PD settings.

### Teacher Professional Development in the United States

Research on PD activities identified that fewer than half of U.S. teachers are involved in sustained collaborative activities (Darling-Hammond, Wei, Andree, Richardson, & Orphanos, 2009). Scotchmer, McGrath, and Coder (2005) found that teacher expertise, experience, and concerns were not considered when teacher PD activities were planned. Teachers report mostly attending traditionally formatted learning activities such as conferences or workshops instead of environmentally enmeshed activities such as mentoring, coaching, and peer observations.

Most PD activities are lecture-type, in which teachers are expected to take in the information dispensed and return to their respective classrooms to implement that information (Ball & Cohen, 1999; McGlinn, Calvert, & Johnson, 2003). Administrators in charge of the trainings often bring in experts without input from teachers about what is needed. This "outside-in" method with an emphasis on information dissemination remains the most common approach to professional learning. Such traditional PD activities typically emphasize procedural learning, resulting in only surface-level implementation of instructional approaches (Butler, Lauscher, & Jarvis-Selinger, 2004), flying in the face of research which identifies the most successful practices for adult learning (Darling-Hammond et al., 2009).

### Professional Book Clubs

The use of professional book studies as a means of providing effective teacher PD has not been widely studied. While research on informal book groups is extensive, research on professional book study as effective PD is relatively rare, and falls into three general categories involving preservice teachers, in-service teachers, and university–public school collaborations. Most research studies took place in the context of a college class with preservice teachers or collaborations between university personnel and in-service teachers in the public school setting (Addington, 2001; Burbank, Kauchak, & Bates, 2010; Hall, 2009; Kooy, 2006b; Lassonde, Stearns, & Dengler, 2005; McGlinn et al., 2003; Mosley & Rogers, 2011; Reilly, 2008; Roberts, Jensen, & Hadjiyianni, 1997). Our research adds to the body of knowledge generated by previous studies of in-service (practicing) teachers.

**Professional Book Studies With In-Service Teachers**

Surprisingly little research involves professional book studies within the school setting as a means of professional development for in-service teachers; however, four studies were particularly applicable to this work (Burbank et al., 2010; Kooy, 2006a; McGlinn et al., 2003; Selway, 2003). Kooy (2006a) compared two independent groups of teachers, one of novice teachers and one of experienced teachers, who participated in a book study of mostly fiction texts centered around teaching or educational experiences. Kooy found that regardless of experience level, participants used the meetings as a chance to share stories of their experiences, often initiated by reflections on the readings. Teachers developed close relationships as result of the social nature of the interactions within the groups. Burbank et al. (2010) compared preservice and in-service teachers participating in book study groups centered around texts about pedagogy, school culture, and/or classroom management. They found that in-service teachers discussed and examined beliefs, values, and teaching practice. Preservice teachers, on the other hand, were more focused on technical aspects of curriculum delivery, the "how to" of teaching. Two studies, McGlinn et al. (2003) and Selway (2003), involved professional book study groups that lasted 3 and 4 years, respectively. McGlinn et al. (2003) formed a 3-year professional book study involving in-service teachers and college professors in a collaborative group. The group read texts they considered to be multicultural young adult novels (e.g., *Bud Not Buddy* by Christopher Paul Curtis and *Tears of a Tiger* by Sharon Draper). The authors of this study believe that professional book studies are a viable alternative to traditional PD because they allow participants to become a community of lifelong learners who support and challenge each other to expand their own learning and improve their classroom practices (McGlinn et al., 2003). Selway (2003) wanted to share her collaborative experiences in college classes by promoting book study groups as professional development in her school. Members of her book study group read texts centered around pedagogy and trends and issues in education by scholars like Linda Darling-Hammond and Johnathan Kozol. As a cofacilitator of a 4-year professional book group with meetings four times per year, she found that some of the participants changed their classroom practices as a result of their book study participation.

*Blanton et al.*

Bonner and Tarner (1999) argued that professional book studies build on practices integral to informal book clubs. Successful professional book studies included voluntary participation and choice of materials, the learning and discussing of new strategies and ideas, development of relationships in a trust-filled social group in which all members play an equal role, and an extended duration of time relative to traditional PD. The book studies were personalized, but the group focused on "professional expertise . . . growing collectively as a group" (Bonner & Tarner, 1999, p. 46).

**Purpose and Significance**

The purpose of this research was to provide a description of the professional book study experience and gain insight into its use to support teacher PD. There have been in-depth studies of a variety of PD delivery activities, but few that specifically investigated the effectiveness of professional book studies from the participants' point of view. For the purposes of this work, a professional book study is defined as any planned group discussion of a particular text or texts as a means of furthering professional understanding of a specific subject or phenomenon. Typically members read books or portions of books or articles in preparation for each meeting (Bach, Hensley Choate, & Parker, 2011). The significance of this work lies in the lack of research that specifically considers the participants' perceptions of book study as a form of PD. In addition, the study provided information about what occurs within professional book study groups to influence teacher learning. This work addresses the question: What are the participants' perceptions of the book study experience as a professional development activity?

## Theoretical Framework

Lave and Wenger's (1991) situated learning theory provides the theoretical framework for our study. Within their theory, Lave and Wenger first described the concept of *communities of practice*. Wenger defines communities of practice as "groups of who share a concern or a passion for something they do and learn how to do it better as they interact regularly" (Wenger, 2004, p. 1). We view the professional book study groups as a specific category of community of practice: a knowledge community.

The term *knowledge community* was first coined by Craig (1995), who identified them in educational settings. In knowledge communities, participants come together because they share an interest or interests and want to learn about a specific topic. According to Craig (2007), teachers also rely on knowledge communities in times of stress to help with solutions to problems and as a source of support. Within knowledge communities, participants use narrative as a way of communicating their thinking (Schultz & Ravitch, 2013), interpreting new information, and gaining knowledge from other members. Vu (2019) suggested that, "Encouraging teachers to engage and reflect in learning communities empowered them to leverage the potential of social networks and to challenge images of teachers as simply vessels for knowledge dissemination" (p. 542). As teachers share stories of their classroom experiences, they incorporate critical reflection and creative problem solving into their dialogue (Craig, 2007).

Knowledge communities do not follow a specific life cycle, but members come together because of a shared interest in learning (Seaman, 2008). The community may become inactive for indefinite periods, then members will come back together when the need for new learning arises.

Carpenter and Linton (2015), in a study of "unconferences" called EdCamps, suggested that PD activities characterized by collaboration, teacher-driven inquiry, and teacher agency have the potential to foster changes in teaching praxis. Trust and Horrocks (2017) further suggested that "well-designed" blended communities of practice, like the book study PD examined in our research, have "transformative potential for shaping teachers' knowledge, skills, attitudes, identities and workplace practices" (p. 658). Thus, we examined the book study and its influence on teachers' beliefs, ideas, and practices through the lens of situated learning (Lave & Wenger, 1991), under the assumption that these book studies served as a knowledge community within the larger construct of a community of practice.

## Methodology

This qualitative case study describes what occurred in professional book study groups through a constructivist lens. As adult learners, when teachers learn new methods, they rely on connections with their own experience and understanding of teaching practices, and on their positions as educators (van den Berg, 2002). This learning takes time and cannot typically be accomplished through traditional forms of professional development.

### Context

The study took place in a rural county in East Tennessee. According to the U.S. Census Bureau (2014), the county had an estimated population of 53,470 and a poverty rate of 14.4% in 2012. It is important to note that the individual schools' poverty rates varied greatly due to the isolation of some communities and the lack of job opportunities. The county's school system has 18 schools which include a preschool/alternative school, 6 elementary schools, 4 middle schools, and 5 high schools.

The professional book studies took place over a 4-year span as part of the special education department's professional development. During this time period, there were 10 book studies with a total of 28 different participants involved. All of the qualifying participants were invited to take part in the research, because their positions, background, and experience had the potential to represent different points of view and a diversity of perspectives.

Of the 28 participants, 12 participated in this research. Four of the participants taught at the elementary level, five taught at the middle and/or high school levels and the remaining three worked in multiple schools or county-wide as professional support personnel. All 12 of the participants are female. Eleven are White, and one is African American. Their years of professional experience range from 7 to 35 years.

**District Book Study Groups**

The original invitation and guidelines for participation in the book study groups were sent out via e-mail. The e-mail outlined that the book studies were voluntary and were open to any of the special education department's professional staff. Participants were required to attend all meetings and write a reflective summary to receive either a $150 stipend or 6 hours in-service credit.

*Book Selection*

The procedures for choosing books for each study evolved over time. For the first two studies, one of the system-wide literacy coaches chose the two books. The Special Education Supervisor selected the next book and a subsequent book for one of the studies during the second spring. The rest were chosen by vote from a list of recommendations compiled by the department's Research and Development Team. The books primarily focused on pedagogy in general (e.g., *Choice Words; Classrooms That Work*), literacy instruction (e.g., *The Fluent Reader; I Read It But I Don't Get It*), school culture (*The Essential 55; Rules in School*) and unique perspectives on education (e.g., *Thinking in Pictures; The Out of Sync Child*). Book study books and the participants in this study are listed in Table 1. The participants chose which study group to attend based on their needs and/or interests. The book study meetings included three face-to-face meeting weeks and one blog week. Two different days of the week were made available for meetings, and participants were asked to choose one. Every participant completed all requirements of the study and was paid the accompanying stipend.

*Data Collection*

Data for this case study came from three sources. It included open-ended interviews of selected participants and two different artifacts collected during the book studies: blog entries and reflective summaries. The use of a variety of methods provided triangulation, as well as within-case and

Table 1
**Book Study Overview**

| Book Study | Book Title | Artifact Collected | Natasha | Ava | Meg | Melody | Hanna | Juanita | Prieta | Lily | Suzie | Ginger | Stella | Crystal |
|---|---|---|---|---|---|---|---|---|---|---|---|---|---|---|
| Spring 1 | *The Fluent Reader* | Reflective essay |  | X |  |  |  |  |  |  |  |  | X | X |
|  |  | Blog entries |  |  |  |  |  |  |  |  |  |  |  |  |
| Spring 1 | *Choice Words* | Reflective essay |  |  |  |  |  | X | X |  | X |  |  | X |
|  |  | Blog entries |  |  |  |  |  |  |  |  |  |  |  |  |
| Winter 1 | *The Out-of-Sync Child* | Reflective essay |  | X | X |  | X | X | X | X | X | X | X | X |
|  |  | Blog entries |  |  |  |  |  |  |  |  |  |  |  |  |
| Spring 2 | *I Read It, but I Don't Get It* | Reflective essay |  | X |  | X |  |  | X |  |  | X | X |  |
|  |  | Blog entries |  |  |  |  |  |  |  |  |  |  |  |  |
| Spring 2 | *New Passages* | Reflective essay |  |  | X |  | X | X |  | X | X |  |  | X |
|  |  | Blog entries |  |  |  |  |  |  |  |  |  |  |  |  |
| Winter 2 | *The Essential 55* | Reflective essay | X | X | X |  | X |  |  | X |  | X |  | X |
|  |  | Blog entries | X |  | X |  | X |  |  | X |  | X |  | X |
| Winter 2 | *Going With the Flow* | Reflective essay |  |  |  | X |  | X |  |  |  |  |  |  |
|  |  | Blog entries |  |  |  |  |  |  |  |  |  |  |  |  |
| Spring 3 | *Thinking in Pictures* | Reflective essay | X | X |  | X | X | X | X | X | X | X |  | X |
|  |  | Blog entries |  |  |  |  |  |  |  |  |  |  |  |  |
| Fall 3 | *Classrooms that Work* | Reflective essay | X | X | X | X | X | X | X | X | X | X | X | X |
|  |  | Blog entries | X | X | X | X | X | X | X | X | X | X | X | X |
| Spring 4 | *Rules in School* | Reflective essay | X | X | X | X | X | X | X | X | X | X | X | X |
|  |  | Blog entries |  |  |  |  |  |  |  |  |  |  |  |  |
| Total book study participation |  |  | 4 | 5 | 6 | 5 | 5 | 7 | 7 | 6 | 6 | 7 | 7 | 7 |

1021

*Blanton et al.*

cross-case comparison of the data (Merriam, 2009). A list of the artifacts associated with each of the participants is included in Table 1.

Each interview was transcribed by the researcher using InqScribe software. After initial transcription, each transcript was analyzed and any comments that correlated with participants' perceptions of book study as professional development were highlighted. Data were analyzed using inductive analysis methods as outlined by Hatch (2002). Themes were created based on relationships within these categories. Salient categories were assigned codes and all data were reviewed again to look for further examples (and/or nonexamples) of these categories. Categories (and related subcategories) were then adjusted or changed based on this analysis. Themes were analyzed both within and across categories.

Several strategies to provide checks and balances were designed to enhance the trustworthiness of the research. Creswell (1998) explained that prolonged engagement in the field helps a researcher to build trust with participants and learn the culture. Before this research took place, one of the coauthors had worked in the book study district for 15 years, the last 5 years as a system-wide literacy coach. There was a concern that her status as a system-wide employee might affect some participants' candor and willingness to communicate their perceptions; however, the research began after she took a new position in a neighboring district, potentially allowing participants more freedom in the sharing of their perceptions. Thus, she took the position of complete participant-observer in this study (Spradley, 1980). She experienced being an insider and outsider simultaneously. Through introspective debriefing with one of the coauthors and her record keeping she worked to maintain explicit awareness of her dual role (Spradley, 1980) in the book study group. The fact that she was already an active participant in the book studies prior to deciding to undertake the research contributes to the unique nature of this work. Her immersion in the context provided her with a level of understanding that is not often available to researchers.

## Results

After analyzing the interview and artifact data and identifying patterns, the researchers identified two major themes: processes and outcomes. The Process Theme includes the structures that were put in place as a requirement of participation and, more importantly, those which developed as a result of the environment that was created by the participants themselves. The Outcomes Theme developed as the coauthor found evidence that the participants made changes in practices, thinking, or beliefs.

### Process Theme

The Process Theme encompasses what occurred within the professional book studies. There are three categories within this theme: Elements of the

Book Study, Perceptions of What Occurred in the Book Studies, and Evolving of a Learning Community. Table 2 shows the organization of the Process Theme.

*Process Theme: Elements of the Book Studies*

This category includes book selection, the discussions, the facilitator's role, and the blog and reflective summaries. The participants' own words are used to describe these elements.

*Choosing the books.* Ten of the 12 participants identified the importance of being able to choose the books they read. In her interview, Hannah explained that, "[It] starts off from the beginning. It gives you . . . an opportunity to express your needs . . ." Before each meeting, participants were asked to read a specific number of pages. There were no other preparation requirements; however, participants demonstrated the intentionality of their preparation for the meetings by annotating the books as they read. Suzie explained, "We would highlight as we read the text so we would know what stood out to us and we would know what to bring up in discussion."

*Discussions.* The group members described the face-to-face meetings as loosely structured, informal discussions. Suzie, the facilitator of one group, recounted the initiating discussion question as, "Who implemented or who tried what we discussed last week and how did that go for you?" Prieta explained: "We took turns speaking . . . We could read little selections and illustrate it with situations from our classrooms, give examples of our own lessons and how we might implement ideas that we gleaned out of the texts."

The members felt that everyone participated equally in the discussions. The sense of story and a nonjudgmental atmosphere allowed everyone to feel that they were part of the group and motivated them to fully participate. Natasha explained that, "Nobody was ever quiet. There was never a lull in the conversation. . . . I think everybody contributed." In fact, the atmosphere that developed within the groups helped the participants open up and share their thinking. Juanita identified overcoming her initial reluctance to speak up in the group as something that eventually helped her build confidence in public speaking, "[A]t first I . . . listened. But after a while, I had ideas and I just couldn't keep my mouth shut . . . Then after that, I just felt more comfortable . . . Actually it helped me be able to discuss things more with people."

Eleven of the 12 case study members believed that the discussions clarified their understanding. Stella explained,

> You could sit there and say, "Okay, well now I can see it. I don't have to talk about it, but so-and-so feels the same way I do." or "Oh, is that

Table 2
## Process Theme Categories and Subcategories

| Category | Subcategories | Participants Involved[a] |
|---|---|---|
| Elements of the Book Studies | Choosing the books | 10 |
| | Discussions | 12 |
| | How discussions unfolded | 10 |
| | The facilitator's role | 9 |
| | Blog | 9 |
| | Stipend | 3 |
| Perceptions of What Occurred in the Book Studies | Comparison to other learning opportunities | 12 |
| | • Comparison to other PD experiences | |
| | • Comparison to college experiences | |
| | Perceptions of the books | 11 |
| | • Perseverance | |
| | Impact of discussions | 10 |
| | • Different points of view | |
| | • Being challenged to think in different ways | |
| | • Sharing ideas with other participants | |
| | Participants validated/applauded each other's ideas | |
| Evolving of a Learning Community | Companionship | 9 |
| | Collegial atmosphere provided a support network | 11 |
| | Building relationships | 9 |
| | • Being accepted unconditionally | |
| | • Motivating | |
| | • Fighting the feeling of isolation and lack of support | |
| | Collaboration | 12 |

[a]Participants who made comments related to the category.

## Professional Development

what it meant?" and hearing people discuss it, you're like . . . "Well, I didn't get that, but now I do."

The participants clearly identify that the spirit of mutual respect helped all members. Stella explained, "There are different perspectives on the books and not everybody [is] going to see things exactly the same way." In fact, 11 of the 12 participants described how they enjoyed the different perspectives that the others brought to the table. Stella explained further:

> You got to see other points of view and you understood that just because you saw it one way didn't mean that was the only way it could be . . . being able to hear other people's thoughts on things sort of made you look at things in different ways.

*The facilitator's role.* Ten of the 12 participants in the case study made comments about the impact of the facilitator. Natasha explained how the lack of rigid expectations allowed everyone to relax and feel comfortable enough to participate while Juanita added that the facilitator did not dominate the discussion; the way the facilitator led the groups added to the positive experiences of the participants. This finding echoes that of Burbank et al. (2010), whose participants indicated a clear preference for instructor-facilitated meetings.

*Blogs and reflective summaries.* Beginning with *The Essential 55* (Clark, 2003), the participants were required to post at least one blog entry in lieu of a face-to-face meeting sometime during each book study. This was the first experience with this type of technology for most of the participants, and it too, became a learning experience for the book study groups. Surprisingly, the fact that the new technology did not always go as expected added to the feeling of community. Suzie elaborated,

> At first, I think it was confusing, We didn't know how to reply or start a new thread . . . and I think it was good in giving people that don't use their technology or don't feel comfortable with it, [it made them] a little more comfortable.

The blog gave the participants the chance to try a new media and the ability to communicate between meetings, adding to their experience while teaching them a new skill.

The reflective summaries provided an opportunity to discover the members' perceptions of the books. Ginger explained that, "I found myself in this book. Along with the others in the book study, I found myself analyzing my teaching style, past and present."

Like all readers, these educators made constant comparisons and connections to their experiences. Melody saw her child in one author's words,

"Her story hit home. . . . As she described her tantrums due to lack of communication skills, I could close my eyes and see my Brady laying in the floor of the kitchen screaming." The participants' initial comprehension of the text was filtered through their life experiences. Reflective summaries revealed, however, that these perceptions changed when they entered the discussions and heard differing interpretations from other members.

*Process Theme: Perceptions of What Occurred in the Book Studies*

The second category in the Process Theme provides understanding of the participants' views and beliefs about the book study. It includes subcategories on perseverance in reading, fighting a feeling of isolation and lack of support, and the diversity of participants' experiences and ideas.

*Comparison to other learning environments.* Every interviewee believed that the professional book studies were superior to traditionally structured professional development programs they had attended.

First, the participants believed that the ability to be actively involved enhanced their experience and learning and many described traditional PD as ineffective. They used phrases such as "You are just sitting in a seat thinking, "When do I get outta here?" (Natasha), "You're just listening or watching something but you really don't have interaction." (Ava), "That's what, I think, lacks in a lot of professional developments is that you don't have an opportunity to respond! It's just, "Okay, what time is it?" (Hannah), and "Just sitting and listening for hours upon hours or a whole day is difficult." (Meg). The participants identified that the interactions and their personal commitment to professional book studies required that they were always involved. Ginger explained that, "I felt like I had some greater insight just from reading the whole book myself and then being able to have our lively discussions and proactive solutions to problems... like a member of a group that was accomplishing something." Crystal's perception of book studies provided a picture of active learning, something she had not experienced in traditional PD, "The book study gave me the opportunity, again, to collaborate and then practice what I read. So it gave me opportunity to actually have more hands-on [learning]."

*Perseverance.* Even though the book did not always turn out to be exactly what they expected, the teachers reported that they persevered in their reading of the book. Suzie wrote in her *Rules in School* (Brady, 2003) reflective essay, "I wasn't very excited about this book being chosen for our book study because I was hoping for a different book. My opinion has totally changed. I have enjoyed every chapter." Stella felt strongly that reading *New Passages* (Sheehy, 1995) changed her thinking as well, "As I stated in my first blog entry, this is not a book I would have chosen for myself. I was, however, enlightened on some things I hadn't thought about before."

*Impact of discussions.* Even though the participants had much in common, they identified the diversity of their experiences and ideas as one of the reasons the group was successful. One third of the participants made reference to this diversity. Ginger believed that the members had a variety of different capabilities, "[It] gives you different perspectives and everybody's from different directions instead of all of us being all resource teachers, you know, or all speech teachers, or all psychologists." In her *Thinking in Pictures* (Grandin, 1996) essay, Natasha wrote, "I have learned how much I don't know and how valuable other professionals will be to me." Lily, the school psychologist, also believed sharing their unique experiences was important, "I like hearing other people's perspectives because it's based on their experiences. Somebody may have a different take on it than I did just because their experiences were different than mine." The participants' examples demonstrate the diversity of the members' professional experiences.

Five of the interviewees made a direct reference to the fact that the discussions added to their reason for coming back. Ava reflected on what she learned in the book studies in light of what she already knew, "Instead of being . . . narrow [minded], with blinders on . . . somebody says, 'Well, but look at this.' . . . I like that because I want to be challenged. If I'm learning the same thing over and over again, why would I be there?" Within the group discussions, the members felt that this give and take provided them with a deeper experience than solo reading.

Participants felt the other members saw their ideas as valid and helpful. Two participants believed that the way the group members treated each other helped them open up. Hannah's perception was that, "It was encouraging . . . when they would say, "Well, that's a good idea!" . . . they'd pat you on the back and say, "Oh, that's a good thing that you're doing!" . . . when a lot of times you don't get many pats on the back." Stella believed that the unconditional acceptance by the other group members grew from the way they treated each other within the discussions, "Over time, it became easier . . . and then we found out later on that it didn't have to be right because . . . it was our take on what we read." This comment is particularly interesting because Stella initially felt reluctant to discuss what she had read, writing, "You start wondering, 'If I write this and they read it, am I really off-base?' and are they going to go, 'Well, she's dumber than a bag of rocks!'" The ways in which members validated and applauded each other's ideas led them to feel more comfortable and more willing to share. These findings are similar to Burbank et al.'s (2010) finding that preservice and in-service teachers reported the advantages of the book club PD model in providing an avenue for safe and across-site teacher dialogues.

*Blanton et al.*

*Process Theme: Evolving Into a Knowledge Community*

The third category in the Process theme provides data that indicate that the book study groups may have developed into a knowledge community. The participants gave evidence that the group provided companionship, a support network, and a space to discuss new ideas collaboratively.

*Companionship.* A feeling of companionship or camaraderie was referenced by three fourths of the participants. Companionship among the participants, however, was not automatic because the members were not well acquainted before the studies. Stella described how her initial reluctance changed over time:

> We became more comfortable with each other and [were] willing to hear what other people had to say. Because we had gotten to know each other a little better and it wasn't like, "They're going to think I'm dumber than a bag of rocks." It was "Well, they know I am, but they love me anyway."

Other participants used phrases such as "sounding board," "respect," "camaraderie," "ownership," "support," "friendship," and "comfortable," to describe their perceptions. These ways of naming interpersonal connections and the evolution of such connections over time supports Kooy's (2006) finding that "[b]y providing social space, interactions, and time (Clark, 2001; Kooy 2003a,b), the book club experience builds relationships (professional and personal) and opens the way for professional learning" (p. 671). McGlinn et al. (2003) reported the emergence of a similar sense of collegiality in the teachers' reading circle they established.

*Support network.* Through participation in the book studies, the members began to see each other as experts in different areas. This expertise was unknown before the book studies. Stella explained how she asked for support from other members, "We [were] more open to be able to say, 'If I don't understand this, I can call so-and-so . . . she may have an insight.'" As they developed this learning community, the participants came to rely on each other to enhance their learning. Hannah reflected on this, "So you're learning not just from the book but from other people's life experiences."

*Fighting isolation and lack of support.* The participants also believed that the discussions provided them with support in ways that added to their desire to continue returning. They felt that the other members challenged them to consider additional ways of thinking because of the different points of view and interpretations of the text, and they accepted and supported each other's ideas. In addition, they were able to hear and share their own experiences through stories. These features of the book study helped diminish the sense of isolation and lack of support often experienced in traditional PD structures.

*Collaboration.* All 12 participants acknowledged that the book studies provided them with the chance to try out new strategies. Melody mentioned that the collaborative atmosphere helped her relax and be open to new ideas, saying, "Well, the first thing that comes to mind is collaborative learning. . . . We learned from the books, but we also learned things from each other."

Suzie, as a system-wide coach, had a regular opportunity to observe teachers' instruction and found the collaboration exciting:

> Just seeing educators come together and want to learn or use [new strategies] in their classroom to help students . . . it was something they wanted to do, not something they had to do. . . . Then you go back to your classroom and implement them and we meet later on and talk to discuss about how they worked and how to modify it, how it worked for different teachers.

Natasha believed that this collaboration empowered her to feel responsible for her own learning, "[Y]ou feel a part of something. You feel some ownership. You feel... like you're in control more than just being told what to do."

The collegial atmosphere, camaraderie, and companionship enabled the participants to build trusting relationships. As they shared ideas and experiences during the discussions and blogs, they came to look to each other for support and expertise. All of these elements combined to create a learning community.

### Outcomes Theme

The Outcomes themes developed as the coauthor found evidence that the participants made a variety of changes in their practices, thinking, and beliefs. These changes were categorized as, *Instructional Practice, Academic Thinking,* and *Personal Beliefs.* Table 3 shows the Outcomes theme's categories and subcategories.

*Outcomes Theme: Instructional Practice*

In many of the book studies, the chosen text was laden with instructional strategies. For these book studies, the facilitators made a conscious effort to empower the participants to choose a strategy to try in the classroom or workplace between meetings. Usually, the members each chose one strategy at the meeting, implemented it in the interval between the meetings, and then reported back about the outcome. Stella explained this in her *The Essential 55* (Clark, 2003) reflection:

Blanton et al.

Table 3
**Outcomes Theme Categories and Subcategories**

| Category | Subcategories | Participants Involved[a] |
|---|---|---|
| Instructional Practice | Implementing new ideas during the book studies | 11 |
|  | Continued use of the books' strategies | 9 |
| Academic Thinking | Academic Learning | 9 |
|  | Application outside of the book study | 9 |
| Personal Beliefs | Changes in thinking about students | 9 |
|  | Changes in thinking about people with disabilities and their abilities | 10 |
|  | Changes in thinking about teaching | 10 |
|  | Changes in thinking about themselves | 9 |

[a]Participants who made comments related to the category.

> She [the facilitator] wants us to pick one of the rules and use it in our classroom and tell how it goes. I find myself saying, "We do this one, and that one, and this one too." My TA [teaching assistant] and I discussed the idea of introducing rules one by one and building social studies/social skills lessons around them. These are working out splendidly for all involved.

All 12 of the participants commented on the changes they made to instructional practices during and after their involvement in the book study groups.

*Implementing new strategies.* Eleven of the participants listed specific strategies they implemented between meetings, including managing student behaviors, empowering students to become their own advocates, and changing reading comprehension instruction in their classroom.

The participants used their chosen strategy with one student, a small group, or the whole class, depending on their specific needs. Juanita chose a single student on whom to focus for the *Choice Words* (Johnston, 2004) book study:

> One boy, "C," in particular comes to mind. He is a sixth-grader on my caseload. He goes through most of his days without anyone speaking to him. Other students are not intentionally cruel. He is not at their same social/maturity level and, therefore, just does not exist [to them]. Adults only speak to him to give directions and to correct his numerous inappropriate behaviors. Throughout the course of this study I decided to make "C" my project and incorporate ideas presented in the book while working with him and the rest of the class.

Natasha implemented a whole-class strategy, as she wrote in her *The Essential 55* (Clark, 2003) essay, "One of the specific rules that I have begun

*Professional Development*

to follow is: during discussion, only one person can have their hand up at a time." In Stella's reflective essay from *Classrooms That Work* (Cunningham & Allington, 2003), she explained how she implemented one of the strategies:

> I really liked the "Reading and Me" activity on page 13. I tried it in my classroom to see where the students saw themselves as readers. Several said that they like to read. The problem though is that they have such difficulty decoding that their comprehension level is very low. I'm hoping that by the end of the year there will be more students wanting to read to read, not just because they have to.

A few of the participants worked in inclusion settings which necessitated collaborating with another teacher in order to implement the strategies. Juanita wrote about the use of a strategy in the inclusion classroom,

> Copying and memorizing definitions still remains a common practice especially in Science and Social Studies classrooms (p. 88). The students will copy definitions that they cannot even read. When introducing the words *archaeologist, anthropologist,* and *geologist,* I suggested [to the general education teacher] that I bring in some of my son's artifacts and tools (pg. 91). I brought in arrowheads, flat chips, fossils and other rocks, a shovel and a sifter. These items in themselves brought about the discussion of the meaning of other related words. Students seemed to do better on this test than they had on the previous one.

Because of the delicacies of collaboration in the inclusion setting, Prieta reflected on how she changed her approaches in more subtle ways in the essay for *Choice Words* (Johnston, 2004):

> I have attempted to model differing opinions and even, with respect and great care, corrected other adults where before I might have "ignored" a co-teacher's mispronunciation or misinterpretation of an academic rule. I tactfully pull a dictionary or text from the shelf to "clarify for myself" to put myself in the position of not knowing "the answer" in order to teach that it is acceptable to disagree or to be corrected when wrong, and that there is an appropriate way for adults to learn from one another . . . just as students need to be open to constructive criticism or need to "cross-check their warrants," and can solve differences by searching for common solutions.

Other participants worked outside the classroom as support services personnel and were also able to use the books' ideas. Lily, a school psychologist, described how she used information from *The Essential 55* (Clark, 2003), saying, "I have used some of Mr. Clark's rules with my after-school groups at two elementary schools." Meg, a speech therapist, related two examples of instructional changes that she was making in communication with students in her reflective essay about *The Out-of-Sync Child* (Kranowitz, 1998):

> There were a couple of other pointers that I especially liked in the book that I have already used with children. To avoid learned helplessness about tying shoes or anything the author suggests saying, "I know it's hard for you to tie your shoes but each time you do it, it will get easier." The other pointer that I really liked and have used was concerning children with hard to understand speech. The author suggests if you catch one word say, "Tell me more about the truck." I used to say, "Something about a truck."

Both of these case study members were able to apply what they learned from the book study to settings outside the classroom environment.

Finally, the participants described how students were responding to new strategies. In her *The Essential 55* (Clark, 2003) blog, Stella wrote,

> It is interesting how, so far, each rule seems to build on or relate to the one before it. My students seem to enjoy talking about how the rules do apply to our little world as well as the school and our community.

Crystal also saw a change in her students after implementing read-alouds in her classroom:

> As a result of being a part of the study group, I began to implement novel reading into my program. Our first novel that we read was "Because of Winn-Dixie." My students loved it! Not only did they like the story, but they enjoyed being read to. It was a great way to begin class or change to another activity. It is not so much a chore to get the students to read when their peers are helping and encouraging them.

Meg described how the use of another strategy changed the atmosphere within her small speech therapy group, "I have also incorporated having everyone in the group clap for other students when they do a good job on something during therapy. I feel this has made therapy more fun for everyone including myself."

In her interview, Juanita remembered the practice of trying out strategies: "That was really the point . . . to try things in the classroom and see if it worked with your students, or make adaptations with it, so it would work." She spoke more about this in her essay for *Classrooms That Work* (Cunningham & Allington, 2003):

> I tried the activity on page 107 for students to self-assess their vocabulary knowledge. It did not go over well with my reading group. They all either responded with a "1 = I have never heard of that word" or a "4 = I know what that word means." I do not feel like these were accurate responses for the majority of the students. The 1's did not want to be called on. The 4s did not know the correct definitions. I simply decided to wait and try it again after the students had gained greater self-confidence.

It was clear that she assessed the failure of the strategy in light of what her students needed instead of the strategy's basic worth. As a result, she began working on building her students' self-confidence. Comments from this category showed that the teachers in the book study used the books' strategies within their own classrooms and in inclusion classrooms, with individual students and whole classes, and in collaboration with other teachers. These findings echo Carpenter and Linton's (2015) finding that teachers who participated in voluntary, democratic "unconference" PD overwhelmingly reported that the experiences had resulted in changes in their instructional practices. McGlinn et al. (2010) also found that teachers in a reading circle shared teaching strategies in their meetings that teachers subsequently implemented with students.

*Continued use of the book strategies.* Nine of the 12 participants directly referred to the continued use over time of the strategies originally learned in the book studies. Ginger explained that she continued to use at least three of the specific strategies she learned. In her interview she said, "I added Reader's Theater. We added Question-Answer-Response . . . I added making my file folders for . . . my sensory integration-disordered children." Other participants commented on the importance of adding the books to their professional library. Crystal, in *The Out of Sync Child* (Kranowitz, 1998) reflective essay explained that "Although the information in the book may not be something that I can use in my classroom today, however, it is information that can be used to help another student or parent along the way." These remarks provide insight into the continued use of strategies discussed in the book studies.

*Academic learning.* The participants found that the book studies provided them with a new understanding of academic thinking and the importance of keeping abreast of current research. Three fourths of the participants made remarks that identified this change in thinking. According to Natasha, "[The professional book studies] helped me learn how to read professional books and interpret them. . . . You just need to stay relevant and [read] things that you're interested in so you will be ready to understand what's going on." Prieta found that the book studies helped her to think differently, "[I]t was very helpful to me and got me to think outside-the-box and learn from authors that I might not have even been aware of." Ava explained, "Whenever things come on that has anything to do with education or the medical community, I watch. I'm not sure when I was younger, I would've cared. Knowing things are changing all the time, makes a huge difference." The participants began looking at research as a source for additional support in their daily teaching, and their remarks reflect the perspective that learning did not end when the professional book studies ceased.

## Outcomes Theme: Academic Thinking

In addition to changing their academic learning habits, the educators made shifts in their academic thinking during their participation in the professional book studies.

*Application of learning outside the book studies.* Eleven of the 12 educators made comments that were relevant to the application of learning after the book studies. First, the members continued reading about the subjects presented in the studies on their own. Additionally, the learning was used to inform others, including educators, parents, and community members.

Prieta reflected on the changes she had made in what she read: "I tend to choose some of these more academics books . . . when I would not have chosen those before. I spend more time reading and looking at the journals . . . giving myself time to sit down and read and think through that information." Participants also explained that the book studies helped them begin the pursuit of additional reading on specific disabilities. In her interview, Juanita remembered how one of the book studies, *Thinking in Pictures* (Grandin, 1996) started her on a journey of research into autism spectrum disorders, "Just at that time, I was becoming really more aware of kids on the Autism Spectrum and it really made me think, and I looked up whole lot more information on how to work with these kids."

Participants also mentioned how they were sharing their academic learning and thinking with fellow teachers and with parents. Meg, a speech therapist less familiar with classroom instructional practices, described sharing with a fellow educator in her *Classrooms That Work* (Cunningham & Allington, 2003) essay, "I shared the ideas that I liked with a friend, who has been a teaching assistant for many years. Some of these ideas were even new to him. I now have a much better idea about how classrooms work." Lily described how she had already used the information she gained from *Thinking in Pictures* (Grandin, 1996), "I have already recommended it as reading for at least five families and any teacher and administrator I have had an opportunity to speak with." Crystal related the way she was able to help someone in her reflective essay for *The Out-of-Sync Child* (Kranowitz, 1998),

> There is a family that goes to my church who has a four-year-old [who has] been diagnosed with being autistic. The other Sunday, when I asked how he was doing, she started telling me how he is beginning to talk and how pleased they were with his progress. She went on to say that he still had a lot of sensory issues. Boy, did I light up like a Christmas tree in December. I got so excited! I began to share with her some information from the book. I told her that I would let her borrow my book. She was so grateful and I felt sooo good. . . . A few months ago I couldn't have offered that parent in my church anything about sensory disorders.

*Professional Development*

Data in this category provided evidence that the case study participants implemented the books' strategies and ideas in a variety of settings. The participants shared their learning with different people at work, at home, and even at church. They were able to relate the books to a variety of situations and individuals. As a result, the members became interested in further learning and the application of the strategies outside of their professional lives.

*Outcomes Theme: Personal Beliefs*

The last category in the Outcomes theme provides insight into the changes the participants felt they made in their thinking and belief systems as a result of the book study experience. The subcategories include changes in thinking about students, about people with disabilities, about teaching practice, and about themselves.

*Changes in thinking about students.* Nine of the 12 participants identified ways in which their thinking about students' abilities and needs had changed. Natasha's beliefs, for instance, about the importance of looking at students in a more complete light changed after the book study experience. In her interview, she said, "I had lost sight of the fact that I was teaching a person instead of a seventh grader. . . . They're a whole person and . . . you're not just teaching them how to read and do math." Stella also reflected that she could now empathize with her students, "I can visualize myself as a student, as a teacher, as a reader. I could put myself in my student's place. I could see it from their point of view." This finding in our study echoes that of Selway (2003), who found that the reading and discussion in her book study group impacted the way participants related to their students.

*Changes in thinking about people with disabilities.* Of 12 participants in the case study, 10 made comments that are relevant to this category. Participants described how their new learning helped them to be less hesitant to work with people with disabilities. Ginger explained how she felt more at ease teaching students with disabilities with whom she had not previously worked, "Well, you just, kind of, know now . . . if [I] get this child, or a child with those tendencies, now I can kind of pick up on those." Melody described how reading Temple Grandin's book gave her the chance to reflect and feel more confident teaching students with autism spectrum disorder, "I would often find myself rereading sections of text in order to try to see things she did. I would try to imagine myself as a visual learner and try to make connections as she described...Wow! That was difficult." Prieta added, "It made me really think much more deeply about my students . . . who've been autistic and how we structured the lessons for them that were probably totally meaningless and how I could have done things . . . much differently."

Comments in this subcategory revealed that participants began to look at their students with a new sense of empathy, enabling change in

*Changes in thinking about teaching practice.* Eleven of the participants provided data for this category. In her *Thinking in Pictures* (Grandin, 1996) essay, Prieta wrote, "I have tried to assess my classroom structure . . . and what may contribute to possible success or . . . failure of my students." In her interview, Hannah said, "It made me see . . . all the kids that come in my room . . . may have certain disabilities and similar diagnoses but all of them are individuals and . . . learn in different ways. You cannot use one way to teach every child." Ginger also described how the book study helped her to think reflectively about her teaching in her *I Read It but I Don't Get It* (Tovani, 2000) essay: "I found myself in this book. Along with the others in the book study, I found myself analyzing my teaching style, past and present." Ginger's comment, in particular, is indicative of the reflective nature of the studies.

*Changes in thinking about themselves.* Nine of the case study participants were represented in this category. Stella wrote about a new need to be self-evaluative before she could begin to look at others, "You really need to know yourself and your place in the universe to understand where others may be in theirs."

Other participants believed that the book studies helped them to become more open to new ideas and willing to change. Ava wrote, "It got me unstuck . . . I've . . . gotten to know people who knew what their strengths were." Ava's transformation echoes that of Selway (2003), who reported "a tremendous amount of growth" (p. 23) after facilitating a teacher book club.

Another insight that occurred was the change in how the participants interacted with others. Juanita described how *New Passages* (Sheehy, 1995) helped her to gain a new understanding of how to cope in stressful situations, "I have learned that the little things that people stress over really do not matter that much in the whole scheme of things."

After participating in the book study groups, the participants described how their thinking transformed them. They felt more open to try new things, more aware of others, and more comfortable with sharing their ideas with others.

## Discussion

The professional book study data provide support for research in two areas: that the book studies offer the components of effective professional development by focusing on the needs of adult learners, and that they can enable the development of a knowledge community.

### Effective Professional Development

This case study shows that the book studies provided elements of effective professional development. According to Lawler and King (2000), six principles of adult learning should be considered in planning professional development activities. The findings of our study suggest that these six principles contributed to professional development that increased teacher buy-in and changed their instructional practices and academic thinking.

First, any PD activity should create a climate of respect. The participants of this case study believed that the freedom to voluntarily participate and to read books directly related to their needs created a climate of respect (Darling-Hammond & Richardson, 2009). After experiencing the trusting atmosphere in which all ideas were accepted, and participants were encouraged to communicate their comprehension of the text, they felt safe enough to participate completely. Even case study members who were hesitant to speak in groups for fear of ridicule eventually became actively involved in the book studies. There is an implication in these data that building such trust takes time (Wei, Andree, & Darling-Hammond, 2009).

Another component of the book studies found in the six principles is the opportunity for members to be actively involved (Darling-Hammond & Richardson, 2009). Active participation occurred through the readings, discussions, implementation of new strategies, and interactive blogs.

Lawler and King (2000) also recommended that any learning be built on participant experiences, acknowledging the importance of the adult learner's extensive knowledge base. During the meetings and through their blog posts, the case study members added their personal reflections and stories to create a more realistic picture of how it meshed with their previous understandings. Book study groups allow for sharing of knowledge, thus, leading to deeper understanding (Darling-Hammond & Richardson, 2009). According to the participants' perceptions, this element added to the value of their experience and facilitated the incorporation of new learning into their workplaces (Darling-Hammond & Richardson, 2009; Wei et al., 2009).

The next principle of adult learning that should be applied to PD is the employment of collaborative inquiry within the activities. The case study members described how the book studies provided them with a support network and a safe environment in which to try out new ideas and get feedback from other members, indicative of research on supportive PD (Hunzicker, 2011). Hoover (1996) recommended that professional development should allow educators "to test their understandings and build new ones" (para. 9). The book study groups provided the participants with this opportunity.

The fifth principle, according to Lawler and King (2000), is the need for any activity to allow participants to apply their learning to their own practice. The book study participants were constantly considering the text they read and the topics of the discussion in light of how it would fit with their particular situation and their students. The book studies also provided a longer duration of time than traditional PD, giving the educators the time they need to try out new ideas (Lee, 2005). The ongoing nature of the book study groups gave the members the chance for experimentation in the implementation of new strategies (Wei et al., 2009).

Finally, any PD activity should empower the participants (Lawler & King, 2000). As a result of their participation, the members described developing confidence and empathy in their understanding of students' special needs, empowering them to advocate and plan for students more effectively.

### Emergence of a Knowledge Community

Participants reflected on their opportunities to collaborate, give advice, provide support to each other, and challenge each other. In knowledge communities, participants come together based on common interests and goals (Craig, 1995, 2007; Seaman, 2008). In addition, there is a desire to enhance personal knowledge, which was the primary reason the participants gave for coming to the book studies again and again. Knowledge communities are formed because adult learners have a need to control their own learning (Gravani, 2012; Gregson & Sturko, 2007; Merriam, 2008).

In the professional book studies, the initial reason for the discussions was to assist the participants in considering how they could incorporate the books they read into their practice. The case study members all reflected on how the discussions accomplished this need. The participants explained that other participants helped fill in their knowledge gaps, enabling them to more deeply comprehend the text.

They also reported that they risked sharing their interpretation of the text even when they were not sure of their level of comprehension, and related their own experiences to the others in the group. The case study members reported that their participation provided them with a feeling of inclusion rather than the isolation they had reported in their workplace.

### Limitations

This research study has limitations that could affect its outcome. First, this work was conducted after the participants had completed the book studies. Thus, the memories of the participants may not be as accurate or strong as if the study had been conducted while the professional book studies were still taking place. However, in order to assist the interviewees in recollection, two artifacts were made available to them during the interviews. First, each participant's reflective essays were provided for perusal prior to and during the interview. In addition, participants' available blog entries were also provided in order to refresh their memories. We believe that returning to these artifacts after some time may also have resulted in "big picture" recollections

*Professional Development*

of the experience, and may account for the lack of reported dissent or dissatisfaction in our findings.

The case study was limited to a purposeful sampling of 12 participants. The participants were chosen because they participated in a minimum of four book studies, representing a period of approximately two school years. Because the book studies were voluntary, this repeated participation was believed to be indicative of a preference for this type of PD activity. This sampling size is believed to provide enough data to reach saturation; however, it is not enough to be generalizable (Merriam, 2009).

Finally, in considering our findings, practitioners may want to consider the realities of teacher attrition in designing book study PD. Our findings reflect the experiences of a group with relative stability over 3½ of the 4 years of the study.

**Implications**

One of the most powerful tenets of knowledge communities is the very idea that new learning is both personally and socially constructed (Seaman, 2008). This case study shows that the participants in the professional book studies had the chance to have an individual voice and shared voice in the discussion groups and in their own professional learning. Adult learners need to have the ability to direct their own learning. Without the buy-in that voice provides, any other professional development activity is a waste of time and energy.

*Individual Voice*

Like all adult learners, the educators brought their own experiences, beliefs, and expectations to their readings and to the book study discussions. The participants read the books on their own, then shared their understanding of the text in the discussions. Within the book study groups, each participant had different interpretations of the reading because of their unique understanding. As they read, they interpreted the text in light of this knowledge. These unique backgrounds are important to understand, acknowledge, and celebrate within PD activities.

One of the underlying elements valued by the participants was choice. Personal choice was built into the studies in several ways. The voluntary nature of participation, the choice of books, and freedom to discuss and implement strategies all provided the learners with freedoms participants claimed were not typically experienced in professional development activities. Research has shown that adult learners need to be able to make their own choices and direct their own learning (Hunzicker, 2011). These educators, in particular, wanted to know that their time would not be wasted, and that the PD would help them solve problems and become more adept at their jobs, all of which are elements of effective professional development.

*Blanton et al.*

*Shared Voice*

Participants also demonstrated a shared voice in the face-to-face discussions and interactive blogs. Through narrative sharing of their experiences, their stories brought the book's information to life and added a realistic and personal interpretation, providing participants with a deeper understanding of the material. Sharing through story helped other members understand how the strategies might work in their own classrooms. Research on narrative learning helps to explain how this sense of story provides support for learning (Kooy, 2006a; Rossiter, 2002) by enabling listeners to better envision how new learning can be put into action.

The need for shared voices has implications for any who are attempting to support adult learners in acquiring new knowledge. Adult learners need to be able to share their thinking in terms of their previous experiences, and they do this through narrative. Open discussions led by the participants provided learners with the opportunity to give voice to ideas. When planning PD activities, teachers should be given the opportunity to have ample time to discuss their ideas in a relaxed setting where the sharing of stories is encouraged as a method of making meaning. In the knowledge community that formed through the professional book stories, the participants found a shared voice.

**Continuing Questions**

Although the participants held many different positions in the district, most of them were similar in at least three ways: age, experience, and gender. Eight of the participants were in their late 50s at the time of the book studies. The other four were in their late 30s. All but two of the participants had worked in education for at least 17 years at the time of the last book studies. Half of the participants had worked in education for 26 years or more. This study examined a teacher book club through the lens of andragogy; Blaschke (2012) suggests that andragogy, or *self-directed* learning, progresses to heutagogy, or *self-determined* learning. As future generations of teachers arrive to the profession, will their needs in book study become more heutagogic? What are the implications for teacher-learners who acquire knowledge through social media or other Web 2.0 tools (Blaschke, 2012)? Selway (2003) began to consider this evolution in learning as well when she considered the question: "It is a challenge to decide what books to read that will meet everybody's needs. Is that even possible?" (p. 21).

In addition to their status as a group of mostly veteran teachers, all of the participants were special educators. How did the peculiarities of their jobs add to or define their experience in the book study? Is this form of professional development particularly meaningful for special educators?

Finally, all of the participants of the book studies were women. While there are all-male and mixed-gender book study groups in the larger body of research, women dominate them (Kooy, 2006a; O'Connor, 1996). What does this say about how women learn? How should this information inform professional development?

## Conclusions

The results from this study highlight how a knowledge community centered around the needs of adult learners can initiate and support changes in teachers' practices, thinking, and beliefs. The book study PD offered participants a unique learning experience which facilitated relationships across school sites as well as ongoing teacher dialogue around instructional strategies and differing views on teaching. The majority of participants reported shifts in their thinking about teaching and about students. As school districts weigh their professional development needs, it is important for researchers, teacher educators, and administrators to continue to examine the essential nature of teachers' needs as adult learners in designing andragogic opportunities to support teacher growth.

## ORCID iD

Amy D. Broemmel https://orcid.org/0000-0003-2373-9564

## References

Addington, A. H. (2001). Talking about literature in university book club and seminar settings. *Research in the Teaching of English, 36,* 212–248. Retrieved from https://www.jstor.org/stable/40171537

Avalos, B. (2011). Teacher professional development in *Teaching and Teacher Education* over ten years. *Teaching & Teacher Education, 27,* 10–20. doi:10.1016/j.tate.2010.08.007

Bach, J., Hensley Choate, L., & Parker, B. (2011). Young adult literature and professional development. *Theory Into Practice, 50,* 198–205. doi:10.1080/00405841.2011.584030

Ball, D., & Cohen, D. G. (1999). Developing practice, developing practitioners: Toward a practice-based theory of professional education. In L. Sykes (Ed.), *Teaching as the learning profession: Handbook of policy and practice* (pp. 3–32). San Francisco, CA: Jossey-Bass.

Blaschke, L. M. (2012). Heutagogy and lifelong learning: A review of heutagogical practice and self-determined learning. *International Review of Research in Open and Distance Learning, 13*(1), 56–71.

Bonner, D., & Tarner, L. (1999). Once upon an HRD book club. *Training and Development, 53*(12), 45–51. Retrieved from https://link.galegroup.com/apps/doc/A58386338/AONE?u=googlescholar&sid=AONE&xid=212d8d6a

Brady, K. (2003). *Rules in school.* Greenfield, MA: Northeast Foundation for Children.

Burbank, M. D., Kauchak, D., & Bates, A. J. (2010). Book clubs as professional development opportunities for preservice teacher candidates and practicing teachers:

An exploratory study. *New Educator, 6,* 56–73. doi:10.1080/1547688X.2010.10399588

Butler, D. L., Lauscher, H. N., & Jarvis-Selinger, S. (2004). Collaboration and self-regulation in teachers' professional development. *Teaching & Teacher Education, 20,* 435–455. doi:10.1016/j.tate.2004.04.003

Carpenter, J., & Linton, J. (2015). Educators' perspectives on the impact of Edcamp unconference professional learning. *Teaching and Teacher Education, 73,* 56–69. doi:10.1016/j.tate.2018.03.014

Chan, S. (2010). Applications of andragogy in multi-disciplined teaching and learning. *Journal of Adult Education, 39*(2), 25–35.

Cizek, G. J. (1999). Give us this day our daily dread: Manufacturing crises in education. *Phi Delta Kappan, 80,* 737–743. Retrieved from https://www.jstor.org/stable/20439557

Clark, R. (2003). *The essential 55: An award-winning educator's rules for discovering the successful student in every child.* New York, NY: Hyperion.

Craig, C. J. (1995). Knowledge communities: A way of making sense of how beginning teachers come to know. *Curriculum Inquiry, 25,* 151. doi:10.1080/03626784.1995.11076175

Craig, C. J. (2007). Illuminating qualities of knowledge communities in a portfolio-making context. *Teachers & Teaching, 13,* 617–636. doi:10.1080/13540600701683564

Creswell, J. W. (1998). *Qualitative inquiry and research design.* Thousand Oaks, CA: Sage.

Cunningham, P. M., & Allington, R. L. (2003). *Classrooms that work: They can all read and write.* Boston, MA: Allyn & Bacon.

Darling-Hammond, L., & Richardson, N. (2009). Teacher learning: What matters? *Educational Leadership, 66*(5), 46–53.

Darling-Hammond, L., Wei, R. C., Andree, A., Richardson, N., & Orphanos, S. (2009). State of the profession: Study measures status of professional development. *Journal of Staff Development, 30*(2), 42–44.

Day, B. W., Lovato, S., Tull, C., & Ross-Gordon, J. (2011). Faculty perceptions of adult learners in college classrooms. *Journal of Continuing Higher Education, 59,* 77–84. doi:10.1080/07377363.2011.568813

Grandin, T. (1996). *Thinking in pictures: And other reports from my life with autism.* New York, NY: Vintage Books.

Gravani, M. N. (2012). Adult learning principles in designing learning activities for teacher development. *International Journal of Lifelong Education, 31,* 419–432. doi:10.1080/02601370.2012.663804

Gregson, J. A., & Sturko, P. A. (2007). Teachers as adult learners: Re-conceptualizing professional development. *Journal of Adult Education, 36*(1), 1–18.

Gulamhussein, A. (2013). *Teaching the teachers: Effective professional development in an era of high stakes accountability.* Arlington, VA: Center for Public Education.

Hall, L. A. (2009). "A necessary part of good teaching": Using book clubs to develop preservice teachers' visions of self. *Literacy Research and Instruction, 48,* 298–317. doi:10.1080/19388070802433206

Hatch, J. A. (2002). *Doing qualitative research in education settings.* Albany: State University of New York Press.

Hirsh, S. (2009). A new definition. *Journal of Staff Development, 30*(4), 10–16.

Hoover, W. (1996). The practice implications of constructivism. *SEDL Letter, 3.* Retrieved from http://www.sedl.org/pubs/sedletter/v09n03/practice.html

Hunzicker, J. (2011). Effective professional development for teachers: A checklist. *Professional Development in Education, 37,* 177–179. doi:10.1080/19415257.2010.523955

Johnston, P. H. (2004). *Choice words: How our language affects children's learning.* Portland, ME: Stenhouse Publishers.

Kooy, M. (2006a). *Telling stories in book clubs: Women teachers and professional development.* New York, NY: Springer.

Kooy, M. (2006b). The telling stories of novice teachers: Constructing teacher knowledge in book clubs. *Teaching & Teacher Education, 22,* 661–674.

Kranowitz, C. S. (1998). *The out-of-sync child: Recognizing and coping with sensory integration dysfunction.* New York, NY: Perigee Books.

Lassonde, C., Stearns, K., & Dengler, K. (2005). What are you reading in book groups? Developing reading lives in teacher candidates. *Action in Teacher Education, 27*(2), 43–53. doi:10.1080/01626620.2005.10463382

Lave, J., & Wenger, E. (1991). *Situated learning: Legitimate peripheral participation.* New York, NY: Cambridge University Press.

Lawler, P. A., & King, K. P. (2000). *Planning for effective faculty development: Using adult learning strategies* (Original ed.). Malabar, FL: Krieger.

Le Fevre, D. M. (2014). Barriers to implementing pedagogical change: The role of teachers' perceptions of risk. *Teaching & Teacher Education, 38,* 56–64. doi:10.1016/j.tate.2013.11.007

Lee, H. (2005). Developing a professional development program model based on teachers' needs. *Professional Educator, 27*(1–2), 39–49.

Lemann, N. (2010, September 27). Schoolwork. *The New Yorker.* Retrieved from https://www.newyorker.com/magazine/2010/09/27/schoolwork

McGlinn, J. M., Calvert, L. B., & Johnson, P. S. (2003). University-school connection: A reading circle for teachers. *Clearing House, 77*(2), 44–49. doi:10.1080/00098650309601227

Merriam, S. B. (2008). Adult learning theory for the twenty-first century. *New Directions for Adult & Continuing Education, 119,* 93–98. doi:10.1002/ace.309

Merriam, S. B. (2009). *Qualitative research: A guide to design and implementation.* San Francisco, CA: Jossey-Bass.

Mitchell, R. (2013). What is professional development, how does it occur in individuals, and how it may be used by educational leaders and managers for the purpose of school improvement? *Professional Development in Education, 39,* 387–400. doi:10.1080/194115257.2012.762721

Mosley, M., & Rogers, R. (2011). Inhabiting the "tragic gap": Pre-service teachers practicing racial literacy. *Teaching Education, 22,* 303–324. doi:10.1080/10476210.2010.518704

Nevills, P. (2003). Cruising the cerebral superhighway. *Journal of Staff Development, 24*(1), 20–23.

O'Connor, S. (1996). Book clubs: Communities of learners. *New Directions for Adult & Continuing Education, 71,* 81-90.

Porter, A. C., Garet, M. S., Desimone, L., Yoon, K. S., & Birman, B. F. (2000). *Does professional development change teaching practice? Results from a three-year study* (Report to the U.S. Department of Education, Office of the Under Secretary on Contract No. EA97001001 to the American Institutes for Research). Washington, DC: Pelavin Research Center.

Reilly, M. A. (2008). Occasioning possibilities, not certainties: Professional learning and peer-led book clubs. *Teacher Development, 12,* 211–221. doi:10.1080/13664530802259230

Roberts, S., Jensen, S., & Hadjiyianni, E. (1997). Using literature study groups in teacher education courses: Learning through diversity. *Journal of Adolescent & Adult Literacy, 41,* 124–133. Retrieved from https://www.jstor.org/stable/40013493

Rossiter, M. (2002). Narrative and stories in adult teaching and learning. *ERIC Digest.* Retrieved from https://files.eric.ed.gov/fulltext/ED473147.pdf

Schultz, K., & Ravitch, S. M. (2013). Narratives of learning to teach: Taking on professional identities. *Journal of Teacher Education, 64,* 35–46. doi:10.1177/0022487112458801

Scotchmer, M., McGrath, D. J., & Coder, E. (2005). *Characteristics of public school teachers' professional development activities: 1999-2000* (NCES Issue Brief No. NCES 2005-030). Washington, DC: National Center for Education Statistics.

Seaman, M. (2008). Birds of a feather? Communities of practice and knowledge communities. *Curriculum & Teaching Dialogue, 10,* 269–279.

Selway, L. G. (2003). *Leading a professional book club: Staff development to build understanding and grapple with difficult issues.* Paper presented at the annual meeting of the American Educational Research Association, Chicago, IL.

Sheehy, G. (1995). *New passages: Mapping your life across time.* New York, NY: Random House.

Spradley, J. P. (1980). *Participant observation.* Orlando, FL: Harcourt College.

Tovani, C. (2000). *I read it but I don't get it: Comprehension strategies for adolescent readers.* Portland, ME: Stenhouse.

Trotter, Y. D. (2006). Adult learning theories: Impacting professional development programs. *Delta Kappa Gamma Bulletin, 72*(2), 8-13.

Trust, T., & Horrocks, B. (2017). "I never feel alone in my classroom": Teacher professional growth within a blended community of practice. *Professional Development in Education, 43,* 645–665. doi:10.1080/19415257.2016.1233507

U.S. Census Bureau. (2014). *State and county quickfacts.* Washington, DC: U.S. Census Bureau. Retrieved from https://www.census.gov/quickfacts/fact/table/US/PST045218

van den Berg, R. (2002). Teachers' meanings regarding educational practice. *Review of Educational Research, 72,* 577–625. doi:10.3102/00346543072004577

Vu, Julie. (2019). Exploring the possibilities for professional learning. *Reading Teacher, 72,* 539–543. doi:10.1002/trtr.1771

Wei, R. C., Andree, A., & Darling-Hammond, L. (2009). How nations invest in teachers. *Educational Leadership, 66*(5), 28–33.

Wenger, E. (2004). Knowledge management as a doughnut. *Ivey Business Journal.* Retrieved from Ivey Business Journal website: https://iveybusinessjournal.com/publication/knowledge-management-as-a-doughnut/

Manuscript received February 4, 2019
Final revision received June 29, 2019
Accepted July 1, 2019

# Bridging the Gap Between Research and Practice: Predicting What Will Work Locally

Kathryn E. Joyce
*University of California, San Diego*
Nancy Cartwright
*University of California, San Diego*
*Durham University*

*This article addresses the gap between what works in research and what works in practice. Currently, research in evidence-based education policy and practice focuses on randomized controlled trials. These can support causal ascriptions ("It worked") but provide little basis for local effectiveness predictions ("It will work here"), which are what matter for practice. We argue that moving from ascription to prediction by way of causal generalization ("It works") is unrealistic and urge focusing research efforts directly on how to build local effectiveness predictions. We outline various kinds of information that can improve predictions and encourage using methods better equipped for acquiring that information. We compare our proposal with others advocating a better mix of methods, like implementation science, improvement science, and practice-based evidence.*

---

KATHRYN E. JOYCE is a PhD candidate in the Department of Philosophy at the University of California, San Diego, and a McPherson fellow at the Center for Ethics & Education, e-mail: *kejoyce@ucsd.edu*. She specializes in social and political philosophy and philosophy of education. Her research focuses on theories of social justice, relational egalitarianism, and educational justice. Her recent projects concern methodological and normative issues associated with education policy.

NANCY CARTWRIGHT is a professor of philosophy at the University of Durham, a Distinguished Professor of Philosophy at the University of California, San Diego, and codirector of Durham's Centre for Humanities Engaging Science and Society (CHESS), 50 Old Elvet, Durham DH1 3HN, UK; e-mail: *nancy.cartwright@durham.ac.uk*. She is a philosopher of natural and social science, with a special interest in causation, objectivity, evidence, and the philosophy of social technology. Her current research investigates how scientific research can inform policy. Her recent books include *Evidence Based Policy: A Practical Guide to Doing it Better* (with Jeremy Hardie, 2012), *Evidence: For Policy and Wheresoever Rigor is a Must* (2013), and *Improving Child Safety: Deliberation, Judgment, and Empirical Research* (with Eileen Munro, Jeremy Hardie, and Eleonora Montuschi, 2016).

*Joyce, Cartwright*

KEYWORDS:  causal claims, evidence-based education, educational research, RCTs, research-informed practice

For nearly two decades, the dominant model for evidence-based education (EBE) has focused on improving schools by researching "what works." Yet anyone familiar with EBE recognizes its relentless adversary: the gap between research and practice (Coburn & Stein, 2010; Farley-Ripple, May, Karpyn, Tilley, & McDonough, 2018; McIntyre, 2005; Nelson & Campbell, 2017; Tseng & Nutley, 2014). The challenge is how educators can use research results to improve their outcomes in practice. Despite efforts to bridge the gap, primarily by more effective dissemination of results from high-quality experimental research, interventions adopted on the basis of recommendations often fail to be effective in practice. Many respond by rejecting EBE in its current form or by denouncing the entire enterprise (e.g., Archibald, 2015; Biesta, 2007, 2010; Hammersley, 2002, 2013; Smeyers & Depaepe, 2007). Rather than opposing the dominant model, in which vast resources have been invested, we propose ways to shift and expand it to improve its performance. We begin with an analysis of why the gap exists and propose constructive advances to help bridge it. We argue that the research-practice gap reflects a gap between the causal claims supported by the experimental research results generally favored in EBE—"It *worked*"—and the causal claims that are relevant to practice—"It *will work* here." Researchers may produce evidence to support the former, but those on the practice side must figure out whether a program can work for them and, if so, what they need to put in place to get it to do so. But so far, EBE has not been centrally concerned with producing and disseminating research that helps with this task. We conclude that addressing the gap requires a major rethinking of the research investigation and theory building needed to support EBE and of the demands on research users.

Drawing on guidance for deliberating in other policy areas, especially child protection, we provide a catalog of some things that can help research users—educators and decision makers in their home sites—make more reliable predictions about what might work for their school, their district, and their students, and how it might do so. Discussing the knowledge *use* side of EBE carries implications for organizing the knowledge *production* side and for knowledge mobilization. Our principal contribution on the knowledge production side is in diagnosing the gap and showing that bridging it requires dramatically expanding the kinds of evidence that are collected and disseminated by EBE and adjusting the methods used to judge its acceptability.

EBE relies on researchers to produce evidence of effectiveness for educators to use in practice. Intermediary organizations like the What Works Clearinghouse (WWC) in the United States, the What Works Network and Educational Endowment Foundation in the United Kingdom, the European

EIPPEE (Evidence Informed Policy and Practice in Education in Europe) Network are supposed to help bridge the gap between research and practice by evaluating research, summarizing results, and advertising interventions that have proven efficacious in rigorous experimental studies, especially randomized controlled trials (RCTs).

Proponents of EBE commonly attribute the gap between outcomes in research and practice to deficiencies in how tasks are performed on one or both sides of the knowledge production/knowledge use divide. Accordingly, plans for addressing it encourage researchers to conduct more relevant research, offer guidance for implementation (Gutiérrez & Penuel, 2014), translate "research-based knowledge into . . . generalized practical suggestions" (McIntyre, 2005, p. 364), and effectively communicate research findings to decision makers in easily digestible formats and in ways that encourage use. For example, Levin (2013) explains that research findings must compete with information from other sources that influence potential users. He suggests using knowledge mobilization strategies that appreciate how experience, organizational practices, and attitudes shape educators' engagement with research. On the practice, or knowledge use, side, strategies generally emphasize the importance of cultivating evidence literacy and exercising professional judgment when choosing and implementing interventions (C. Brown, Schildkamp, & Hubers, 2017; Bryk, 2015). By contrast, we are concerned with the *information and reasoning* needed to bridge the gap between the kind of causal claim supported by research—*causal ascriptions*—and the kinds of claims that are relevant to practice—*effectiveness generalizations* and *effectiveness predictions*.

RCT results, along with meta-analyses and systematic reviews of them, are taken across evidence-based policy communities as "gold-standard" evidence for "what works." There is much debate about this concentration on RCTs (see the section below). Our concerns are with sloppy talk. Currently, EBE is plagued by casual use of language that is imprecise about what kinds of causal claims are at stake and also about what kind of work it takes to warrant the kinds needed in practice. We will look carefully and critically at just what kinds of claims can (in the ideal) be warranted by RCTs and contrast that with the kinds of claims educators need to know. Then we will turn to *the argument theory of evidence* and the related *material theory of induction* to lay out what it takes, beyond the much-discussed RCT, to provide evidence for claims that a policy works generally or that it will work *here*.

We shall defend the claim that RCTs can provide evidence for *causal ascriptions*—that is, the intervention *worked* in the study population in the study setting. But, as the argument theory shows, it takes a great deal of additional knowledge besides an RCT result to warrant the claims educators need—namely, *effectiveness generalizations* and *effectiveness predictions*. Yet the WWC, like other organizations driving EBE, claims that "high-quality research" from RCTs can show "what works in education" and support "evidence-based decisions" (WWC, n.d.). Similarly, Connolly, Biggart, Miller, O'Hare, & Thurston (2017) assert that "RCTs within education offer the possibility of developing a cumulative body of knowledge and an evidence base around the effectiveness of different practices, programmes, and policies" (p. 11). Like other philanthropic organizations involved in social policy, the Gates Foundation (n.d.) calls for schools to use methods "grounded in data and evidence." Their programs focus on "identifying new, effective approaches that can be replicated in other schools" and "using evidence-based interventions and data-driven approaches to support continuous learning."

As these statements indicate, EBE aims to support predictions concerning how an evidence-based intervention will perform in a new educational setting *indirectly* by establishing effectiveness *generalizations*. While generalizations would naturally justify predictions about specific cases, we argue that they are unnecessary. This is a good thing since there are few useful, reliable general effectiveness claims to be had. The kinds of claims that can be expected to hold widely in education, or even over restricted domains characterized by a handful of descriptors (inner city, free school lunches, ESL [English as a second language], high-achievers, etc.), are usually too abstract to guide practice—for example, the generalization that children learn better when they are well nourished, have a secure environment, read at home, and have adequate health care.

Instead of just trying to support local plans and predictions indirectly by establishing "what works," we argue that EBE should focus on research that supports these *directly* by producing evidence that research users can employ to make effectiveness predictions locally. We should note at the start that because context does significantly affect effectiveness, it will seldom be possible to replicate results by moving a program as implemented in study sites to new settings. If the program is to work in a context it will have to be *fitted* to that context, and this seldom involves just tinkering around the edges or implementing it well. Rather, it requires getting all the right features in place that will allow the program to work *here* and guarding against features that can derail it *here*. So the work is harder than one might expect: If the program is to be effective, a context-local program plan must be built. And to build a program plan that you can expect to be effective for the goals you want to achieve in a specific context, you need to know what facts make it likely that the program will work *here*. Since just what these facts are can vary from context to context, often dramatically, there's no telling in advance what they will be. But we can tell what *kinds* of facts matter. To this end, we will outline a variety of kinds of information that can guide local planning and make local predictions about the effectiveness of proposed plans more reliable and more useful.

The kinds of research that can produce the requisite information, locally or more generally, often in coproduction, require a mix of methods well

## Bridging the Gap Between Research and Practice

beyond those listed in current evidence hierarchies. This suggestion resonates with other recent calls for more mixed-method research in education,[1] but our reasons are different from the usual ones. The standard reasons for mixing methods in EBE are to aid implementation (Gorard, See, & Siddiqui, 2017) and to make general effectiveness claims more reliable (Connolly et al., 2017; Bryk, 2015). We, by contrast, encourage mixed methods because reliable and useful effectiveness predictions require a variety of different kinds of information relevant to determining how an intervention will perform in a specific setting that different kinds of research help uncover. While local effectiveness predictions will never be certain, incorporating this information can improve them.

The next section situates our project within the general intellectual setting and clarifies the contribution we wish to make. The section Diagnosing the Problem provides an account of how the current research emphases in EBE contribute to the research-practice gap. Building on a framework[2] that distinguishes different kinds of causal claims prevalent in evidence-based policy, we assess the role that RCTs can play for each, underlining that RCTs are poor evidence for what educators need to know: Will this intervention work here in our school or classroom for these students? The subsequent section titled Bridging the Gap offers a catalog of the kinds of information that can help educators make local predictions that are reliable and useful. We urge developing and using a broader panoply of research methods than are generally endorsed in EBE to help uncover these kinds of information. The penultimate section, Complementary Calls for Expanding the Research Agenda in EBE, compares other approaches that call for a better, broader mix of methods, like "implementation science," "improvement science," and "practice-based evidence," with our own.

### Framing Our Project

We acknowledge that debates about education research methodologies often stem from fundamental disagreements about the epistemology and ontology underpinning various approaches (e.g., Bridges, 2017; Bridges & Watts, 2008; Crotty, 1998; Howe, 2009; Phillips, 2007; Smeyers & Depaepe, 2007). In particular, many critics of EBE target the positivist approach they take it to represent. Our project does not contribute to these debates. Disagreements about the meaning of broad positions like positivism abound; yet differences between the relevant ontological and epistemological positions are difficult to pin down, as are their precise consequences for debates concerning EBE. Additionally, interlocutors on all sides disagree about whether the disputed positions bear a *necessary* relation to EBE.[3] Although they continue to favor RCTs for investigating causal questions, many education researchers involved in EBE now take their background theories to be compatible with a wide range of methodologies (e.g., Connolly et al., 2017; Gorard et al., 2017). Instead of attempting to navigate this familiar, fraught landscape, we articulate the specific problems underlying the research-practice gap without relying on controversial terminology or attempting to identify their ideological origins. Still, readers will find that our arguments reflect and address some of the prevalent criticisms of EBE driving these debates—namely, the tendency to neglect the importance of context, of professional judgment, and of what teachers, students, and parents can contribute (Biesta, 2007; Smeyers & Depaepe, 2007).

However one classifies the dominant EBE model, it is part of a larger evidence-based movement across areas of social policy and is enshrined in U.S. federal education policy (Eisenhart & Town, 2003; Education Sciences Reform Act, 2002; No Child Left Behind Act, 2002; Every Student Succeeds Act, 2015). We start from their assumption that research can help improve students' educational outcomes and experiences and hunt for ways to build and improve on the widespread efforts in this direction that are already in place. As we understand it, this assumption need not imply that social phenomena yield to the same study methods as natural phenomena, nor does it entail an instrumentalized view of education, a narrow conception of causality as deterministic, or denying that education involves agents who engage in socially developed normative and cultural processes (Bridges & Watts, 2008; Biesta, 2007, 2010; Smeyers & Depaepe, 2007). On the contrary, we hold that effects issue from a plurality of context-local causal factors that contribute to them, including the actions, attitudes, norms, and habits of students, teachers, parents, school administration, and the wider neighborhoods in which these are embedded. Because we often cannot identify, untangle, or measure all of the factors that contribute to an outcome, our causal claims are never certain. Still, the warrant for them can be better or worse.

We recognize that assessing effectiveness is only one part of a larger decision-making process that involves considering values, setting goals, and mapping local assets (Brighouse, Ladd, Loeb, & Swift, 2018). Educational policies and practices should be fair, compassionate, and effective; and it is difficult to predict if any of these, let alone all three, will be true for a policy we consider using *here*, as we are planning to implement it *here*, given the complicated set of interacting factors that are relevant *here*. We cannot tackle all three in this article. We offer positive suggestions on the third—effectiveness—which is EBE's central concern. Expectations can always go wrong, both ethically and epistemically, but care in deliberation can improve predictions on both sides. Our suggestions here focus on the epistemic side, more specifically, on whether policies as planned for the local context will achieve their targeted aims, setting aside important ethical questions regarding choice of the target and who is advantaged or disadvantaged by those choices.

Because we situate our arguments within the existing EBE framework, they ought to be of interest to proponents seeking to improve its performance. At the same time, our diagnosis captures some of the familiar concerns commonly raised by antipositivist critics. Like them, we reject the intervention-centered approach to EBE, which focuses exclusively on "what works." Instead, we endorse a context-centered approach, which starts from local problems in local settings, with their own local values, aims, resources, and capabilities to provide tools that help predict what will work there and what students, teachers, parents, and school staff can contribute to success.[4] Framing our project this way allows readers with differing commitments to seriously consider and even accept our arguments, albeit they may do so for different reasons.

## Diagnosing the Problem

### Getting Clear on Just What Claims Are Being Made

Clarifying our concept of evidence helps distinguish distinct causal claims, which are often conflated without note to bad effect. Evidence is always evidence for (or against) *something*. The question is not "What makes a result good evidence?" but rather "What makes a result good evidence for *a particular claim*?" Using the concept of *evidence* imprecisely within EBE generates confusion, which can undercut its success (Joyce & Cartwright, 2018; Kvernbekk, 2016; Spillane & Miele, 2007). A fact counts as evidence for a specified claim when it speaks to the truth of that claim. The *argument theory of evidence* provides a good way to capture this idea (Cartwright, 2010, 2013, 2017; Reiss, 2013; Scriven, 1994). According to the argument theory, something is evidence for a claim when it serves as a premise in a sound argument for that claim. Sound arguments are composed of trustworthy premises that jointly imply the conclusion. To be trustworthy, each premise must be supported by good reasons. These can include, among other things, empirical facts established by research, observations, and credible theory.[5]

The argument theory is a close cousin of the philosopher John Norton's highly regarded *material theory of induction*. Norton (2003) points out that there are serious problems with all the standard attempts to articulate a theory of inductive inference that depend on form alone (as in the case of induction by simple enumeration, which we discuss later). What really does the work, he maintains, are material facts—encoded in substantive claims that connect the evidence with the hypothesis to be evidenced—showing just *why* the putative evidence is evidence for that hypothesis. For this theory too, a research result is evidence *relative to* a target hypothesis and to a set of additional claims describing material facts about the world, including often general truths.

By contrast, some scholars use *research findings* or *research, knowledge*, and *evidence* interchangeably (e.g., C. Brown, 2014; Nutley, Walters,

*Joyce, Cartwright*

& Davies, 2007). Their usage implies that when a study meets the espoused standards, its results count as evidence. This does not make sense. We must know what the claim is to decide whether and how—or under what further assumptions—some finding counts as evidence for it. Labeling research results as "evidence" obfuscates differences in the claims for which EBE needs evidence.

**What Kind of Claim Can RCTs Support?**

Much of the literature advocating RCTs stresses the benefits of randomly assigning subjects to intervention and control groups in order to balance the other causal factors that might affect the outcome, other than the intervention (e.g., Borman, 2009; Connolly et al., 2017; Gorard et al., 2017; Shavelson, Towne, & National Research Council, 2002; Slavin, 2002). There are two problems with this.

First, even if randomization did achieve balance, it only balances other factors at the time of selection. Correlation with other causal factors can arise postrandomization. For example, getting new materials may encourage effort or confidence that aids performance among students in the intervention group. Where possible, experiments use blinding to reduce this problem. Ideally, research subjects, those delivering interventions, those measuring outcomes, and those doing the statistical analysis are all unaware of who received which treatment. But, at best, blinding works poorly in educational experiments. Teachers, students, and administrators know who is receiving the intervention and who is in the comparison group. This knowledge alone can impact outcomes, as can the training and support for teachers delivering the intervention.

Second, even supposing there is random assignment and successful blinding, we should not expect to get two groups that are the same with respect to all causal factors affecting the outcome other than the intervention. Getting a balance like that in a study of, say, 100 units is like getting 50 heads in 100 flips of a fair coin. Of course, generally, the larger the sample, the closer the results on a single run can be expected to be to the true average. This is one of the reasons why the WWC encourages multiple, large RCTs.

What we can expect if randomization, blinding, and so on are successful is this:

> RCT conclusion: The measured outcome—the difference between the observed mean in the treatment group and that in the control group—gives
> a. the average treatment effect[6]
> b. of the intervention provided in the treatment wing of the study compared to that provided in the control wing
> c. in expectation – across a hypothetical infinite sequence of runs of the experiment with a new randomization on each run
> d. for the population enrolled in the study.

*Bridging the Gap Between Research and Practice*

We need to be attentive to each of these:

a. We only learn average results. Individuals in the study will generally have differed in their responses to the intervention, and some may even have been harmed by it.
b. Exactly what the control group received matters significantly to the size of the measured treatment effect; the effect will seem much bigger if the control group received an intervention that performs badly than if it received one that performs well.
c. The estimate is, in technical language, *unbiased*. This has nothing to do with how close it is to the truth.[7]
d. The conclusion can at best be a causal *ascription*.

It is the last point that we want to discuss in some detail.

Studies—any study, RCT or otherwise—can only support results about the things studied. Conclusions about things not studied must depend on assumptions outside the range of the study. To play the role EBE assigns them, RCTs would have to be almost sufficient by themselves for effectiveness claims. But what a positive result from a very well-conducted RCT can directly support is just a claim of the form we have labeled "RCT conclusion," which is about the study population. Indeed, the WWC (2017b) explicitly states that its standards "focus on the causal validity within the study sample (*internal* validity)" (p. 1). It should be obvious that a causal ascription in a study population cannot directly evidence a general effectiveness claim or an effectiveness prediction. The fact that an intervention *worked* somewhere cannot show that the intervention *works* or that it *will work* in some other target.

### Using Results Outside the Study Population

Assume for the moment that the WWC succeeds in vetting RCTs for what is called "internal validity," that is, to ensure the RCTs really do support causal ascriptions in the study populations. Educators are concerned with whether the tested intervention can positively affect outcomes in *their setting*. The central question for them is not what is usually called "external validity": Will the *same results* hold here? Generally, the answer to that is "probably not," given that the local factors that determine an effect size vary from setting to setting. Rather, the central question is "*What* must be in place here for the intervention to deliver good enough results for us?" And then, "Are the gains worth the expected costs?" For guidance, the WWC combines study results that meet the standards for internal validity to provide an effectiveness rating and an "improvement index" for each intervention. In both cases, it explicitly uses evidence supporting causal ascriptions and effect sizes within study groups as evidence for general effectiveness claims. For example, the WWC rates Read 180, a multifaceted literacy program, as having a positive effect on comprehension. The effectiveness rating rates the strength of evidence supporting the claim that Read 180 positively affected outcomes within a domain (e.g., comprehension) based on "the quality of research, the statistical significance of findings, the magnitude of findings, and the consistency of findings across studies" (WWC, 2016a). Interventions, including Read 180, merit the highest effectiveness rating when these factors provide "strong evidence that an intervention *had* a positive effect on outcomes [italics added]" (WWC, n.d.).

The WWC calculates the *effect size* for each individual study by dividing the observed average treatment effect (the observed difference in mean outcomes between treatment and control) by the standard deviation of the outcome in the pooled treatment and control groups (WWC, 2017b). This means that all studies are recorded in standard deviation units, which, it is hoped, provides some sensible way to compare sizes for studies that use different scales for outcomes. Then, it averages those (standardized) effect sizes to produce the effect size for the domain. For Read 180, the effect size, averaging across all six qualifying studies, is .15. This average provides the basis for the improvement index, which depicts "the size of the effect from using the intervention" in a way that is supposed to help educators "judge the practical importance of an intervention's effect" (WWC, 2017b, p. 14). That is, it is supposed to represent the expected change in outcome for "an average comparison group student if the student *had* received the intervention [italics added]" (p. 14). Based on the expected improvement for an average comparison group student, the improvement index for Read 180 is +6. Explaining its significance, the intervention report for Read 180 says that if given the intervention, the average student *"can"* be expected to improve their percentile rank by 6% (WWC, 2016b).

The presentation of these findings implies that the stronger the evidence for causal ascriptions (as represented by effectiveness ratings), the more credible is a general effectiveness claim or an effectiveness prediction. Educators are thus led to believe that "strong" evidence and large effect sizes indicate that the intervention will be highly effective, producing significant effects for them.[8] For instance, because Read 180 earns the highest effectiveness rating, educators can reasonably expect their students to improve by 6%, on average. Apparently, using RCTs justifies the inference from the internal validity in each study and consistent results across multiple studies (or faring well in a meta-analysis of study results) of this conclusion about what can be expected in general.

Clearly this is a mistake. It is only consistency across studies that is an indicator that the result might hold generally or at my school for my students; and without more ado—much more ado—it is a weak indicator at that. Adding up causal ascriptions, as the EBE literature seems to recommend, amounts to induction by simple enumeration, which has long been condemned as a weak form of inference.[9] When it *does* work, as it sometimes may in education, it is because the feature being generalized is projectable across the domain in question—*not* because the inference is based on multiple RCTs or combined findings from multiple RCTs. Any education examples we cite will likely prove controversial, so instead to make the point clear, we illustrate with the example of electric charge: A negative charge of $1.602177 \times 10^{-19}$ coulombs measured on a single electron is projectable across all electrons because we have strong reasons from theory and a multitude of different kinds of empirical results to believe that all electrons have the same charge. The same form of inference yields a false conclusion if it generalizes a feature that does *not* project onto the target population. Consider the oft-cited case of the color of swans. Multiple samples consisting of only swans in London's Regents Park do not license the inference that all swans are white. Likewise, even if multiple RCTs show positive results, predictions need further premises showing that the causal relations being generalized to other students or schools projects onto them, as in the case of electrons. Without such premises, we risk drawing false conclusions, as with the swans.

What kind of reasons can support projectability in educational contexts? Imagine that a study has produced a good estimate of an average treatment effect in the study population and you want to argue that that estimate should hold for some target population. This would be warranted if you could argue that the students in the study sample are representative of the target. One way to warrant that is to draw those students randomly from the target, where the target could be either a broad range of U.S. students or the students in a particular district or school. If the study group genuinely *represents* the target, then whatever probabilities are true of the study population (like the average effect of an intervention) are true of the target—that's what it means to be representative. But, of course, just as random assignment should not be expected to produce a balance of other factors in any one run of an experiment, random sampling should not be expected to produce a representative study population on any one draw of a sample for study. The probabilities of the study sample and the parent only match in expectation—over a hypothetical infinite series of trials. For any one study, the average treatment effect may be accurate for the study group but far from it for the parent.[10]

Even without this worry, within education, transferring results from the study to the target population usually cannot be justified on these grounds because samples are seldom randomly selected from the target population. Sometimes though, a sample is drawn randomly from the students in a specific school or district, often with the plan to scale up the intervention to the whole school or district if the study results are positive. But even if you draw your study students randomly, there are notorious difficulties in scaling up – the study population may not be representative of the whole population since people behave differently when a few receive the treatment than when all do.

Nor is it easier to construct study populations that are representative of a broad range of U.S. students. Researchers work within constraints that impose criteria for selecting study populations. For instance, researchers might be confined to a school district. Within the district, they must conduct the study within schools that are willing and able to participate. Even if all schools in the district are obliged to participate if randomly selected, that district becomes the parent population. The district itself was not randomly selected, so the experiment cannot support generalizing average effect sizes beyond the district.

A second problem is that random sampling from the target is not enough to ensure that the study results are representative of the target facts since much can happen to differentiate the two after sampling. Just as in an RCT, what happens to the sample after it is drawn may change the probabilities for relevant factors, so that the sample is no longer representative of the parent. For instance, monitoring both groups may impact student performance independently of the intervention itself, which inflates effect sizes. Moreover, educational contexts are dynamic systems (Reynolds et al., 2014). Changes may occur that have no connection to the implementation of the intervention but may still impact the reliability and generalizability of the results.

The alternative to random sampling from the target is to argue that the target and the study populations are the same, or similar enough, in the ways that affect the results to be projected (e.g., the size or the direction of the effect). This indeed can be, and often is, the case. Then the result from the study is evidence for the conclusion drawn about the target. But as the argument theory demands, it is only evidence relative to the additional assumption that the two populations are alike in the relevant respects. For example, if students' background knowledge or out-of-school resources impact the intervention's performance in the study sample, then students in the target population must have relevantly similar background knowledge or resources. For *credible* evidence-based policy or practice, the assumption that populations are alike in the relevant ways itself needs to be backed up by good reasons (see Joyce, 2019), which can come from both theory and other empirical results.

This is important in education because, as we have repeatedly remarked, it cannot be taken for granted that differences between educational contexts are negligible. Nuthall (2004) observes that "teaching one specific class of students is different from teaching any other class of students" because "engaging the students in relevant learning activities requires a unique understanding of each student's interests and relationships with classroom peers" (p. 294). Thus, Nuthall concludes that "what works in one classroom or with one student will not necessarily work in other

## Bridging the Gap Between Research and Practice

contexts" (p. 294). Nuthall's insight that students within and across educational contexts have different needs to which educators must be responsive in practice seems especially important in contexts of inequality, where there are significant disparities in resources among students, schools, or districts.

Research on school effectiveness and improvement conducted over the past few decades demonstrates the complex, dynamic nature of learning environments (e.g., Reynolds et al., 2014). Researchers in these fields argue that various factors within schools and broader education systems and outside school impact students' learning outcomes (Lareau, 2011; McLeod & Kaiser, 2004; Sarason, 1990). For example, such research highlights the importance of school culture (Freiberg, 1999; Hargreaves, 1995; MacNeil, Prater, & Busch, 2009; Sarason, 1996). As Halsall, Carter, Curley, and Perry (1998) observe, "One of the most consistent messages from the school improvement literatures is that school culture has a powerful impact on any change effort" (p. 175). Those who study its impact aim for "understandings of sociocultural or organizational factors at the school level that facilitate or impede school improvement" (Schoen & Teddlie, 2008, p. 148). This literature indicates that even schools that share broad superficial characteristics like population density, range of socioeconomic statuses, or urbanicity may differ in ways that bear on an intervention's effectiveness. Regardless of whether or not one accepts the conclusions from school effectiveness or improvement research, the assumption that differences beyond superficial characteristics are irrelevant or inconsequential for assessing representation is big enough to require serious justification.[11]

Judging when study and target settings are similar enough in the right ways requires *theory*—lots of it and of very different kinds. To learn just what it is about two different settings that allows them to support similar causal pathways from intervention to outcome requires a wealth of knowledge. This includes not only the theory of the process by which that intervention is supposed to bring about the intended result (sometimes called "the logic model" or the "theory of change"of the intervention) but also what contextual factors in the setting can support that process: For example, what are the psychological and sociological mechanisms at work, and what can we expect them to do in this setting? What kinds of things in the economic and material structure can facilitate and what kinds can hinder the process, and how can these be expected to interact in the target setting? None of this is to be learned from RCTs, even bigger and better ones. Still, it seems that conducting more large-scale RCTs remains the primary EBE strategy for improving evidence for effectiveness claims.

There is good reason to think that focusing narrowly on RCT results without considering contextual factors has sometimes invited harmful consequences, especially for students from marginalized groups. Helen Ladd (2012) argues that EBE and associated reform strategies (e.g., test-based accountability systems) are potentially harmful because they "pay little attention to meeting the social needs of disadvantaged children" (p. 204). Attempts to improve school quality ignore the contextual factors associated with poverty "that directly impede student learning" (p. 219; see also Duncan & Murnane, 2011, 2014). One particular harm concerns the unequal distribution of highly credentialed teachers. The idea that interventions work when implemented with fidelity de-emphasizes the importance of quality teachers by suggesting that disparities in teachers' abilities across schools are not a problem as long as those in schools serving disadvantaged students are good enough to implement effective interventions faithfully.

Another harm stems from the fact that "instructionally focused interventions pay insufficient attention" to students' "socioemotional-learning needs" (Rowan, 2011, p. 534). Rowan points out that benefitting from even high-quality instruction usually requires certain socio-emotional skills, which correspond to social advantage (Becker & Luthar, 2002; McLeod & Kaiser, 2004; Rowan, 2011). Failure to recognize this can conceal the need to provide students with socio-emotional learning opportunities and support. At the same time, overestimating the impact of students' behavior or social skills on learning (which seems sometimes to stem from implicit biases) can lead to overemphasizing discipline (e.g., some high-commitment "No Excuses" charter schools), to the detriment of socially disadvantaged students (see Brighouse & Schouten, 2011; Curto, Fryer, & Howard, 2011; Tough, 2013). Further, thinking only about the effectiveness of interventions risks neglecting the quality of students' experiences, which can bear on outcomes but are also important in their own right. While here we focus on the general question of how to bridge the research-practice gap, these unintended side effects underscore the importance of shifting toward a context-centered approach to EBE.

## Bridging the Gap

For EBE to help improve educational outcomes, it must generate evidence that supports effectiveness predictions: "This intervention will work here, as we plan to use it." One indirect route to this is via establishing a *general* effectiveness claim: "If it almost always works, within a broad range of contexts and ways of implementing it, it will probably work here." But establishing this is a demanding task. Supporting such claims requires showing that an intervention has a *stable causal capacity*. For example, aspirin has a relatively stable capacity to relieve headaches: It reliably does so across wide-ranging circumstances and populations. It is unclear whether educational interventions can have stable capacities and, if at least some of them do, how we could arrive at them given the nature of educational contexts. As Berliner (2002) remarks, educational researchers conduct their investigations "under conditions that physical scientists find intolerable" (p. 19). However, establishing general effectiveness is not necessary for making reliable predictions of the sort educators need.

We suggest less emphasis on general effectiveness claims, which are hard to come by, in favor of research that produces evidence educators can use to make causal predictions locally. After all, for educators, a successful intervention is one that contributes to a positive effect for them regardless of whether it can do so elsewhere. Incorporating a prediction phase into the EBE model means educators and policymakers must do more than choose and implement interventions that are deemed to work. To plan policy and predict effectiveness, they must construct arguments with multiple premises supported by various types of information, which will necessarily involve theory as well as empirical results of various kinds. Because their arguments require complex local knowledge and judgment, educators themselves can have much of the information needed to make predictions for their settings. But most premises need to be warranted by other types of evidence, some of which researchers can supply. Namely, researchers can provide information to help educators recognize just what facts may be relevant for predictions regarding particular interventions. And many of these facts will be theoretical in nature. It is no good just piling up study results. If we want research to be useful to practice, there is no way to avoid the heavy work of detailing the theory of change of the intervention; of developing new concepts like metacognition, meta-affective awareness, stereotype threat, and Bloom's taxonomy (and its revision); and of generating and vetting general claims and mechanisms well beyond those of the form "It works."

We can take a lead here from recent work on evidence and child protection. Munro, Cartwright, Hardie, and Montuschi (2016) outline three kinds of information that contribute to reliable local effectiveness predictions about an intervention:

1. Knowledge that the local social, economic, cultural, and physical *structure* can afford the necessary causal pathways for the intervention to lead to the outcome
2. Knowledge of what the *support factors* are for that intervention to work in the local setting and whether these are available or can be made available
3. Knowledge of what *derailers* might interfere along the way to diminish or entirely halt progress toward the outcome

### Structure

Can the intervention work here, or do local conditions simply not afford the steps that it takes, start-to-finish (which are hopefully outlined in the theory of change of the intervention), to produce the outcome? For instance, no amount of tinkering with the reward structures for parents within resource constraints will get more children vaccinated in places where there are no vaccination clinics within reach and there is a strong cultural resistance to vaccination. Similarly, creating systems to incentivize teachers or hold them accountable for students' performance is unlikely to improve outcomes in schools where teachers are already doing their best and low performance is primarily attributable to out-of-school factors. Likewise, active-learning strategies like in-class peer review may only contribute to better essays in classes where students have similar writing skills and have the time to receive feedback from multiple classmates, revise their essays to accommodate the feedback, and receive supplementary guidance from the teacher.

A successful RCT result can play a role in helping to warrant a prediction that the intervention tested will work here since a positive study result indicates that the intervention *can* produce the effect at least under some set(s) of circumstances. There is still much to add, however, since the whole point is that the same intervention can perform differently in settings with different underlying social structures. Thus, far more research effort needs to be devoted to understanding which kinds of social/economic/cultural/material structures can afford success with an intervention and which cannot. (For example, see our discussion of Posey-Maddox's study below.) This way, decision makers will have better resources for choosing the intervention that is best for them or for assessing and adjusting structures to accommodate new practices. Additionally, this kind of information can help educators who may want to adapt an intervention to allow for better integration with other practices or simply to capitalize on the potential it has within their setting.

*Support factors* are all those usually less obvious features that need to operate with the intervention for it to achieve its intended effects in the local setting.[12] Causes, including interventions, rarely work alone. They require support from various factors that are easy to overlook if we focus narrowly on isolating and measuring the effects of interventions. For example, available technology and computer literacy may be support factors for educational software. Values and norms within the school and community also count as support factors. *Derailers* are those disruptive factors that can undermine the intervention. Geography might serve as a derailer for interventions that assign substantial amounts of homework or require students to work with classmates outside school. If many students commute from rural areas, they might not have time to do the homework and may not have access to their classmates. Even if all support factors are present, a significant and ineliminable detracting factor may count against choosing an intervention—especially if the potential derailer is a valuable school asset we would not wish to eliminate. For instance, a thriving collaborative environment that facilitates interaction and group work might be a derailer relative to an intervention that mostly requires working independently.

To make plans and evaluate predictions about how effective they will be, decision makers need to decide whether their setting is at all right for the intervention and then gauge whether the requisite support factors are in place in their setting or can be put into place with reasonable cost and effort, as well as what the chances are of derailers occurring down the line to disrupt the process. What then is needed from the research community is more theorization and a far broader body of empirical work that can help with these issues, even if the information supplied is not as certain as the RCT results that currently take center stage.

Researchers can also help by identifying the underlying causes that contributed to the problem in the study populations, so that educators can consider whether their problem has the same causes as it did in the study population where the intervention worked to provide a solution. Sometimes an intervention may be an effective solution to a problem with different underlying causes, and sometimes the causes may not be relevant. But we cannot rely on that to hold generally.

Consider, for example, practices that involve ability grouping—assigning students to classrooms or groups according to their perceived level of abilities. Creating more homogeneous groups is supposed to solve the problems teachers face when their classes contain students with varying abilities and needs. Students at all levels are thought to benefit from instruction that caters to their abilities and from learning alongside peers who are similar in that respect (Colangelo, Assouline, & Gross, 2004; Gentry, 2014; Steenbergen-Hu, Makel, & Olszewski-Kubilius, 2016). Such practices have many advocates who laud their significant positive effects on student achievement across ability groupings, while many critics argue that they are harmful to students who are socially disadvantaged (Carbonaro, 2005; Carbonaro & Gamoran, 2002; Ladd, 2012; Oakes, 2005).

Plausibly, the different outcomes stem in part from the fact that wide variation in performance among students in the same grade is a problem with different underlying causes. In places where variance grows out of natural differences among students, ability grouping may affect students' experiences and outcomes differently than in schools where social disadvantage significantly contributes to the variation. In the latter case, sorting mechanisms—no matter how objective they are intended to be—will likely be unreliable because social disadvantage affects both talent development and performance. Also, implicit biases may unfairly affect the process by influencing expectations and interpretations of performance. For these reasons, assessments of ability may be inaccurate, and focusing on ability may actually undermine educational goals. Targeting underlying causes may not always be feasible or desirable, but understanding them can inform predictions about which solution will be most effective for aiding achievement or pursuing other aims.

Many of these issues can be tackled by educational theory and research that is currently marginalized within the dominant EBE framework. Qualitative methods like case studies and ethnographies can illuminate social structures and analyze their interactions with processes or programs.

They can study the causes of local problems and identify factors that may support or detract from improvement efforts.

For example, in her ethnographic case study of Morningside Elementary in Northern California, Posey-Maddox (2014) examines the role of parental involvement in an urban school. As a participant-observer, she intensively collected data over the course of one school year, using a range of qualitative methods including interviews with parents, teachers, and community members; observation; anonymous surveys; and analysis of relevant documents. Posey-Maddox collected further data for 1 year after her time at Morningside Elementary, returning for follow-up interviews and observation. Two years later, she returned for additional follow-up to assess the longer-term effects of the parental involvement programs at Morningside (Posey-Maddox, 2014, Appendix B).

Posey-Maddox uses a variety of theoretical tools and frameworks to analyze the data she collected during her time at Morningside. She describes unintended consequences stemming from the school's reliance on parental volunteers to fill resource gaps and identifies factors that likely contributed to those outcomes, like shifting student demographics and disparities in parents' social capital. Additionally, she distinguishes between various types of parental contributions and explores the ways in which each of them shaped the dynamics within the school. Her analysis of the context, processes, and effects associated with parental involvement draws on social theory, including Bourdieu's theories of social power structures (e.g., Bourdieu & Passeron, 1990; Bourdieu & Nice, 1977).

In addition to providing conceptual tools to aid interpretation, using theory allows researchers like Posey-Maddox to develop plausible, albeit defeasible, explanations and explore possible implications beyond the case at hand. Her analysis of Morningside Elementary reveals factors that could affect interventions and identifies the circumstances in which they might do so. It thus alerts educators to the potential costs and benefits of the various options for designing programs, like opportunities for parental involvement. More broadly, this kind of study demonstrates how particular norms, beliefs, values, and dynamics that may be present elsewhere bear on general processes within schools.

Although these findings do not speak directly to what will happen in other cases and thus do not supply *conclusions*, they identify *premises* that are relevant to predictions about similar programs in similar contexts. Unlike RCTs, ethnographies and case studies attempt to understand which factors made a difference and how they did so. This information can help others determine whether their settings are similar in ways that affected outcomes in the study setting. Relatedly, they can offer insight into how costs and benefits were distributed across students and the factors that influenced that distribution.

*Bridging the Gap Between Research and Practice*

For instance, Posey-Maddox attributes the negative consequences she observed to race- and class-based inequalities within the larger community that affected the nature and outputs of parental involvement. These consequences may not occur at a school serving a racially and socioeconomically homogeneous population, even if parents become involved in the same ways. What her ethnography illustrates is the potential for race and socioeconomic status to detract from efforts to fill resource gaps by involving parents in certain environments. More generally, it suggests that social structures outside of the school are relevant to the success of such programs and to their distribution of benefits and burdens. Ignoring these factors can undermine improvement efforts. Equally, assuming that they are always relevant without considering when and how they matter might lead decision makers to dismiss interventions that could work for them. Credible theory examining the relationships between education and external social structures, especially those involving social power, could help draw out the significance of these findings for other cases (e.g., Horvat, Weininger, & Lareau, 2003; Lareau, 2011).

To be sure, educational theory and non-experimental research have value beyond the role they could play in effectiveness predictions. Indeed, academic communities engage in this work and numerous academic journals are dedicated to publishing it. However, these enterprises usually run parallel to educational research for EBE. Their potential to contribute to the EBE model as we are suggesting remains underappreciated. As a result, information from these literatures is not prepared or mobilized for use by educational decision-makers. Intermediaries like the WWC emerged in part because proponents of EBE recognize that academic journals generally do not highlight practical implications and educational decision makers rarely consult them (Gorard et al., 2017; Nuthall, 2004; Phillips, 2007).

Existing work from these disciplines may be useful for identifying necessary premises and finding evidence for effectiveness predictions. Applying these methods within EBE can expand its relevance. Researchers could investigate questions concerning context, social structure, support factors, and derailers alongside RCTs. Similarly, they could study interventions that have been tested by RCTs as they are implemented in new settings or analyze failed attempts to use them. Learning about successful adaptations and failures can inform predictions and local planning.

Some researchers have undertaken such projects, but their efforts are largely directed at aiding implementation rather than prediction. Consider the study of Success for All (SFA) conducted by Datnow and Castellano (2000). After experimental studies found positive effects on students' literacy across multiple sites, Datnow and Castellano investigated "how teachers respond to SFA and how their beliefs, experiences, and programmatic adaptations influence implementation" (p. 777). Their findings yield multiple suggestions that can help decision makers predict what will happen if they implement SFA in their own sites. For instance, nearly all teachers adapted the program in some way as a response to local factors or the perceived deficiencies of SFA. Considering the most common adaptations, the teachers' reasons for implementing them, and how they affected outcomes could help decision makers assess, among other things, the need for and likelihood of adaptations in their own settings and evaluate their estimated impact. In these ways, their findings provide guidance for applying local knowledge and professional judgment within the predictions. While some of their other findings are more useful for implementation than for prediction, designing the project with predictions in mind could lead to results that more directly support them.

When considering interventions tested by RCTs, it is useful for educators to evaluate how well the study population and setting represent their own along relevant dimensions. Information about which demographics are relevant to that particular intervention and how they affect it can be of great help to them here. Currently, the WWC provides the same demographic information for all interventions (e.g., minority status, qualification for the National School Lunch Program) without specifying which are likely to affect effectiveness.

Considering only those interventions with populations and contexts that are representative in terms of all observed characteristics will likely leave research users with few choices. Moreover, the categories the WWC uses are too broad to be useful in many cases. For example, there are surely many differences among students who qualify for the National School Lunch Program and among those with minority status. How does low socioeconomic or minority status bear on the intervention in question? Can it be expected to have a similar impact on other interventions? We cannot expect all low-socioeconomic status communities or households to have the same assets, support factors, and derailers. Socioeconomic status may be a better indicator of family dynamics and parental behavior in some places than in others, and those behaviors and dynamics may vary qualitatively, for instance (Furstenberg, 2011). Additionally, some interventions depend on these factors more than others. Without some understanding of why and how the results emerged, decision makers may hastily dismiss interventions that have failed in broadly similar settings without asking whether they could have been effective for them, perhaps with some adaptation and improvement. For these reasons, research that helps educators assess representativeness along the relevant dimensions could be especially useful.

Educational decision makers must also consider costs and benefits. Effectiveness predictions play an important role in a larger, all-things-considered decision-making process (see Brighouse et al., 2018). Educators need to know what effects they can expect so they can decide whether those benefits outweigh the expected costs. The average treatment effect documents the difference in effects between intervention and comparison groups. Recall point (b) in the "RCT conclusion" formula. For educators, estimating

## Bridging the Gap Between Research and Practice

the average effect they can expect requires comparing their current curriculum with what was used in the study's comparison group. If they are using a much better literacy program than the comparison group, for example, they should prima facie expect a smaller average effect or even a negative one. Likewise, if their program is much worse, the effect might be greater. Again, recall point (a) in the "RCT conclusion."

Often, cost-benefit calculations at the local level concern individual students, not averages. Knowing how an intervention affected particular individuals or students with certain characteristics can usefully inform predictions about who is likely to benefit and who may bear the costs. This information is especially important given the persistent achievement gaps between racial and socioeconomic groups. Knowing average effects is insufficient for choosing interventions that will help (or at least not harm) underserved students.

Let us return to the case of ability grouping. Meta-analyses of multiple RCTs and second-order meta-analyses consistently show positive effects on average (Steenbergen-Hu et al., 2016), leading some to recommend ability grouping (e.g., Gentry, 2014). However, these averages do not document the effects on students from particular groups. A particularly strong endorsement claims that, because "talent cuts across all demographics: ethnicity, gender, geography, and economic background," ability grouping is particularly beneficial to talented students from disadvantaged groups (Colangelo et al., 2004, p. 7). But this is just an assumption—the data are disaggregated only by ability groups, not by social groups. Even if they were, the averages for subgroups do not account for intersectionality, accuracy of sorting, different applications, or quality of experience. On average, students of color may benefit because the most advantaged among them improve dramatically while those with the lowest socioeconomic status are negatively impacted.

Research that can shed light on these issues should be considered, even if its findings are less certain than RCT results. For instance, recall Posey-Maddox's (2014) observation that students from middle-class families gained greater benefits from parental involvement programs than more disadvantaged students and her examination of the factors that likely contributed to the uneven distribution of benefits. Although qualitative studies do not document effect sizes or precisely quantify benefits, they provide some relevant evidence for premises that concern costs and benefits. Addressing the harms stemming from educational inequalities continues to be a central goal for education policy and practice in the United States. Given its urgency, there is a pressing need for relevant information to inform predictions concerning individual students.

While we have offered some suggestions, we are calling on the research community to seriously investigate just what can be done at the research level to help local decision makers identify and find the facts they need to predict if an intervention is likely to work in their setting, what it would take to get it to do so, and whom it might help and whom it might harm. In this enterprise, the best should not be the enemy of the good. It is no use insisting that the information supplied be information that one can be fairly certain is correct—as in policing and reporting causal ascriptions supported by well-done RCTs—when it is not the information educators need. A far more ambitious, and riskier, program of research, theorizing, and reporting needs to be undertaken if evidence is really going to help improve educational outcomes.

## Complementary Calls for Expanding the Research Agenda in EBE

Our proposal resembles some others that suggest diversifying research approaches and incorporating educators' local knowledge, judgment, and expertise into the decision-making process (e.g., Bridges, Smeyers, & Smith, 2009; C. Brown et al., 2017; Bryk, 2015; Hammersley, 2013; McIntyre, 2005; Smeyers & Depaepe, 2007). But none focus on the source of the research-practice gap that we identify: unwarranted effectiveness claims. Some think general effectiveness claims can be established more quickly and reliably by supplementing evidence from research with evidence from practice. Others advise using these resources to connect the two "communities" or "worlds" of research and practice by translating general effectiveness claims into practical suggestions. Still others focus on improving implementation protocols. Additionally, some recent policy initiatives encourage place-based or practice-based interventions.[13] The idea is to identify promising interventions created within educational contexts as responses to problems and to figure out how to scale them up. While some of them propose recasting the relationship between research and practice communities as bidirectional (e.g., Farley-Ripple et al., 2018), these new suggestions for connecting research and practice tend to preserve the division of labor that assigns establishing causal claims to researchers and implementation to educators. From our perspective, these suggestions get something right that will help address the gap that concerns us—though not intentionally for that reason—and are thus potentially useful within the adjusted, context-centered EBE approach we endorse.

### Research-Practice Partnerships

Research-practice partnerships (RPPs) are gaining momentum as a strategy for bridging the research-practice gap (Coburn & Penuel, 2016).[14] RPPs are collaborations between educators and educational researchers. They aim for research with greater relevance to practice and to improve the use of research in decision making and practice. The driving idea is that research outcomes will be more applicable to practice if educators influence the research agenda. Educators share pressing problems with researchers,

*Bridging the Gap Between Research and Practice*

who then study interventions targeting those issues. Researchers can directly help educators interpret the findings and decide how to use them in practice.

For example, Stanford University's School of Education has partnered with the San Francisco Unified School District to "help Stanford researchers produce more useful research and to help San Francisco administrators use research evidence to inform their decisions" (Wentworth, Mazzeo, & Connolly, 2017, p. 244). One of their projects studied the outcomes of an ethnic studies course the district was piloting in some of its schools. Researchers used a regression discontinuity design to study the effects of the program on five school-year cohorts from three schools in the district. Ninth graders whose GPA (grade point average) was below 2.0 in the previous year were automatically enrolled in the course, while students with GPAs at or above 2.0 were not. Those enrolled could opt out, and others could opt in. The study compares students just below the 2.0 threshold with students just above it because students in these two groups are taken to be similar (Dee & Penner, 2016). The researchers found that the course positively affected GPAs and attendance for students assigned to the course, who were identified as at risk of dropping out (Dee & Penner, 2016; Wentworth, Carranza, & Stipek, 2016). The district leaders used these findings to decide whether to implement the course throughout the district.

In this case, the school administrators needed data to answer a specific question, and the researchers designed a study to obtain such data. The researchers then helped the administrators to interpret the data, highlighting what they mean for the decision they needed to make. While the research questions differ, this case exemplifies the typical relationship between practitioners and researchers involved in RPPs. The partnership is supposed to be mutually beneficial in that researchers obtain results with broader significance that they can publish in academic journals and practitioners get data directly relevant to them.

Organized this way, RPPs support the standard division of labor between educators and researchers. Practice guides research in the sense that it influences the research question. Beyond that, the interaction primarily involves researchers helping to interpret results, drawing out their implications for practice. Researchers still produce information to serve as evidence in decisions about practice, while educators focus on implementation.

RPPs pinpoint the *relevance* of evidence as a key step in bridging the gap. We too underline the need for research to produce relevant evidence. However, RPPs generally aim for results relevant to particular *learning outcomes*, whereas we urge evidence relevant to *effectiveness predictions*. Researching interventions that target local problems does not reduce the need for predicting effectiveness. In addition to aligning research projects with school districts' goals, RPPs seek generalizable results, which they suggest can be achieved using experimental or quasi-experimental methods (sometimes with the help of statistics), contrary to what we have argued. The research projects Stanford conducts in partnership with the San

Francisco Unified School District are supposed to meet "high standards of validity and generalizability" so they are relevant to other districts aiming for similar learning outcomes (Wentworth et al., 2016, p. 68). To serve our purposes, RPPs would need to address a wider array of questions. As it is, they are motivated by the idea that "education research does not influence policy because it takes too long to produce, is too expensive, is not applicable to a specific context of interest, and is not disseminated in a clear and direct manner" (López Turley & Stevens, 2015, p. 6S).

From our point of view, university-partnered, single-school RPPs, like those pioneered by four University of California (UC) campuses, take a more promising approach.[15] Researchers and educators collaborate at all levels of school design and practice. Quartz et al. (2017) describe their distinctive RPPs as "multidimensional and multilevel problem-solving ecologies" that are committed to "democratic participation" wherein "researchers and practitioners . . . bring their knowledge to the table and together 'build the plane while flying'" (pp. 144–145). The university-partnered, single-school design departs from standard RPPs by integrating researchers and practitioners. Instead of working with educators to identify problem areas and test potential solutions, help them implement best practices, or replicate an existing, evidence-based solution, researchers become familiar with available causal pathways and, in collaboration with educators, design solutions that are likely to be effective within that particular context.

For example, researchers in the Education Department at UC Berkeley partnered with Aspire Public Schools to prepare students for college. After working together for more than 10 years, Aspire and UC Berkeley designed and cofounded CAL Prep, a public charter school based on local knowledge and "community-engaged scholarship."[16] Their strategies were "developed through careful and systematic study of the conditions that allowed all students to meet high academic standards" (Quartz et al., 2017, p. 144–145). While, according to Quartz et al. (2017), these RPPs strive to create knowledge that generalizes beyond their settings, their approach equips them to share information about support factors, the underlying causes of the problems they respond to, and individual students, which, we have argued, is important for local predictions and planning.

## Implementation Science

Implementation science has emerged as another response to the research-practice gap. The National Implementation Research Network (NIRN) describes it as "the science related to *implementing* [evidence-based] programs, with high fidelity, in real-world settings" (Active Implementation Hub, n.d.) According to prominent implementation scientists Blasé (Blasé & Fixsen, 2013; Blasé, Fixsen, & Duda, 2011) and Fixsen (Fixsen & Blasé, 2013; Fixsen, Blasé, Duda, Naoom, & Van Dyke, 2010; Fixsen, Blasé, Metz, & Dyke,

2013; Fixsen, Blasé, Naoom, & Wallace, 2009), the way to bridge the gap between research and practice is to create the infrastructure for successful implementations within education sites. The University of North Carolina–Chapel Hill hosts the State Implementation and Scaling-Up of Evidence-Based Practices Center, funded by the U.S. Department of Education. This is a project within the NIRN that creates resources for educators. It presents implementation as an active, recursive process of "making it happen," instead of simply letting or helping an intervention succeed (Fixsen et al., 2009, p. 532; NIRN, 2016). Implementation scientists have developed an infrastructure that affords the capacity to implement effective interventions or innovations. It is made up of five integrated frameworks called active implementation frameworks (AIFs).

Roughly, AIFs encourage and support high-fidelity implementations of evidence-based interventions within particular local contexts (Blasé, Fixsen, Sims, & Ward, 2015; Carroll et al., 2007; Fixsen et al., 2009; Fixsen, Blasé, Metz, & Van Dyke, 2015). While implementation capacities are considered to be standard across educational settings, AIFs encourage users to take the characteristics of their own sites into account at various stages. Implementation begins by exploring options using the Hexagon Tool to assess the extent to which available interventions and their expected effects fit with local needs, priorities, and existing programs (Metz & Louison, 2018).

Once they decide to implement an intervention, teams install it by making practical arrangements and "developing the knowledge, skills and abilities of teachers and administrators" through training and coaching (Active Implementation Hub, n.d.). From there, they build toward full implementation. In part, this involves establishing and sustaining implementation drivers—core components that ensure competence among those engaging in implementation efforts, develop supports for the program, and assign leadership roles. These drivers capture the "common features that exist among many successfully implemented programs and practices" (Active Implementation Hub, n.d.). The additional frameworks concern implementation teams, which monitor the implementation infrastructure and employ AIFs, and improvement cycles, which offer tools for identifying and solving the problems that occur during implementation (Blasé et al., 2015).

This brief description oversimplifies AIF, but it sketches the contours well enough to contrast it with our proposal. Although it recognizes that local context can significantly impact implementation and eschews the idea that educators can simply apply research findings in practice, implementation science still adopts some of the problematic assumptions driving the dominant EBE model. In particular, it appears to assume that "the integrity of the research presented stands alone on its scientific grounds; . . . context becomes only an issue once we move to consider policy implementation, focusing on particular target/risk groups and the way we can deliver to them what we know works" (Seckinelgin, 2017, p. 132). Implementation science focuses on scaling up evidence-based programs by helping schools cultivate the capacities for successful implementation because it accepts that interventions that produce a positive outcome in a handful of good RCTs can generally be expected to do so elsewhere if implemented with fidelity, barring positive reasons to the contrary. As Fixsen puts it, the aim is to "take these good ideas that work in some places and get them to work in all places" (Fixsen & Blasé, 2013). Like other proposals to bridge the research-practice gap, it neglects the question of whether an intervention *can* work in a specific target when implemented, whether with or without fidelity.

Using AIFs cannot ensure that an intervention can or will be effective in a particular setting. The Hexagon Tool for selecting interventions directs attention to local context but focuses on whether it allows high-fidelity implementation. It does not ask whether the intervention *can* work in that setting as it did in the study settings. In other words, it leaves the prediction phase out of the selection process. This is a problem because assessing the feasibility of implementation without predicting effectiveness could lead educators to choose an intervention that can be implemented in their setting but that cannot work in that setting. To see this, consider a simple example: a computer program that produced positive effects for study populations when students used it in 2-hour increments three times per week. Implementing the program in their own site requires educators to schedule 2-hour time slots for students to use the program three times per week. Successful implementation, however, does not guarantee that it will produce positive effects there. There are many reasons why it might not: By making time for it, they could inadvertently eliminate support factors for the program or activities that contribute to morale; their students might stop concentrating on it after just half an hour for any number of reasons; or 6 hours per week may be too much or too little time given their students' skill sets. There is an important difference between the question of whether the local context affords a causal pathway through which the intervention can make a positive contribution and the question of whether it will allow educators to implement the intervention with fidelity.

Evaluating evidence is part of the Hexagon process, but it invites users to take general effectiveness for granted if the research meets certain criteria. It prompts users to evaluate the strength of evidence according to the number of studies that have been conducted using experimental methods with diverse populations, much like the WWC (Metz & Louison, 2018). It does encourage users to compare their setting and population with the study and replication sites. These are factors relevant to predictions. But recall that an RCT does not identify what population or contextual characteristics are important for projectability. Even granting projectability, unless one replicates the whole set of postallocation differences with exact fidelity, there is no evidence that similar effects will occur. [17]

*Bridging the Gap Between Research and Practice*

We want to reiterate that perfect fidelity is not possible. Educational settings are so complex that reporting how interventions were implemented in a study likely leaves out details that could be relevant. More importantly, aiming for fidelity is very often not the best way to implement an intervention. Given the significant differences across students and contexts, even if the target closely resembles the study along visible dimensions, producing positive effects in new settings will very likely require adjustments. Appreciating this, Nuthall (2004) observed that "the contextual details that [have] been eliminated from these studies in order to make the results generalizable are what teachers needed to know in order to . . . apply the results" (p. 286).

The Hexagon Tool attempts to address these issues by advocating interventions with clearly defined components and a logic model or theory of change that can help educators make only "safe" adjustments and assess progress. Although descriptions of the intervention components and implementation are likely to be incomplete, we agree that understanding the key components and how they are supposed to produce outcomes would be helpful during implementation. However, prior to that they should inform predictions, which ask if an intervention *can* work in the target setting as it would be implemented there.

Note that, just as they do not identify relevant population and contextual characteristics, RCTs do not show which components of interventions are essential to the causal process, nor do they provide theories of change. Those who designed the interventions may offer theories, but they are seldom tested, and, again, RCTs cannot be used to verify them. Positive RCT results do not show that a theory of change is accurate. Calling for this kind of information, then, signals the need for evidence from different sources to support different claims. Even if it were available, though, neither the Hexagon Tool nor the other AIFs offer guidance for evaluating the kind of research and theory that could evidence these claims or speak to alternative pathways through which the intervention could make a positive contribution. Importantly, to be relevant to predictions, theories of change must account for support factors and derailers specific to the local target. While they provide some helpful information, even well-supported theories of change conceptualized by researchers or developers cannot identify these in advance. We suggest that far more effort be dedicated to providing the resources educators need to choose interventions that are likely to work for them and to make an implementation plan that is best for their setting.

## Improvement Science, Networked Improvement Communities, and Practice-Based Evidence

Another solution encourages complements to RCTs under the rubric of *improvement* science. Improvement science investigates the variation in educational outcomes and devises strategies for addressing the sources of variation so interventions can work more effectively across contexts. Instead of building capacities for implementing programs with fidelity, as implementation science does, improvement science builds "capacity within the organization to understand the factors that shape improvement" and "to notice and learn from variation" (Lewis, 2015, p. 59). Whereas implementation science attempts to avoid modifications that undermine the intervention by clearly specifying its components and monitoring fidelity at the implementation site, improvement science does not recommend fidelity. Instead, it attempts to avoid modifying interventions in a way that undermines them by monitoring the indicators predicted by theories of change and contextual factors that are taken to shape improvement (Lewis, 2015).

For example, Bryk (2015; Bryk, Gomez, & Grunow, 2011) proposes a collaborative effort among educators to investigate why effects from the same intervention vary across educational settings. This would involve looking for the kinds of information we outlined in the previous section. Understanding what causes variation in effectiveness, he argues, allows educators to identify the contextual factors that affect results. Sharing their findings widely through networked improvement communities (NIC) provides "practice-based evidence" that educators can use for better implementation (Campbell, Pollock, Briscoe, Carr-Harris, & Tuters, 2017).

In a similar vein, C. Brown and colleagues (2017) suggests integrating scientific research with locally collected quantitative and qualitative data that educators use to identify goals and inform action for reaching them. Gutiérrez and Penuel (2014) argue that establishing that an intervention "works" requires qualitative research from educators, who can provide information about how they used interventions within their settings. These proposals highlight the need for practical knowledge and action research to supplement experimental research.

Like us, Bryk (2015) identifies an important, but neglected, distinction within EBE. He agrees that research shows only that an intervention *can* work, but he claims that what educators really need is "knowledge of how to actually *make it work* reliably over diverse contexts and populations [italics added]" (p. 469). Contrast this with our claim that what educators need to know is that the intervention *will* work *here, in their setting*. Bryk assumes that evidence-based interventions *can* work across educational settings if the intervention or setting is properly adjusted. He advocates figuring out how to replicate results in different settings to produce "quality student outcomes reliably at scale" (p. 475). While they concern more specific target settings, these suggestions still aim for general effectiveness claims that can justify predictions. We definitely agree that if there are reasonably reliable general claims about the kinds of populations and settings an intervention works for and how it can be used effectively in these, this is important information to secure. But we fear that there will often be too few of these kinds of reliable claims to provide much help. Looking for this sort of information is important, but it cannot replace the need for helping educators to find the necessary information and to piece it together to make local effectiveness predictions as we are suggesting.

If we treat the information from practice as evidencing general effectiveness claims, then these proposals are unlikely to improve significantly on the existing model. Using the argument theory, we can more easily see what practice-based evidence can be evidence for and how it can be used responsibly in policy deliberations. Instead of (or in addition to) using practice-based evidence to improve implementation or gain evidence of general effectiveness, educators should use it for their predictions. Reporting on variation in outcomes and trying to determine which local variables made the difference can be very useful for learning what conditions affect the effectiveness of particular interventions. But we cannot use knowledge from practice to neutralize the sources of variability, making interventions work reliably across contexts. Nor can we rely on the accumulated knowledge as an inductive base that warrants conclusions about other targets on its own. A large and varied evidence base cannot warrant a prediction without further premises supporting the claim that the results will travel. Even then, induction provides weak evidence compared with premises that speak to the local structure and the support factors needed there.

The upshot is that collecting and disseminating results and practical knowledge through NICs and other knowledge mobilization networks can help to close the gap between research and practice if educators use the evidence to make predictions rather than continuing to abide by the standard division of labor. Whether it comes from educators or researchers, predictions require information about the conditions that affect effectiveness no matter which intervention is under consideration. Of course, not all information obtained this way will be equally reliable. Responding to the suggestion that individual craft knowledge or knowledge from action research can be coordinated and compiled to provide a useful, evidence-based body of professional knowledge that educators can use to improve their practice, Nuthall (2004) points out that "what is going on in a classroom that leads to student learning is more complex and difficult to disentangle than a teacher has time to record, analyze, and interpret" (p. 292). Additionally, educators' interpretations may be influenced by biases, leading them to pass on inaccurate, or potentially harmful, information.

Causal ascriptions and explanations can be true or false, and the reasoning offered in support can be better or worse. Well-warranted predictions require premises that are supported by strong evidence. The relevant information is difficult to assess rigorously. As such, to make good use of alternative research types and reports from practice, we need mechanisms for evaluating the claims.

It seems that the WWC and similar databases could be well positioned to evaluate and disseminate these resources. Currently, the WWC includes only original research from RCTs or quasi-experimental studies. Much of what we and others are recommending is *secondary* research, including conceptualization and theorizing. For example, reports about successful and failed attempts to use an intervention in practice do not currently qualify as evidence by their standards. Reports about individual cases and various alternative forms of educational research can be found on blogs, on websites, or in academic journals. These are ineffective channels both because they are inconsistently accessed and because some lack credibility. Even if NICs provide better avenues for sharing results, knowledge transmitted that way could be misleading. Databases could evaluate these materials and organize them according to research type. They could also communicate how particular resources might be useful and how they should *not* be used in predictions. The last thing we want to do is bombard educators with more information without guidance on how to use it alongside experimental research and their own local knowledge in deliberations about policy and practice.

## Conclusion

We attribute the persistent gap between what worked in research and what works in practice in part to lack of support for the effectiveness planning and prediction central to the standard EBE model. We distinguish between three kinds of causal claims: causal ascriptions, general effectiveness claims, and effectiveness predictions. We argue that, at best, educational RCTs evidence causal ascriptions, which, without further assumptions, are irrelevant to general effectiveness claims and effectiveness predictions. Because general effectiveness claims are not essential for predictions and are difficult to establish, we propose a serious rethinking of the EBE model to figure out how better to produce evidence and theory relevant to effectiveness predictions directly. Recognizing the sort of considerations that are necessary to support local predictions suggests a far broader, context-centered research agenda. Additionally, materials for decision makers should highlight local planning and prediction as an indispensable step.

Examining other recent strategies for addressing the research-practice gap, we find that they can be helpful for facilitating our proposal but are not, on their own, enough to bridge the gap. If the planning and prediction phase remains invisible, educators invited to collaborate with researchers are unlikely to request the information most relevant to their predictions. Educators can surely influence research agendas by identifying widespread problems. Research that investigates those problems will be relevant to practice in a topical sense. But bridging the gap between research and practice requires more than topically relevant research or more detailed plans for implementation and adaptation—it requires research that is relevant to local effectiveness predictions.

## Bridging the Gap Between Research and Practice

### Notes

We are grateful to Adrian Simpson for helpful feedback on an early draft of this paper and to Harry Brighouse for his thoughtful input. Funding for this project was provided to Nancy Cartwright and Kathryn Joyce by the Center for Ethics & Education. For Cartwright, this material is based on research supported by the National Science Foundation under Grant No. 1632471 and the European Research Council under the European Union's Horizon 2020 research and innovation program (Grant Agreement No. 667526 K4U). It is acknowledged that the content of this work reflects only the authors' views and that the European Research Council is not responsible for any use that may be made of the information it contains.

[1] For example, C. Brown et al. (2017) suggest integrating local knowledge with rigorous evidence of effectiveness; Bryk (2015) calls for networks wherein educators share the information they learn in the course of implementing evidence-based interventions; and Jinfa Cai et al. (2017) urge researchers to "offer information on effective ways to implement" effective interventions (p. 345). In the section titled Complementary Calls for Expanding the Research Agenda in EBE, we consider these proposals in more detail.

[2] See Cartwright and Hardie (2012) for the general structure of this frame; for essential components, see Rothman (1976) on "support" factors and Bechtel & Abrahamsen (2005) on underlying structure.

[3] EBE, in its current form, may have arisen from disputed positivist doctrines, but that doesn't mean that all versions of EBE owe what justification they have to those doctrines. Thus, instead of targeting the doctrines taken to motivate EBE, as many have done, we directly address the problems present within EBE. See Kvernbekk (2016) for an excellent discussion distinguishing the *necessary* attributes of EBE as a basic concept from the attributes attached to particular conceptions of EBE, like the dominant "what works" model.

[4] For a general discussion of context-centered versus intervention-centered approaches, see Cartwright (2019) or the extended discussion of context centering on the area of international HIV AIDS policies in Seckinelgin (2017), much of which applies to education.

[5] We cannot account for what makes theory credible with any precision in the abstract. However, academic standards and expertise on the part of theorists in relevant fields could be used to assess theory.

[6] This is the average of the "individual treatment effects" for the individuals in the study population, that is, how much difference the intervention would make to the individual supposing all other causes of the outcome were the same. Amazingly, RCTs allow for an estimate of the average of individual treatment effects, even though we cannot measure these counterfactual values themselves. For more on this, see Rubin (1974).

[7] More technically, if randomization, blinding, and other post–random assignment policing succeed in ensuring the intervention is probabilistically independent in the mean from the net effect of all other causes of the outcome, the difference in means between the intervention and control groups will be an *unbiased estimate* of the average intervention effect in the study population—which can be far from a correct (or "precise") estimate. For more on unbiasedness versus precision in RCTs, see Deaton and Cartwright (2018).

[8] Simpson (2017) noted that "strong evidence of positive effects" is commonly misinterpreted to mean "evidence of strong positive effects."

[9] As Francis Bacon taught in 1620, "Induction by simple enumeration is puerile" (Bacon, 2000, *NO* I:105).

[10] As with the estimates of average treatment effect in an RCT on a study population, the estimate should be better as the size of the sample increases, but the problem never goes away entirely.

[11] School effectiveness research has been subject to some criticism, especially by advocates of using research evidence to inform education policy (see, e.g., S. Brown, Duffield, & Riddell, 1997; Coe & Fitz-Gibbon, 1998; Goldstein & Woodhouse, 2000). We do not rely on this research in a substantial way, nor are we claiming that it should inform policy. Rather, we take the enterprise and its general observations to indicate that schools, although not wholly independent of one another and the broader education system, differ

in ways that *may* bear on the performance of educational interventions in some cases. Thus, we should not assume that differences between study and target schools are negligible when assessing generalizability. For more on assumptions about representativeness, see Joyce (2019).

[12]The distinction between structural features, on the one hand, and support factors and derailers, on the other, is not a hard and fast one. But it is useful to separate factors that are deeply entrenched and difficult for the relevant educators to change—which Munro et al. (2016) label "structural"—from ones that educators can more readily change or substitute for (e.g., by an after-school homework club where students from distant homes can work together)—which they label "support factors" and "derailers."

[13]The Every Student Succeeds Act of 2015 created programs to support innovations developed locally by educators. For example, the Education Innovation and Research Program (sec. 4611) provides funding for "evidence-based, field-initiated innovations" that can be scaled up to help more students.

[14]The U.S. Institute for Education Sciences has introduced programs that encourage RPPs.

[15]These are UC Los Angeles, UC Berkeley, UC San Diego, and UC Davis.

[16]See the Center for Educational Partnerships website: cep.berkeley.edu/cal-prep.

[17]As Adrian Simpson noted in commenting on a draft of this paper (personal communication, April 11, 2018).

## ORCID iDs

Kathryn E. Joyce https://orcid.org/0000-0003-4535-8471
Nancy Cartwright https://orcid.org/0000-0002-0873-6966

## References

Active Implementation Hub (n.d.). *Module 1: An overview of active implementation frameworks*. Retrieved from https://implementation.fpg.unc.edu/module-1

Archibald, T. (2015). "They just know": The epistemological politics of "evidence-based" non-formal education. *Evaluation and Program Planning, 48*, 137–148. doi:10.1016/j.evalprogplan.2014.08.001

Bacon, F. (2000). *The new organon* (L. Jardine & M. Silverthorne, Eds.). New York, NY: Cambridge University Press.

Bechtel, W., & Abrahamsen, A. (2005). Explanation: A mechanist alternative. *Studies in the History and Philosophy of the Biological and Biomedical Sciences, 36*, 421–441.

Becker, B. E., & Luthar, S. S. (2002). Social-emotional factors affecting achievement outcomes among disadvantaged students: Closing the achievement gap. *Educational Psychologist, 37*, 197–214. doi:10.1207/S15326985EP3704_1

Berliner, D. C. (2002). Comment: Educational research: The hardest science of all. *Educational Researcher, 31*(8), 18–20.

Biesta, G. (2007). Why "what works" won't work: Evidence-based practice and the democratic deficit in educational research. *Educational Theory, 57*, 1–22. doi:10.1111/j.1741-5446.2006.00241.x

Biesta, G. J. J. (2010). Why "what works" still won't work: From evidence-based education to value-based education. *Studies in Philosophy and Education, 29*, 491–503. doi:10.1007/s11217-010-9191-x

Blasé, K., & Fixsen, D. (2013). *Core intervention components: Identifying and operationalizing what makes programs work* (ASPE Research Brief). Washington,

DC: Office of the Assistant Secretary for Planning and Evaluation, Office of Human Services Policy, U.S. Department of Health and Human Services.

Blasé, K. A., Fixsen, D., & Duda, M. (2011, February). *Implementation science: Building the bridge between science and practice*. Invited presentation to the Institute of Education Sciences, U.S. Department of Education, Washington, DC.

Blasé, K. A., Fixsen, D. L., Sims, B. J., & Ward, C. S. (2015). *Implementation science: Changing hearts, minds, behavior, and systems to improve educational outcomes*. Oakland, CA: The Wing Institute. Retrieved from https://fpg.unc.edu/node/7729

Borman, G. D. (2009). The use of randomized trials to inform education policy. In G. Sykes, B. Schneider, & D. Plank (Eds.), *Handbook of education policy research* (pp. 129–138). New York, NY: Routledge. doi:10.4324/9780203880968.ch11

Bourdieu, P., & Passeron, J. C. (1990). *Reproduction in education, society, and culture*. Newbury Park, CA: Sage, in association with Theory, Culture & Society, Department of Administrative and Social Studies, Teesside Polytechnic.

Bourdieu, P., & Nice, R. (1977). *Outline of a theory of practice*. Cambridge, England: Cambridge University Press. doi:10.1017/CBO9780511812507

Bridges, D. (2017). *Philosophy in educational research*. Cham, Germany: Springer International. doi:10.1007/978-3-319-49212-4

Bridges, D., Smeyers, P., & Smith, R. (Eds.). (2009). *Evidence-based education policy: What evidence? What basis? Whose policy?* Malden, MA: Wiley-Blackwell.

Bridges, D., & Watts, M. (2008). Educational research and policy: Epistemological considerations. *Journal of Philosophy of Education, 42*, 41–62. doi:10.1111/j.1467- 9752.2008.00628.x

Brighouse, H., Ladd, H. F., Loeb, S., & Swift, A. (2018). *Educational goods: Values, evidence, and decision making*. Chicago, IL: University of Chicago Press.

Brighouse, H., & Schouten, G. (2011). Understanding the context for existing reform and research proposals. In G. J. Duncan & R. J. Murnane (Eds.), *Whither opportunity? Rising inequality, schools, and children's life chances*. New York, NY: Russell Sage Foundation.

Brown, C. (2014). *Making evidence matter: A new perspective for evidence-informed policy making in education*. London, England: Institute of Education Press.

Brown, C., Schildkamp, K., & Hubers, M. D. (2017). Combining the best of two worlds: A conceptual proposal for evidence-informed school improvement. *Educational Research, 59*, 154–172. doi:10.1080/00131881.2017.1304327

Brown, S., Duffield, J., & Riddell, S. (1997). School effectiveness research: The policy maker's tool for school improvement? In N. Bennett, A. Harris, & M. Preedy (Eds.), *Organizational effectiveness and improvement in education* (pp. 138–146). Philadelphia, PA: Open University Press.

Bryk, A. S. (2015). 2014 AERA distinguished lecture: Accelerating how we learn to improve. *Educational Researcher, 44*, 467–477. doi:10.3102/0013189X15621543

Bryk, A. S., Gomez, L. M., & Grunow, A. (2011). Getting ideas into action: Building networked improvement communities in education. In M. T. Hallinan (Ed.), *Frontiers in sociology of education* (pp. 127–162). Dordrecht, Netherlands: Springer. doi:10.1007/978-94-007-1576-9_7

Cai, J., Morris, A., Hohensee, C., Hwang, S., Robison, V., & Hiebert, J. (2017). Making classroom implementation an integral part of research. *Journal for Research in Mathematics Education, 48*, 342–347. doi:10.5951/jresematheduc.48.4.0342

Campbell, C., Pollock, K., Briscoe, P., Carr-Harris, S., & Tuters, S. (2017). Developing a knowledge network for applied education research to mobilise evidence in and for educational practice. *Educational Research, 59*, 209–227. doi:10.1080/00131881.2017.1310364

Carbonaro, W. (2005). Tracking, students' effort, and academic achievement. *Sociology of Education, 78*, 27–49. doi:10.1177/003804070507800102

Carbonaro, W. J., & Gamoran, A. (2002). The production of achievement inequality in high school English. *American Educational Research Journal, 39*, 801–827. doi:10.3102/00028312039004801

Carroll, C., Patterson, M., Wood, S., Booth, A., Rick, J., & Balain, S. (2007). A conceptual framework for implementation fidelity. *Implementation Science, 2*(1). doi:10.1186/1748-5908-2-40

Cartwright, N. (2010). What are randomised controlled trials good for? *Philosophical Studies, 147*, 59–70. doi:10.1007/s11098-009-9450-2

Cartwright, N. (2013). Evidence, argument and prediction. In V. Karakostas & D. Dieks (Eds.), *EPSA11 perspectives and foundational problems in philosophy of science* (pp. 3–17). doi:10.1007/978-3-319-01306-0_1

Cartwright, N. (2017). Single case causes: What is evidence and why. In H.-K. Chao & J. Reiss (Eds.), *Philosophy of science in practice* (pp. 11–24). doi:10.1007/978-3-319-45532-7_2

Cartwright, N. (2019). *Nature, the artful modeler: Lectures on laws, science, how nature arranges the world and how we can arrange it better*. Chicago, IL: Open Court.

Cartwright, N., & Hardie, J. (2012). *Evidence-based policy: A practical guide to doing it better*. New York, NY: Oxford University Press.

Coburn, C. E., & Penuel, W. R. (2016). Research–practice partnerships in education: Outcomes, dynamics, and open questions. *Educational Researcher, 45*, 48–54. doi:10.3102/0013189X16631750

Coburn, C. E., & Stein, M. K. (Eds.). (2010). *Research and practice in education: Building alliances, bridging the divide*. Lanham, MD: Rowman & Littlefield.

Coe, R., & Fitz-Gibbon, C. T. (1998). School effectiveness research: Criticisms and recommendations. *Oxford Review of Education, 24*, 421–438. doi:10.1080/0305498980240401

Colangelo, N., Assouline, S. G., & Gross, M. U. M. (2004). *A nation deceived: How schools hold back America's brightest students* (Vol. 1; The Templeton National Report on Acceleration). Retrieved from http://www.accelerationinstitute.org/Nation_Deceived/ND_v1.pdf

Connolly, P., Biggart, A., Miller, S., O'Hare, L., & Thurston, A. (2017). *Using randomised controlled trials in education* (1st ed.). Thousand Oaks, CA: Sage.

Crotty, M. (1998). *The foundations of social research: Meaning and perspective in the research process*. Thousand Oaks, CA: Sage.

Curto, V. E., Fryer, R. G., Jr., & Howard, M. L. (2011). It may not take a village: Increasing achievement among the poor. In G. J. Duncan & R. J. Murnane (Eds.), *Whither opportunity? Rising inequality, schools, and children's life chances* (pp. 483–506). New York, NY: Russell Sage Foundation/Spencer Foundation.

Datnow, A., & Castellano, M. (2000). Teachers' responses to Success for All: How beliefs, experiences, and adaptations shape implementation. *American Educational Research Journal, 37*, 775–799. doi:10.3102/00028312037003775

Deaton, A., & Cartwright, N. (2018). Understanding and misunderstanding randomized controlled trials. *Social Science & Medicine, 210*, 2–21. doi:10.1016/j.socscimed.2017.12.005

Dee, T., & Penner, E. (2016). *The causal effects of cultural relevance: Evidence from an ethnic studies curriculum* (CEPA Working Paper No. 16-01). Retrieved from http://cepa.stanford.edu/wp16-01.

Duncan, G. J., & Murnane, R. J. (Eds.). (2011). *Whither opportunity? Rising inequality, schools, and children's life chances*. New York, NY: Russell Sage Foundation/Spencer Foundation.

Duncan, G. J., & Murnane, R. J. (2014). *Restoring opportunity: The crisis of inequality and the challenge for American education*. Cambridge, MA: Harvard Education Press.

Eisenhart, M., & Towne, L. (2003). Contestation and change in national policy on "scientifically based" education research. *Educational Researcher, 32*(7), 31–38.

Farley-Ripple, E., May, H., Karpyn, A., Tilley, K., & McDonough, K. (2018). Rethinking connections between research and practice in education: A conceptual framework. *Educational Researcher, 47*, 235–245. doi:10.3102/0013189X 18761042

Fixsen, D., Blasé, K., Metz, A., & Van Dyke, M. (2013). Statewide implementation of evidence-based programs. *Exceptional Children, 79*, 213–230. doi:10.1177/ 001440291307900206

Fixsen, D., Blasé, K. A., Metz, A., & Van Dyke, M. (2015). Implementation science. In J. Wright (Ed.), *International encyclopedia of the social & behavioral sciences* (2nd ed., Vol. 11, pp. 695–702). Oxford, UK: Elsevier.

Fixsen, D. L., Blasé, K., Duda, M., Naoom, S., & Van Dyke, M. (2010). Sustainability of evidence-based programs in education. *Journal of Evidence-Based Practices for Schools, 11*, 30–46.

Fixsen, D. L., & Blasé, K. A. (2013). *An overview of scaling-up and active implementation* [Video]. Chapel Hill, NC: State Implementation and Scaling-Up Evidence-Based Practices Center. Retrieved from http://sisep.fpg.unc.edu/

Fixsen, D. L., Blasé, K. A., Naoom, S. F., & Wallace, F. (2009). Core implementation components. *Research on Social Work Practice, 19*, 531–540. doi:10.1177/ 1049731509335549

Freiberg, H. J. (Ed.). (1999). *School climate: Measuring, improving, and sustaining healthy learning environments*. Philadelphia, PA: Falmer Press.

Furstenberg, F. F. (2011). The challenges of finding causal links between family educational practices and schooling outcomes. In G. J. Duncan & R. J. Murnane (Eds.), *Whither opportunity? Rising inequality, schools, and children's life chances*. New York, NY: Russell Sage Foundation/Spencer Foundation.

Gates Foundation. (n.d.). *K–12 education strategy overview*. Retrieved from https:// gatesfoundation.org/What-We-Do/US-Program/K-12-Education

Gentry, M. L. (2014). *Total school cluster grouping and differentiation: A comprehensive, research-based plan for raising student achievement and improving teacher practices* (2nd ed.). Waco, TX: Prufrock Press.

Goldstein, H., & Woodhouse, G. (2000). School effectiveness research and educational policy. *Oxford Review of Education, 26*(3–4), 353–363. doi:10.1080/ 713688547

Gorard, S., See, B. H., & Siddiqui, N. (2017). *The trials of evidence-based education: The promises, opportunities and problems of trials in education*. New York, NY: Routledge.

Gutiérrez, K. D., & Penuel, W. R. (2014). Relevance to practice as a criterion for rigor. *Educational Researcher, 43*, 19–23. doi:10.3102/0013189X13520289

Halsall, R., Carter, K., Curley, M., & Perry, K. (1998). School improvement: The case for supported teacher research. *Research Papers in Education, 13*, 161–182. doi:10.1080/0267152980130204

Hammersley, M. (2002). *Educational research, policymaking and practice*. London, England: Paul Chapman.

Hammersley, M. (2013). *The myth of research-based policy and practice*. Thousand Oaks, CA: Sage.
Hargreaves, D. H. (1995). School culture, school effectiveness and school improvement. *School Effectiveness and School Improvement, 6*, 23–46. doi:10.1080/0924345950060102
Horvat, E. M., Weininger, E. B., & Lareau, A. (2003). From social ties to social capital: Class differences in the relations between schools and parent networks. *American Educational Research Journal, 40*, 319–351. doi:10.3102/00028312040002319
Howe, K. R. (2009). Positivist dogmas, rhetoric, and the education science question. *Educational Researcher, 38*, 428–440. doi:10.3102/0013189X09342003
Joyce, K. E. (2019). The key role of representativeness in evidence-based education [Online]. *Educational Research and Evaluation*, 1–20. doi:10.1080/13803611.2019.1617989
Joyce, K. E., & Cartwright, N. (2018). Meeting our standards for educational justice: Doing our best with the evidence. *Theory and Research in Education, 16*, 3–22. doi:10.1177/1477878518756565
Kvernbekk, T. (2016). *Evidence-based practice in education: Functions of evidence and causal presuppositions*. New York, NY: Routledge.
Ladd, H. F. (2012). Education and poverty: Confronting the evidence. *Journal of Policy Analysis and Management, 31*, 203–227. doi:10.1002/pam.21615
Lareau, A. (2011). *Unequal childhoods: Class, race, and family life* (2nd ed.). Berkeley: University of California Press.
Levin, B. (2013). To know is not enough: Research knowledge and its use. *Review of Education, 1*, 2–31. doi:10.1002/rev3.3001
Lewis, C. (2015). What is improvement science? Do we need it in education? *Educational Researcher, 44*, 54–61. doi:10.3102/0013189X15570388
López Turley, R. N., & Stevens, C. (2015). Lessons from a school district-university research partnership: The Houston Education Research Consortium. *Educational Evaluation and Policy Analysis, 37*(1 Suppl.), 6S–15S. doi:10.3102/0162373715576074
MacNeil, A. J., Prater, D. L., & Busch, S. (2009). The effects of school culture and climate on student achievement. *International Journal of Leadership in Education, 12*, 73–84. doi:10.1080/13603120701576241
McIntyre, D. (2005). Bridging the gap between research and practice. *Cambridge Journal of Education, 35*, 357–382. doi:10.1080/03057640500319065
McLeod, J. D., & Kaiser, K. (2004). Childhood emotional and behavioral problems and educational attainment. *American Sociological Review, 69*, 636–658. doi:10.1177/000312240406900502
Metz, A., & Louison, L. (2018). *The hexagon tool: Exploring context*. Chapel Hill, NC: National Implementation Research Network.
Munro, E., Cartwright, N., Hardie, J., & Montuschi, E. (2016). *Improving child safety: Deliberation, judgment and empirical research* (ISSN 2053-2660). Retrieved from https://www.dur.ac.uk/resources/chess/ONLINE_Improvingchildsafety-15_2_17-FINAL.pdf
National Implementation Research Network. (2016). *Active implementation practice and science*. Chapel Hill, NC: Author. Retrieved from https://nirn.fpg.unc.edu/sites/nirn.fpg.unc.edu/files/resources/NIRN-Briefs-1-ActiveImplementationPracticeAndScience-10-05-2016.pdf
Nelson, J., & Campbell, C. (2017). Evidence-informed practice in education: Meanings and applications. *Educational Research, 59*, 127–135. doi:10.1080/00131881.2017.1314115

Norton, J. D. (2003). A material theory of induction. *Philosophy of Science, 70*(, 647–670. doi:10.1086/378858

Nuthall, G. (2004). Relating classroom teaching to student learning: A critical analysis of why research has failed to bridge the theory-practice gap. *Harvard Educational Review, 74*, 273–306. doi:10.17763/haer.74.3.e08k1276713824u5

Nutley, S. M., Walter, I., & Davies, H. T. O. (2007). *Using evidence: How research can inform public services.* Bristol, England: Policy Press.

Oakes, J. (2005). *Keeping track: How schools structure inequality* (2nd ed.). New Haven, CT: Yale University Press.

Phillips, D. C. (2007). Adding complexity: Philosophical perspectives on the relationship between evidence and policy. *Yearbook of the National Society for the Study of Education, 106*, 376–402. doi:10.1111/j.1744-7984.2007.00110.x

Posey-Maddox, L. (2014). *When middle-class parents choose urban schools: Class, race, and the challenge of equity in public education.* Chicago, IL: University of Chicago Press.

Quartz, K. H., Weinstein, R. S., Kaufman, G., Levine, H., Mehan, H., Pollock, M., . . . Worrell, F. C. (2017). University-partnered new school designs: Fertile ground for research-practice partnerships. *Educational Researcher, 46*, 143–146. doi:10.3102/0013189X17703947

Reiss, J. (2013). What's wrong with our theories of evidence? *Theoria, 29*, 283–306. doi:10.1387/theoria.10782

Reynolds, D., Sammons, P., De Fraine, B., Van Damme, J., Townsend, T., Teddlie, C., & Stringfield, S. (2014). Educational effectiveness research (EER): A state-of-the-art review. *School Effectiveness and School Improvement, 25*, 197–230. doi:10.1080/09243453.2014.885450

Rothman, K. J. (1976). Causes. *American Journal of Epidemiology, 104*, 587–592.

Rowan, B. (2011). Intervening to improve the educational outcomes of students in poverty: Lessons from recent work in high-poverty schools. In G. J. Duncan & R. J. Murnane (Eds.), *Whither opportunity? Rising inequality, schools, and children's life chances* (pp. 523–538). New York, NY: Russell Sage Foundation/Spencer Foundation.

Rubin, D. B. (1974). Estimating causal effects of treatments in randomized and nonrandomized studies. *Journal of Educational Psychology, 66*, 688–701. doi:10.1037/h0037350

Sarason, S. B. (1990). *The predictable failure of educational reform: Can we change course before it's too late?* (1st ed.). San Francisco, CA: Jossey-Bass.

Sarason, S. B. (1996). *Revisiting "The Culture of the School and the Problem of Change."* New York, NY: Teachers College Press.

Schoen, L. T., & Teddlie, C. (2008). A new model of school culture: A response to a call for conceptual clarity. *School Effectiveness and School Improvement, 19*, 129–153. doi:10.1080/09243450802095278

Scriven, M. (1994). The final synthesis. *Evaluation Practice, 15*, 367–382. doi:10.1016/0886-1633(94)90031-0

Seckinelgin, H. (2017). *The politics of global aids: Institutionalization of solidarity, exclusion of context.* New York, NY: Springer.

Shavelson, R. J., Towne, L., & National Research Council (Eds.). (2002). *Scientific research in education.* Washington, DC: National Academies Press.

Simpson, A. (2017). The misdirection of public policy: Comparing and combining standardised effect sizes. *Journal of Education Policy, 32*, 450–466. doi:10.1080/02680939.2017.1280183

Slavin, R. (2002). Evidence-based education policies: Transforming educational practice and research. *Educational Researcher, 31*(7), 15–21.

Smeyers, P., & Depaepe, M. (Eds.). (2007). *Educational research: Why "what works" doesn't work*. Dordrecht, Netherlands: Springer. doi:10.1007/978-1-4020-5308-5

Spillane, J. P., & Miele, D. B. (2007). Evidence in practice: A framing of the terrain. *Yearbook of the National Society for the Study of Education, 106*, 46–73.

Steenbergen-Hu, S., Makel, M. C., & Olszewski-Kubilius, P. (2016). What one hundred years of research says about the effects of ability grouping and acceleration on K–12 students'· academic achievement: Findings of two second-order meta-analyses. *Review of Educational Research, 86*, 849–899. doi:10.3102/0034654316675417

Tough, P. (2013). *How children succeed: Grit, curiosity, and the hidden power of character*. Boston, MA: Mariner Books.

Tseng, V., & Nutley, S. (2014). Building the infrastructure to improve the use and usefulness of research in education. In K. S. Finnigan & A. J. Daly (Eds.), *Using research evidence in education* (pp. 163–175). Cham, Germany: Springer International. doi:10.1007/978-3-319-04690-7_11

Wentworth, L., Carranza, R., & Stipek, D. (2016). A university and district partnership closes the research-to-classroom gap. *Phi Delta Kappan, 97*(8), 66–69. doi:10.1177/0031721716647024

Wentworth, L., Mazzeo, C., & Connolly, F. (2017). Research practice partnerships: A strategy for promoting evidence-based decision-making in education. *Educational Research, 59*, 241–255. doi:10.1080/07391102.2017.1314108

What Works Clearinghouse. (n.d.). *Find What Works homepage*. Retrieved from https://ies.ed.gov/ncee/wwc/FWW

What Works Clearinghouse. (2016a). *Read 180®: Adolescent literacy*. Retrieved from https://ies.ed.gov/ncee/wwc/EvidenceSnapshot/665

What Works Clearinghouse. (2016b). *Read 180®: WWC summary of evidence for this intervention*. Retrieved from https://ies.ed.gov/ncee/wwc/Intervention/742

What Works Clearinghouse. (2017a). *Procedures handbook: Version 4.0*. Retrieved from https://ies.ed.gov/ncee/wwc/Docs/referenceresources/wwc_procedures_handbook_v4.pdf

What Works Clearinghouse. (2017b). *Standards handbook: Version 4.0*. Retrieved from https://ies.ed.gov/ncee/wwc/Docs/referenceresources/wwc_standards_handbook_v4.pdf

Manuscript received June 12, 2018
Final revision received June 11, 2019
Accepted July 1, 2019

# Parent Engagement and Satisfaction in Public Charter and District Schools

### Zachary W. Oberfield
### Haverford College

*Using nationally representative parent surveys over a 10-year period, this article asks if there were differences in parent engagement and satisfaction at public charter and district schools. It then examines whether any such differences persisted when accounting for observable school and family characteristics, including whether parents conducted a school search prior to selecting their child's school. It finds that charter parents volunteered more but, in aggregate, were not more engaged in school-related activities, relative to district parents. In contrast, charter parents reported higher levels of satisfaction than district parents throughout the period. These differences persisted even when accounting for observable ways in which these families and schools differed.*

KEYWORDS: charter schools, educational policy, parent engagement and satisfaction

Public charter schools (hereafter "charter schools") are funded with public revenues but primarily function outside the authority and rules of a school district (Finn, Jr., Manno, & Wright, 2016). Perhaps as a result, they have emerged as one of the most controversial public education reforms of the past half-century. Proponents argue that charter schools decentralize decision making, set the stage for innovation, and enable more dynamic, nimble public schools (Osborne, 2017). Critics argue that charter schools are chaotic and unaccountable and that they draw precious resources from public district schools (hereafter "district schools") (Ravitch,

---

ZACHARY W. OBERFIELD is an associate professor of political science at Haverford College, 370 Lancaster Avenue, Haverford, PA 19041, USA; e-mail: *zoberfie@haverford.edu*. His research interests include schools, leadership, and street-level bureaucracy. He is the author of two books: *Are Charters Different? Public Education, Teachers, and the Charter School Debate*, Harvard Education Press (2017), and *Becoming Bureaucrats: Socialization at the Front Lines of Government Service*, University of Pennsylvania Press (2014), winner of the 2015 Best Book Award from the Public and Nonprofit Division of the Academy of Management. He received his PhD in political science from the University of Wisconsin–Madison.

2013). In response to this controversy, scholars have sought to determine how charter schools perform relative to district schools. Most of this research focuses on standardized achievement tests or graduation rates (Betts & Tang, 2014; Epple, Romano, & Zimmer, 2016; Wohlstetter, Smith, & Farrell, 2013).

In addition, it is important to study how stakeholders experience charter schools. In particular, parents are central to claims that charter schools function differently. Borrowing the language and logic of the market, charter school advocates imagine parents as consumers whom schools must satisfy in order to maintain market share (Bulkley & Fisler, 2003; Miron & Nelson, 2002). Charter school critics portray this experiment in very different terms: Echoing general critiques of privatization and neoliberalism (Schram, 2015), they suggest that these schools have the potential to increase the precarity of historically marginalized people and communities (Buras, 2014; Ravitch, 2013; Wells, 2002).

Despite these conflicting views about what charter schools mean for parents, few works compare the experiences of parents with a child in a charter school with those of parents with a child in a district school. Research that does focus on parents tends to look at limited geographic areas (usually a single state) at a particular time. Because the charter school sector has evolved over the years, and differs significantly from state to state, it is important to look over a wide area and sustained period of time.

This article contributes by using nationally representative data over a 10-year period to compare levels of parent engagement and satisfaction in charter and district schools. The article begins by further elaborating the role of parents in the theory of charter schools and how this fits with the privatization movement and research in public administration. It then reviews empirical works that have examined whether charter schools foster higher levels of parent engagement and satisfaction relative to district schools. The findings from this literature generally show that charter schools have higher levels of parent engagement and satisfaction. However, because students are not randomly assigned to charter or district schools, it is difficult to know what role self-selection plays in any of these observed differences. Thus, this article considers how a novel set of variables, aimed at capturing differences in family characteristics, experiences, and school-selection processes, may account for charter-district differences in parent engagement and satisfaction. It then specifies its method and data and presents its findings.

In summary, the article shows that charter parents reported volunteering more than district parents. However, in aggregate, there was little evidence that charter schools fostered higher levels of parent engagement relative to those found in district schools. Despite approximately equal levels of parent engagement, charter parents did report higher levels of school satisfaction. Further analysis reveals that these differences in volunteering and satisfaction persisted even when controlling for observable differences in parent self-selection. The article closes by considering what these findings mean for our understanding of charter schools and suggesting directions for future research.

## Parents and the Argument for Charter Schools

In the latter part of the 20th century, there was a growing sense among American policymakers and the public that the nation's public schools were failing (National Commission on Excellence in Education, 1983). Many of the reforms that sought to respond to this perceived crisis—like charter schools and school voucher programs—aimed to improve public education by decentralizing decision making and harnessing the forces of the market (Ravitch, 2013). Initially, the idea for charter schools was that they would serve as temporary experiments for the betterment of district schools (Budde, 1988). Rather quickly, however, charter schools became understood as an alternative, semipermanent type of public school that would compete with district schools for resources and students (Kolderie, 1990; Mintrom, 2003; Ravitch, 2013). In doing so, proponents suggested, charter schools would jolt district schools into change, improving the entire public education system.

In turning to markets and decentralization to solve the education crisis, these early charter supporters were tapping into the emerging—and arguably still dominant—effort to use privatization to alter how government operates (Donahue, 1989). From this perspective, government is inefficient and, without competition, has monopolistic tendencies. Thus, public policy can be improved by contracting out core government services to private-sector actors who, in responding to the demands of the market, will deliver better services to citizens (Donahue, 1989; Verkuil, 2007). Put concisely, government should endeavor to "steer rather than row" (Osborne & Gaebler, 1993, p. 28).

Although the rhetoric of the privatization movement flattens some important nuances about how public sector organizations function, public policy and administration scholars agree that public organizations are constrained relative to their private-sector counterparts (Meier & O'Toole, 2011; Rainey & Chun, 2005). For example, because they must answer to elected officials from varied political institutions, and pay close attention to public opinion, public organizations may be less responsive to change, more focused on processes than outcomes, and less efficient (Bozeman, 2004). In addition, public organizations are thought to be constrained because public leaders and managers have a harder time hiring and firing problematic employees, cannot set intermediate goals, and work in environments with higher levels of red tape (Feeney & Bozeman, 2009; Meier & O'Toole, 2011; Rainey & Chun, 2005).

Born during a surge of interest in privatization—and buttressed by the aforementioned scholarly research on public organizations—it is unsurprising that charter schools, like many recent education reforms, seek to use the logic of the market to solve perceived problems. The task in front of scholars now is making sense of how these experiments have played out in practice. Most efforts to reckon with the performance of charter schools have relied on test scores and graduation rates (Betts & Tang, 2014; Epple et al., 2016; Wohlstetter et al., 2013). However, there has also been interest in comparing the experiences of various stakeholders—such as teachers and principals—in charter and district schools (Gawlik & Bickmore, 2017; Oberfield, 2017).

This article contributes by looking at parents, the "consumers" of the charter school marketplace. Since the dawn of the charter sector, parents have been central to charter theorists' views of why these schools would be different. Kolderie (1990), one early charter proponent, argued that removing the monopoly that districts had on public schools would prevent them from taking their "customers for granted" (p. 54). In other words, by generating competition among schools, the needs and wants of parents—the people in the best position to assess their child's well-being (Hill, Lake, & Celio, 2002)—would be elevated. Similar to Kolderie, Shanker (1988) envisioned a system in which, "Parents could choose which charter school to send their children to, thus fostering competition," driving system-wide school improvement and bottom-up accountability. Nathan (1996), another early proponent, also expected that charter schools would foster tight connections between families and schools based on frequent meetings between teachers and parents. In this way, they would improve student well-being and learning.

Today, parents remain central to arguments in favor of charter schools. For instance, the National Alliance for Public Charter Schools (2018), one of the most prominent charter school advocacy organizations, argues that charter schools "provide families with options in public education, allowing parents to take a more active role in their child's education." Also, various states have written their state charter laws with the express intent of fostering higher levels of parent involvement. For example, the Tennessee charter law is intended to "afford parents substantial meaningful opportunities to participate in the education of their children" (as quoted by Smith, Wohlstetter, Kuzin, & De Pedro, 2011, p. 75).

This article is devoted to exploring whether gains in parent engagement and satisfaction have been realized, even when accounting for the possibility that charter school parents may differ from district school parents in particular ways. The next section advances toward this goal by asking why parent engagement and satisfaction are important educational outcomes. Following that, it reviews prior works that assess levels of parent engagement and satisfaction in charter and district schools and articulates two of the hypotheses that this article will test.

## Parent Engagement and Satisfaction

Among scholars and policymakers, there is a consensus that parent engagement—the working together of parents and school staff to support and improve student learning, development, and health (Centers for Disease Control and Prevention, 2018)—is strongly related to a host of educational and social outcomes (Epstein, 2010; Henderson & Mapp, 2002; Jeynes, 2012; U.S. Department of Education, 1994; Zimmer & Buddin, 2007). Parent involvement is thought to benefit students by enabling more sophisticated coordination between parents and teachers—drawing parents into the life of the school—and giving parents and teachers more power to monitor student well-being and learning.

Looking across the United States in the 2011-2-12 school year, over 80% of parents attended a general school meeting, over 70% attended a parent-teacher meeting, and around 40% volunteered or served on a committee at their child's school (Child Trends, 2013). However, parent engagement is not distributed equally across demographic groups: Wealth, Whiteness, and parent education are positively associated with parent engagement (Child Trends, 2013). One explanation for these findings is that parent engagement, like civic engagement more generally, is enabled by resources and their attendant privileges (Verba, Schlozman, & Brady, 1995).

Although it has not received as much attention, parent satisfaction, the level of contentment that parents have with their child's school, has recently become a more prominent interest for scholars and policymakers (Chambers & Michelson, 2016). This increased attention is coincident with the rise of the school choice movement and growing interest in understanding how parents choose and experience their children's schools (Buckley & Schneider, 2007). As researchers have begun to explore parent satisfaction, they have found that it is strongly linked to school performance (Gibbons & Silva, 2011). Put simply, parents' perceptions of schools are a useful indicator of academic quality.

Research also shows that satisfaction is associated with family and student characteristics. For example, Black and Hispanic parents appear to be less satisfied with their children's schools relative to White and Asian parents (Friedman, Bobrowski, & Geraci, 2006); similarly, married parents are more satisfied than single parents (Fantuzzo, Perry, & Childs, 2006). Parent satisfaction also appears to be linked to student characteristics. For example, the parents of special education students (those with an individualized education plan or IEP) typically report being less satisfied with their child's school relative to the parents of students without an IEP (Beck, Maranto, & Lo, 2014). Though family and child characteristics are correlated with parent satisfaction, the programs and policies that schools implement also appear to affect parent satisfaction (Bailey, Scarborough, & Hebbeler, 2003).

*Oberfield*

As noted in the prior section, parent engagement and satisfaction are central to arguments in support of charter schools. With higher levels of autonomy and accountability, charter schools are expected to, among other things, involve parents more, improve educational quality, and generate higher levels of parent satisfaction (Bulkley & Fisler, 2003; Smith et al., 2011). Next, the article evaluates this expectation by reviewing empirical research investigating parent engagement and satisfaction in charter and district schools.

**Parent Engagement in Charter and District Schools**

Nathan (1996), in an early tour of the charter sector, notes that charter schools are achieving the goal of enhanced parent involvement. Citing research in California charter schools, he finds that charter teachers are discussing strategies for engaging hard-to-reach parents, giving parents ideas about what they can do at home to engage students, and assigning homework that involves parents. In addition, Nathan notes that charter schools are actively getting parents involved in the upkeep and advancement of schools. Another early view on this aspect of the charter experiment, provided by Finn, Manno, and Vanourek (2000), also suggests important gains in parent engagement. In fact, they argue, "If charter schools can declare any clear-cut victory today, it is in the battle against adult disengagement. One of the secrets of these schools' success is their knack for tapping vast resources of parental involvement" (p. 93).

Becker, Nakagawa, and Corwin (1997) contribute by comparing parent engagement in charter and district schools in California in the mid-1990s. Using a survey of principals, they document higher levels of parent involvement in charter schools. For instance, charter parents were more likely to help or teach in their child's classroom and work on a school committee or governance board. Mintrom (2003) takes a similar approach by comparing parent engagement in district and charter schools in Michigan in the late 1990s. Using a survey of principals, he finds higher levels of parent volunteering and meeting attendance in charter schools; however, he shows no differences in parent-teacher conference attendance. Zimmer and Buddin (2007) also use principal surveys to assess levels of parent engagement in charter and district schools. Focusing on California, they find that charter principals in elementary, middle, and high schools report higher levels of parental engagement than comparable principals in district schools.

The Becker et al., Mintrom, and Zimmer and Buddin studies focus on only one state; because state charter laws may affect parent engagement (Wohlstetter et al., 2013) and, subsequently, principals' experiences, Bifulco and Ladd (2006) make a helpful contribution by examining a nationwide survey of principals in charter and district schools. Their findings match with these other studies and suggest that charter schools foster a higher level of parent engagement, especially in primary and middle schools.

Differences in parent engagement have also been studied by looking at the experiences of teachers and parents. Oberfield (2017) uses nationally representative teacher survey data from four times between the 1999–2000 and 2011–2012 school years. The study shows that teachers in charter schools report significantly higher levels of parent support and involvement than teachers in district schools. A working paper from Buckley (2007) focuses on parents' experiences in charter and district schools. Drawing from a nationally representative survey of parents, fielded in the 2003–2004 school year, the paper compares levels of engagement in public schools that were assigned and chosen; this latter category included charter schools, magnet schools, and other schools of choice. It measures engagement in terms of the number of activities parents engaged in and the hours they spent in their child's school. When the study controls for location, school, and family characteristics, it finds that parents in public schools of choice reported volunteering more hours than parents in assigned schools; however, these two groups were indistinguishable in terms of the number of activities that they reported.

On balance, this review suggests support for the expectation that charter schools foster higher levels of parent engagement. However, it also reveals some important limitations in the literature: Most of the studies look at one moment in time, focus on a particular state, and look at principals and teachers rather than parents. The Buckley paper, with its focus on parents, is helpful; however, it cannot separate charter schools from other public schools of choice and uses data from over 15 years ago. Another limitation, raised often by charter school critics and observers (Miron & Nelson, 2002; Ravitch, 2013), is that any differences in parent engagement result from the types of parents who select into charter schools. Most obviously, perhaps parents who seek out information on a number of schools, and end up choosing a charter school, are the types of parents who would be more likely to get involved in school life. This possibility, and how the article tries to deal with it, is discussed further below. However, based on the above literature review, and theoretical accounts suggesting that charter schools will engage with parents more, this article tests the following hypothesis:

*Hypothesis 1:* Parents with a child in a charter school will report higher levels of engagement relative to parents with a child in a district school.

**Parent Satisfaction in Charter and District Schools**

In addition to getting parents more involved in school life, early theorists expected that charter schools would foster higher levels of parent satisfaction: By building an education marketplace, and empowering parents to make choices in it, parents would be happier with their chosen public schools (Kolderie, 1990; Shanker, 1988). To examine whether this expectation has been realized, this section reviews empirical studies that have examined parent satisfaction in charter and district schools.

To begin, a number of studies have sought to understand the charter experiment by studying parent satisfaction in charter schools. Miron and Nelson (2002) and Miron, Nelson, and Risley (2002) find that charter school parents in Michigan and Pennsylvania reported high levels of satisfaction in curriculum and instruction. Specifically, charter parents were likely to agree that their school had high expectations, accountable teachers and leaders, and, generally speaking, a bright future. Another study, by Finn et al. (2000), notes that two thirds of charter school parents in a multistate sample indicate that their child's charter school is better than their previous school with regard to class size, school size, attention from teachers, and curriculum. Wohlstetter, Nayfack, and Mora-Flores (2008) examine surveys of charter school parents in California and find high levels of satisfaction with, among other aspects of school life, academic programs, support services, and teachers. These studies are helpful in describing charter parent satisfaction, but they provide little sense of how it compares with district parent satisfaction.

Thus, Buckley and Schneider (2006) contribute by examining charter and district parent satisfaction in Washington, D.C. They find that charter parents grade their schools higher relative to district parents. In response to concerns that this result is driven by self-selection into charter schools, they conduct a propensity score–matching analysis using a model that includes family background characteristics and other potentially relevant control variables. The results of their analysis suggest that self-selection is not driving differences in parent satisfaction. Complicating the picture, they also look at how parent satisfaction evolves over time. Although charter parents are more satisfied initially, they show that these gains recede over a number of years such that, at the end of a 4-year period, there were no differences in satisfaction.

Jochim, DeArmond, Gross, and Lake (2014) examine parent satisfaction in eight cities with high levels of school choice. In general, they find no statistically significant difference in satisfaction between parents with children in a neighborhood district school (i.e., assigned) and those with children in a nonneighborhood school (i.e., a school of choice). Barrows, Cheng, Peterson, and West (2017) examine satisfaction among charter and district parents using two nationally representative surveys. Controlling for various background and demographic characteristics, they find that parents with a student in a charter school are more satisfied than parents with a child in an assigned-district school and around equally satisfied as parents with a child in a chosen-district school.

Gleason, Clark, Tuttle, and Dwoyer (2010) look at oversubscribed charter middle schools that held admission lotteries to allot student seats in 36 schools across 15 states. They find that charter schools positively affected lottery winners' parent satisfaction. Specifically, winners were 33 percentage points more likely to rate their child's school as excellent relative to lottery losers. Finally, Tuttle et al. (2013) examine parent satisfaction using admission lotteries into KIPP (Knowledge Is Power Program) schools, a network of charter schools that has achieved impressive levels of student achievement. They find that lottery winners were 6 percentage points more satisfied.

Although it is by no means unanimous, the empirical literature generally supports the claim that parents with a child in a charter school report higher levels of satisfaction relative to parents with a child in a district school. However, there are a number of ways in which the literature is incomplete. Many of the early studies provide information on parents in charter schools but lack a counterfactual or comparison case (i.e., parents in district school). Those that came later satisfy this need, but they tend to focus on a small geographic area or a limited number of cities. The Gleason et al. (2010) and Tuttle et al. (2013) pieces make a major contribution, by zeroing in on the causal effect of particular charter schools, but may not be externally generalizable to the full set of charter schools due to focusing only on schools that were oversubscribed (i.e., popular). Finally, Barrows et al. (2017) use national data to look at parent satisfaction and control for a variety of potential covariates of parent satisfaction—such as parent race, educational attainment, household income, and homeowner status. However, they do not control for differences in parent behavior that might be linked to self-selection (like considering other schools), school-level covariates, and the state in which parents live. This article contributes by remedying some of these deficiencies in the literature while testing the following hypothesis:

*Hypothesis 2:* Parents with a child in a charter school will report higher levels of satisfaction relative to parents with a child in a district school.

## Are Differences in Parent Engagement and Satisfaction Driven by Self-Selection?

The prior section suggests that it is reasonable to expect charter schools to foster higher levels of parent engagement and satisfaction relative to district schools. This article's first goal is assessing whether there is evidence to support these expectations. If there is, its second goal is explaining why such differences may exist. The most prominent explanation in the literature is that higher levels of parent engagement and satisfaction are artifacts of self-selection or the process of choosing a school, rather than any actions taken by the schools (Buckley & Schneider, 2006). This section fleshes out that argument and concludes with a related hypothesis.

Many observers have argued that enhanced parent engagement in charter schools results from self-selection (Finn et al., 2000; Hamlin, 2017; Miron & Nelson, 2002; Ravitch, 2013). Because parents with children in a charter school made a choice to send them there, the argument goes, they are more likely to be the type of parents who want to be involved in their children's schooling. For example, Kahlenberg and Potter (2014) argue,

> Charter schools generally have a leg up in parent involvement. As opposed to district schools that enroll all students within their boundaries, charter schools have only families where someone—a parent, grandparent, aunt, social worker, or friend—was involved enough in a child's education to apply for the school. (p. 143)

If this argument is right, different levels of parent engagement are driven less by school or teacher practices and more by which families choose charter schools.

Similarly, scholars have suggested that differences in satisfaction between charter and district school parents may derive from the mere fact that charter parents made an affirmative choice (Bosetti, 2004; Buckley & Schneider, 2006). For example, Rothstein (1998) argues, "Because charter school parents and students have chosen to be there, we should also expect surveys to show a predisposition in favor of the charter school, even with no difference in quality." More specifically, scholars have argued that in choosing a charter school, parents associate themselves with an oppositional identity (Wells, 2002). By defining themselves against district schools, which are old and static, charter parents see their chosen schools as unique, innovative, and special. Another possible reason why self-selection could lead to higher levels of satisfaction in charter schools is an *ex post* rationalization made by parents (Teske & Schneider, 2001). Since charter parents invested time and energy in the decision-making process, they feel compelled to report high levels of school satisfaction as a way to justify this expenditure.

Finally, there is evidence that in choosing among schools, parents seek student populations that they favor (Buckley & Schneider, 2007). Specifically, parents care intensely about the racial dynamics of the students in the schools that they consider and work to find environments that they think are suitable for their children. By exercising some agency over who their children attend school with, charter parents may have higher levels of satisfaction. However, this satisfaction gain would not be caused by, for example, the practices of teachers and principals.

If accounts like this are true, differences in parent engagement and satisfaction between charter and district schools result from the process of choosing a school, or parent and student self-selection, rather than any programs or policies that the schools are implementing. Without an experiment, it is difficult to know for sure how self-selection affects parent engagement and satisfaction. Nonetheless, this article advances our understanding of the role played by self-selection by controlling for various family characteristics—such as parent-school selection processes—that may distinguish district and charter families. As such, the article tests the following hypothesis:

> *Hypothesis 3:* Higher levels of parent engagement and satisfaction in charter schools are attributable to the characteristics of the students and families who select into them.

## Data and Method

To study parent engagement and satisfaction in district and charter schools, this article uses data from the Parent and Family Involvement (PFI) surveys, conducted by the U.S. Department of Education's National Center for Education Statistics (NCES).[1] Because aspects of charter school life may have changed over time—as schools matured and standard operating procedures were implemented (Buckley & Schneider, 2007; Rothstein, 1998)—it is important to look at parent engagement and satisfaction over a number of years. As such, the article examines PFI survey data from 2007, 2012, and 2016. These cross-sectional (i.e., not panel) surveys, conducted as part of the National Household Education Survey, randomly chose thousands of respondents in each cycle.

In 2007, the NCES administered the survey using computer-assisted telephonic interviewing; in 2012 and 2016, a self-administered questionnaire was sent to respondents via the U.S. mail. The response rates and sample sizes for the survey were 39.1% and 14,080, respectively, in 2007; 57.6% and 17,560, respectively, in 2012; and 49.3% and 10,680, respectively, in 2016.[2] Since the PFI randomly selects respondents—and weights the data to account for sample nonresponse—the surveys provide a nationally representative view of parents' experiences. More specifically, the PFI can be used to make inferences about U.S. families with a child 20 years of age or younger who is enrolled in kindergarten through 12th grade. This article focuses on families with a child in a public school (i.e., it excludes families that homeschool or have a child in a private school).

Throughout this time period, charter schools were a relatively small, but growing, portion of the national public school universe. Using the NCES's Common Core of Data, which yearly captures data on all public schools, in the 2007–2008 school year charter schools accounted for less than 5% of all public schools; by 2015–2016 school year, this had risen to around 8%. In 2007–2008, around 3% of the public school population attended charter schools; by 2015, this figure had risen to around 6%. Mirroring this reality, the PFI includes mostly district student parents; however, over the years (as seen in Table 1), a growing proportion of respondents had children in a charter school.

## Table 1
### Survey Means

|  | All 2007 | All 2012 | All 2016 | District 2007 | District 2012 | District 2016 | Charter 2007 | Charter 2012 | Charter 2016 |
|---|---|---|---|---|---|---|---|---|---|
| **Dependent variables** | | | | | | | | | |
| Attend event | 0.73 | 0.72 | 0.77 | 0.73 | 0.72 | 0.76 | 0.73 | 0.79* | 0.80 |
| Volunteer | 0.37 | 0.35 | 0.36 | 0.37 | 0.34 | 0.35 | 0.54* | 0.52* | 0.53* |
| General meeting | 0.86 | 0.82 | 0.84 | 0.86 | 0.82 | 0.84 | 0.88 | 0.84 | 0.84 |
| PTO meeting | 0.48 | 0.43 | 0.46 | 0.48 | 0.43 | 0.45 | 0.56 | 0.52* | 0.57* |
| PT conference | 0.74 | 0.72 | 0.73 | 0.74 | 0.72 | 0.73 | 0.83 | 0.81* | 0.84* |
| Fundraising | 0.61 | 0.55 | 0.56 | 0.61 | 0.55 | 0.56 | 0.62 | 0.59 | 0.53 |
| Committee | 0.13 | 0.10 | 0.10 | 0.13 | 0.10 | 0.10 | 0.19 | 0.13 | 0.09 |
| Guidance | 0.39 | 0.35 | 0.35 | 0.39 | 0.35 | 0.35 | 0.37 | 0.39 | 0.39 |
| Number of activities | 7.94 | 6.73 | 7.08 | 7.93 | 6.70 | 7.09 | 8.79 | 7.61 | 6.70 |
| Satisfaction | 3.40 | 3.40 | 3.42 | 3.39 | 3.39 | 3.42 | 3.63* | 3.56* | 3.55* |
| **Independent variable** | | | | | | | | | |
| Charter | 0.01 | 0.03 | 0.05 | 0.00 | 0.00 | 0.00 | 1.00 | 1.00 | 1.00 |
| **Control variables** | | | | | | | | | |
| Nonassigned school | 0.15 | 0.08 | 0.13 | 0.15 | 0.09 | 0.13 | 0.00 | 0.00 | 0.00 |
| First choice | 0.82 | 0.78 | 0.79 | 0.82 | 0.77 | 0.79 | 0.87 | 0.85* | 0.84 |
| Considered other | 0.29 | 0.29 | 0.25 | 0.29 | 0.28 | 0.24 | 0.50* | 0.56* | 0.55* |
| Grades | 3.22 | 3.29 | 3.27 | 3.22 | 3.29 | 3.27 | 3.38 | 3.41* | 3.31 |
| Moved | 0.27 | 0.19 | 0.19 | 0.28 | 0.19 | 0.20 | 0.12* | 0.10* | 0.12* |
| Suspended | 0.15 | 0.14 | 0.12 | 0.15 | 0.14 | 0.12 | 0.16 | 0.14 | 0.13 |
| Hispanic | 0.20 | 0.24 | 0.25 | 0.20 | 0.24 | 0.25 | 0.37* | 0.33* | 0.38* |
| Asian | 0.03 | 0.07 | 0.08 | 0.03 | 0.07 | 0.08 | 0.04 | 0.07 | 0.07 |
| Black | 0.19 | 0.19 | 0.18 | 0.19 | 0.18 | 0.18 | 0.20 | 0.32* | 0.30* |
| White | 0.72 | 0.70 | 0.69 | 0.72 | 0.70 | 0.69 | 0.63 | 0.53* | 0.54* |
| Female | 0.48 | 0.49 | 0.48 | 0.48 | 0.49 | 0.48 | 0.52 | 0.56 | 0.52 |
| English at home | 0.90 | 0.86 | 0.83 | 0.90 | 0.86 | 0.83 | 0.81* | 0.79* | 0.78 |
| Parent education | 2.92 | 2.94 | 3.03 | 2.92 | 2.94 | 3.04 | 2.86 | 2.88 | 2.99 |
| Two-parent family | 0.70 | 0.72 | 0.72 | 0.70 | 0.72 | 0.72 | 0.60 | 0.65 | 0.71 |
| Welfare | 0.33 | 0.35 | 0.36 | 0.33 | 0.35 | 0.36 | 0.37 | 0.43* | 0.46* |
| Family income | 9.37 | 5.95 | 6.25 | 9.38 | 5.96 | 6.25 | 9.24 | 5.60 | 6.20 |
| Neighborhood poverty | 1.90 | 2.12 | 2.18 | 1.90 | 2.10 | 2.16 | 2.23* | 2.65* | 2.69* |
| Disability | 0.24 | 0.17 | 0.17 | 0.24 | 0.17 | 0.17 | 0.21 | 0.16 | 0.15 |
| Urban | 0.30 | 0.29 | 0.31 | 0.30 | 0.28 | 0.30 | 0.55* | 0.52* | 0.56* |
| Percent lunch | 3.48 | 3.69 | 3.74 | 3.48 | 3.69 | 3.74 | 3.20* | 3.69 | 3.75 |
| Percent Black | 2.53 | 2.55 | 2.56 | 2.53 | 2.54 | 2.55 | 2.70 | 2.88* | 2.83* |
| Percent Hispanic | 2.53 | 2.91 | 3.02 | 2.52 | 2.91 | 3.00 | 3.01* | 2.98 | 3.23* |
| Elementary school | 0.32 | 0.41 | 0.41 | 0.32 | 0.41 | 0.40 | 0.36 | 0.48 | 0.50* |
| Middle school | 0.29 | 0.27 | 0.26 | 0.29 | 0.26 | 0.26 | 0.32 | 0.34* | 0.28 |
| High school | 0.39 | 0.33 | 0.33 | 0.40 | 0.33 | 0.34 | 0.32 | 0.18* | 0.22* |
| N | 6350 | 13100 | 10330 | 6250 | 12720 | 9890 | 100 | 380 | 440 |

*Note.* Attend event, volunteer, general meeting, PTO (parent-teacher organization) meeting, PT (parent-teacher) conference, fundraising, committee, and guidance are binary variables; *number of activities* is a continuous variable; *satisfaction* is an index variable with a range from 1 = *very dissatisfied* to 4 = *very satisfied*. Observations rounded to the nearest 10 per Department of Education guidelines. Asterisks indicate statistically significant charter-district differences at the $p < .05$ level.
*Source.* Parent and Family Involvement surveys.

**Dependent Variables**

*Parent Engagement*

The first outcome examined in this article is parent engagement, the working together of parents and school staff to support and improve student learning, development, and health (Centers for Disease Control and Prevention, 2018). To study parent engagement, this article examines eight types of parent-school interactions drawn from a set of questions that asked parents:

> Since the beginning of this school year, has any adult in this child's household done any of the following things at this child's school? (1) attended a school or class event, such as a play, dance, sports event, or science fair; (2) served as a volunteer in this child's classroom or elsewhere in the school; (3) attended a general school meeting, for example, an open house, or back-to-school night; (4) attended a meeting of the parent-teacher organization or association; (5) gone to a regularly scheduled parent-teacher conference with this child's teacher; (6) participated in fundraising for the school; (7) served on a school committee; and (8) met with a guidance counselor in person.[3]

Using responses to these questions, the article examines the dichotomous variables—*attend event, volunteer, attend meeting, attend PTO (parent-teacher organization), PT (parent-teacher) conference, fundraising, committee,* and *guidance*—coded 0 if the response was "No" and 1 if the response was "Yes."

The second way in which it studies parent engagement is by examining the total number of activities reported by parents over a school year. Specifically, it draws from a question in which parents were asked, "During this school year, how many times has any adult in the household gone to meetings or participated in activities at this child's school?" In response, they could indicate a number from 0 to 99. Responses to this question were coded as indicated by parents and used to create the *number of activities* variable.

*Parent Satisfaction*

The article's second outcome of interest is parent satisfaction, the level of contentment that parents have with their child's education. Satisfaction is measured with a latent variable, *satisfaction*, drawn from five questions on the PFI. Specifically, parents were asked,

> How satisfied or dissatisfied are you with each of the following?: (1) the school this child attends this year; (2) the teachers this child has this year; (3) the academic standards of the school; (4) the order and

discipline at the school; and (5) the way that school staff interacts with parents.[4]

For each question, parents could respond, *very dissatisfied*, coded 1; *somewhat dissatisfied*, coded 2; *somewhat satisfied*, coded 3; or *very satisfied*, coded 4. The latent variable adds these responses and divides them by five (to maintain the 4-point scale) such that 1 is the minimum and 4 is the maximum. Across these years, the latent variable had an average scale reliability coefficient (Cronbach's alpha) of .88.

## Estimation and Models

The article's first goal is to isolate the relationship between school type (charter vs. district) and parent engagement and satisfaction. To do so, the article uses survey-weighted ordinary least squares (OLS) for the dependent variables *number of activities* and *satisfaction*, which are continuous, and survey-weighted logistic regression for the *attend event, volunteer, general meeting, PTO meeting, PT conference, fundraising, committee,* and *guidance* variables, which are dichotomous.[5] To test Hypotheses 1 and 2, the article estimates the following equation:

$$Y_i = \alpha_i + \beta_1 charter_i + \beta_2 X_i + \beta_3 Z_i + \varepsilon_i \tag{1}$$

where *i* refers to an individual parent and *Y* refers to the parent's reported engagement or satisfaction. *Charter*, the independent variable, is dichotomous and indicates whether the child's school is identified by the NCES as a charter school. *Z* refers to a set of dummy variables indicating the state in which each parent lived (state fixed effects).

*X* refers to a vector of school-level controls that prior research suggests have a meaningful relationship with the educational experiences of children and their parents (Epple et al., 2016; Wohlstetter et al., 2013). As such, including them in Equation 1 helps show how *charter* is related to the article's dependent variables. *Urban* indicates whether a school is located in an urban area; *pct. lunch* indicates the percentage of students in the school who qualified for free or reduced school lunch (coded such that 1 = less than 1%, 2 = 1% to 5%, 3 = 5% to 25%, and 4 = 25% or more); *pct. Black* indicates the percentage of students in the school identified as Black or African American (coded such that 1 = less than 1%, 2 = 1% to 5%, 3 = 5% to 25%, and 4 = 25% or more); *pct. Hispanic* indicates the percentage of students in the school identified as Hispanic (coded such that 1 = less than 1%, 2 = 1% to 5%, 3 = 5% to 25%, and 4 = 25% or more); *elementary school* and *middle school* measure the grades taught at the school according to the following scheme: Schools offering grades K–5 were coded as an *elementary school*, schools offering grades 6–8 were coded as a *middle school*. The variable *high school*, which denotes a school in which grades

9–12 were offered, serves as the reference category for these last two variables. Finally, this vector includes *nonassigned*, which is coded 1 if a district school is a magnet, vocational, continuing, alternative, special education, or other unassigned school, and 0 if it is a geographically assigned district school (or charter school). Because nonassigned schools often involve an admissions process of some sort, it is important to include this variable to distinguish geographically and nongeographically assigned district-operated schools.

After establishing whether there are differences in parent engagement and satisfaction, while controlling for school and state characteristics, the article then explores whether differences in parent engagement and satisfaction result from parent self-selection (Hypothesis 3). To examine this explanation, the article adds variables that measure family characteristics to its existing model and reports whether any associations between *charter* and parent engagement and satisfaction remain significant. More specifically, it examines the *charter* variable using Equation 2:

$$Y_i = \alpha_i + \beta_1 charter_i + \beta_2 W_i + \beta_3 X_i + \beta_4 Z_i + \varepsilon_i \tag{2}$$

This equation is the same as Equation 1 except that it adds *W*, a vector of family and student-level variables that help account for parent demographic and behavioral characteristics. First, it includes the variable *first choice*, created based on parent responses to the following question: "Is the school this child attends your first choice, that is, the school you wanted most for him/her to attend?"[6] This variable, coded 0 if parents answered "No" and 1 if parents answered "Yes," is useful as it identifies (1) district parents who have a child in a first-choice school, even if it is a geographically assigned school and (2) parents who selected a charter school that was not their most preferred school. Second, it includes the variable *considered other*, which captures parents' responses to the question: "Did you consider other schools for this child?" This variable—coded 0 if parents answered "No" and 1 if parents answered "Yes"—is helpful as it captures a parent behavior that is thought to distinguish some parents: education market savvy or a willingness to explore a variety of different educational options.

In addition, *W* includes a host of other variables that measure student and family differences and are thought to be associated with parents' school experiences (Child Trends, 2013; Smith et al., 2011; Wohlstetter et al., 2008, 2013). As with the school-level controls, the inclusion of these variables helps focus our understanding on how *charter* is related to the article's dependent variables. *Grades* indicates the parent's response when asked what grades the student received during the school year (coded from 1 = *mostly Ds and lower* to 4 = *mostly As*); *moved* indicates whether the family changed residences to attend the child's school; *suspended* captures whether the student had ever received an in- or out-of-school suspension; *Hispanic*,

Oberfield

*Asian, Black,* and *White* indicate the race and ethnicity of the child; *female* indicates the child's sex; *English at home* indicates whether the child predominantly speaks English at home; *parent education* indicates the level of education (1 = less than high school, 2 = high school or GED, 3 = some college or technical training, 4 = college graduate, and 5 = graduate or professional degree); *two parent* indicates whether the child lived in a two-parent household; *welfare* indicates whether the family received assistance from the following programs in the past year (TANF, state welfare program, WIC, food stamps, Medicaid, CHIP, and Section 8); *family income* indicates the household's total income in the past year; *neighborhood poverty* indicates the percentage of families with children under the poverty line in the family's zip code (coded such that 1 = 4% or less, 2 is 5% to 9%, 3 is 10% to 19%, and 4 is 20% and above); and *disability* indicates whether the child currently has a disability.[7]

## Findings

Table 1 presents the survey-weighted means for the article's dependent, independent, and control variables broken down by all respondents, district respondents, and charter respondents.[8]

The table shows that more charter parents indicated volunteering at their child's school and, in 2 of the 3 years, attending a parent-teacher conference and a PTO meeting. In addition, in 1 of the 3 years, more charter parents attended a school event. However, there were no statistically significant charter-district differences (at the $p < .05$ level) in *general meeting, fundraising, committee, guidance,* or *number of activities* in any of the years examined here.

The *satisfaction* means tell a different story: In each year, charter parents reported higher levels of *satisfaction* than district parents. However, we also see that, as the charter sector expanded from 2007 to 2016, there was an uptick in district parent satisfaction and a downtick in charter parent satisfaction. If these trends continue, we may see a convergence in district and charter parent satisfaction in the near future.

A look at the control variables in Table 1 reveals important differences between the families that attend charter and district schools. For example, in each year examined here, over half of the charter parents sampled indicated that they had considered other schools before choosing their child's current school; in contrast, the percentage of district parents who shopped around never went above 29%. This suggests an important way that district and charter parents differ, outside the standard socioeconomic characteristics that scholars attempt to control for. Nonetheless, the control data also highlight the importance of these more traditional control variables. For example, more charter families lived in neighborhoods with higher levels of poverty, received welfare, and identified as Hispanic or Black. We also

## Table 2
**Parent Engagement and Satisfaction: Subgroup Analysis (Pooled)**

|  | White | Non-White | Urban | Nonurban | College | Noncollege |
|---|---|---|---|---|---|---|
| District school: Number of activities | 8.06 | 5.40 | 6.44 | 7.67 | 8.97 | 6.27 |
| Charter school: Number of activities | 9.02* | 5.85 | 6.84 | 8.77* | 9.57 | 6.21 |
| District school: Satisfaction | 3.45 | 3.38 | 3.41 | 3.44 | 3.49 | 3.40 |
| Charter school: Satisfaction | 3.65* | 3.50* | 3.57* | 3.61* | 3.64* | 3.55* |

*Note. Number of activities* is a continuous variable; *satisfaction* is an index variable with a range from 1 = *very dissatisfied* to 4 = *very satisfied*. Asterisks indicate statistically significant charter-district differences at the $p < .05$ level (within a column). See the text for information about statistically significant differences between subgroups within a school type (i.e., across rows).
*Source.* Parent and Family Involvement surveys.

see some differences in school characteristics. In particular, charter schools were more likely to be located in urban areas and have higher percentages of Black and Hispanic students.

To dig deeper into these descriptive statistics, we now turn to a subgroup analysis of parent engagement and satisfaction. Specifically, Table 2 presents findings from an analysis that pooled all respondents over these 3 years and then calculated the *number of activities* and *satisfaction* for subgroups within and between these schools.[9]

The top row focuses on parent engagement in district schools. Starting on the left we see that White district parents reported engaging in approximately two-and-a-half more school activities per year than non-White district parents. We also see that nonurban district parents reported around one more school activity per year (relative to urban district parents), and college-educated district parents reported over two more school activities per year (relative to non–college-educated district parents). All of these differences were statistically significant at the $p < .05$ level. The second row, which looks among charter school parents, reveals different point estimates but similar trends. As with the district school point estimates, all of these differences were statistically significant at the $p < .05$ level.

The third row examines satisfaction in district schools. It shows that White, nonurban, and college-educated district parents were, respectively, 0.07, 0.03, and 0.09 of a point more satisfied with their child's school relative to non-White, urban, and non–college-educated district parents. All of these differences were statistically significant at the $p < .05$ level but were not large enough to shift respondents one category on the 4-point *satisfaction* scale. The fourth row looks at satisfaction among subgroups within charter schools. The difference in satisfaction between White and non-White charter parents was statistically significant; however, there were no statistically significant differences in satisfaction between urban and nonurban charter parents and college-educated and non–college-educated charter parents.

Next, we look within subgroups across different school settings; statistically significant charter-district differences (at the $p < .05$ level) are denoted with an asterisk. The first column from the left shows higher levels of engagement among White parents with a child in a charter school relative to White parents with a child in a district school. The fourth column from the left shows higher levels of engagement among nonurban charter parents. However, in the other categories (non-White, urban, non-college educated, and college educated), there were no statistically significant differences in engagement between parents in charter and district schools.

Finally, the table shows that charter parents reported higher levels of satisfaction within all subgroups. For example, White charter parents were 0.20 of a point more satisfied than White district parents, and non-White charter parents were 0.12 of a point more satisfied than non-White district parents. Though it is difficult to know how to think of these differences substantively, it is worth pointing out that none were large enough to shift a respondent from one category to another on the 4-point satisfaction scale. At the same time, it is noteworthy that the average non-White charter school parent was more satisfied than the average White district school parent; similarly, the average noncollege charter parent was more satisfied with their child's school than the average college district parent.

The examination of parent engagement and satisfaction in Tables 1 and 2 is helpful, but it is important to recall that it does not account for school-level differences, including the racial and class demographics of students attending the school. Thus, we now turn to a multivariate analysis (Equation 1) that examines whether parents with a child in a charter school reported higher levels of engagement and satisfaction, relative to parents with a child in a district school, when controlling for school-level differences. One important point as readers seek to interpret these findings: Equation 1 includes a variable (*nonassigned*), which distinguishes between children who were in a geographically assigned school and those that were not. Therefore, when the below text alludes to charter-district differences, it is referring to differences between parents with a child in a charter school and parents with a child in an assigned district school.

Figure 1 displays the findings from the analysis of the eight types of parent engagement measured on the PFI (for full results, see Appendix Table A.1). Because logistic regression coefficients are difficult to interpret, the figure presents odds ratios; these ratios can be interpreted as gauging the likelihood of an engagement difference between charter and district parents, controlling for all other model variables. The lines extending from the

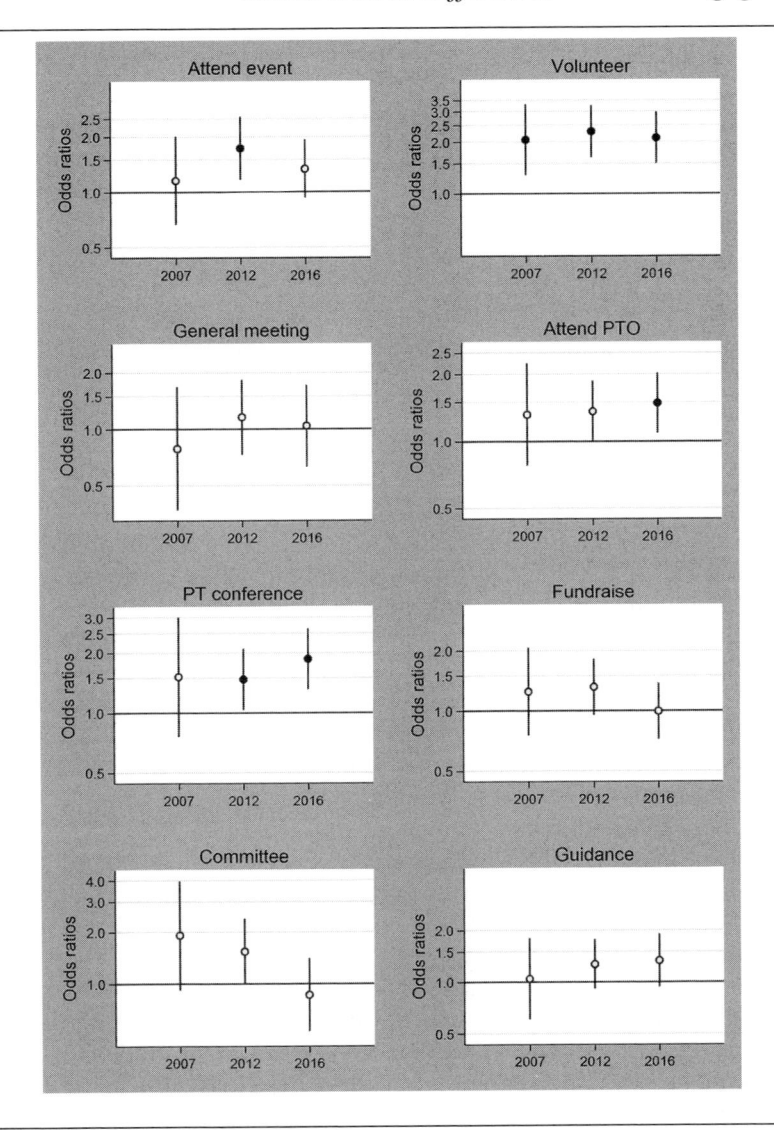

*Figure 1.* **Comparing charter parents with district parents: Differences in types of parent engagement (*without* self-selection controls).**
*Note.* Circles indicate odds ratios—indicating the likelihood of a difference between charter and district parents—derived from logistic regression coefficients; darkened circles denote statistical significance at the $p < .05$ level.
*Source.* Parent and Family Involvement surveys.

markers indicate the 95% confidence intervals, and darkened markers indicate that a difference was statistically significant at the $p < .05$ level.

Figure 1 reveals that in all three years, charter parents were more likely to volunteer in their child's school. It also shows that in two of the three years, charter parents were more likely to attend a parent-teacher conference. In one year, charter parents were more likely to attend a school event (2012) and attend a meeting of the PTO (2016). However, in half of the areas of engagement (*general meeting, fundraising, committee*, and *guidance*), there was no evidence of charter-district differences.

Figure 2 shows charter-district differences in the aggregate number of school activities engaged in by parents (top panel) and differences in satisfaction (bottom panel). Because the variables *number of activities* and *satisfaction* are continuous, the figure presents regression coefficients (for full results see Appendix Table A.1). As with Figure 1, darkened markers indicate that a charter-district difference was statistically significant at the $p < .05$ level.

The top panel shows that in 2012 charter parents reported engaging in approximately one-and-a-half more school activities relative to district parents. To put this in context, this difference is around half of the engagement gap between college and noncollege district parents and White and non-White district parents (seen in Table 2). In the other 2 years examined in Figure 2, there were no differences in the number of activities reported by district and charter parents.[10] The bottom panel reveals that, in each year examined here, charter parents reported higher levels of satisfaction relative to district parents. Specifically, charter parents were, on average, 0.18 of a point more satisfied.

How meaningful is this satisfaction difference? On the one hand, this difference is not enough to shift a respondent from one category to another on the 4-point *satisfaction* scale. However, in Table 2 we saw that non-White district parents reported 0.07 of a point less *satisfaction* than White parents and that non-college-educated district school parents reported 0.09 of a point less satisfaction than college-educated district school parents. Thus, the average charter-district *satisfaction* difference in Figure 2 is around twice as large as the average White-non-White and college–non–college-educated district satisfaction gaps.

These gains are impressive, but it is important to recall that they do not account for any of the family-level differences, identified in Table 1, between charter and district school parents. Thus, we now shift to consider how the inclusion of these variables alters our view of any charter-district parent engagement and satisfaction differences. To do so, Figures 3 and 4 were generated using the model shown in Equation 2. The analytic strategy here was to (1) add the variables measuring family-level differences to the prior model (Equation 1) and (2) determine if charter-district differences in Figures 1 and

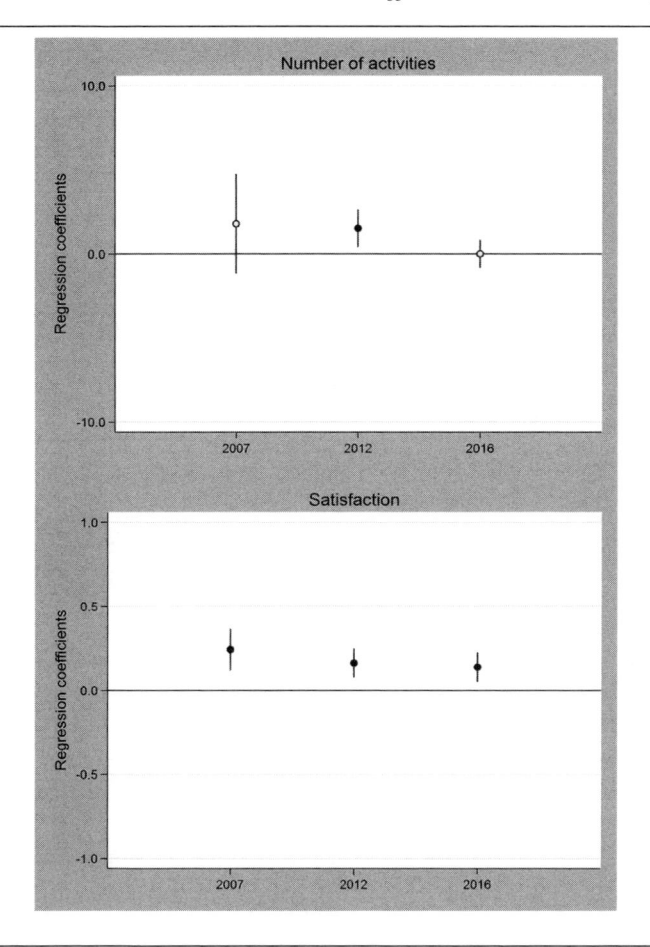

*Figure 2.* **Comparing charter parents with district parents: Differences in the number of parent engagement activities and satisfaction (*without* self-selection controls).**
*Note.* Circles indicate regression coefficients; darkened circles denote statistical significance at the $p < .05$ level.
*Source.* Parent and Family Involvement surveys.

2 persist. Full results from these analyses can be found in Appendix Tables A.2-A.4.

Figure 3 shows that even when controlling for the observable ways in which these two populations differed, charter school parents were more likely to report volunteering. Adding these family-level controls did not change the finding that charter parents were more likely to attend a school

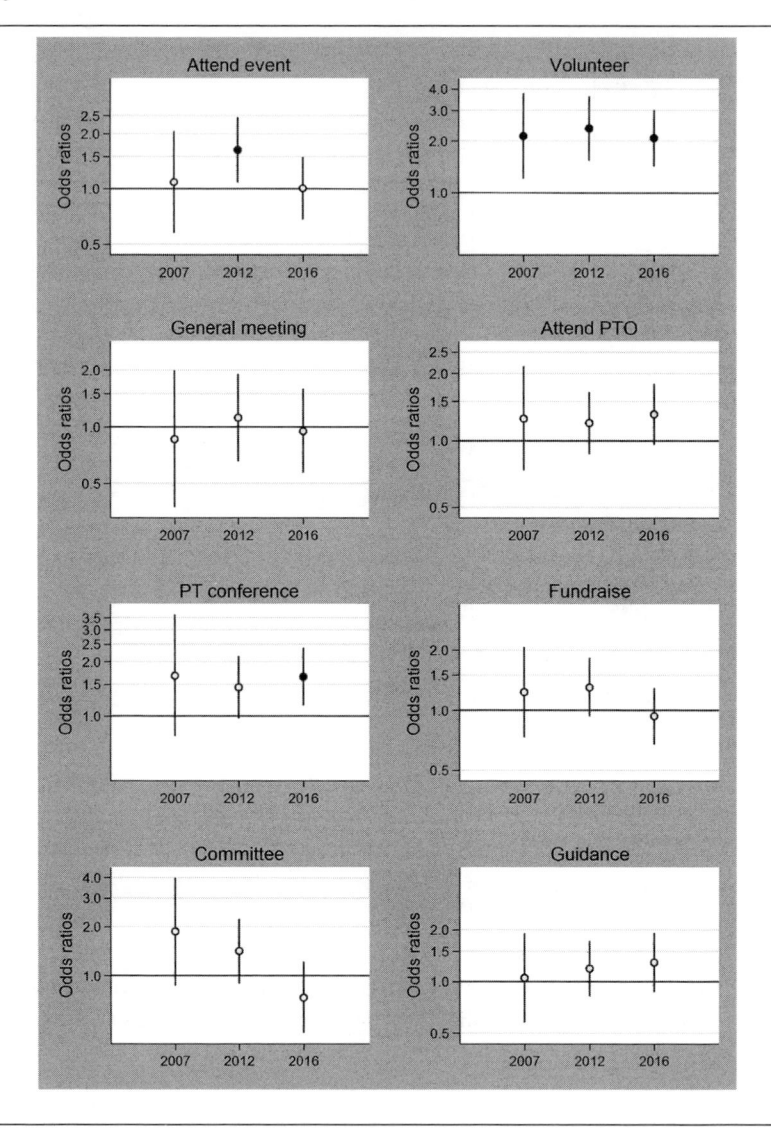

*Figure 3.* **Comparing charter parents with district parents: Differences in types of parent engagement (*with* self-selection controls).**
*Note.* Circles indicate odds ratios—indicating the likelihood of a difference between charter and district parents—derived from logistic regression coefficients; darkened circles denote statistical significance at the $p < .05$ level.
*Source.* Parent and Family Involvement surveys.

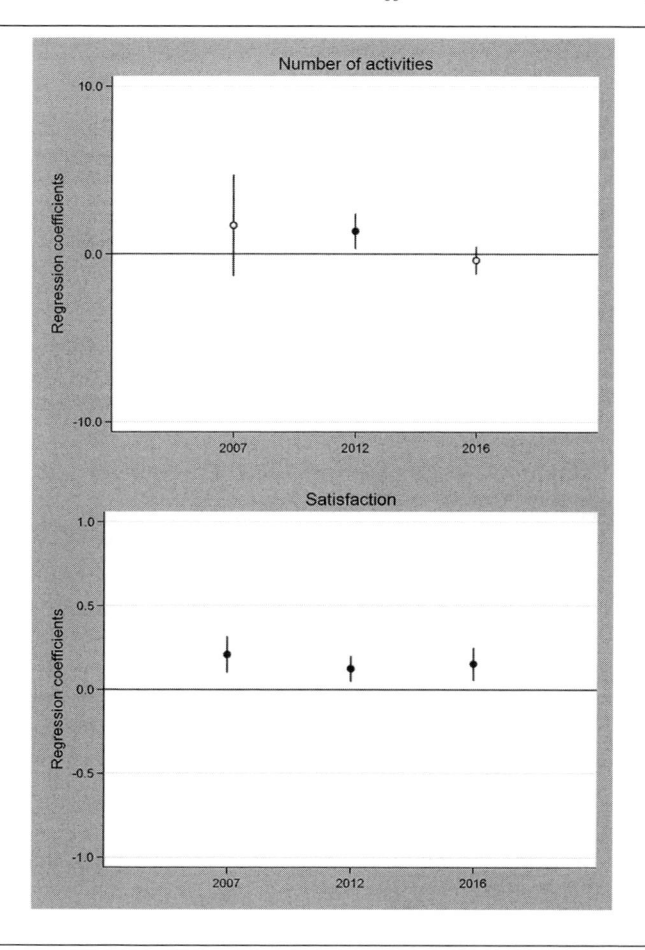

*Figure 4.* **Comparing charter parents with district parents: Differences in the number of parent engagement activities and satisfaction (*with* self-selection controls).**

*Note.* Circles indicate regression coefficients; darkened circles denote statistical significance at the $p < .05$ level.
*Source.* Parent and Family Involvement surveys.

event, in 2012, or a parent-teacher conference, in 2016. However, the inclusion of these controls did render the difference in PTO meeting attendance, in 2016, and a parent-teacher conference, in 2012, statistically insignificant. As in Figure 1, there were no significant differences between charter and district parents regarding meeting attendance, fundraising, committee work, or guidance counselor meeting.

*Oberfield*

Turning to the examination of *number of activities* and *satisfaction*, we see that the findings in Figure 4 largely match those of Figure 2. Nevertheless, the inclusion of the family-level variables reduced the size of the *number of activities* coefficient from 1.5 to 1.4; similarly, the inclusion of these variables reduced the size of the average *satisfaction* coefficient from 0.18 to 0.16. As such, this analysis suggests that family characteristics did not play a large role in driving the findings in Figure 2. In other words, the self-selection characteristics measured here did not appear to be the sole reason for the charter-district gap in parent engagement and satisfaction.

## Discussion

As readers reckon with these findings, it is important to consider a few of the article's limitations. First, its engagement measures rely on self-reported behaviors, which, due to various survey response biases (Podsakoff, MacKenzie, Lee, & Podsakoff, 2003), may not reflect reality. Second, although the article tries to account for self-selection, its independent variable was not randomly assigned. Therefore, it cannot say for sure whether any associations between *charter* and parent engagement and satisfaction result from self-selection, what schools are doing, or other unobserved but nonrandom factors. Third, although the surveys in each year include thousands of respondents, the response rate in 2007 was quite a bit lower relative to other years. This may be responsible for some of the year-to-year volatility in the article's engagement findings. Finally, although it looks over a 10-year period, the article relies on cross-sectional data. As such, it cannot track individual-level changes over time.

These are important limitations. However, by using nationally representative parent survey data over a 10-year period, this article makes an important contribution to the literature: It provides a temporal comparison of the experiences of parents with a child in charter school with the experiences of parents with a child in a geographically assigned district school. This section discusses the findings here and considers their implications.

To begin, the article showed that parent volunteering occurred more frequently at charter schools. This difference persisted even controlling for observed self-selection characteristics. On its face, this is a positive finding for charter schools: Parent volunteering saves money and builds a stronger school community. However, it is possible that differences in volunteering stemmed from higher rates of mandated parent volunteering in charter schools (Mommandi & Welner, 2018; Nathan, 1996; Smith et al., 2011; Weiler & Vogel, 2015; Wells, 2002). Because wealthier families have more time to spend on volunteering, parent volunteering contracts may discourage applications from—or push out—families with lower means (Hammel, 2014; Kahlenberg & Potter, 2014). Put differently, parent-volunteering contracts, if they are more common in charter schools, could generate selection differences between charter and district school parents. Unfortunately, there was no recent data about the existence of parent volunteering contracts in charter and district schools,[11] and the PFI did not ask parents about whether they had entered into a parent-volunteering contract with their child's school. As a result, it is not clear whether this gain in volunteering indicates selection differences or something that charter schools are doing to encourage parents to get more involved in their children's schools.

Aside from volunteering, the article found that—in most ways and in most years—charter and district parents were equally engaged in their children's schools. Though it is impossible to fully explain a nonfinding, it is worth asking why there were not higher levels of parent engagement in charter schools. One explanation is that there is significant state-to-state variation in parent engagement, related to what state charter laws say about parent involvement (Wohlstetter et al., 2013). As noted earlier, some states have gone to greater lengths to make parent engagement a core part of their approach to charter schools. If this leads to national variation—wherein some states foster charter schools with greater emphasis on parent engagement than others—nationally there may be no uniform trend. Put differently, this null finding could be driven by charter policy differences across the states.

Alternatively, perhaps this nonfinding results from the vast diversity of the charter sector and the mission-driven nature of these schools (Carpenter, 2006). In other words, because there are so many different kinds of charter schools, and some are driven to pursue missions that do not place parent engagement at their center, there would be no meaningful national difference with district schools. If this is true, it means that the charter label, on its own, tells us little about how parents are likely to engage with their child's school.

The article's analysis of parent satisfaction revealed a clearer picture. Across all 3 years, charter parents reported higher levels of satisfaction with their child's school relative to district parents. These differences persisted even when controlling for variables that measured observable differences in family demographics and behavior, such as the consideration by parents of other schools prior to enrollment. To put the size of these gains in context, they were much larger than the non-White–White and college–non-college district school parent satisfaction gaps. It is notable that these satisfaction findings align with Gleason et al. (2010) and Tuttle et al. (2013), which compared the satisfaction of charter school lottery winners and losers, and Barrows et al. (2017), which examined two nationally representative parent surveys. As such, the evidence is accumulating that charter schools are associated with higher levels of parent satisfaction.

Although we cannot know for sure why these differences are manifest, it is worth revisiting some of the theories considered in the early parts of this article. Drawing inspiration from the larger privatization movement, charter school theorists and advocates suggested that these schools would have a greater incentive to treat parents as "customers" and ensure that they are satisfied (Kolderie, 1990). As such, perhaps these gains in satisfaction result from steps that charter schools are taking to ensure that parents and children are having positive schooling experiences. In fact, some in-depth, qualitative work suggests that charter schools are devising innovative strategies to connect with parents and create the conditions for student success (Kahlenberg & Potter, 2014; Wohlstetter et al., 2013).

However, it is also possible that these satisfaction differences are artifacts of some unmeasured aspects of self-selection or that they are inherent to the process of school selection—perhaps winning a lottery or choosing a school, outside a parent's and student's experiences with the school, conditions a positive feeling. Supporting this possibility, the appendix tables reveal that parents with a child in a nonassigned district school also had higher levels of satisfaction relative to parents in a geographically assigned district school. Because the variable includes a diverse set of schools (see the Data and Method section for a list of possible schools captured by this variable), it is difficult to know what is driving these relationships. Still, it is important to note that parents with a child in a charter or nonassigned school—both of which indicate departures from the traditional, geographically assigned school—reported higher levels of satisfaction.

## Conclusion

Advocates and critics have opined about what effects charter schools have on parents. However, relatively little is known about parent-reported differences in engagement and satisfaction in charter and district schools. This article contributes by examining these claims empirically over a number of years in the first two decades of the 21st century. In short, its findings suggest that charter schools may be making real gains in satisfaction even if parents are not more engaged in their children's schools.

These findings raise a number of questions and topics for future study. To start, prior research suggests that teachers and principals in charter schools report higher levels of parent engagement. However, this article was unable to produce much evidence that parents in charter schools report a more diverse array, or higher levels, of engagement. Thus, further research is needed to understand the discrepancy between charter teachers', principals', and parents' experiences with parent engagement.

Another area for future inquiry relates to the contours of the charter sector: Although educational management organizations have centralized operations in a variety of cities, the charter sector remains incredibly diverse. Thus, it will be important for future work to look at parent engagement and satisfaction within the charter sector. Do educational management organizations foster higher or lower levels of parent engagement and satisfaction? What about nonprofit versus for-profit charter schools? There has been some compelling, geographically-limited work in this area (Hamlin,

2017); however, moving forward, we will need nationally representative research examining what explains differential levels of engagement and satisfaction inside the charter sector. To spur this research along, it would be helpful if the NCES collected, maintained, and reported information about the profit and franchise status of the nation's charter schools in its Common Core of Data. It would also be helpful if the NCES included a question about the existence of parent volunteering contracts on the PFI.

In addition, we need further research on why we are observing higher levels of parent satisfaction in charter schools. In particular, it would be helpful to better explore parents' experiences with school leaders and teachers. School leaders are thought to have one of the largest impacts on school success (Kelley, Thornton, & Daugherty, 2005), and teachers are thought to have a major effect on student learning (Hanushek & Rivkin, 2006). Although charter proponents suggest that school leaders and teachers operate in a significantly different environment relative to their district counterparts, differences with district schools are contingent and mixed (Oberfield, 2017). Future work could contribute by comparing how district and charter parents experience the teachers and leaders who run their child's school and how this is connected to their engagement and satisfaction.

More specifically, there are signs that charter schools are pioneering advances in school communication (Barrows et al., 2017; Mintrom, 2003; Oberfield, 2017; Smith et al., 2011). In particular, some charter schools are finding innovative approaches to contacting hard-to-reach parents and increasing the frequency of school-parent communication. These advances are important because school communication is theorized to affect parent engagement, parent-student interactions, and student achievement (Kraft & Rogers, 2015; Peterson, 2009). Specifically, Epstein (1995) argues that when schools conduct high-quality communications, they get communities, schools, and parents to interact and collaborate more closely. As this process plays out, student learning is enhanced, and parents have easier, more rewarding interactions with their child's schools. As such, future research may advance our understanding of the charter-district satisfaction gap by looking at the role played by school communication.

Finally, aside from suggesting new avenues for research, this article's results have implications for the tenor of the nationwide charter school debate, which often devolves into caricature and hardline position taking. Despite the heated rhetoric and the polarized nature of the debate, whether we look at in-school processes or student outcomes, the evidence suggesting that charter schools are different or better than district schools is mixed. This article's findings mirror and contribute to this larger literature. As these results accumulate, perhaps they can encourage policymakers and stakeholders to ratchet down the rhetoric and engage in more generative conversations. In doing so, we can deepen our understanding of how charter and district schools compare and what they can learn from one another.

Table A.1
Charter-District Differences in Types of Parent Engagement and Satisfaction (Without Self-Selection Controls)

|  | Attend Event |  |  | Volunteer |  |  | Attend Meeting |  |  | Attend PTO |  |  |
| --- | --- | --- | --- | --- | --- | --- | --- | --- | --- | --- | --- | --- |
|  | 2007 | 2012 | 2016 | 2007 | 2012 | 2016 | 2007 | 2012 | 2016 | 2007 | 2012 | 2016 |
| Charter | 1.16 | 1.73* | 1.33 | 2.06* | 2.30* | 2.12* | 0.79 | 1.16 | 1.04 | 1.32 | 1.36 | 1.48* |
|  | (0.33) | (0.35) | (0.25) | (0.50) | (0.41) | (0.37) | (0.30) | (0.27) | (0.27) | (0.36) | (0.22) | (0.23) |
| Nonassigned | 1.13 | 1.57* | 1.13 | 1.23 | 1.78* | 1.27* | 0.98 | 1.52* | 1.34* | 1.29* | 1.12 | 1.14 |
|  | (0.14) | (0.16) | (0.13) | (0.14) | (0.16) | (0.12) | (0.16) | (0.18) | (0.17) | (0.14) | (0.10) | (0.11) |
| Urban | 0.78* | 0.76* | 0.78* | 0.75* | 0.76* | 0.94 | 0.91 | 0.81* | 0.69* | 1.08 | 1.26* | 1.04 |
|  | (0.08) | (0.06) | (0.07) | (0.07) | (0.05) | (0.08) | (0.11) | (0.07) | (0.08) | (0.10) | (0.08) | (0.08) |
| Percent lunch | 0.79* | 0.69* | 0.63* | 0.78* | 0.69* | 0.60* | 0.65* | 0.59* | 0.58* | 0.91 | 0.97 | 0.88* |
|  | (0.05) | (0.05) | (0.05) | (0.04) | (0.03) | (0.04) | (0.05) | (0.04) | (0.05) | (0.05) | (0.04) | (0.05) |
| Percent Black | 0.95 | 0.84* | 0.85* | 0.90* | 0.87* | 0.87* | 1.12 | 0.96 | 1.03 | 1.07 | 1.02 | 1.15* |
|  | (0.05) | (0.03) | (0.04) | (0.04) | (0.03) | (0.04) | (0.07) | (0.04) | (0.08) | (0.05) | (0.03) | (0.05) |
| Percent Hispanic | 0.88* | 0.85* | 0.77* | 0.83* | 0.83* | 0.93 | 1.00 | 0.96 | 0.93 | 0.99 | 1.06 | 1.11 |
|  | (0.05) | (0.05) | (0.05) | (0.04) | (0.03) | (0.05) | (0.07) | (0.06) | (0.07) | (0.05) | (0.04) | (0.06) |
| Elementary | 2.50* | 2.10* | 2.01* | 3.28* | 3.02* | 2.70* | 2.96* | 2.72* | 2.20* | 2.05* | 2.15* | 2.02* |
|  | (0.26) | (0.16) | (0.19) | (0.30) | (0.19) | (0.20) | (0.37) | (0.22) | (0.24) | (0.18) | (0.13) | (0.15) |
| Middle | 1.33* | 1.25* | 1.18 | 1.26* | 1.21* | 1.05 | 1.84* | 1.83* | 1.97* | 1.50* | 1.24* | 1.34* |
|  | (0.13) | (0.09) | (0.10) | (0.12) | (0.08) | (0.08) | (0.24) | (0.16) | (0.21) | (0.13) | (0.08) | (0.10) |
| N | 6340 | 13080 | 10290 | 6340 | 13100 | 10330 | 6340 | 13080 | 10330 | 6340 | 13100 | 10330 |

*(continued)*

Table A.1 (continued)

|  | PT Conference | | | Fundraising | | | Committee | | | Guidance Meeting | | |
|---|---|---|---|---|---|---|---|---|---|---|---|---|
|  | 2007 | 2012 | 2016 | 2007 | 2012 | 2016 | 2007 | 2012 | 2016 | 2007 | 2012 | 2016 |
| Charter | 1.51 | 1.47* | 1.86* | 1.25 | 1.32 | 1.00 | 1.91 | 1.54 | 0.86 | 1.05 | 1.27 | 1.33 |
|  | (0.53) | (0.27) | (0.33) | (0.32) | (0.22) | (0.17) | (0.71) | (0.34) | (0.21) | (0.29) | (0.21) | (0.24) |
| Nonassigned | 1.34* | 1.33* | 1.21 | 1.00 | 1.39* | 1.20 | 1.31* | 1.51* | 1.13 | 1.04 | 1.23* | 1.01 |
|  | (0.17) | (0.13) | (0.13) | (0.11) | (0.12) | (0.12) | (0.18) | (0.18) | (0.15) | (0.11) | (0.11) | (0.10) |
| Urban | 1.07 | 0.91 | 0.91 | 0.76* | 0.71* | 0.67* | 0.87 | 0.87 | 1.11 | 1.13 | 1.00 | 1.05 |
|  | (0.11) | (0.07) | (0.08) | (0.07) | (0.05) | (0.05) | (0.10) | (0.08) | (0.13) | (0.10) | (0.07) | (0.09) |
| Percent lunch | 1.01 | 0.99 | 0.90 | 0.84* | 0.81* | 0.71* | 0.80* | 0.77* | 0.76* | 0.94 | 1.03 | 1.00 |
|  | (0.07) | (0.05) | (0.05) | (0.05) | (0.04) | (0.04) | (0.05) | (0.04) | (0.06) | (0.05) | (0.05) | (0.06) |
| Percent Black | 1.00 | 1.00 | 1.11* | 0.94 | 0.86* | 0.88* | 0.94 | 0.92 | 0.90 | 1.19* | 1.12* | 1.06 |
|  | (0.05) | (0.04) | (0.05) | (0.04) | (0.03) | (0.04) | (0.05) | (0.04) | (0.06) | (0.06) | (0.04) | (0.05) |
| Percent Hispanic | 1.14* | 0.96 | 1.09 | 0.83* | 0.86* | 0.86* | 0.85* | 0.92 | 0.94 | 1.06 | 1.02 | 1.00 |
|  | (0.07) | (0.05) | (0.07) | (0.05) | (0.04) | (0.05) | (0.06) | (0.05) | (0.09) | (0.06) | (0.05) | (0.05) |
| Elementary | 6.00* | 6.10* | 7.31* | 1.94* | 2.49* | 1.88* | 1.16 | 1.38* | 1.12 | 0.28* | 0.38* | 0.40* |
|  | (0.76) | (0.50) | (0.72) | (0.18) | (0.16) | (0.14) | (0.13) | (0.12) | (0.12) | (0.03) | (0.02) | (0.03) |
| Middle | 2.16* | 1.85* | 2.08* | 1.26* | 1.31* | 1.13 | 0.83 | 1.02 | 0.90 | 0.57* | 0.62* | 0.57* |
|  | (0.21) | (0.12) | (0.17) | (0.12) | (0.08) | (0.08) | (0.10) | (0.10) | (0.11) | (0.05) | (0.04) | (0.04) |
| N | 6340 | 13100 | 10330 | 6340 | 13100 | 10330 | 6320 | 13100 | 10320 | 6340 | 13100 | 10330 |

*(continued)*

Table A.1 (continued)

|  | Number of activities ||| Satisfaction |||
|---|---|---|---|---|---|---|
|  | 2007 | 2012 | 2016 | 2007 | 2012 | 2016 |
| Charter | 1.80 | 1.51* | 0.01 | 0.24* | 0.16* | 0.14* |
|  | (1.51) | (0.57) | (0.43) | (0.06) | (0.04) | (0.04) |
| Nonassigned | 0.40 | 1.64* | 0.55 | 0.14* | 0.07* | 0.09* |
|  | (0.50) | (0.35) | (0.32) | (0.03) | (0.03) | (0.03) |
| Urban | −1.34* | −0.71* | −0.42 | −0.04 | 0.01 | −0.01 |
|  | (0.35) | (0.19) | (0.25) | (0.03) | (0.02) | (0.02) |
| Percent lunch | −0.83* | −0.49* | −0.53* | −0.06* | −0.12* | −0.11* |
|  | (0.24) | (0.17) | (0.18) | (0.02) | (0.01) | (0.02) |
| Percent Black | −0.61* | −0.60* | −0.42* | −0.06* | −0.05* | −0.06* |
|  | (0.18) | (0.11) | (0.14) | (0.01) | (0.01) | (0.01) |
| Percent Hispanic | −0.98* | −0.83* | −0.55* | −0.01 | 0.03* | 0.01 |
|  | (0.24) | (0.16) | (0.16) | (0.01) | (0.01) | (0.02) |
| Elementary | −0.65 | −0.13 | −0.46 | 0.24* | 0.25* | 0.21* |
|  | (0.42) | (0.21) | (0.25) | (0.03) | (0.02) | (0.02) |
| Middle | −1.65* | −1.10* | −1.22* | 0.08* | 0.07* | 0.04 |
|  | (0.40) | (0.21) | (0.26) | (0.03) | (0.02) | (0.02) |
| N | 6350 | 13100 | 10330 | 6350 | 13100 | 10330 |

*Note.* PTO = parent-teacher organization. Cells for "Attend event," "Volunteer," "Attend meeting," "Attend PTO," "PT conference," "Fundraising," "Committee," and "Guidance meeting" report odds ratios. Cells for "Number of activities" and "Satisfaction" report regression coefficients. Standard errors in parentheses. Observations rounded to the nearest 10 per Department of Education guidelines; estimates include state fixed effects. Asterisks denote statistical significance at the $p < .05$ level.
*Source.* Parent and Family Involvement surveys.

Table A.2
**Charter-District Differences in Types of Parent Engagement (With Self-Selection Controls)**

|  | Attend Event 2007 | Attend Event 2012 | Attend Event 2016 | Volunteer 2007 | Volunteer 2012 | Volunteer 2016 | Attend Meeting 2007 | Attend Meeting 2012 | Attend Meeting 2016 | Attend PTO 2007 | Attend PTO 2012 | Attend PTO 2016 |
|---|---|---|---|---|---|---|---|---|---|---|---|---|
| Charter | 1.09 (0.36) | 1.63* (0.34) | 1.01 (0.20) | 2.13* (0.62) | 2.36* (0.52) | 2.08* (0.40) | 0.86 (0.37) | 1.12 (0.30) | 0.95 (0.25) | 1.26 (0.35) | 1.20 (0.20) | 1.31 (0.21) |
| Nonassigned | 1.01 (0.14) | 1.23 (0.14) | 0.89 (0.11) | 1.12 (0.13) | 1.44* (0.14) | 1.12 (0.12) | 0.90 (0.15) | 1.26 (0.16) | 1.21 (0.16) | 1.23 (0.14) | 1.03 (0.10) | 1.09 (0.10) |
| First choice | 1.36* (0.16) | 1.39* (0.11) | 1.40* (0.14) | 1.28* (0.15) | 1.12 (0.08) | 1.29* (0.13) | 1.02 (0.15) | 1.23* (0.11) | 1.01 (0.12) | 1.35* (0.15) | 1.23* (0.08) | 1.08 (0.10) |
| Considered other | 1.25* (0.13) | 1.31* (0.10) | 1.48* (0.15) | 1.39* (0.13) | 1.29* (0.09) | 1.33* (0.11) | 1.35* (0.18) | 1.22* (0.11) | 1.09 (0.12) | 1.31* (0.12) | 1.27* (0.08) | 1.25* (0.09) |
| Grades | 1.41* (0.08) | 1.35* (0.06) | 1.38* (0.07) | 1.41* (0.08) | 1.34* (0.05) | 1.31* (0.07) | 1.34* (0.08) | 1.25* (0.06) | 1.32* (0.08) | 1.17* (0.06) | 1.01 (0.04) | 1.07 (0.05) |
| Moved | 1.02 (0.10) | 0.98 (0.09) | 0.87 (0.09) | 1.28* (0.11) | 1.14 (0.08) | 1.15 (0.09) | 1.35* (0.19) | 1.01 (0.11) | 0.98 (0.12) | 1.30* (0.11) | 1.13 (0.08) | 1.23* (0.09) |
| Suspended | 0.89 (0.11) | 0.94 (0.09) | 0.78 (0.10) | 0.90 (0.12) | 1.00 (0.10) | 0.77 (0.10) | 0.85 (0.13) | 0.88 (0.09) | 0.82 (0.12) | 0.97 (0.12) | 0.93 (0.08) | 1.09 (0.12) |
| Hispanic | 0.92 (0.12) | 0.88 (0.09) | 0.98 (0.12) | 0.89 (0.11) | 0.75* (0.07) | 0.80* (0.08) | 0.97 (0.18) | 1.08 (0.12) | 0.92 (0.13) | 1.18 (0.15) | 1.07 (0.09) | 1.11 (0.11) |
| Asian | 0.66 (0.15) | 0.56* (0.08) | 0.69* (0.13) | 0.77 (0.17) | 0.53* (0.08) | 0.71* (0.10) | 0.92 (0.31) | 0.53* (0.08) | 0.46* (0.08) | 0.94 (0.20) | 0.93 (0.11) | 0.64* (0.09) |
| Black | 1.08 (0.19) | 1.30* (0.16) | 1.08 (0.16) | 1.04 (0.18) | 0.91 (0.11) | 0.91 (0.14) | 1.31 (0.28) | 1.19 (0.17) | 1.11 (0.18) | 1.34 (0.22) | 1.08 (0.11) | 1.23 (0.16) |
| White | 0.91 (0.13) | 1.33* (0.13) | 1.15 (0.15) | 1.08 (0.16) | 1.13 (0.12) | 1.00 (0.12) | 0.79 (0.14) | 1.26* (0.14) | 1.16 (0.16) | 0.91 (0.12) | 0.79* (0.07) | 0.62* (0.07) |

*(continued)*

Table A.2 (continued)

|  | Attend Event ||| Volunteer ||| Attend Meeting ||| Attend PTO |||
| --- | --- | --- | --- | --- | --- | --- | --- | --- | --- | --- | --- | --- |
|  | 2007 | 2012 | 2016 | 2007 | 2012 | 2016 | 2007 | 2012 | 2016 | 2007 | 2012 | 2016 |
| Female | 1.39* | 1.24* | 1.20* | 1.20* | 1.10 | 1.09 | 0.90 | 1.00 | 1.10 | 1.02 | 0.95 | 1.03 |
|  | (0.12) | (0.08) | (0.10) | (0.09) | (0.06) | (0.07) | (0.10) | (0.07) | (0.10) | (0.08) | (0.05) | (0.06) |
| English home | 1.66* | 1.32* | 1.49* | 2.62* | 1.52* | 1.42* | 1.26 | 1.19 | 1.19 | 0.85 | 0.70* | 0.82 |
|  | (0.26) | (0.14) | (0.19) | (0.48) | (0.17) | (0.18) | (0.26) | (0.14) | (0.18) | (0.13) | (0.07) | (0.09) |
| Parent education | 1.22* | 1.42* | 1.32* | 1.13* | 1.27* | 1.27* | 1.26* | 1.39* | 1.37* | 0.98 | 1.07* | 1.03 |
|  | (0.05) | (0.05) | (0.05) | (0.04) | (0.04) | (0.05) | (0.08) | (0.05) | (0.06) | (0.04) | (0.03) | (0.03) |
| Two-parent | 1.21 | 1.41* | 1.07 | 1.54* | 1.35* | 1.21* | 1.21 | 1.24* | 1.29* | 1.23* | 0.99 | 1.45* |
|  | (0.13) | (0.11) | (0.10) | (0.16) | (0.10) | (0.11) | (0.15) | (0.11) | (0.14) | (0.11) | (0.07) | (0.12) |
| Welfare | 0.75* | 0.75* | 1.12 | 0.78* | 0.79* | 0.92 | 0.96 | 0.91 | 1.04 | 0.93 | 1.03 | 1.08 |
|  | (0.08) | (0.06) | (0.12) | (0.09) | (0.06) | (0.10) | (0.13) | (0.10) | (0.13) | (0.10) | (0.08) | (0.10) |
| Income | 1.06* | 1.02 | 1.12* | 1.07* | 1.05* | 1.06* | 1.11* | 1.04 | 1.05* | 1.00 | 1.01 | 0.99 |
|  | (0.02) | (0.02) | (0.02) | (0.02) | (0.02) | (0.02) | (0.02) | (0.02) | (0.03) | (0.01) | (0.01) | (0.02) |
| Poverty | 1.01 | 0.96 | 1.03 | 1.04 | 0.99 | 0.97 | 0.87* | 0.92 | 0.97 | 1.00 | 1.05 | 0.98 |
|  | (0.06) | (0.04) | (0.06) | (0.06) | (0.04) | (0.05) | (0.06) | (0.04) | (0.06) | (0.05) | (0.04) | (0.04) |
| Disability | 0.86 | 0.92 | 0.79* | 1.04 | 0.94 | 1.07 | 1.15 | 1.17 | 1.35* | 1.08 | 1.19* | 1.08 |
|  | (0.09) | (0.08) | (0.08) | (0.10) | (0.07) | (0.10) | (0.16) | (0.11) | (0.16) | (0.10) | (0.08) | (0.09) |
| Urban | 0.96 | 0.98 | 0.91 | 0.98 | 0.96 | 1.10 | 1.16 | 1.05 | 0.85 | 1.07 | 1.16* | 1.00 |
|  | (0.10) | (0.08) | (0.09) | (0.09) | (0.07) | (0.10) | (0.14) | (0.09) | (0.09) | (0.10) | (0.08) | (0.08) |
| Percent lunch | 0.91 | 0.94 | 0.93 | 0.91 | 0.88* | 0.85* | 0.80* | 0.78* | 0.82* | 0.92 | 0.99 | 0.88* |
|  | (0.06) | (0.06) | (0.07) | (0.05) | (0.04) | (0.05) | (0.06) | (0.05) | (0.08) | (0.05) | (0.05) | (0.05) |
| Percent Black | 0.99 | 0.84* | 0.92 | 0.93 | 0.89* | 0.94 | 1.14* | 0.99 | 1.09 | 1.05 | 0.97 | 1.07 |
|  | (0.06) | (0.03) | (0.05) | (0.04) | (0.03) | (0.04) | (0.07) | (0.04) | (0.07) | (0.05) | (0.03) | (0.05) |

*(continued)*

Table A.2 (continued)

|  | Attend Event ||| Volunteer ||| Attend Meeting ||| Attend PTO |||
| --- | --- | --- | --- | --- | --- | --- | --- | --- | --- | --- | --- | --- |
|  | 2007 | 2012 | 2016 | 2007 | 2012 | 2016 | 2007 | 2012 | 2016 | 2007 | 2012 | 2016 |
| Percent Hispanic | 0.97 | 0.96 | 0.83* | 0.92 | 0.93 | 1.03 | 1.09 | 1.05 | 1.02 | 0.97 | 1.04 | 1.08 |
|  | (0.06) | (0.05) | (0.06) | (0.05) | (0.04) | (0.06) | (0.08) | (0.06) | (0.08) | (0.05) | (0.04) | (0.06) |
| Elementary | 2.52* | 1.96* | 2.02* | 3.43* | 3.00* | 2.74* | 2.71* | 2.54* | 2.17* | 1.89* | 2.09* | 1.96* |
|  | (0.29) | (0.16) | (0.20) | (0.33) | (0.20) | (0.22) | (0.37) | (0.23) | (0.24) | (0.18) | (0.13) | (0.15) |
| Middle | 1.37* | 1.15 | 1.16 | 1.26* | 1.11 | 1.00 | 1.85* | 1.72* | 1.99* | 1.49* | 1.22* | 1.33* |
|  | (0.14) | (0.09) | (0.11) | (0.12) | (0.08) | (0.08) | (0.24) | (0.15) | (0.22) | (0.13) | (0.08) | (0.10) |
| N | 6340 | 13080 | 10290 | 6340 | 13100 | 10330 | 6340 | 13080 | 10330 | 6340 | 13100 | 10330 |

*Note.* PTO = parent-teacher organization. Cells report odds ratios derived from logistic regression coefficients. Standard errors in parentheses. Observations rounded to the nearest 10 per Department of Education guidelines; estimates include state fixed effects. Asterisks denote statistical significance at the $p < .05$ level.
*Source.* Parent and Family Involvement surveys.

Table A.3
**Charter-District Differences in Types of Parent Engagement (With Self-Selection Controls)**

|  | PT Conference ||| Fundraising ||| Committee ||| Guidance Meeting |||
|---|---|---|---|---|---|---|---|---|---|---|---|---|
|  | 2007 | 2012 | 2016 | 2007 | 2012 | 2016 | 2007 | 2012 | 2016 | 2007 | 2012 | 2016 |
| Charter | 1.68 | 1.44 | 1.65* | 1.23 | 1.30 | 0.94 | 1.86 | 1.41 | 0.73 | 1.05 | 1.19 | 1.29 |
|  | (0.66) | (0.29) | (0.31) | (0.33) | (0.22) | (0.16) | (0.73) | (0.33) | (0.19) | (0.32) | (0.23) | (0.26) |
| Nonassigned | 1.29 | 1.16 | 1.13 | 0.89 | 1.15 | 1.07 | 1.22 | 1.19 | 0.98 | 0.99 | 1.13 | 1.01 |
|  | (0.18) | (0.12) | (0.13) | (0.11) | (0.11) | (0.12) | (0.18) | (0.14) | (0.13) | (0.11) | (0.11) | (0.10) |
| First choice | 0.98 | 1.16 | 0.93 | 1.55* | 1.31* | 1.02 | 1.14 | 1.35* | 1.32 | 0.77* | 1.03 | 1.06 |
|  | (0.13) | (0.09) | (0.09) | (0.18) | (0.09) | (0.09) | (0.18) | (0.15) | (0.21) | (0.08) | (0.07) | (0.10) |
| Considered other | 1.28* | 1.23* | 1.44* | 1.27* | 1.12 | 1.11 | 1.12 | 1.38* | 1.15 | 1.35* | 1.41* | 1.29* |
|  | (0.14) | (0.09) | (0.13) | (0.13) | (0.07) | (0.09) | (0.14) | (0.13) | (0.13) | (0.12) | (0.09) | (0.10) |
| Grades | 0.84* | 0.91* | 0.87* | 1.19* | 1.35* | 1.34* | 1.35* | 1.38* | 1.30* | 0.78* | 0.78* | 0.70* |
|  | (0.05) | (0.04) | (0.04) | (0.06) | (0.05) | (0.06) | (0.11) | (0.08) | (0.10) | (0.04) | (0.03) | (0.03) |
| Moved | 1.29* | 0.96 | 1.14 | 1.17 | 1.16 | 0.99 | 1.19 | 1.33* | 0.91 | 1.17 | 1.14 | 1.28* |
|  | (0.13) | (0.08) | (0.10) | (0.11) | (0.09) | (0.08) | (0.14) | (0.12) | (0.09) | (0.10) | (0.09) | (0.11) |
| Suspended | 1.01 | 1.04 | 1.06 | 0.82 | 0.81* | 0.69* | 0.68 | 0.91 | 1.22 | 1.41* | 1.50* | 1.26* |
|  | (0.13) | (0.10) | (0.13) | (0.10) | (0.07) | (0.08) | (0.14) | (0.15) | (0.32) | (0.16) | (0.13) | (0.14) |
| Hispanic | 1.27 | 1.19 | 0.95 | 0.85 | 0.85 | 0.80* | 1.03 | 0.95 | 1.24 | 1.11 | 1.02 | 1.03 |
|  | (0.18) | (0.12) | (0.11) | (0.11) | (0.07) | (0.08) | (0.18) | (0.13) | (0.20) | (0.14) | (0.10) | (0.12) |
| Asian | 0.64 | 0.92 | 0.78 | 0.69 | 0.54* | 0.71* | 0.70 | 0.49* | 0.59* | 0.74 | 0.68* | 0.91 |
|  | (0.15) | (0.14) | (0.12) | (0.17) | (0.07) | (0.10) | (0.22) | (0.11) | (0.13) | (0.16) | (0.09) | (0.16) |
| Black | 1.18 | 1.29* | 1.10 | 1.10 | 1.12 | 1.11 | 1.31 | 0.95 | 1.06 | 1.49* | 1.40* | 1.47* |
|  | (0.22) | (0.17) | (0.18) | (0.19) | (0.12) | (0.15) | (0.31) | (0.17) | (0.22) | (0.24) | (0.16) | (0.20) |
| White | 0.81 | 1.01 | 0.89 | 1.10 | 1.14 | 1.36* | 1.12 | 0.98 | 1.02 | 0.99 | 1.04 | 1.10 |
|  | (0.12) | (0.11) | (0.11) | (0.15) | (0.10) | (0.16) | (0.23) | (0.15) | (0.19) | (0.13) | (0.10) | (0.13) |

(continued)

Table A.3 (continued)

|  | PT Conference ||| Fundraising ||| Committee ||| Guidance Meeting |||
|---|---|---|---|---|---|---|---|---|---|---|---|---|
|  | 2007 | 2012 | 2016 | 2007 | 2012 | 2016 | 2007 | 2012 | 2016 | 2007 | 2012 | 2016 |
| Female | 0.94 | 0.93 | 0.90 | 1.10 | 1.03 | 1.09 | 0.90 | 0.99 | 1.22* | 1.08 | 0.95 | 0.91 |
|  | (0.08) | (0.06) | (0.06) | (0.09) | (0.06) | (0.07) | (0.09) | (0.07) | (0.12) | (0.08) | (0.05) | (0.06) |
| English home | 0.75 | 1.18 | 1.45* | 1.66* | 1.92* | 1.84* | 1.05 | 1.38* | 1.03 | 1.24 | 0.87 | 0.91 |
|  | (0.13) | (0.14) | (0.20) | (0.25) | (0.19) | (0.21) | (0.27) | (0.22) | (0.21) | (0.20) | (0.10) | (0.13) |
| Parent education | 1.19* | 1.20* | 1.15* | 1.19* | 1.29* | 1.19* | 1.33* | 1.23* | 1.19* | 1.12* | 1.09* | 1.13* |
|  | (0.05) | (0.04) | (0.04) | (0.05) | (0.04) | (0.04) | (0.07) | (0.05) | (0.07) | (0.04) | (0.03) | (0.04) |
| Two-parent | 1.29* | 0.97 | 1.04 | 1.24* | 1.29* | 1.09 | 1.38* | 1.42* | 1.24 | 1.03 | 0.95 | 0.83* |
|  | (0.14) | (0.09) | (0.10) | (0.12) | (0.09) | (0.09) | (0.21) | (0.16) | (0.18) | (0.10) | (0.07) | (0.08) |
| Welfare | 1.10 | 1.03 | 1.12 | 0.91 | 0.78* | 1.09 | 0.70 | 0.80 | 1.00 | 0.98 | 1.27* | 0.92 |
|  | (0.13) | (0.10) | (0.11) | (0.09) | (0.06) | (0.10) | (0.14) | (0.09) | (0.17) | (0.11) | (0.11) | (0.09) |
| Income | 1.01 | 1.03 | 1.02 | 1.05* | 1.05* | 1.10* | 1.07* | 1.06* | 1.11* | 1.00 | 1.02 | 0.97 |
|  | (0.02) | (0.02) | (0.02) | (0.01) | (0.02) | (0.02) | (0.03) | (0.02) | (0.04) | (0.02) | (0.02) | (0.02) |
| Poverty | 1.01 | 1.04 | 0.99 | 0.95 | 1.01 | 0.98 | 1.17* | 1.00 | 1.01 | 1.08 | 1.05 | 1.01 |
|  | (0.06) | (0.04) | (0.05) | (0.06) | (0.04) | (0.04) | (0.08) | (0.05) | (0.07) | (0.06) | (0.04) | (0.05) |
| Disability | 1.24* | 1.64* | 1.76* | 1.03 | 1.12 | 0.86 | 1.27* | 1.01 | 1.19 | 1.78* | 2.04* | 2.26* |
|  | (0.14) | (0.14) | (0.19) | (0.10) | (0.08) | (0.07) | (0.15) | (0.11) | (0.16) | (0.17) | (0.15) | (0.20) |
| Urban | 1.03 | 0.90 | 0.91 | 0.97 | 0.90 | 0.80* | 1.03 | 1.05 | 1.23 | 1.01 | 0.88 | 0.99 |
|  | (0.11) | (0.07) | (0.08) | (0.09) | (0.06) | (0.06) | (0.13) | (0.11) | (0.15) | (0.10) | (0.06) | (0.08) |
| Percent lunch | 1.04 | 1.05 | 0.99 | 0.98 | 1.05 | 1.00 | 0.93 | 0.97 | 0.94 | 0.92 | 0.96 | 0.97 |
|  | (0.07) | (0.05) | (0.06) | (0.06) | (0.05) | (0.07) | (0.06) | (0.06) | (0.08) | (0.05) | (0.05) | (0.06) |
| Percent Black | 0.96 | 0.95 | 1.05 | 0.98 | 0.86* | 0.94 | 0.97 | 0.94 | 0.95 | 1.06 | 1.02 | 0.96 |
|  | (0.05) | (0.04) | (0.05) | (0.05) | (0.03) | (0.04) | (0.06) | (0.05) | (0.06) | (0.05) | (0.04) | (0.05) |

(continued)

Table A.3 (continued)

|  | PT Conference ||| Fundraising ||| Committee ||| Guidance Meeting |||
| --- | --- | --- | --- | --- | --- | --- | --- | --- | --- | --- | --- | --- |
|  | 2007 | 2012 | 2016 | 2007 | 2012 | 2016 | 2007 | 2012 | 2016 | 2007 | 2012 | 2016 |
| Percent Hispanic | 1.11 | 1.01 | 1.19* | 0.92 | 1.00 | 0.96 | 0.88 | 1.01 | 0.94 | 1.08 | 1.06 | 1.05 |
|  | (0.07) | (0.05) | (0.08) | (0.05) | (0.05) | (0.06) | (0.07) | (0.06) | (0.09) | (0.06) | (0.05) | (0.06) |
| Elementary | 6.19* | 6.41* | 8.01* | 1.88* | 2.38* | 1.87* | 0.97 | 1.24* | 1.10 | 0.31* | 0.40* | 0.42* |
|  | (0.78) | (0.55) | (0.80) | (0.19) | (0.16) | (0.15) | (0.12) | (0.12) | (0.13) | (0.03) | (0.03) | (0.03) |
| Middle | 2.14* | 1.85* | 2.20* | 1.26* | 1.21* | 1.09 | 0.76* | 0.91 | 0.88 | 0.55* | 0.62* | 0.59* |
|  | (0.21) | (0.13) | (0.18) | (0.12) | (0.08) | (0.09) | (0.10) | (0.09) | (0.11) | (0.05) | (0.04) | (0.05) |
| N | 6340 | 13100 | 10330 | 6340 | 13100 | 10330 | 6320 | 13090 | 10320 | 6340 | 13100 | 10330 |

*Note.* PT = parent-teacher. Cells report odds ratios derived from logistic regression coefficients. Standard errors in parentheses. Observations rounded to the nearest 10 per Department of Education guidelines; estimates include state fixed effects. Asterisks denote statistical significance at the $p < .05$ level.

*Source.* Parent and Family Involvement surveys.

## Charter-District Differences in Parent Engagement

### Table A.4
**Charter-District Differences in the Number of Parent Activities and Satisfaction (With Self-Selection Controls)**

|  | Number of Activities |  |  | Satisfaction |  |  |
| --- | --- | --- | --- | --- | --- | --- |
|  | 2007 | 2012 | 2016 | 2007 | 2012 | 2016 |
| Charter | 1.70 | 1.35* | −0.38 | 0.21* | 0.12* | 0.15* |
|  | (1.55) | (0.54) | (0.43) | (0.06) | (0.04) | (0.05) |
| Nonassigned | 0.00 | 0.96* | 0.20 | 0.12* | 0.08* | 0.09* |
|  | (0.50) | (0.35) | (0.32) | (0.03) | (0.02) | (0.03) |
| First choice | 0.58 | 0.45* | 0.29 | 0.56* | 0.45* | 0.42* |
|  | (0.39) | (0.19) | (0.25) | (0.04) | (0.02) | (0.03) |
| Consider. other | 0.98* | 0.61* | 0.54* | −0.09* | −0.09* | −0.14* |
|  | (0.40) | (0.20) | (0.24) | (0.03) | (0.02) | (0.02) |
| Grades | 0.73* | 0.77* | 0.50* | 0.15* | 0.17* | 0.15* |
|  | (0.19) | (0.10) | (0.14) | (0.01) | (0.01) | (0.01) |
| Moved | 0.57 | 0.54* | 0.28 | 0.06* | 0.06* | 0.07* |
|  | (0.38) | (0.23) | (0.31) | (0.02) | (0.02) | (0.02) |
| Suspended | −1.50* | −0.02 | −0.43 | −0.18* | −0.11* | −0.13* |
|  | (0.41) | (0.23) | (0.32) | (0.04) | (0.02) | (0.03) |
| Hispanic | −0.02 | −0.98* | −0.54 | 0.06 | 0.03 | −0.00 |
|  | (0.63) | (0.22) | (0.30) | (0.04) | (0.02) | (0.03) |
| Asian | −2.80* | −2.59* | −1.70* | 0.01 | −0.05 | −0.09* |
|  | (0.66) | (0.30) | (0.44) | (0.05) | (0.03) | (0.04) |
| Black | −0.00 | 0.14 | 0.48 | −0.01 | −0.03 | −0.03 |
|  | (0.61) | (0.32) | (0.39) | (0.05) | (0.03) | (0.03) |
| White | 1.10* | 0.52* | 0.86* | −0.00 | −0.07* | −0.03 |
|  | (0.40) | (0.24) | (0.33) | (0.04) | (0.02) | (0.03) |
| Female | −0.55 | −0.11 | 0.01 | −0.04 | −0.01 | −0.06* |
|  | (0.33) | (0.17) | (0.20) | (0.02) | (0.01) | (0.02) |
| English at home | 2.22* | 0.33 | 0.31 | −0.07 | −0.02 | −0.08* |
|  | (0.46) | (0.21) | (0.39) | (0.04) | (0.02) | (0.03) |
| Parent education | 0.97* | 0.69* | 0.60* | −0.01 | −0.02* | −0.01 |
|  | (0.16) | (0.08) | (0.10) | (0.01) | (0.01) | (0.01) |
| Two-parent | 1.30* | 0.54* | 0.49* | 0.01 | 0.02 | −0.01 |
|  | (0.39) | (0.20) | (0.23) | (0.03) | (0.02) | (0.02) |
| Welfare | −0.55 | −0.45* | −0.01 | 0.02 | −0.00 | −0.03 |
|  | (0.40) | (0.21) | (0.27) | (0.03) | (0.02) | (0.02) |
| Family income | 0.12* | 0.19* | 0.21* | 0.00 | 0.00 | −0.00 |
|  | (0.06) | (0.04) | (0.05) | (0.00) | (0.00) | (0.00) |
| Neighborhood poverty | 0.31 | 0.01 | 0.33* | −0.00 | −0.02* | −0.00 |
|  | (0.20) | (0.11) | (0.15) | (0.01) | (0.01) | (0.01) |
| Disability | 0.64 | 0.41 | 0.55 | 0.01 | 0.04 | −0.02 |
|  | (0.36) | (0.23) | (0.28) | (0.02) | (0.02) | (0.02) |

*(continued)*

Table A.4 **(continued)**

|  | Number of Activities |  |  | Satisfaction |  |  |
|---|---|---|---|---|---|---|
|  | 2007 | 2012 | 2016 | 2007 | 2012 | 2016 |
| Urban | −0.54 | −0.03 | −0.07 | 0.03 | 0.05* | 0.01 |
|  | (0.36) | (0.19) | (0.25) | (0.02) | (0.02) | (0.02) |
| Percent lunch | −0.30 | 0.24 | 0.13 | −0.04* | −0.07* | −0.07* |
|  | (0.24) | (0.17) | (0.20) | (0.02) | (0.01) | (0.01) |
| Percent Black | −0.31 | −0.54* | −0.28 | −0.01 | −0.02 | −0.01 |
|  | (0.19) | (0.11) | (0.16) | (0.01) | (0.01) | (0.01) |
| Percent Hispanic | −0.69* | −0.45* | −0.32* | −0.01 | 0.03* | 0.01 |
|  | (0.25) | (0.16) | (0.16) | (0.02) | (0.01) | (0.02) |
| Elementary | −1.25* | −0.41 | −0.58* | 0.13* | 0.19* | 0.17* |
|  | (0.45) | (0.21) | (0.26) | (0.02) | (0.02) | (0.02) |
| Middle | −1.81* | −1.38* | −1.29* | 0.05* | 0.04* | 0.03 |
|  | (0.40) | (0.21) | (0.26) | (0.03) | (0.02) | (0.02) |
| N | 6,350 | 13,100 | 10,330 | 6,350 | 13,100 | 10,330 |

*Note.* Cells report regression coefficients. Standard errors in parentheses. Observations rounded to the nearest 10 per Department of Education guidelines; estimates include state fixed effects. Asterisks denote statistical significance at the $p < .05$ level.
*Source.* Parent and Family Involvement surveys.

## ORCID iD

Zachary W. Oberfield https://orcid.org/0000-0001-9418-3693

## Notes

[1] The instruments used to create the PFI database can be found online on the NCES website https://nces.ed.gov/nhes/
[2] Observations rounded to nearest 10 as per Department of Education guidelines.
[3] The question prompt was slightly different in 2007: "Since the beginning of this school year, (have/has) (you/any adult in your household)." In addition, the attend event question included "because of [this child]?" at the end. These differences may have had some minor effect on year-to-year response differences. However, they do not affect the article's main goal of comparing the responses of charter and district parents within particular years.
[4] The question prompt in 2007 was, "Would you say that you are very satisfied, somewhat satisfied, somewhat dissatisfied or very dissatisfied . . ." followed by the five aspects of school satisfaction. This difference may have had some minor effect on year-to-year response differences. However, it does not affect the article's main goal of comparing the responses of charter and district parents within particular years.
[5] The variable *number of activities* is a count variable. A supplementary analysis, which employed a Poisson regression and this same model, resulted in identical findings as those from the OLS analysis. OLS findings are reported here for ease of interpretation.
[6] The question prompt in 2007 was, "Is the school [this child] attends the one you wanted most for (him/her), that is, your first choice?" This difference may have had some minor effect on year-to-year response differences. However, it does not affect the

article's main goal of comparing the responses of charter and district parents within particular years.

[7] To ensure that the model did not suffer from multicollinearity—whereby variables are so strongly correlated that they distort our understanding of the relationships between independent and dependent variables—a variance influence factor test was conducted. This analysis revealed no concerns about overlaps among the model's variables (average VIF [variance inflation factor] = 2.06).

[8] District respondents include parents with children in geographically assigned and non–geographically assigned (i.e., magnet, vocational, continuing, alternative, or special education school) district schools. Excluding non–geographically assigned district students did not change the table's findings regarding district-charter differences.

[9] As with Table 1, district respondents include parents with children in geographically assigned and non–geographically assigned district schools. Excluding non–geographically assigned district students did not change the table's findings regarding district-charter differences.

[10] A supplementary analysis revealed that when the data were divided by school level—elementary, middle, and high—there were no charter-district engagement differences at any level in 2007 and 2016. This analysis revealed that the 2012 *number of activities* charter-district difference was likely driven by a higher level of engagement among elementary charter school parents.

[11] Becker et al. (1997) found parent volunteering contracts in 74% of charter schools and 8% of district schools. A more recent analysis by Mintrom (2003) found these contracts in 37% of charter schools and 22% of district schools.

## References

Bailey, D., Scarborough, A., & Hebbeler, K. (2003). *Families' first experiences with early intervention*. Menlo Park, CA: SRI International.

Barrows, S., Cheng, A., Peterson, P., & West, M. (2017). *Parental perceptions of charter schools: Evidence from two nationally representative surveys of us parents* (PEPG 17-01). Cambridge, MA: Harvard Kennedy School.

Beck, D., Maranto, R., & Lo, W. (2014). Determinants of student and parent satisfaction at a cyber charter school. *Journal of Educational Research, 107*, 209–216.

Becker, H., Nakagawa, K., & Corwin, R. (1997). Parent involvement contracts in California's charter schools: Strategy for educational improvement or method of exclusion? *Teachers College Record, 98*, 511–536.

Betts, J., & Tang, Y. (2014). *The effect of charter schools on student achievement: A meta-analysis of the literature*. Seattle, WA: Center on Reinventing Public Education.

Bifulco, R., & Ladd, H. (2006). Institutional change and coproduction of public services: The effect of charter schools on parental involvement. *Journal of Public Administration Research and Theory, 16*, 553–576.

Bosetti, L. (2004). Determinants of school choice: Understanding how parents choose elementary schools in Alberta. *Journal of Education Policy, 19*, 387–405.

Bozeman, B. (2004). *All organizations are public: Bridging public and private organizational theories*. Washington, DC: Beard Books.

Buckley, J. (2007). *Choosing schools, building communities? The effect of schools of choice on parental involvement*. Fayetteville: University of Arkansas, Department of Education Reform. Retrieved from https://eric.ed.gov/?id=ED508943

Buckley, J., & Schneider, M. (2006). Are charter school parents more satisfied with schools? Evidence from Washington, DC. *Peabody Journal of Education, 81*(1), 57–78.

Buckley, J., & Schneider, M. (2007). *Charter schools: Hope or hype?* Princeton, NJ: Princeton University Press.

Budde, R. (1988). *Education by charter: Restructuring school districts.* Andover, MA: Regional Laboratory for Educational Improvement of the Northeast and Islands.

Bulkley, K., & Fisler, J. (2003). A decade of charter schools: From theory to practice. *Educational Policy, 17,* 317–342.

Buras, K. (2014). *Charter schools, race, and urban space: Where the market meets grassroots resistance.* New York, NY: Routledge.

Carpenter, D. (2006). Modeling the charter school landscape. *Journal of School Choice, 1,* 47–82.

Centers for Disease Control and Prevention. (2018). Parent engagement. Retrieved from https://www.cdc.gov/healthyyouth/protective/parent_engagement.htm

Chambers, S., & Michelson, M. (2016). School satisfaction among low-income urban parents. *Urban Education,* 1–23. doi:10.1177/0042085916652190

Child Trends. (2013). *Parental involvement in schools.* Washington, DC: Author. Retrieved from https://www.childtrends.org/indicators/parental-involvement-in-schools/

Donahue, J. (1989). *The privatization decision: Public ends, private means.* New York, NY: Basic Books.

Epple, D., Romano, R., & Zimmer, R. (2016). Charter schools: A survey of research on their characteristics and effectiveness. In E. Hanushek, S. Machin, & L. Woessmann (Eds.), *Handbook of the economics of education* (Vol. 5, pp. 139–208). New York, NY: Elsevier.

Epstein, J. (1995). School/family/community partnerships: Caring for the children we share. *Phi Delta Kappan, 92*(3), 81–96.

Epstein, J. (2010). *School, family, and community partnerships: Preparing educators and improving schools.* Boulder, CO: Westview Press.

Fantuzzo, J., Perry, M., & Childs, S. (2006). Parent satisfaction with educational experiences scale: A multivariate examination of parent satisfaction with early childhood education programs. *Early Childhood Research Quarterly, 21,* 142–152.

Feeney, M., & Bozeman, B. (2009). Stakeholder red tape: Comparing perceptions of public managers and their private consultants. *Public Administration Review, 69,* 710–726.

Finn, C., Jr., Manno, B., & Vanourek, G. (2000). *Charter schools in action: Renewing public education.* Princeton, NJ: Princeton University Press.

Finn, C., Jr., Manno, B., & Wright, B. (2016). *Charter schools at the crossroads: Predicaments, paradoxes, possibilities.* Cambridge, MA: Harvard Education Press.

Friedman, B., Bobrowski, P., & Geraci, J. (2006). Parents' school satisfaction: Ethnic similarities and differences. *Journal of Educational Administration, 44,* 471–486.

Gawlik, M., & Bickmore, D. (Eds.). (2017). *Unexplored conditions of charter school principals: An examination of the issues and challenges for leaders.* New York, NY: Rowman & Littlefield.

Gibbons, S., & Silva, O. (2011). School quality, child wellbeing and parents' satisfaction. *Economics of Education Review, 30,* 312–331.

Gleason, P., Clark, M., Tuttle, C., & Dwoyer, E. (2010). *The evaluation of charter school impacts: Final report.* Washington, DC: National Center for Education Evaluation and Regional Assistance.

Hamlin, D. (2017). Parental involvement in high choice deindustrialized cities: A comparison of charter and public schools in Detroit. *Urban Education.* doi:10.1177/0042085917697201

Hammel, H. (2014). *Charging for access: How California charter schools exclude vulnerable students by imposing illegal family work quotas.* San Francisco, CA: Public Advocates.

Hanushek, E., & Rivkin, S. (2006). Teacher quality. In E. Hanushek & F. Welch (Eds.), *Handbook of the economics of education* (Vol. 2, pp. 1051–1078). Amsterdam, Netherlands: North-Holland.

Henderson, A., & Mapp, K. (2002). *A new wave of evidence: The impact of school, family, and community connections on student achievement* (Annual Synthesis 2002). Austin, TX: National Center for Family and Community Connections with Schools.

Hill, P., Lake, R., & Celio, M. (2002). *Charter schools and accountability in public education.* Washington, DC: Brookings Institution Press.

Jeynes, W. (2012). A meta-analysis of the efficacy of different types of parental involvement programs for urban students. *Urban Education, 47,* 706–742.

Jochim, A., DeArmond, M., Gross, B., & Lake, R. (2014). *How parents experience public school choice.* Seattle, WA: Center for Reinventing Public Education.

Kahlenberg, R., & Potter, H. (2014). *A smarter charter: Finding what works for charter schools and public education.* New York, NY: Teachers College Press.

Kelley, R., Thornton, B., & Daugherty, R. (2005, Fall). Relationships between measures of leadership and school climate. *Education, 126*(1), 17–23.

Kolderie, T. (1990). *Beyond choice to new public schools: Withdrawing the exclusive franchise in public education* (Policy Report No. 8.) Washington, DC: Progressive Policy Institute.

Kraft, M., & Rogers, T. (2015). The underutilized potential of teacher-to-parent communication: Evidence from a field experiment. *Economics of Education Review, 47,* 49–63.

Meier, K., & O'Toole, L. (2011). Comparing public and private management: Theoretical expectations. *Journal of Public Administration Research and Theory, 21*(Suppl. 3), i283–i299.

Mintrom, M. (2003). Market organizations and deliberative democracy: Choice and voice in public service delivery. *Administration & Society, 35,* 52–81.

Miron, G., & Nelson, C. (2002). *What's public about charter schools? Lessons learned about choice and accountability.* Thousand Oaks, CA: Corwin Press.

Miron, G., Nelson, C., & Risley, J. (2002). *Strengthening Pennsylvania's charter school reform: Findings from the statewide evaluation and discussion of relevant policy issues. Year five report.* Kalamazoo, MI: Evaluation Center.

Mommandi, W., & Welner, K. (2018). Shaping charter enrollment and access. In I. Rotberg & J. Glazer (Eds.), *Choosing charters: Better schools or more segregation?* (pp. 61–81). New York, NY: Teachers College Press.

Nathan, J. (1996). *Charter schools: Creating hope and opportunity for American education.* San Francisco, CA: Jossey-Bass.

National Alliance for Public Charter Schools. (2018). What is a charter school? Retrieved from https://www.publiccharters.org/about-charter-schools/what-charter-school

National Commission on Excellence in Education. (1983). *A nation at risk: The imperative for education reform.* Washington, DC: U.S. Department of Education.

Oberfield, Z. (2017). *Are charters different? Public education, teachers, and the charter school debate.* Cambridge, MA: Harvard Education Press.

Osborne, D. (2017). *Reinventing America's schools: Creating a 21st century education system.* New York, NY: Bloomsbury.

Osborne, D., & Gaebler, T. (1993). *Reinventing government: How the entrepreneurial spirit is transforming the public sector.* New York, NY: Plume.

Peterson, P. (2009). Voucher impacts: Differences between public and private schools. In M. Berends, M. Springer, D. Ballou, & H. Walberg (Eds.), *Handbook of research on school choice* (pp. 249-266). New York, NY: Routledge.

Podsakoff, P., MacKenzie, S., Lee, J., & Podsakoff, N. (2003). Common method biases in behavioral research: A critical review of the literature and recommended remedies. *Journal of Applied Psychology, 88*, 879–903.

Rainey, H., & Chun, Y. (2005). Public and private management compared. In E. Ferlie, L. Lynn Jr., & C. Pollitt (Eds.), *The Oxford handbook of public management* (pp. 72–102). New York, NY: Oxford University Press.

Ravitch, D. (2013). *Reign of error: The hoax of the privatization movement and the danger to America's public schools*. New York, NY: Knopf.

Rothstein, R. (1998, July-August). Charter conundrum. *The American Prospect*. Retrieved from http://prospect.org/article/charter-conundrum

Schram, S. (2015). *The return of ordinary capitalism: Neoliberalism, precarity, occupy*. New York, NY: Oxford University Press.

Shanker, A. (1988, July 10). Where we stand: Convention plots new course; a charter for change. *The New York Times*.

Smith, J., Wohlstetter, P., Kuzin, C., & De Pedro, K. (2011). Parent involvement in urban charter schools: New strategies for increasing participation. *School Community Journal, 21*(1), 71–94.

Teske, P., & Schneider, M. (2001). What research can tell policymakers about school choice. *Journal of Policy Analysis and Management, 20*, 609–631.

Tuttle, C. C., Gill, B., Gleason, P., Knechtel, V., Nichols-Barrer, I., & Resch, A. (2013). *KIPP middle schools: Impacts on achievement and other outcomes. Final report*. Princeton, NJ: Mathematica Policy Research.

U.S. Department of Education. (1994). *Strong families, strong schools: Building community partnerships for learning*. Washington, DC.

Verba, S., Schlozman, K., & Brady, H. (1995). *Voice and equality: Civic voluntarism in American politics*. Cambridge, MA: Harvard University Press.

Verkuil, P. (2007). *Outsourcing sovereignty: Why privatization of government functions threatens democracy and what we can do about it*. New York, NY: Cambridge University Press.

Weiler, S., & Vogel, L. (2015). Charter school barriers: Do enrollment requirements limit student access to charter schools? *Equity & Excellence in Education, 48*, 36–48.

Wells, A. (Ed.). (2002). *Where charter school policy fails: The problems of accountability and equity*. New York, NY: Teachers College Press.

Wohlstetter, P., Nayfack, M., & Mora-Flores, E. (2008). Charter schools and "customer" satisfaction: Lessons from field testing a parent survey. *Journal of School Choice, 2*, 66–84.

Wohlstetter, P., Smith, J., & Farrell, C. (2013). *Choices and challenges: Charter school performance in perspective*. Cambridge, MA: Harvard Education Press.

Zimmer, R., & Buddin, R. (2007). Getting inside the black box: Examining how the operation of charter schools affects performance. *Peabody Journal of Education, 82*, 231–273.

Manuscript received May 21, 2018
Final revision received June 24, 2019
Accepted July 15, 2019

# Pedagogy and Profit? Efforts to Develop and Sell Digital Courseware Products for Higher Education

Matthew D. Regele
Xavier University

*The individual economic benefits of higher education are largely determined by what students learn in the process of obtaining their degrees. Increasingly, for-profit companies that develop and sell digital courseware products influence what college students learn. Employees' pedagogical expertise, content knowledge, and understanding of organizational goals are likely to affect product characteristics and outcomes associated with the use of those products. This study draws on 15 months of ethnographic data to examine one organization's efforts to develop and sell courseware for use in higher education. The data suggest organization members' interpretations of educational access and quality support product development and sales efforts consistent with profit aims, but that may promote credentialism, negatively affect learning, and exacerbate quality differences across institutions.*

KEYWORDS: courseware, higher education, learning materials, credentialism, product development

Although academics have long debated the purpose and value of a higher education degree, it is clear that benefits related to economic opportunity and career growth have emerged as a key focus for students, parents, and policymakers (e.g., P. Brown, Lauder, & Ashton, 2011; Haas & Fischman, 2010; Labaree, 1997; Olssen & Peters, 2005). Furthermore, although academics continue to debate whether this largely neoliberal, individualistic view of education is normatively appropriate or desirable (e.g., Hursh, 2007; Labaree, 1997), its growing prominence outside the academy has encouraged—and to a certain extent, been supported by—significant

---

MATTHEW D. REGELE is an assistant professor in the Department of Management and Entrepreneurship in the Williams College of Business at Xavier University, 3800 Victory Parkway, Cincinnati, OH 45207, USA; e-mail: *regelem@xavier.edu*. Dr. Regele's current research interests include entrepreneurship, innovation, and organizational change in higher education and other sectors.

research.[1] In particular, the question of who derives economic benefits from higher education, as well as how and why these benefits accrue has attracted a great deal of attention from economists, sociologists, and other scholars (Abel & Deitz, 2014; Brand & Xie, 2010; Hout, 2012). Since the gains largely depend on what students actually learn in the process of obtaining a higher education, curricula issues have been a key area of debate. Much of the discussion has been about what students *should* learn. This includes arguments about the types of content that should be covered, how this content should be taught, and how these factors may (or should) vary across institutions and contexts (e.g., Bok, 2013).

Yet, even as research and debate has generated valuable insights, the discussion remains markedly incomplete. Much of the debate occurs among and between academics and policymakers, sometimes taking into account the interests of students, parents, and employers. However, such debate largely fails to consider another critical set of stakeholders: for-profit companies that develop and sell learning materials. These organizations' influence has largely paralleled the rise of economically focused, neoliberal views of education, in part because advocates of these views believe that market competition will lead to more efficient, desirable outcomes (Labaree, 1997; Olssen & Peters, 2005). Although some research has examined the impact of learning material providers and other for-profit service organizations on K–12 education (e.g., Apple & Christian-Smith, 1991; Heinrich & Nisar, 2013; Roberts-Mahoney, Means, & Garrison, 2016; Spring, 2011), the influence of such businesses on higher education has been largely overlooked. This oversight is especially troubling given the increasing popularity of digital learning products (e.g., courseware solutions), the extent to which such products are increasingly embedded in teaching and learning processes, and the outsize role that such businesses play in their development and sale.

This study begins to address this significant gap in our collective knowledge. Specifically, through an inductive examination of one company's efforts to develop and sell digital courseware solutions, I consider three questions: (1) What do organization members believe is the organization's appropriate role in the higher education ecosystem, particularly with regard to what and how students learn? (2) How do these understandings affect the organization's product development and sales efforts? (3) What potential implications do these efforts have for higher education outcomes, both overall and in terms of how they vary across students and institutions? Ethnographic observation, interviews, and archival materials collected while embedded within the organization for a 15-month period allow me to triangulate on organization members' understanding of higher education, pedagogy, and their own work; how these understandings guide their actions; and how these actions affect product design and sales.

My data suggest that, although organization members consistently express beliefs that they are improving educational quality (i.e., ensuring each student learns "more") and access (i.e., ensuring more individuals are able to pursue a higher education), these beliefs are sustained by conceptions of quality and access that are narrow even among neoliberal perspectives of higher education. In addition, I find these conceptualizations, coupled with the organization's increasing influence over content and pedagogy, encourage the development of products that make it easier for students to obtain a college diploma, but that also may diminish what students learn in the process of acquiring their degree. As a result, the products support outcomes consistent with theories of human capital signaling and credentialism (e.g., Arrow, 1973; D. K. Brown, 2001; Collins, 1979) but may undermine human capital development (e.g., Becker, 1993). My findings also suggest that for-profit businesses' efforts to sell these digital courseware solutions may exacerbate differences in educational quality across institutions.

## Theoretical Background

### The Individual Economic Benefits of Higher Education

As neoliberal perspectives of education have become increasingly dominant, researchers have asked if and how individual students benefit from higher education, especially whether and in what ways obtaining a degree affects subsequent occupational and career prospects. In general, there is consensus that a college degree improves job opportunities, job satisfaction, and lifetime earnings (e.g., Becker, 1993; Brand & Xie, 2010; Mincer, 1974; Olneck, 1979); however, there is less agreement about how or why those benefits materialize. There are at least three alternative explanations.[2] The first, and perhaps most dominant, is that higher education is valuable because it develops human capital that makes graduates more productive employees (e.g., Becker, 1993). The second explanation suggests the primary value of a higher education degree may be the signal provided by the diploma. This line of argument suggests that a diploma serves as a signal of the individual's inborn capabilities and natural intelligence, as well as his or her diligence and work ethic, rather than particular knowledge or skills developed in the process of obtaining the degree (Arrow, 1973; Spence, 1974; Thurow, 1975). Finally, a third explanation, sometimes referred to as credentialism, argues that the diploma is not a signal of any particular knowledge or capability, but rather a sign of cultural fit.[3] In other words, the diploma indicates the individual is socially and culturally suited for a particular job or occupation (D. K. Brown, 2001; Collins, 1979). Although this suitability may, in part, reflect the individual having acquired some technical knowledge or ability to "speak the language" of an occupation during his or her degree program, such knowledge does not necessarily imply the individual will be more productive or effective in a particular role, or that he or she is more talented than those without the credential (D. K. Brown, 2001; Collins, 1979).

In addition to theorizing about individual educational outcomes, academics have, to a degree, attempted to examine them empirically. Some scholars have focused on students (and their families), including the expectations they have about higher education and its potential benefits, how these expectations influence how they allocate time and effort while obtaining a degree, and the post-graduation implications of their efforts (e.g., Arum & Roksa, 2010; Bleemer & Zafar, 2018; Domina, Conley, & Farkas, 2011; Dominitz & Manski, 1994; Nielsen, 2015). Others have focused on employers' evaluation of academic credentials and whether they prioritize, for example, institutional prestige, major, or just the diploma itself (e.g., Deterding & Pedulla, 2016; Rivera, 2011). Such research offers valuable insights about the links between expectations and outcomes in higher education. Yet the focus on expectations and outcomes in some ways fails to address another central issue: the education process itself. In particular, such research may not consider what, if anything, students actually learn in college. This is problematic because it is what students actually learn in college that *determines* whether they graduate with newly acquired human capital, a signifier of cultural fit, or a signal of preexisting capabilities.

## Curricula Debates

To argue that we have an incomplete understanding of what students learn in college is not to suggest that no one has considered what students *should* learn. Academics, students, parents, employers, and policymakers interested in individual-level economic outcomes have long debated college curricula. For example, a frequently raised question is whether curricula should focus on traditional liberal arts topics (e.g., "Great Books" courses) or more vocational, occupational-specific content (e.g., Bok, 2013).

Although such curricula questions are certainly important, they also miss key elements of the learning process likely to influence individual economic outcomes. Specifically, such questions consider what information students are exposed to, but do not necessarily consider *how* they engage with that material or what capabilities and skills they develop in the process. For example, a "Great Books" curriculum could either be taught in a way that focuses on content coverage and reading comprehension or a way that focuses on analysis and the transferability of lessons to other domains (e.g., Barber, 2012; Hannaway, 1992). Each approach seems more consistent with a different interpretation of the economic value of a college degree. For example, outcomes consistent with signaling and credentialist perspectives are possible even if students only superficially cover content and quickly obtain a diploma, whereas developing capabilities and skills may require deeper, more time-consuming engagement with material (e.g., Sandberg &

Barnard, 1997; van Gelder, 2005). In other words, what students actually "do" in college is inextricably linked to whether their diploma is primarily a signal of innate ability or cultural fit, represents newly acquired human capital, or none of the above. What students actually "do" is, in turn, at least partly influenced by the learning materials they use and the assignments they complete. This raises important questions about how such materials are designed and used. Yet such questions have received surprisingly little attention from researchers.

### The Role of For-Profit Learning Material Providers

In part, scholars may not have paid systematic attention to learning materials because individual instructors are commonly viewed as having significant latitude in selecting (or creating) materials and assignments. Indeed, such control is often viewed as a key component of academic freedom (e.g., Schrecker, 2010). In other words, researchers may expect to find few systematic links between learning materials and educational outcomes because there is so much variation in the content and assignments that instructors choose. Yet there is significant reason to question such assumptions.

First, at least since the introduction of the famous McGuffey Readers in the mid-19th century (Vail, 1911), textbook publishers and other businesses have sought to profit by selling (supposedly) high-quality, standardized learning materials. In recent decades, a few dominant players have produced a large proportion of these materials. For example, in 2008 the four-firm concentration ratio in textbook publishing was 87.5, making it one of the most concentrated manufacturing industries in the United States (Koch, 2013).

Second, in recent years, the potential influence of such organizations has increased significantly, as they have shifted their focus from print textbooks to digital learning products. From a business perspective, this reflects publishers' efforts to reinvent themselves in the face of declining revenues from print textbooks (Thompson, 2005). In particular, these organizations have sought to offer "courseware" solutions that extend beyond the textual content students consume to encompass all the major components of the course, including graded assignments and even exams (Tyton Partners, 2014). By linking their products to all aspects of the course, the companies hope to force every enrolled student to purchase subscription-based access to the courseware (Tyton Partners, 2014). Pedagogically, if the shift is successful, it will make these for-profit companies increasingly central to teaching and learning processes (Alavi & Leidner, 2001; Reiser, 2001; Roberts-Mahoney et al., 2016). Specifically, whereas textbooks shape what content students consume, digital products also have the potential to shape *how* students engage with that content. In part, this is because assignments may increasingly be completed—and associated grades determined—*within* the digital learning products (e.g., Roberts-Mahoney et al., 2016). This is particularly true of courseware solutions, in which students may complete virtually all of their coursework, often using the companies' proprietary content and tools. This includes not only graded homework but also other assignments traditionally created or curated by faculty, such as case studies, group projects, and exams. In addition to reshaping who curates content and designs assignments, digital delivery of learning materials is also likely to influence how students receive and engage with the material. For example, studies have repeatedly shown that reading is more tiring and difficult online and may result in lower levels of comprehension and retention (e.g., Baron, 2015; Mangen, Walgermo, & Brønnick, 2013). This encourages digital learning material providers to revisit how material is presented; for example, to move away from traditional chapter formats and toward much shorter "bites" of information.

Not surprisingly, companies developing digital learning products argue that these products are beneficial for students and faculty. In particular, they have argued the products will improve individual educational (and economic) outcomes by increasing educational *access* and *quality*. By access, I refer to claims that technology can increase the number of people to whom a postsecondary education is available. Proponents argue that technology can lower the cost of an education—or even make it free (Amiel & Reeves, 2008; Bok, 2004). Advocates also suggest that technology—especially when used to support online classes—can offer flexibility in when students with jobs, families, or other commitments complete their coursework (e.g., Moore & Kearsley, 2012). By quality, I refer to arguments that technology can improve what and how students learn. For example, technology proponents claim that digital products enable greater personalization in the content students consume, as well as how that content is presented (e.g., Christensen, Horn, & Johnson, 2008). Advocates also argue that personalization promotes engagement and information retention (e.g., Christensen et al., 2008). Furthermore, they suggest technology can facilitate frequent, targeted assessment, helping faculty and students to pinpoint and address learning gaps (e.g., Christensen et al., 2008).

Publishers' claims about the benefits of their digital learning products may be valid; it is possible these products are increasing the quality and accessibility of teaching and learning. Yet research also suggests some skepticism may be warranted. First, it's not clear that the goals of increasing access and quality are well aligned, or how they might influence individual educational benefits and outcomes. For example, improving access to a higher education can be interpreted as reducing the barriers (e.g., cost, time, etc.) students face in completing a degree. This may entail minimizing the time and effort required to obtain a diploma, which might in turn reduce the cost of obtaining a credential or sending a signal to employers about one's capabilities. However, educators have also long recognized that learning capabilities and skills (i.e., developing human capital) is often inherently difficult and may require time-consuming struggle with difficult ideas and problems (e.g., Sandberg & Barnard, 1997). At a certain point, minimizing the time and effort necessary to obtain the signal/credential may begin to reduce the human capital that is developed in the process.

Second, it is not clear the extent to which or how goals related to increasing educational quality and access correlate with profit aims. For example, maximizing the number of individuals who have access to education seems to imply minimizing the cost they incur in obtaining a degree; however, from a financial perspective, maximizing the number of people for whom a higher education is affordable would suggest making learning materials as inexpensive as feasibly possible. This might be achieved by minimizing profits, or at least doing so beyond the level necessary to support research and development and to fund growth. In addition, if students are primarily interested in obtaining a degree with minimal effort, they may resist paying for products that require them to complete time-consuming or difficult assignments. This may incentivize companies to offer assignments that are fast and easy, but result in little actual learning.

More generally, existing research suggests mixing educational goals and profit motives could create perverse incentives for publishers or encourage marketing claims that are (perhaps unintentionally) misleading. For example, scholarship on organizational hybridity demonstrates that efforts to combine social (e.g., educational) and economic goals often result in "mission drift," or one mission being subordinated to the other (Battilana & Lee, 2014; Battilana, Sengul, Pache, & Model, 2015; Dees & Anderson, 2003). In fact, researchers have uncovered evidence of mission drift in education. They have found, for example, that competition can cause education providers to focus resources on consumer-oriented marketing, rather than improving educational processes (Jacob, McCall, & Stange, 2013; Lubienski, 2005). In addition, regardless of organizations' stated goals, scholars have long recognized that organization members' own knowledge, beliefs, and interests shape if and how they pursue those goals (March & Simon, 1958). As a result, it is important to consider who actually works in these organizations, as well as their understandings of education, the organization's role in facilitating education, and what motivates them personally.

In summary, for-profit developers of learning materials increasingly influence what and how students learn in college, which is likely to directly shape the extent to which and how graduates individually benefit from the degrees they earn. These companies' product development efforts will be shaped by the organizations' own interests and capabilities, as well as those of their members. To date, there has been little research on such efforts or their potential impact on individual-level educational and economic outcomes, including whether students leave higher education with a degree that represents newly acquired human capital or that is primarily a signal of innate personal characteristics or cultural fit. Furthermore, we know little about how digital courseware products might influence variance in outcomes across students or institutions. This study begins to explore these issues through an inductive, ethnographic investigation of one organization's efforts to develop digital courseware products.

## Method

### Setting

EduTech[4] has published print textbooks for over a century. It remains one of the largest companies operating in the industry. Like its competitors, EduTech has struggled in recent years due to the emergence of national markets for used and rental textbooks, from which it makes no revenue. Also like its competitors, the company has responded by attempting to reinvent itself as an educational technology company. Rather than focusing on marketing and selling print textbooks, EduTech has increasingly sought to offer digital learning tools, particularly courseware solutions, that both deliver and allow students to engage with content (e.g., through graded assignments). The hope is to establish a subscription-based business model and to reinvigorate profitability and growth.

In order to pursue its new strategic direction, EduTech needed to develop the new digital learning products it was going to sell. Initially, I entered EduTech in an authoring role and quickly became interested in the strategic change efforts underway.[5] At the time of my entry, EduTech's leaders had begun (and continued) to communicate the organization's shift to educational technology and employees had begun (and continued) to develop and sell the digital courseware products that would enact this strategic shift. I spent the next 15 months ethnographically observing these efforts as they unfolded, while collecting archival documents extending back to the initial introduction of the new strategy. These data allowed me to inductively derive the study's key findings. In the remainder of this section, I describe my data and analyses in more detail.

### Data

My data span a 2-year period starting at the beginning of 2015, when leaders first began to communicate EduTech's new strategic direction (i.e., the shift to digital products), through the end of 2016. The data were collected during 15 months of ethnographic fieldwork conducted in 2015 and 2016 as part of my larger doctoral thesis project. Throughout this period, I spent an average of 40 hours per week in EduTech's offices, observing organizational dynamics and collecting archival documents (see Figure 1).

I collected three types of data from every EduTech division involved with developing, selling, or servicing technology products for higher

*Pedagogy and Profit?*

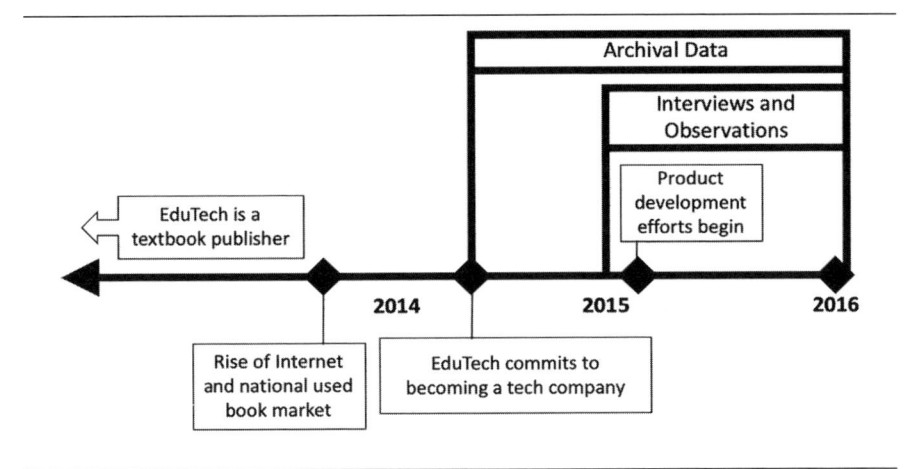

*Figure 1.* **Research timeline.**

*Table 1*
**Data Summary**

|  | Interviews | Observations | Archival Documents |
|---|---|---|---|
| Product | 74 | 187 | 782 |
| Sales and marketing | 23 | 15 | 90 |
| Customer/tech support | 9 | 4 | 2 |
| Other/multiple | 11 | 29 | 113 |
| Total | 117 | 235 | 987 |

education. These data allowed me to triangulate on how organizational leaders communicated about education and technology, how organization members understood these communications, and how their understandings influenced the development and sale of digital learning products. I continued to collect data until I reached theoretical saturation, or the point where additional data collection failed to generate any novel perspectives or information (Charmaz, 2014; Glaser & Strauss, 1967). The data are summarized in Table 1 and described in more detail below.

*Interviews*

I conducted 117 semistructured interviews with EduTech members from all areas of the organization, including product, sales and marketing, and customer and technology support. The interviews spanned organizational levels, ranging from entry-level positions to senior vice president roles. The interviews were recorded and transcribed verbatim. Each interview

*Regele*

followed the same protocol, which attempted to avoid informant bias by helping informants distinguish between what they felt or believed and objective events, behaviors, and facts (Golden, 1992; Miller, Cardinal, & Glick, 1997). The interviews largely followed this protocol, but were semistructured, which allowed me to adjust in real time based on participant responses.

The interviews covered four topics relevant to the present study.[6] First, participants were asked to describe their background and how they ended up in their current roles at EduTech. This line of questioning was important because extant research has highlighted the importance of prior experience in the construction of individual-level identity, which in turn influences how individuals understand their work (e.g., Ashforth & Mael, 1989; Cerulo, 1997; Dutton, Roberts, & Bednar, 2010). Second, participants were asked about their understanding of EduTech's shift to educational technology—for example, what they saw as the purpose(s) of the shift and what successful digital learning products would achieve. Third, participants were asked to describe their own roles and how they fit into the organization. This included a discussion of if and how they saw their roles as contributing to new focus on educational technology and digital learning. Fourth, participants were asked how and the extent to which the organization was fulfilling the promises associated with educational technology. Follow-up interviews ($n$ = 5) were conducted in cases where my analysis indicated gaps or raised additional questions.

The process of seeking theoretical saturation resulted in my interviews being disproportionately concentrated across 12 product teams in the product division of the organization. This concentration was due to members of the product division being responsible for developing and building the new digital learning products. This entailed conducting market research to understand faculty and student preferences and needs related to teaching and learning, including expectations about how products might affect the quality and accessibility of teaching and learning. Product division employees then had to make decisions about how to address these interests through product design. Furthermore, the product division was tasked with setting product prices. As a result, members in this area had to determine how to balance concerns related to increasing profit and increasing educational quality and access. In contrast, sales representatives had to communicate the claimed benefits of the products, but were not forced to make decisions about which interests to address. As a result, these individuals experienced the product development issues less directly and in fewer ways, making fewer interviews necessary to reach theoretical saturation outside of the product division (i.e., in sales and marketing and in customer and technology support).

*Observation*

Observation was conducted throughout the 15-month research period. Whereas interviews provided data about individuals' understandings and

## Pedagogy and Profit?

retrospective accounts of their experiences, observation allowed me to see how they acted out those understandings in real time. This helped address the shortcomings of each data type by disentangling relationships between how people think and act (Leonard-Barton, 1990).

While observing, I took extensive notes on a laptop, capturing as much of the dialogue as possible. In most cases immediately following the observation, but always within 24 hours, I reviewed my notes, expanding shorthand and adding detail. In total, I observed 235 discrete events. These observations included everything from sitting in town hall sessions led by the Chief Executive Officer, to shadowing a sales representative as he made sales calls on campus, to joining product teams during on-campus market research efforts. I continued to seek out observations until theoretical saturation was reached. Like my interviews (and for the same reason), my observations were largely concentrated in the product division.

### Archival Data

For the 15-month data collection period, I had an EduTech-issued laptop, email address, and access to the organization's internal IT systems. I also automatically received all company-wide e-mails. Ultimately, I collected 987 archival documents. These documents included everything from marketing communications, to training materials, to market research results. The documents ranged from the introduction of the shift to educational technology in early 2015 until I reached theoretical saturation and exited the field at the end of 2016.

In presenting my findings, I primarily use direct quotes from interviews and observations to demonstrate how organization members understood EduTech's product development and sales efforts, as well as its broader role in the higher education ecosystem. Nevertheless, although not necessarily quoted directly, archival materials were critical to my analysis. In particular, such materials allowed me to explore and verify how organization members' understandings were reflected in product development efforts (e.g., what market research was conducted), product characteristics, and sales efforts (e.g., which institutions and faculty were targeted and how).

**Analysis**

My data analysis followed an iterative, grounded theory approach. First, I read through, catalogued, and took summary notes on the data to develop an overarching understanding of the material. Second, I used NVivo to analyze the data on a sentence-by-sentence basis, following an open-coding approach (Corbin & Strauss, 2014; Gioia, Corley, & Hamilton, 2013). For example, in one town hall meeting, the Chief Product Officer said,

> Listen, I didn't go to a school where they sit around discussing philosophy. I went to a good, honest state school. . . . One time I was talking to an Ivy League history professor that was saying that he would never use digital because it would take away his connection to his students. And I asked, "Well how many students do you have in a class?" And he said, "Oh, 18." And I was like, "Yeah, I am interested in classes with 200 or 300 students. Not the Ivy League."

During my open-coding effort, I assigned this comment the codes "supporting large courses" and "supporting nonelite schools." Third, I organized initial codes into a series of categories and began to identify the connections between categories (Corbin & Strauss, 2014). For example, the code above was included in the category "types of schools/students served." As part of my analysis of member interpretations, I revisited the raw interview data with the relevant second-level categories and identified connections between members' interpretations of EduTech's goals, understandings of teaching and learning, and their decisions and actions. For example, the quote—along with others he made—demonstrated that the chief product officer saw EduTech's focus as serving large numbers of students at once. He suggested the organization should design products that supported large class sizes, not that enhanced the learning process itself. As part of this exercise, I interrogated the category descriptions and made adjustments to account for aspects that did not capture the interview data. This meant broadening the category descriptions to include a wider range of interpretations. Together, these steps allowed for the induction of conceptual insights that translated into the study's key findings (Charmaz, 2014).

## Findings

### Increasing Involvement in the Learning Process

Although EduTech has long attempted to make a profit by producing and selling higher education learning materials, the company's role in shaping what and how students learn is dramatically changing as it moves from print to digital products. Specifically, the company is attempting to move from marketing and selling content created by leading academics to the in-house development of content and pedagogical tools.

Like those of its competitors, EduTech's textbooks have long been used in college courses and been a key source of the content that students consume (see Watt, 2007, for an overview of the industry's history). As a result, the company has long played some role in shaping what students learn. However, historically, its royalty-based business model muted this role. Under this model, which prevailed across the industry, EduTech was essentially a sales and marketing organization and focused on providing a variety of content in formats matching various faculty preferences and teaching approaches:

*Pedagogy and Profit?*

> When I became an Acquisitions Editor, "acquisitions" was in the title. My job was mostly about acquiring new authors to write content for course areas where there were opportunities. And then to make sure authors were writing to curriculum standards and trends in the market, so we'd do a lot of market research. Really sort of weighing what the revenue opportunity was versus where we were already selling and making sure that the list filled all the different holes. I mean that was really what my job was about. And then working closely with my marketing counterparts to communicate the story to the sales reps and to the customers that I was responsible for. (Product Manager [4])[7]

As this quote suggests, EduTech (and higher education publishers more generally) sought out leaders in particular fields and encouraged them to write textbooks. The company, in turn, marketed these books to other faculty for use in their courses. Students purchased the books their professors selected. The publisher attempted to identify market trends, but authors were largely responsible for determining the specific content that was included in the books. The authors also bore much of the risk associated with a particular book:

> When you look at the risk in print publishing, the risk is just do you have the right people going out and finding the product. Because in print when you look at the model of when we start paying real money for a product, you have Professor Smith who has for the last five years been writing this manuscript, on their own dime. And you're like, "Okay great, so would you mind if I sent a couple chapters out for review?" So, you might have like $750 that you spent to like float that idea. . . . The author is doing everything . . . as a company, you haven't really spent any money testing out that idea. . . . At that point, you are like, "Okay let's sign a contract." They give you the whole manuscript. You review that, so that's maybe another $1,000, maybe. Everybody says, "Oh, God, I want to use this now, it's so great." You've barely spent $2,000 and you have a pretty good idea that you are onto something. (Director of Content [5])

Under this model, the authors might receive a small advance for writing the manuscript, but were primarily paid through royalties, which they received only if the book sold well. This royalty-based model also meant that it cost relatively little for publishers to introduce new books and that they benefited from assembling a large portfolio of books. Essentially, they sought to have an offering that suited every course and teaching style.

Textbook publishers' (both EduTech and its competitors) business model has shifted substantially as they have moved from the provision of print books to offering digital products. First, whereas textbooks influence what information students consume, courseware solutions and other digital learning products increasingly shape how students engage with that information. For example, students increasingly complete assignments—including those that count for a grade—in the products themselves. This change has shifted the function of the publisher, which now plays a more direct role in developing the technology and tools through which these assignments are delivered,

> What has really happened . . . is when we moved to digital, all of a sudden . . . we were developing all of . . . these other things [in terms of assignments]. . . . All of a sudden our role as [a company] . . . got amped up. Because now the authors didn't know what we were developing here. We had to figure that stuff out. Even our best attempts at getting the authors to help us work on that stuff has not gone well [because they are still focused on the textual content]. (Product Manager [15])

> [Moving to digital] absolutely changes what the products are. In the past, we were in the business of putting information out there. When you have a textbook, you are just delivering information to people. Now, we're engaged in the whole learning process. We really aren't putting out textbooks anymore. We're putting out complete classes. (Instructional Designer [27])

Second, the move to digital products has shifted product economics. Digital products are more expensive to develop and maintain. Interactive content and assignments require extensive software coding and programming work. There are also ongoing costs related to hosting and backing up the associated data on servers. These costs—and the associated risk—cannot easily be pushed onto authors:

> In a digital product, you have to spend like $5,000 just to get the manuscript. . . . The money up front for people to even test a product to see if they're going to like it, you are probably going to have to spend $10-20,000 upfront. So, just funding that is very different. The financial models are different from a company standpoint. The risk is a lot more. (Director of Content [5])

Recouping the higher development, hosting, and other costs requires products that scale. Rather than assembling a portfolio of options, standardization is prioritized:

> One of my biggest frustrations is as we move to this digital environment, the story of the book that you could articulate via the print world goes away. . . . Everything has become kind of generic. Even in the e-book, it is highly templatized. (Product Manager [4])

These shifting roles and changing economics have forced employees to revisit EduTech's and their own role in the educational process, as well determine what those roles imply with regard to product features and sales efforts.

*Pedagogy and Profit?*

**Interpretations of Educational Outcomes**

In general, EduTech employees expressed a strong desire to improve educational quality and access. Seventy-one percent (80) of the 112 employees I interviewed discussed improving education as a key motivating factor in their jobs, and a benefit of the new digital focus:

> The for-purpose and for-profit don't have to be mutually exclusive and I really believe that a lot of people here are here because they believe in education, they believe in learning. . . . They believe that they're helping people do that better . . . [People] feel like they're actually making a difference. They're contributing to something good. They're not just going and selling something or creating something to be sold, they're making a difference in people's lives. (Content Developer [101])

> This company wants to be as profitable as possible, but I do believe that what we've been doing and how we're trying to improve what we produce and how we . . . deliver things to students has been good with good intentions. With the intentions of helping the learning experience for people and making education more accessible to students through a digital platform. (Content Developer [19])

Although members' intentions largely appeared genuine and spanned the entire organization, these intentions were associated with interpretations of quality and access that were somewhat narrow, consistent with making a profit, and ultimately—if unintentionally—self-serving.

*Singular Focus on Earning a Credential*

Organization members typically equated access to higher education with access to a higher education *credential* (i.e., a diploma).[8] In general, employees rarely discussed what students actually learned in the process of acquiring the credential.[9] Rather, they focused on the credential itself, which they viewed as a pathway to a better life. At all levels of the organization, members discussed how a diploma could open up new—mostly economic—opportunities. Both organizational leaders and rank-and-file members argued digital products could open such opportunities to a greater share of the U.S. population:

> In 2013, [state university] was $135,000 in [digital product sales]. In 2016, it's $1.6 million [in digital sales]. (applause) And according to [EduTech sales representative], it will be at $2 million by the end of next year. So, that's a great story, isn't it? It's a great corporate story. But I think the really important story is that [sales representative] has 8,500 kids . . . who are having a better experience, are breaking a cycle. . . . That's . . . because she has a story about how [our digital products] improve peoples' lives. And when I look at all of you, this is your story. Are you salespeople? Sure, you're salespeople. Do you get

a quota? Sure you do. But to me, what your job is about is how you improve kids' and students' lives. (Excerpt from a speech by EduTech's chief product officer)

[EduTech's digital products are] life changing. We're starting to see that in some of the videos and things that we're seeing about how [the products have] changed a person's life because . . . they get through . . . they don't fail . . . they end up graduating. . . . With so many first generation students, [EduTech digital products] can be the difference, that's a life changer there. If they stay in college, graduate, I mean so to me, not to sound corny, but that stuff's now starting to be a part of everybody's lives here. And I don't think it was that way, you didn't see that with print textbooks. (Product Manager [15])

This interpretation of educational access was also framed by members at all levels of the organization as clearly aligned with EduTech's profit goals. The logic was that students would be willing to purchase products that increased their chances of obtaining a diploma:

If we can create products that . . . help students perform better and help faculty achieve their goals . . . that raises the value of our products in the eyes of our customers. From a company standpoint [that] puts us in a better position to hit profits because students and faculty say, "Yes, I will pay for this if it moves me closer to my goal." Which we talked about in the beginning [as being a diploma]. (Product Director [14])

### Focus on Grades, Rather Than Learning

Members at all levels and from all divisions also spoke about how the new digital products would improve the quality of education, or what and how students learned. However, when pressed on what they believed the digital products would change about learning and teaching processes, members mostly did not discuss such processes directly. Rather, they focused on learning outcomes, or more precisely, a single measure of learning outcomes. That measure was grades:

At the end of the day, if the students' grades improve because of our product, that's the win that we're looking for. That's the win that the instructors are looking for and that's what the students want, so we have to find a product that they're going to use and then we have to let the instructors know that they will use it, and convince them. (Product Manager [15])

The grade is the ultimate fruit for them, they want to have that. . . . The grade is the ultimate thing. . . . So, I think that the grade is the ultimate thing because that is the certificate of yes, you learned. (Software Development Manager [10])

*Pedagogy and Profit?*

Organization members repeatedly expressed the idea that if EduTech's courseware products improved grades, then those products would be driving more, deeper learning. Yet they had a difficult time explaining how or why this would be the case. Furthermore, they tended to argue improvements in the learning process itself were difficult—and unnecessary—to identify or measure. As long as grades were improving, that was enough:

> Engagement could be a lot of things. It could be that [students] go and talk to their instructor more, that they go to class and they pay attention, that they read the book, or that they study with their friends. It's very difficult for us to measure that in any kind of unified way. I think it's just really a catchall type term. (Product Manager [43])

> I think with [courseware products] what we could do ideally is help students prepare to have successes . . . on the test, the homework, those types of things. (Product Director [14])

Members also made it clear that tying products to improved grades was consistent with the desire to sell the products for a profit. They consistently expressed a belief that students would be willing to pay for better grades:

> If we can get to the point where we . . . can say, if you are willing to do this, most people who do this get an A. If an A isn't important to you . . . to get a B, here is what people do. . . . If you want a small coffee, you are going to pay this much. [Students] live in a world where they understand that if you buy a standard version of a car it's not going to have a sunroof. But, you know, I don't need a sunroof. So, I feel like that's where if we could get to that. . . . I think that that would be super powerful. (Director of Content [5])

> One of the questions I used to ask . . . was through a series of questions, like, "Hey, how much would you pay for an A in this class?" "You know, I'd give five hundred bucks." "You wouldn't have to come tomorrow, or anything, you'd just get an A." "Yeah, I'll give you five hundred." "What about for a C?" "Three hundred." You know, [often] students really only just want to get that grade. (Vice President of Product [28])

Indeed, organization members recognized making assignments completed in the digital products a part of students' grades as a way to compel them to purchase those products:

> If the professor requires technology as part of the grade, then the students have to buy an access code. . . . Students either have to buy it or depending on how much they care that it's a part of their grade, they may say, "I don't care, I'm still not going to buy it. I'll take a C in the class and just fail the [courseware] piece." I don't think many people are willing to gamble like that. (Sales Representative [81])

*Regele*

Members frequently spoke of a "magical 20%" number. They argued students were unwilling to sacrifice 20% of their grade; so as long as faculty made the digital assignments count for at least that proportion of the course grade, students would purchase those products:

> We're hoping [the digital product] gets to 20% of the grade. That's kind of a magic number for us. If an instructor assigns 20% of the grade to [the digital product], there's a really good chance the student's going to buy the [product] and use it and then actually improve their grade. (Product Manager [15])

**Impact on Product Characteristics**

EduTech members' perspectives on educational access and quality guided their product development efforts, encouraging the development of products consistent with signaling or credentialist interpretations of higher education. Specifically, they developed products that sought to make a diploma easier to achieve or lowered the hurdles to obtaining a particular grade, as opposed to improving teaching or learning processes. The result was products focused on exposing students to information and requiring them to demonstrate a basic level of comprehension, rather than engagement or higher level thinking. In the rest of this section, I discuss how interpretations of educational access and quality affected products in more detail.

*Maximizing Access to Credentials*

Organization members' diploma-focused perspective of educational access encouraged product characteristics that maximized the number of students that could be served by a particular instructor and that minimized the effort needed to complete one's coursework and obtain a diploma. Maximizing the number of students that a faculty member could serve was accomplished by minimizing the amount of time and effort the faculty member had to spend teaching or evaluating each student. Members argued that this, in turn, would reduce the cost of providing an education. The implication was that these savings would be passed on to students, making a degree affordable for a larger number of students:

> [Digital products can] allow schools that teach hundreds of students at the same time . . . to get rid of some instructors because now one instructor can [teach more students]. She used to teach 200 students and now she can teach 500 students using our online tools. (Instructional Designer [85])

Products that provided self-contained lessons (i.e., that both delivered content and assessed the learning), as well as auto-grading capabilities, helped achieve this goal:

> Auto-grading is a really big [selling point for instructors] . . . because the classes can be very, very large. It's not uncommon for someone to have a 200 or 300 student class. . . . Usually people hear digital and they think, "I don't have to grade." I think that's just synonymous [with digital]. (Instructional Designer [34])

> Our [courseware] products, everything is pre-built for [instructors]. They don't have to create the assignments. They don't have to find videos. It's all there in one place for them. I think that's a big selling feature and then the time saving is huge. . . . The fact that this stuff can be graded for them. [That] everything automatically flows to the gradebook is huge. (Sales Representative [58])

From the student perspective, organization members saw many hurdles that could prevent students from achieving high marks in a course and, ultimately, from obtaining a diploma. In order to maximize the number of students that were able to complete a degree, they sought to develop products that circumvented or minimized those hurdles. Organization members consistently suggested that the single greatest hurdle for students was a lack of time: Students were busy and so needed solutions that allowed them to complete their coursework and obtain good grades quickly and efficiently:

> Nowadays, students, they may have a couple of jobs. They may have a family. The net has been cast pretty wide, and students are leading very complex lives, and a lot of instructors are still teaching the way that [courses were] being taught 30 years ago. In actuality, students are looking for more focus. They're looking for more efficiencies because this is just one part of a very busy day for them, getting their work done. What we've been doing a lot of is crafting products that actually speak to how students actually live their lives. (Marketing Manager [33])

> [Students] want an A. They want a good grade. They don't want to spend forever doing the coursework and getting that A. If they learn something along the way, long term, and not just regurgitating it and learning it to spit it back out on a test, great, but it's really just graduating through grades and balancing the social life and everything at the same time. (Content Developer [101])

> If we could give students material that would help them get an A easily, we would definitely do it, but what is that material? What is the material that would do that for them? That's what we're always trying to find out. It's not as simple as it sounds. (Content Developer [92])

To address such student concerns, members focused on developing products that fit students' preferred ways of working. This included minimizing the time and effort spent completing assignments, especially time spent learning material not directly addressed in a homework problem or test question:

*Regele*

> I'm not of the mindset . . . that says, "Students are lazy now," or "Students, they don't even know how to do basic math. I can't believe everything's with a calculator." Well, students are more efficient now. . . . I just think . . . we need to meet them where they are. A student's not going to want to read a 20-page, 25-page chapter. That's not the way they're built. Students do not do that. (Content Developer [108])

> We see research with students too that they're not sitting down and reading a 30-page chapter and then okay, now let me go do my homework or whatever. They start with the homework. Go find what they need to find and they go do that. And . . . I kind of look at it like they're trying to do it in the most effective and efficient way possible. . . . We shouldn't knock students for that. . . . Our products need to . . . meet the way that students are really working these days. (Product Director [17])

> Students benefit from information in short bits, being very brief and prescriptive. Prescriptive is saying, "Here's what you need to know and here are your next steps." (Marketing Manager [25])

In fact, members often suggested students should not be asked to complete any work for which they would not receive some objective credit, which typically meant a grade:

> [We won't have achieved our goals] until we've got some kind of mobile device that's spinning off data and says, "Just study these three things because you're just about to go into class. This is what you don't know, and this will help you get the A." (Vice President of Product [28])

*Maximizing Grades, Not Learning*

Organization members' grade-focused perspective on educational quality encouraged the development of products that led directly to higher grades. Specifically, members suggested products could drive higher grades in three ways. First, members argued that auto-grading reduced the challenges associated with grading work. As a result, instructors could require students to complete graded assignments, which—since students' primary concern (at least in organization members' eyes) was grades—would force them to complete the work. Members were confident this would result in more learning:

> [Faculty] felt like students were not engaged, they weren't keeping up with the content so they started making them watch these videos, and [we created] quizzes tied to [the videos] and that sort of thing. . . . It forces them to keep up with the content. (Sales Representative [58])

> I think [instructors'] initial appeal to the digital solution is that the digital solves some sort of problem that they have that the print world

doesn't do for them. And a good example of that is that my students don't come to class prepared. I can use digital to require my students to complete some activity so that now when they get to class they've done something that they weren't doing before. (Product Manager [4])

Second, organization members attempted to build products and assignments with very clear, explicit pathways to a particular grade. Members expressed a checklist-like view of learning, suggesting if students completed a distinct set of tasks, they would receive a particular grade, which would be a clear demonstration of their learning:

> I think it's a hard argument to say that it's not a good idea if you tell a student. . . . "If you spend four hours a week and you watch these videos, you work through these problems, you go to class, these are the things you have to do and you are going to end up getting the grade that you want." I mean, who can argue against that? We are helping them. It's like you tell your kids, "You do these chores and you get an allowance." I mean this is like behavior-reward. (Director of Content [5])

> When they get the product, then they should have something that says, "Hey, try this out. A, B, C, and D, and now you're smarter." (Product Manager [43])

In addition to making the path to a particular grade clear, members also sought to give students multiple chances to demonstrate their "learning," by allowing them repeated attempts at an assignment. Auto-grading also helped facilitate giving students multiple chances at an assignment, as instructors did not need to grade each of these attempts:

> It's kind of how the instructor sets it up too. If an instructor is a little more harsh on the settings like, "Okay you guys only get one try and that's it and it's X percent of your grade" [students are unhappy], whereas some instructors will say, "You guys get unlimited attempts." Then they're like, "Oh, cool. I get as many tries as I want so I can make sure I get that high grade." It doesn't seem like as much pressure. It's more of a positive experience. (Sales Representative [58])

Third, to a certain extent, EduTech's grade-based interpretation of quality was consistent with broader arguments about accountability observable at all levels of education (e.g., Koretz, 2017). Accountability advocates suggest student learning can be improved through frequent assessment of what students know and how this changes throughout a course or degree program (e.g., Koretz, 2017). EduTech members believed frequent, auto-graded assignments were a way to hold instructors and institutions accountable:

> [Frequent] assessments [are a big benefit of our courseware products]. Accountability is huge. How can we show increases in learning from

week one to week 16 or from year one to year two? These are questions that universities are increasingly having [to answer]. Now we're trying to say, "Well look, we have solutions for that. We can help you with those kinds of things." Every one of those is a gradual release of responsibility [from the] teacher . . . and more [responsibility] on our [end]. (Content Developer [92])

We hope that [instructors] are committed to their students' success. We hope. If we want to assume that that's the teacher we're talking to, then they are concerned about their students having the right material to achieve the course objectives, and . . . . being able to report on specific outcomes for the department, or for the administration . . . or for accreditation, and there could be different criteria that they have to meet for those different sources. (Sales Representative [86])

In short, although in many cases grades might be an effective way to evaluate learning, the ways EduTech members conceptualized the link between learning and grades corrupted this relationship. Typically, we would expect a student to develop knowledge and skills that would be demonstrated on an assessment and reflected in a grade. At EduTech, however, members to a large extent reversed this relationship. Specifically, they focused on increasing grades, which they presumed would mean students had learned more. This logic is flawed because they were also determining the criteria on which a grade was determined. As a result, they might not be generating more learning, but rather altering the amount of learning that was required to earn a particular grade. In this case, the outcome seems likely to be less learning, especially given their efforts to tie grades to the completion of a task, as opposed to demonstration of mastery or skill.

**Counterfactuals**

Thus far, I have emphasized the prevailing interpretations of educational access and quality at EduTech, and how these interpretations influenced product characteristics. Although dominant, such views were not universal. Thirteen percent (15) of my 112 interview participants, all of whom had prior teaching experience, expressed views of higher education more consistent with a human capital development perspective. They viewed learning as being about developing critical thinking abilities and transferable skills, rather than demonstrating basic comprehension of specific content:

The products that we're providing, if it is helping a student get a degree that's going to help them improve their life in some way, then that's a good thing. At the same time, it may not necessarily be education. If students really want to [learn], and really . . . retain this understanding, there are probably better ways . . . than the approach that our product might take . . . [To truly learn,] they aren't just checking things off. It's a conflict. It's constantly in conflict. (Content Developer [38])

> [Most EduTech employees] don't see knowledge as an accretion; they don't see that building knowledge is like making pearls. They think [students] are just collecting sand and if they get enough of them then zip zap, they've got their beach built and they can move on. Instead you take that one little sand granule and you layer and you layer and you hone it, and if you really work at it and cultivate it you can end up with a really cool pearl. Otherwise it's one of those irregular things that looks like somebody ought to stomp and take out of existence. (Instructional Designer [93])

In several cases, these individuals attempted to incorporate their views of learning into products, proposing features that used technology to directly enhance the learning experience. For example, one product manager developed a mockup of a business simulation in which students would need to analyze a business problem in real time. The simulation would provide various forms of data (e.g., videos of meetings and spreadsheets of financial data), which students would need to analyze to reach a decision. Another product manager proposed a simulation that built on a "privilege walk" exercise often used in sociology courses. The idea was not focused on covering particular material or testing students on what they knew, but on developing an appreciation of the complexity of inequality and the many challenges it creates.

Yet, although individuals occasionally proposed these types of product ideas, they rarely, if ever, made it beyond a rudimentary prototype. Most often, the product ideas were seen as too expensive or complicated to build, which would make them unprofitable:

> In terms of innovation, [the business simulation] is a brilliant idea, but in terms of the company's ability to answer that need . . . it requires a huge infrastructure that we don't have. . . . We don't have the capability to build it. . . . The use of it cannot sustain the cost of building it. So, in terms of building a digital product that meets the needs of the market, I mean we were showing wireframes to people and they were lining up, saying, "When will it be ready, we want it now." And, you don't often get that. I haven't seen a response like that in other things that we've been doing. (Product Manager [6])

My data suggests not only that there were few organization members who held these divergent views of learning in the first place but also that they were becoming increasingly less likely to express these views. First, these members continually indicated that they felt marginalized and treated as if they were troublemakers:

> It seems like I'm saying, "This is where I can be most useful," but I don't have any power. They have decision-making meetings without me. I don't have any power or respect, in general. I don't have any prestige in the organization. I can get invested in something, and then I get pulled off. . . . They don't necessarily tell me what's

happening. I just will find out that I've been put on another project. I feel very disempowered. (Instructional Designer [79])

I have to be real careful because my job is to write for [products that have been built]. I haven't come up with a way, I can't . . . let my own passion for [learning] really come out. I wish someday that I could. When . . . we could write content that would explore beyond just the mantras of the past kind of stuff. But . . . I'll never live that long. (Instructional Designer [93])

Second, turnover (i.e., exit from the organization) was much higher among individuals who held such divergent views than among other organization members. Specifically, turnover was 47% among informants that expressed such views versus 15% for everyone else I spoke with.

**Impact on Product Sales Efforts**

Members' interpretations of access and quality also affected how and to whom the digital products were sold. Specifically, sales representatives were incentivized to target large courses at nonelite colleges and universities. Individuals associated with such courses also tended to be the most receptive to their sales pitches:

A lot of the classes that we sell [to] are going to be the lower level classes, the intro level classes. Those are the ones that we have the most digital on. (Sales Representative [65])

One of the things is that the online classes are growing, especially in our community colleges. . . . For those instructors, especially the ones that teach online, they more than see the value because they are not going to be able to get in front of their students to lecture. . . . I think sometimes the best thing to do is kind of convince the online faculty and get them to love it. (Sales Representative [81])

This focus was recognized and encouraged by the rest of the organization as well:

There are some instructors . . . adjuncts . . . [as an adjunct] I want something more short and easy where I'm not making a commitment to long-term. . . . There are the instructors who are really caring for the students to learn and they want to impart all this knowledge. Those ones I find are a bit more hesitant . . . because with digital, things are getting standardized and are getting put in a cookie-cutter mold. . . . So, that is the one side. The other side is, which there seems to be more of an increase in, is more the adjuncts. . . . For them, whatever gets the work done quickest, fastest, and in a repeatable manner is something that they will be willing to try. There are more open to [digital products]. (Software Development Manager [10])

*Pedagogy and Profit?*

> A lot of the instructors in community colleges, they have jobs. They have day jobs. This is their nights and weekends, augmenting salary, being an adjunct professor, something like that. They don't have time to do a lot of prep work. They don't have time to pore over some text. They don't have time to do a lot of grading. They don't have time to prepare exams and homework assignments. They're much more likely to grab [a digital product], a turnkey solution. They just got to show up, they've got their slides, they talk a little bit, tell them, "Okay your homework's due three weeks from today," but the assignment's already created by us, so they're able to go. They're not having to do the big time investment. . . . If you see a lot more enrollment in a community college for a specific discipline . . . you'll see a much bigger adoption on the digital side. (Software Development Director [44])

The focus on large courses and nonelite institutions is consistent with the product characteristics discussed above. First, from the faculty perspective, large courses make auto-grading and prebuilt lessons especially valuable:

> [Driving] the [use of] digital . . . is that the lecture sizes are getting bigger. From a class management perspective, [instructors have] more and more students. In the past, they were like, "Well I can handle this on my own, with a few [teaching assistants]." Now they're like, "Gosh, I've got 800, or I've got 400, [students]. I've got to have something like a [courseware product] to help me manage that." I think it's probably more that than anything that's really [driving the shift to digital]. (Product Manager [56])

Second, from a student perspective, organization members argued that these are often the schools, courses, and students for which the primary focus is especially likely to be obtaining a diploma, not the learning itself:

> Community college students . . . they generally show up expecting sort of a service relationship. They just pay their tuition, or maybe their financial aid came through, and they're registered in the class, and they sort of have an expectation that they're starting out with an A+ and the teacher is a service provider. . . . I think very often we hear teachers dumbing down their content and approach because they need [better] outcomes, and students just won't do the work. (Sales Representative [86])

> I wish that I had more upper level courses to know what students who actually have an interest in the course would think about the difference between [digital] products and traditional products, but I don't [target] a lot of upper level [courses] . . . [instead] we've always [focused on] hitting those main student pain points of, "I don't want to be in this course anyhow. Just give me exactly what I need in the briefest amount of time to get whatever grade I've already predecided that I'm happy with. If that's a C, a B, or an A. Give it to me for the lowest price possible." We've always hit on most of those marks. (Product Manager [30]s)

## Discussion

Although normative questions remain about the desirability and appropriateness of this trend, neoliberals' growing influence over higher education has increasingly raised questions about how a higher education degree might influence individual-level economic outcomes (Abel & Deitz, 2014; P. Brown et al., 2011; Labaree, 1997). This study investigated a category of stakeholders often overlooked in conversations about such issues: for-profit providers of learning materials.

By examining one organization's efforts to improve individual-level educational and economic outcomes—while also generating a profit—through the development and sale of digital courseware products, this study demonstrates how profit concerns and limited understanding of pedagogy can generate learning materials and tools consistent with signaling or credentialist views of higher education (D. K. Brown, 2001; Collins, 1979, 2002; Spence, 1974). These products may do little to develop human capital or advance what and how students learn, particularly with regard to acquiring new capabilities and skills, and instead reinforce traditional teaching and learning practices, or even detract from learning (Ringstaff & Kelley, 2002; Wenglinsky, 1998). Perhaps even more troubling is the possibility that even if the products increase the accessibility of a postsecondary diploma, they may exacerbate differences in the quality and value of the diplomas, and how these factors vary across students and institutions (e.g., Brewer, Eide, & Ehrenberg, 1999; Collins, 1979, 2002; Rivera, 2011).

Credentialist and human capital signaling perspectives of higher education focus on the employment opportunities created by the possession of a postsecondary diploma, as opposed to the knowledge and skills presumably developed while acquiring that diploma (D. K. Brown, 2001; Collins, 1979, 2002; Spence, 1974). I find that EduTech members' understanding of educational access was largely consistent with such perspectives. This understanding is congruent with much recent policy discussion about higher education, as well as academic studies that measure and tout the financial returns associated with a postsecondary diploma, including the types of degrees offered by institutions in EduTech's target market (e.g., Brewer et al., 1999; Dadgar & Trimble, 2015; Diprete & Buchmann, 2006). Organization members largely saw the company's role in improving educational access as making the provision and acquisition of postsecondary diplomas as frictionless as possible. They focused on developing products that achieved this goal. Members believed this was consistent with generating a profit because students and universities would embrace products that made it easier to acquire an educational credential. This logic is potentially problematic with regard to individual-level educational and economic outcomes for at least four reasons.

*Pedagogy and Profit?*

First, the more successful EduTech members are in making a postsecondary diploma easy and cheap to obtain, the more they lower the value of that diploma as a labor market signal. If everyone is able to obtain a diploma, the credential becomes less of a differentiator (Collins, 1979, 2002). Instead, the degree might become "table stakes" for being considered for a position. This may even extend to positions for which the skills and knowledge associated with the degree are largely irrelevant (Collins, 1979, 2002; Thurow, 1975). In fact, empirical evidence suggests such dynamics may already exist in the U.S. labor market. For example, 70% of high school graduates enroll in postsecondary education (U.S. Bureau of Labor Statistics, 2017) even as data suggest only about a third of jobs currently require this level of education (U.S. Bureau of Labor Statistics, 2018). Yet, at the same time, possession of a college degree is increasingly used as a first level filter for a wide range of jobs (e.g., Cappelli, 2012; Grubb & Lazerson, 2004). EduTech's product development and sales efforts may unintentionally contribute to this type of degree inflation (Collins, 1979, 2002; Thurow, 1975).

Second, in focusing on the signal of the diploma itself, academic research and policy discussions often fail to adequately consider the human capital and productivity gains that are supposedly captured by this signal (e.g., Becker, 1993; Schultz, 1961). This human capital—what capabilities and skills students have developed—is the (admittedly difficult to measure) outcome that determines, from an economic perspective, the true quality and value of an education. Perhaps even more worrying than their impact on the value of a diploma as a signal, organization members' efforts to develop products that make obtaining a degree easier and faster may unintentionally hinder the development of this human capital and, as a result, lower the real economic value of an education, both to the individual and society (Means, 2018). This risk exists because time and effort—and the associated struggle with difficult ideas and problems—underpin learning and human capital development (e.g., Arum & Roksa, 2010; Sandberg & Barnard, 1997; van Gelder, 2005).

Digital courseware products also may reduce learning by allowing courses to be delivered with less direct interaction between students and faculty (e.g., through standardized, online courses or assignments) (Sims, 2017). Such personal interactions have been shown to be a key predictor of learning, particularly when these interactions mediate the use of technology (Paino & Renzulli, 2013). As a result, even if EduTech is successful in developing products that help students acquire a degree—and this degree helps them obtain a job—the products may inadvertently limit their preparation for that job.

Third, organization members' obsession with grades may further limit learning and human capital development. Particularly problematic is the fact that the grades are increasingly endogenously determined by assignments completed *within* the digital products. Indeed, sales representatives

1151

encouraged instructors to make such assignments as large a proportion of the course grade as possible. Making product use part of the course grade forced students to purchase that product, which increased EduTech's revenues and profits. At the same time, though, EduTech was attempting to increase educational access by making a diploma easier to obtain. One way that EduTech did this was by making required assignments quicker and easier to complete, and high grades easier to achieve. Since, as already discussed, learning is often inherently time-consuming and difficult, such product features may reduce learning (Sandberg & Barnard, 1997; van Gelder, 2005). A focus on frequent assessment exacerbated this issue by encouraging members to prioritize auto-grading, which typically relied on simple close-ended problems and evaluated basic comprehension, rather than complex critical thinking or skill development (van Gelder, 2005).

Finally, organization members' focus on degrees as labor market signals—or, more precisely, product development and sales efforts based on this perspective—may have negative, if unintended, effects on inequality within the broader higher education system. This argument builds on findings at the K–12 level, which demonstrate that market competition can have dissimilar effects on different types of educational institutions (Davies & Quirke, 2007). At the higher education level, for-profit educational technology companies' focus on community colleges, large state schools, and other institutions that disproportionately serve first-generation college students and individuals from underprivileged backgrounds may exacerbate differences in educational quality between these schools and more elite, selective alternatives.

EduTech members described the organization's focus on nonelite institutions as both an effort to increase higher education access and consistent with profit goals; however, this also means these institutions may be disproportionally affected by the negative aspects of the courseware solutions discussed above. Indeed, thus far, courseware solutions generally seem to have had less affect on elite universities (Allen & Seaman, 2007). Such schools continue to hire more knowledgeable, experienced, and skilled instructors and to emphasize low student-to-faculty ratios and small class sizes. In part, this is likely because such measures remain key components of many college rankings, such as those compiled by U.S. News and World Report (Morse, Brooks, & Mason, 2017). These factors encourage better pedagogy, more personal interaction, and more complex assignments, and lower the likelihood that EduTech-type courseware products are adopted. Furthermore, even if courseware products are adopted in these more elite institutions, faculty members' greater pedagogical knowledge means those products are likely to be used more effectively than in nonelite institutions. Indeed, this latter prediction is consistent with prior research at the K–12 level, which finds technology is used less effectively in schools serving less privileged students than in wealthier institutions (Wenglinsky, 1998).

To the extent that such patterns exist and persist, courseware solutions and other digital products are likely to exacerbate disparities in the extent to which students at elite and nonelite institutions develop human capital as they complete their degrees. This is particularly troubling given that, whereas students at elite institutions might also develop social and cultural capital, the (supposed) private benefits associated with attending a nonelite institution are almost entirely tied to human capital gains (e.g., Anyon, 1980; Bourdieu & Passeron, 1977; Rivera, 2011).[10]

Scholars have paid surprisingly little attention to the growing use of digital learning products developed by for-profit businesses at all levels of education, even as such products have become increasingly central to what and how students learn. This study takes a step forward in addressing this important issue by considering how members of such businesses understand their own work and that of their organization, as well as how these interpretations influence the development and sale of their organization's products. Because so little was known about such issues, an inductive, qualitative research design was most appropriate; however, like any research design, this approach also has its limitations.

Triangulated, ethnographic data spanning more than a year provided a depth of understanding about the organization's product development and sales efforts that would not be possible with other methodologies; however, this approach also potentially limits the generalizability of my findings. At a minimum, I am confident my results would extend to other organizations making a similar transition from print textbook publishing to educational technology. Indeed, such organizations often make very similar claims in marketing materials and press releases and their products often contain very similar features. Furthermore, there is a great deal of employment mobility between these companies. Many of my interview subjects—and many of the company's senior executives—had prior experience working for close competitors and described similar dynamics in those organizations. Nevertheless, additional work is needed to systematically test and quantify the dynamics I discuss. In particular, perhaps the most interesting arguments to emerge from the study relate to broad issues of educational access and quality, and how digital learning products may affect access and quality across different types of higher education institutions. The study suggests for-profit businesses' product development and sales efforts may have dramatic implications for such issues; however, my research design does not allow me to measure those implications directly. Thus, additional work is needed to examine such possibilities, both from the perspective of organizations such as EduTech and the higher education institutions themselves.

**ORCID ID**

Matthew D. Regele https://orcid.org/0000-0003-0304-7886

## Notes

[1] I use the term *neoliberal* to broadly refer to market-based perspectives that conceptualize higher education primarily as a private good and form of capital that imparts economic benefits on those who complete a degree. For example, see (Pusser, 2006) for a useful discussion of these perspectives and how they compare to conceptualizations of education as a public good. I also wish to emphasize that this article is not intended to endorse neoliberal perspectives. Rather, my goal is to examine how their spread—including to the development and provision of educational materials—may (negatively) affect the economic outcomes those perspectives prioritize.

[2] Consistent with Note 1 and the expressed purpose of this article, all three of these explanations are consistent with neoliberal conceptions of higher education (i.e., education as a "private good"). As indicated above, I am not arguing that such a conceptualization is normatively appropriate. Rather, my goal is to understand the implications of neoliberal conceptualizations when they are applied to educational decision making, in this case the development, sale, and use of learning materials. This includes whether private education-related economic benefits are realized, the specific nature of those benefits, and who realizes them.

[3] At times, both academics and laypeople conflate human capital signaling and credentialist theories, or use them interchangeably. In such cases, these individuals seem to mostly be referring to human capital signaling arguments. The distinction and language I use here follows those of Brown (2001) and Collins (e.g., see Collins, 1979).

[4] To maintain anonymity, EduTech and all other names in the article are pseudonyms.

[5] I was officially an employee of EduTech throughout the research period, which occurred during my doctoral studies. Although the organization utilized my authoring work, my employment was largely the research sponsor's way of overcoming hurdles that would have limited my access to EduTech's IT systems and buildings. My responsibility as an employee was to author content for various EduTech products; however, I was not responsible for designing or making decisions about the products or their characteristics. I had no managerial responsibility. As a result, although study participants knew I was technically an employee, it was clear (and I emphasized) that my primary role was as a researcher. I also emphasized that I had no ability or desire to evaluate their actions or decisions from a performance perspective and made it clear that anonymity would be preserved in any findings communicated to EduTech leadership. Senior management agreed to these terms when they approved the research project and guaranteed complete freedom in publishing my results. Finally, it was made clear to senior management that I would sever my employment at the end of my data collection.

[6] As indicated above, the data were collected as part of my larger doctoral thesis project, which investigated organizational change more broadly. As a result, other portions of the interviews considered organizational issues that extended beyond the education-related aspects of the change discussed in this article.

[7] Numbers in brackets (e.g., [4]) are unique identifiers for each informant.

[8] I consciously use the terms "credential" and "diploma" interchangeably in this context. At the time of my study, EduTech was, at most, just beginning to discuss certificates, microcredentials, and other nondiploma-based credentials. It was not currently offering any products used in such programs nor were there any ongoing or planned product development efforts focused on such credentials. Rather, its products were almost entirely designed for "traditional" college degree programs (i.e., bachelor's and associate's degrees).

[9] Of course, the small fraction of employees who actually wrote content (e.g., homework questions, tests, and other assessments) had to consider what students "should learn." For the most part, though—consistent with the discussion of product characteristics below—these individuals were expected to create close-ended questions that closely mirrored the content in existing texts. They were actively discouraged from raising concerns about pedagogical effectiveness. Indeed, as discussed in the section on "counterfactuals" below, dissenting views were typically marginalized or ignored. Thank you to an anonymous reviewer for raising this issue and highlighting the need for clarification.

[10] Thank you to an anonymous reviewer for raising this point.

## References

Abel, J. R., & Deitz, R. (2014). Do the benefits of college still outweigh the costs? *Current Issues in Economics and Finance, 20*(3), 1–9.

Alavi, M., & Leidner, D. E. (2001). Research commentary: Technology-mediated learning: A call for greater depth and breadth of research. *Information Systems Research, 12*, 1–10.

Allen, I. E., & Seaman, J. (2007). *Online nation: Five years of growth in online learning*. Retrieved from http://www.onlinelearningsurvey.com/reports/online-nation.pdf

Amiel, T., & Reeves, T. C. (2008). Design-based research and educational technology: Rethinking technology and the research agenda. *Journal of Educational Technology & Society, 11*, 29–40.

Anyon, J. (1980). Social class and the hidden curriculum of work. *Journal of Education, 162*(1), 67–92.

Apple, M. W., & Christian-Smith, L. K. (Eds.). (1991). *The politics of the textbook*. New York, NY: Routledge.

Arrow, K. J. (1973). Higher education as a filter. *Journal of Public Economics, 2*, 193–216.

Arum, R., & Roksa, J. (2010). *Academically adrift: Limited learning on college campuses*. Chicago, IL: University of Chicago Press.

Ashforth, B. E., & Mael, F. (1989). Social identity theory and the organization. *Academy of Management Review, 14*, 20–39.

Barber, J. P. (2012). Integration of learning: A grounded theory analysis of college students' learning. *American Educational Research Journal, 49*, 590–617.

Baron, N. S. (2015). *Words onscreen: The fate of reading in a digital world*. Oxford, England: Oxford University Press.

Battilana, J., & Lee, M. (2014). Advancing research on hybrid organizing: Insights from the study of social enterprises. *Academy of Management Annals, 8*, 397–441.

Battilana, J., Sengul, M., Pache, A.-C., & Model, J. (2015). Harnessing productive tensions in hybrid organizations: The case of work integration social enterprises. *Academy of Management Journal, 58*, 1658–1685.

Becker, G. S. (1993). *Human capital: A theoretical and empirical analysis, with special reference to education* (3rd ed.). Chicago, IL: University of Chicago Press.

Bleemer, Z., & Zafar, B. (2018). Intended college attendance: Evidence from an experiment on college returns and costs. *Journal of Public Economics, 157*, 184–211.

Bok, D. (2004). *Universities in the marketplace: The commercialization of higher education*. Princeton, NJ: Princeton University Press.

Bok, D. (2013). *Higher education in America*. Princeton, NJ: Princeton University Press.

Bourdieu, P., & Passeron, J.-C. (1977). *Reproduction in education, society, and culture* (2nd ed.). Thousand Oaks, CA: Sage.

Brand, J. E., & Xie, Y. (2010). Who benefits most from college? Evidence for negative selection in heterogeneous economic returns to higher education. *American Sociological Review, 75*, 273–302.

Brewer, D. J., Eide, E. R., & Ehrenberg, R. G. (1999). Does it pay to attend an elite private college? Cross-cohort evidence on the effects of college type on earnings. *Journal of Human Resources, 34*, 104–123.

Brown, D. K. (2001). The social sources of educational credentialism: Status cultures, labor markets, and organizations. *Sociology of Education, 74*(Extra Issue: Current of Thought), 19–34.

Brown, P., Lauder, H., & Ashton, D. (2011). *The global auction: The broken promises of education, jobs, and income.* Oxford, England: Oxford University Press.

Cappelli, P. (2012). *Why good people can't get jobs: The skills gap and what companies can do about it.* Philadelphia, PA: Wharton Digital Press.

Cerulo, A. K. (1997). Identity construction: New issues, new directions. *Annual Review of Sociology, 23,* 385–409.

Charmaz, K. (2014). *Constructing grounded theory* (2nd ed.). Thousand Oaks, CA: Sage.

Christensen, C. M., Horn, M. B., & Johnson, C. W. (2008). *Disrupting class: How disruptive innovation will change the way the world learns.* New York, NY: McGraw-Hill.

Collins, R. (1979). *The credential society: An historical sociology of education and stratification.* New York, NY: Academic Press.

Collins, R. (2002). Credential inflation and the future of universities. In S. Brint (Ed.), *The Future of the City of Intellect* (pp. 23–46). Stanford, CA: Stanford University Press.

Corbin, J., & Strauss, A. L. (2014). *Basics of qualitative research: Techniques and procedures for developing grounded theory* (4th Ed.). Thousand Oaks, CA: Sage.

Dadgar, M., & Trimble, M. J. (2015). Labor market returns to sub-baccalaureate credentials: How much does a community college degree or certificate pay? *Educational Evaluation and Policy Analysis, 37,* 399–418.

Davies, S., & Quirke, L. (2007). The impact of sector on school organizations: Institutional and market logics. *Sociology of Education, 80,* 66–89.

Dees, J. G., & Anderson, B. B. (2003). Sector-bending: Blurring lines between nonprofit and for-profit. *Society, 40*(4), 16–27.

Deterding, N. M., & Pedulla, D. S. (2016). Educational authority in the "open door" marketplace: Labor market consequences of for-profit, nonprofit, and fictional educational credentials. *Sociology of Education, 89,* 155–170.

Diprete, T. A., & Buchmann, C. (2006). Gender-specific trends in the value of education and the emerging gender gap in college completion. *Demography, 43*(1), 1–24.

Domina, T., Conley, A., & Farkas, G. (2011). The link between educational expectations and effort in the college-for-all era. *Sociology of Education, 84,* 93–112.

Dominitz, J., & Manski, C. F. (1994). *Eliciting student expectations of the returns to schooling* (NBER Working Paper Series No. 4936). Cambridge, MA.

Dutton, J. E., Roberts, L. M., & Bednar, J. (2010). Pathways for positive identity construction at work: Four types of positive identiy and the building of social resources. *Academy of Management Review, 35,* 265–293.

Gioia, D. A., Corley, K. G., & Hamilton, A. L. (2013). Seeking qualitative rigor in inductive research: Notes on the Gioia methodology. *Organizational Research Methods, 16,* 15–31.

Glaser, B. G., & Strauss, A. L. (1967). *The discovery of grounded theory: Strategies for qualitative research.* Chicago, IL: Aldine.

Golden, B. R. (1992). The past is the past-or is it? The use of retrospective accounts as indicators of strategy. *Academy of Management Journal, 35,* 848–860.

Grubb, W. N., & Lazerson, M. (2004). *The education gospel: The economic power of schooling.* Cambridge, MA: Harvard University Press.

Haas, E., & Fischman, G. (2010). Nostalgia, entrepreneurship, and redemption: Understanding prototypes in higher education. *American Educational Research Journal, 47,* 532–562.

Hannaway, J. (1992). Higher order skills, job design, and incentives: An analysis and proposal. *American Educational Research Journal, 29,* 3–21.

Heinrich, C. J., & Nisar, H. (2013). The efficacy of private sector providers in improving public educational outcomes. *American Educational Research Journal, 50,* 856–894.

Hout, M. (2012). Social and economic returns to college education in the United States. *Annual Review of Sociology, 38*, 379–400. doi:10.1146/annurev.soc.012809.102503

Hursh, D. (2007). Assessing No Child Left Behind and the rise of neoliberal education policies. *American Educational Research Journal, 44*, 493–518.

Jacob, B., McCall, B., & Stange, K. M. (2013). *College as country club: Do colleges cater to students' preferences for consumption?* (NBER Working Paper Series No. 18745). Cambridge, MA. Retrieved from https://www.nber.org/papers/w18745.pdf

Koch, J. V. (2013). *Turning the page: An economic analysis of textbooks*. Retrieved from https://files.eric.ed.gov/fulltext/ED497026.pdf

Koretz, D. (2017). *The testing charade: Pretending to make schools better*. Chicago, IL: University of Chicago Press.

Labaree, D. F. (1997). Struggle over educational goals. *American Educational Research Journal, 34*, 39–81.

Leonard-Barton, D. (1990). A dual methodology for case studies: Synergistic use of a longitudinal single site with replicated multiple sites. *Organization Science, 1*, 248–266.

Lubienski, C. (2005). Public schools in marketized environments: Shifting incentives and unintended consequences of competition-based educational reforms. *American Journal of Education, 111*, 464–486.

Mangen, A., Walgermo, B. R., & Brønnick, K. (2013). Reading linear texts on paper versus computer screen: Effects on reading comprehension. *International Journal of Educational Research, 58*, 61–68.

March, J. G., & Simon, H. A. (1958). *Organizations*. New York, NY: Wiley.

Means, A. J. (2018). *Learning to save the future: Rethinking education and work in an era of digital capitalism*. New York, NY: Routledge.

Miller, C. C., Cardinal, L. B., & Glick, W. H. (1997). Retrospective reports in organizational research: A reexamination of recent evidence. *Academy of Management Journal, 40*, 189–204.

Mincer, J. (1974). *Schooling, experience, and earnings*. New York, NY: Columbia University Press.

Moore, M., & Kearsley, G. (2012). *Distance education: A systems view of online learning* (3rd ed.). Belmont, CA: Wadsworth.

Morse, R., Brooks, E., & Mason, M. (2017). *How U.S. News calculated the 2018 best colleges rankings*. Retrieved from https://www.usnews.com/education/best-colleges/articles/how-us-news-calculated-the-rankings

Nielsen, K. (2015). "Fake it 'til you make it": Why community college students' aspirations "hold steady." *Sociology of Education, 88*, 265–283.

Olneck, M. (1979). The effects of education. In C. Jencks (Ed.), *Who gets ahead? The determinants of economic success in America* (pp. 159–160). New York, NY: Basic Books.

Olssen, M., & Peters, M. A. (2005). Neoliberalism, higher education and the knowledge economy: From the free market to knowledge capitalism. *Journal of Education Policy, 20*, 313–345.

Paino, M., & Renzulli, L. A. (2013). Digital dimension of cultural capital: The (In)visible advantages for students who exhibit computer skills. *Sociology of Education, 86*, 124–138.

Pusser, B. (2006). Reconsidering higher education and the public good. In W. G. Tierney (Ed.), *Governance and the public good* (pp. 11–28). Albany: State University of New York Press.

Reiser, R. A. (2001). A history of instructional design and technology: Part I: A history of instructional media. *Educational Technology, Research and Development*,

49(1), 53–64. Retrieved from http://www.capella.edu/idol/HistoryofIDTPartI.pdf

Ringstaff, C., & Kelley, L. (2002). *The learning return on our educational technology investment: A review of findings from research*. San Francisco. Retrieved from http://rtecexchange.edgateway.net/learningreturn.pdf

Rivera, L. A. (2011). Ivies, extracurriculars, and exclusion: Elite employers' use of educational credentials. *Research in Social Stratification and Mobility, 29*(1), 71–90.

Roberts-Mahoney, H., Means, A. J., & Garrison, M. J. (2016). Netflixing human capital development: Personalized learning technology and the corporatization of K-12 education. *Journal of Education Policy, 31*, 405–420.

Sandberg, J., & Barnard, Y. (1997). Deep learning is difficult. *Instructional Science, 25*(1), 15–36.

Schrecker, E. (2010). *The lost soul of higher education: Corporatization, the assault on academic freedom, and the end of the American university*. New York, NY: The New Press.

Schultz, T. W. (1961). Investment in human capital. *American Economic Review, 51*(1), 1–17.

Sims, C. (2017). *Disruptive fixation: School reform and the pitfalls of techno-idealism*. Princeton, NJ: Princeton University Press.

Spence, A. M. (1974). *Market signaling: Informational transfer in hiring and related screening processes*. Cambridge, MA: Harvard University Press.

Spring, J. (2011). *The politics of American education*. New York, NY: Routledge.

Thompson, J. B. (2005). *Books in the digital age: The transformation of academic and higher education publishing in Britain and the United States*. Cambridge, England: Polity Press.

Thurow, L. C. (1975). *Generating inequality*. New York, NY: Basic Books.

Tyton Partners. (2014). *Time for class: Lessons for the future of digital courseware in higher education*. Boston, MA. Retrieved from http://tytonpartners.com/tyton-wp/wp-content/uploads/2015/03/EGA011_Course2_WP_Rd6.pdf

U.S. Bureau of Labor Statistics. (2017). *Sixty-nine percent of 2016 high school graduates enrolled in college in October 2016*. Retrieved from https://www.bls.gov/opub/ted/2017/69-point-7-percent-of-2016-high-school-graduates-enrolled-in-college-in-october-2016.htm

U.S. Bureau of Labor Statistics. (2018). *Employment projections*. Retrieved from https://www.bls.gov/emp/tables/education-summary.htm

Vail, H. H. (1911). *A history of the McGuffey Readers*. Cleveland, OH: Burrows Brothers.

van Gelder, T. (2005). Teaching critical thinking: Some lessons from cognitive science. *College Teaching, 53*, 41–46. doi:10.3200/CTCH.53.1.41-48

Watt, M. G. (2007). Research on the textbook publishing industry in the United States of America. *International Association for Research on Textbooks and Educational Media, 1*(1), 1–17. Retrieved from http://files.eric.ed.gov/fulltext/ED498713.pdf

Wenglinsky, H. (1998). *Does it compute? The relationship between educational technology and student achievement*. Princeton, NJ: Policy Information Center.

Manuscript received October 18, 2018
Final revision received July 3, 2019
Accepted July 17, 2019

# "Dear Future President of the United States": Analyzing Youth Civic Writing Within the 2016 Letters to the Next President Project

Antero Garcia
Amber Maria Levinson
Emma Carene Gargroetzi
*Stanford University*

*This article investigates the civic writing practices of more than 11,000 students writing letters to the next president in the lead up to the 2016 U.S. election. We analyze how letter topics are associated with socioeconomic factors and reveal that 43 topics—including ones prevalent among students such as immigration, guns, and school costs—were significantly associated with socioeconomic and racial majority indicators. Furthermore, we conducted a qualitative analysis of the kinds of arguments and evidence developed in letters from five schools serving predominantly lower income students and/or students of color in different regions of the country. Student arguments and types of evidence used were site dependent, suggesting the importance of teacher instruction. This analysis expands previous conceptions of youth civic learning.*

---

ANTERO GARCIA, PhD, is an assistant professor in the Graduate School of Education at Stanford University, 520 Galvez Mall, Stanford, CA 94305; e-mail: *antero.garcia@stanford.edu*. His research explores youth civic identities and contemporary literacy practices through the use of games and technology. Prior to completing his PhD, Antero was a high school English teacher. Based on his research, Antero codesigned the Critical Design and Gaming School—a public high school in South Central Los Angeles.

AMBER MARIA LEVINSON, PhD, is a research associate at Stanford Graduate School of Education. Levinson conducts research at the intersection of learning, equity, and technology in both formal and informal settings. Her past research has included studies of language minority families and their use of technology for learning, as well as students as creators of digital media. She holds a PhD in Learning Sciences at Technology Design from Stanford University.

EMMA CARENE GARGROETZI is a doctoral candidate in Stanford Graduate School of Education in the areas of race, inequality, and language in education and curriculum and teacher education. Her research examines the production and negotiation of sociopolitical identities in conjunction with academic and disciplinary identities in classrooms and schools.

*Garcia et al.*

KEYWORDS: civic education, classroom instruction, mixed methods, writing, youth identity

Two weeks before Donald J. Trump was elected the 45th president of the United States, Luis, a high school student in Ohio, published a letter to whomever would be elected in the national civic event. "Dear Future President," he begins, before diving into an issue that is of particular importance to him:

> We need to stop police brutality. We have cops killing our people and people killing our cops. This country we say is so called a free country has a war going on within it. We have fathers, mothers and children being killed everyday. Why must we lose our loved ones because someone else killed them. Everyone should die a natural way, not by being shot to death. All the racism has to stop. All the killings have to stop. All the funerals have to stop. We can't do this anymore. We have cops killing unarmed civilians. We have cops killing our children our brother/sisters. I don't want to have to deal with one of my loved ones dead, some of us have been through enough.

Luis's letter continues for two more paragraphs in which he describes personal experiences with violence ("I know what it's like to live on a battleground") and he connects the issues of violence with race and racism in America. Focusing his writing on this singular topic and the need for policy changes, Luis concludes his letter with a two-word, capitalized request: "Please Help." Sharing tacit knowledge as well as ethical and emotional appeals, Luis's letter to the then future president is a personal plea for executive action and leadership. His is an impassioned epistle on a crisis that continues to shape the political landscape of activism and protest in the United States today, more than 2 years after it was written. Most tellingly, as a high school student aware of a public audience reading his open letter, Luis voices the concerns of a generation of young people that are on the cusp of inheriting a political landscape embroiled in disagreement, distrust, and enmity.

Luis is one of more than 11,000 students from 321 sites across the United States that participated in the 2016 Letters to the Next President (LTNP) project. His letter was published alongside his classmates' and remains publicly available for anyone to read. His is also one of 1,065 letters that teen authors labeled as related to gun issues, emphasizing that Luis is not alone in his wish to see legislative leaders proactively address the issues of gun violence, police brutality, and protection of the lives of young people within schools and local communities. Written as a classroom activity, Luis's letter echoes the powerful forms of out-of-school activism that youth throughout the country lead. As potent and visible as they may be, these forms of activism also call into question how civic learning opportunities within schools have kept up with students mobilizing via hashtags, organizing distributed protests, and conducting other forms of savvy civic activities across the country. In contrast to existing literature on civic education's emphasis on gaps between the civic learning opportunities of students based on socioeconomic differences, voices like Luis's offer a glimpse at how forms of civic identity can be fostered and developed within classrooms for all students. In this study, we set out to understand what topics students chose to write about within the LTNP project and how those topics varied by school demographics. In addition, we explored how students developed civic arguments as well as the kinds of evidence they relied on in constructing their arguments. Together, these two interlinked lines of inquiry provide an understanding of how more than 11,000 students, most of them in middle and high school classrooms, express civic thought during the 2016 U.S. election.

Recognizing the many different demands, hopes, and fears voiced by youth within these letters, our analysis explores what issues were most important to the diverse students that took part in this widespread participatory civics project implemented in classrooms across the United States. Analyzing the topics of all the letters within the project and exploring how these topics differed based on socioeconomic indicators, we looked for what topics students intentionally engaged with and how these topics speak back to existing scholarship and assumptions about youth civic education. As perhaps the largest teacher-curated collection of youth writing tied to U.S. politics in recent years, the LTNP authors offer a substantial opportunity to understand youth civic perspectives and to explore how these projects *ready* students for participation in the political and civic landscape outside of schools (e.g., Allen, 2016).

The data also offer a unique look at civic issues through the eyes of middle and high school students; surveys such as Harvard University's Institute of Politics survey of "young Americans" (2016) collect responses from 18- to 29-year-old adults, but more research is needed to understand the viewpoints of teens across diverse contexts. While these letters do not constitute a nationally representative sample, they illustrate youth-understood boundaries of civic action. By looking at what topics students count as civically important and how students rhetorically frame these issues, we attempt to highlight the responsibilities of educators within a new landscape of *participatory readiness* (Allen, 2016, p. 27). Our findings call for civic education that listens to and centers the voices of Luis and his peers as emerging civic actors.

## Letters to the Next President Background

Cosponsored by the National Writing Project (NWP) and by public radio station KQED, LTNP served as a national opportunity for young people to share the topics that were most important to them in the midst of a media-saturated focus on politics and partisan debate. Alongside a visual database that allows visitors to search, find, and read every letter from the project, the LTNP website (https://letters2president.org) explains, "While candidates and media concentrated on issues that mattered to voters in the 2016 election season, teachers and students in our nation's schools concentrated on issues that mattered to the next generation of voters."

At the same time that Luis and his peers author impassioned, researched, and powerful civic arguments in their letters, they are oftentimes cast by media and academic literature as aloof and disengaged from the processes that govern them. A recent *New York Times* article looking at efforts to increase youth voter participation was strikingly titled "Wasted Ballots" (Gonchar, 2018); frequent representations of youth in contemporary media depict them as bemused slacktivists at best and politically toothless and inept in some instances. However, looking across the entire corpus of letters collected as part of LTNP, the voices of young people paint an important picture of what issues are most pressing to them and articulate actions they expect to be taken by their elected leaders. Representing classrooms from 47 states and Washington, D.C., the topics and content of these letters illustrate what students in late 2016 felt was part of the civic domain on which the next president should act. In this sense, these letters are markers of the contours of civic identity, civic engagement, and civic learning in schools today. In addition, a closer look at *how* students wrote about these issues connects the intersections of academic writing and youth civic agency within schools. These letters constitute one event, arguably an important one, in a long history of youth civic engagement both in schools and beyond.

## What Is Civic Engagement and Who Is It For?

Though widely seen as a topic covered in a high school civics course focused on the branches of the U.S. government, civic education is also, if not largely, attained in out-of-school contexts, through the social networks that youth interact with and through their interpretation of popular media and dominant news coverage. Educational research is increasingly exploring how civic learning must transcend the assumed disciplinary and grade-level boundaries of where and who teaches in-school civics (e.g., Mirra, 2018; Payne & Journell, 2019). In addition, civics are also learned explicitly and implicitly across students' experiences in classrooms, the role of content standards, and the "grammar" of schooling (e.g., Garcia & Mirra, 2019; McDonnell, 2000, Tyack & Cuban, 1995). These school-based lessons are not the only ways that youth learn about civic responsibility and engagement. Exemplified by Luis's letter, which ties school-based writing to broader activist movements, students can be powerfully engaged in civic learning that transcends what typically transpires within classroom. The definitions of civic engagement and the ways civics are taught in schools must be reviewed and reimagined. Occluding this reimagining today is that the public too frequently perceives schools as politically neutral sites that do not incorporate the forms of youth activism and political participation that are present in participatory and networked contexts today (e.g., Cohen, Kahne, Bowyer, Middaugh, & Rogowski, 2012; Literat, Kligler-Vilenchik, Brough, & Blum-Ross, 2018).

*The Civic Mission of Schools* (Carnegie Corporation of New York & Center for Information and Research on Civic Learning and Engagement [CIRCLE], 2003) defines civic education broadly, connecting the term in U.S. contexts to preparation of informed citizens: "Civic education should help young people acquire and learn to use the skills, knowledge, and attitudes that will prepare them to be competent and responsible citizens throughout their lives" (p. 4). The report further explains that this definition includes guiding youth toward informed decision making, meaningful participation in community activities, development of foundational knowledge and skills in political participation, and empathetic development for the rights of others. The enactment of instruction around these principles can differ substantially, leading to various perspectives of what citizenship looks like and *means* today (Westheimer & Kahne, 2004).

Civic engagement scholarship grounded in the above civic "mission" of schools often emphasizes how civic instruction is frequently tied to forms of service learning, debate of political issues, and instruction on formal structures of governance (Gould, 2011; Hess & McAvoy, 2015). Collectively, these different approaches guide students toward particular "characteristics of citizenship" (Flanagan & Levine, 2010, p. 161) that include activities such as reading newspapers, volunteering, voting, and engaging in local groups such as clubs, political parties, or religious services (Mirra & Garcia, 2017). Importantly, these traditional markers of civic participation often highlight substantial differences in youth participation along socioeconomic and racial lines (Wray-Lake & Hart, 2012). One perspective of these differences describes a "civic empowerment gap" (M. Levinson, 2010, 2012) that aligns with similar research suggesting that youth of color are surrounded by fewer civic resources and therefore lesser potential to "acquire" civic identities (Atkins & Hart, 2003, p. 159). This framing of a civic learning "gap" is in concert with research that civic engagement increases with levels of education (Nie, Junn, & Stehlik-Barry, 1996). These differences emphasize not simply the "opportunities" for civic learning in the United States (Kahne & Middaugh, 2008) but also how civic education is enacted and for whom.

While there are numerous explorations of the positive characteristics that could be encouraged in youth identity such as honesty, fairness, and mentorship (e.g., Damon, 2001; Schwartz, Chan, Rhodes, & Scales, 2013), other perspectives question why historically marginalized youth would "buy into a system where they feel excluded" (Watts & Flanagan, 2007, p. 781).

Garcia et al.

Exploring this tension, Kirshner (2015) notes "structural" contradictions vis-à-vis youth civic participation and activism in which

> low-income youth of color are exhorted to work hard and fulfill their responsibilities to go to college, but for many this is a remote possibility, either because of failing schools, economic barriers to higher education, or citizenship laws that block children of immigrants from legal employment or financial aid. (p. 6)

Civic expectations of marginalized youth to participate in systems that historically replicate inequality remain a key dilemma for framing meaningful civic participation; these approaches too frequently obscure "the differences in perspective that comes with social diversity" (Watts & Flanagan, 2007, p. 781). Such contradictions are compounded by the fact that these youth are not provided opportunities for participation despite the fact that "structural issues and inequities are rarely mentioned" in measuring youth political engagement (p. 799).

Despite civic literature showing declining civic participation from low-income youth of color leading to widening gaps in civic learning and empowerment, Watts and Flanagan (2007) critique how such findings treat youth attitudes and dispositions as fundamental to the "problem" (pp. 799–780). These disparities between portrayal of low-income youth of color in civic literature in comparison to new directions of engagement and participation suggest that previous, fixed definitions of civics must be challenged. Taking into account the myriad forms of youth civic activism occurring outside of schools, we make use of the construct of youth civic *readiness* to better contextualize the writing and actions of students like Luis in the LTNP project.

## Participatory Readiness

Considering the disparities in how civic engagement is taken up and whom it empowers, educators, researchers, and policymakers alike must engage in a critical "civic interrogation" (Mirra & Garcia, 2017, p. 139) of how youth participate in the ever-roiling political world they are inheriting today. Civic education must not only contend with the historic mechanisms that guide participation in existing U.S. structures but also consider who might be excluded from these structures and why. As one example, the notion of "citizenship" frequently undergirds the assumptions of what is learned (and who should learn it) in civic education (e.g., Carnegie Corporation of New York & CIRCLE, 2003; Westheimer & Kahne, 2004). Such framing is particularly fraught considering that many students in schools today are not recognized as U.S. citizens and that the 2016 election placed immigration as a divisive topic (both nationally and within the letters in the LTNP corpus).

At the same time that the 2016 election sparked a renewed interest in civic education (Tripodo & Pondiscio, 2017), it is also increasingly apparent that traditional boundaries of civics are unable to account for a system in which youth fear for their safety in schools, children are being kept in cages, and individuals are being legislated "out of existence" (Green, Benner, & Pear, 2018). Considering these dehumanizing markers of civic life in the United States today in tandem with the powerful forms of political engagement that youth are a part of—from Black Lives Matter to DREAMers to the Parkland-inspired March for Our Lives—we explore the data in this study with the recognition of new civic "innovation" occurring today (Mirra & Garcia, 2017).

We frame civic learning in this study as a form of *participation* and explore the contexts, audiences, and approaches to civic participation found throughout the LTNP data. We base this framework on Danielle Allen's (2016) call for an education of "participatory readiness" that finds students prepared not only to engage in a "political community but also that of intimate and communitarian relationships" (p. 27). Pedagogical implications of participatory readiness require expanding learning beyond the mechanics of government ("tactical knowledge"). Allen (2016) suggests that civic education must also include *"verbal empowerment* and *democratic knowledge"* (p. 40). Verbal empowerment, according to Allen, includes skills fundamental for interpreting and communicating; we detail the historic role of writing as a civic practice in the section below. Additionally, recognizing that democratic knowledge covers a large span of historical and theoretical ground, Allen emphasizes the critical role of *relational* components of participating in a democracy (p. 41). While Allen's work is largely framed as a way to direct the meaningful support of educators to encourage the civic participation of young people, we instead take student letters as evidence of, rather than an intervention on behalf of, youth's verbal empowerment, democratic knowledge, and participatory readiness. With this in mind, we explore youth civic writing within the context of the 2016 election, one that heightened hate-related violence (Southern Poverty Law Center, 2016), distrust in "fake" news reporting (Knight Foundation, 2018), and a general "increase in uncivil political discourse" (Costello, 2016, p. 4).

### Civic Engagement Enacted Through Civic Writing

Writing practices are socioculturally bounded; the meaning of words is tied to the contexts in which they are authored, the perspectives from which they are interpreted, and the ways they convey aspects of an author's identity (e.g., Bakhtin, 1986; Ivanič, 1998). From this perspective, writing is an important component of civic action and participatory readiness.

Text-based literacy has acted as a guiding force for shaping national identities (McLuhan, 1962), and writing has functioned as a domain through which civic beliefs are developed, contested, and reified over time. Likewise, activist efforts to resist oppressive policies and structures are often rooted in traditional forms of writing and argumentation. From the freedom schools during the U.S. Civil Rights movement (Hale, 2016) to online participatory writing using hashtags like #BlackLivesMatter (Jenkins, Shresthova, Gamber-Thompson, Kligler-Vilenchik, & Zimmerman, 2016), civic writing remains a key means for recruitment, communication, and organization. Letter writing, as a genre at the heart of this study, is also a substantial vein of civic action and acculturation (e.g., King, 1994; Laskas, 2017). At the same time, literacy has also functioned as a barrier for accessing equal civic opportunities; insidious efforts to suppress Black voters through literacy tests, for example, demonstrate literacy's historic role stifling civic participation within the United States (Perman, 2001).

There have been recent shifts to measure, assess, and support civic literacy practices in schools (e.g., National Assessment Governing Board, 2014; NWP, 2017). Research on civic literacy practices particularly in English Language Arts contexts demonstrate the possibilities for debate, dialogue, and empathy in classrooms (e.g., Garcia, 2017; Mirra, 2018). Generally, youth civic writing practices often conform to familiar, Western modes of argumentation and rhetoric (e.g., Burke, 1966). Likewise, Hess and McAvoy (2015) detail how some teachers "work to activate natural political disagreement amongst students" to foster *political literacy* (p. 92). These instructional practices hint at how English Language Arts classrooms and writing instruction can buttress pedagogies of participatory readiness. In the context of the Digital Youth Network, a program designed to support learning and creating with technology among students in an inner city middle school, Barron, Gomez, Pinkard, and Martin (2014) describe efforts to cultivate the "critical disposition" in students, positioning them to critically understand the media around them as well as the "social disposition," encouraging students to advocate for change in their local communities and beyond.

Considering the varied approaches to civic writing instruction, we examine LTNP to consider how a diverse group of students across hundreds of schools engaged in youth civic writing practices. While students may practice forms of civic writing in schools, civic literacy practices are particularly attuned to preparing youth for consuming and interpreting the messages they receive (e.g., Hobbs, 2011; Wineburg, 2018). As Monaghan and Saul (1987) note, society is "much more interested in children as receptors than as producers of the written word" (p. 91). To be clear, our understanding of civics and its enactment in U.S. schools frames it as a "productive and generative" activity that builds relationships and connections across individuals of differing backgrounds (Boyte, 2003). As "generative" activity that engaged thousands of youth, we explore the kinds of topics that students took up and the rhetorical approaches they leveraged to convey their arguments; this

# Youth Civic Writing

work illustrates a redefinition of the meaning of civics in schools at a moment when our country was engaged in a highly contentious presidential election.

## Methodology

### About the Letters to the Next President

As noted above, the LTNP project was a collaboration between the NWP and KQED, and educators working with young people across the country were invited to participate. Though we do not know the specific directions that teachers provided to students, guidelines for participating in this project are posted on the LTNP website:

> Writers are asked to address their letter to the future US President, whomever that person may be. We ask that writers do not address their letter to a specific candidate or party or advocate simply for a specific candidate or party. We welcome multimedia letters as well as text-based letters.

Visiting the website (letters2president.org) today, it is clear that these letters were meant to be easily navigated and read. On loading the page, a random letter is displayed (Figure 1) with a map promoting the thousands of authors that participated in the project. Furthermore, every letter is searchable by name, topic, region, and school; each letter, too, has a unique URL.

Unlike other online platforms, LTNP is intentionally designed as a "safe and supportive" environment. Student letters were only made visible after being approved by teachers and there is no commenting ability within the site; teachers essentially acted as moderators for a community that did not foster comment-driven dialogue. Furthermore, the website features instructional suggestions, sample lesson plans, and professional development resources for teachers.

Submissions of individual letters included several pieces of information that students chose and are publicly available—the name and location of the school site, a student name, the entirety of the letter, a title, a summary (usually a sentence long), and up to five topics ("tags") associated with the letter. With assistance from NWP and KQED, we aggregated and analyzed these data for every letter published to the LTNP website on or before the U.S. presidential election on November 8, 2016, comprising 11,035 letters. Our analysis and research design were independent of NWP and KQED, and we accessed the data only after the conclusion of the 2016 election. We did not participate in conceptualizing or implementing the project. Particularly in light of ongoing questions about how youth learn civically within schools during one of America's most partisan elections in recent history, our research focused on two specific questions:

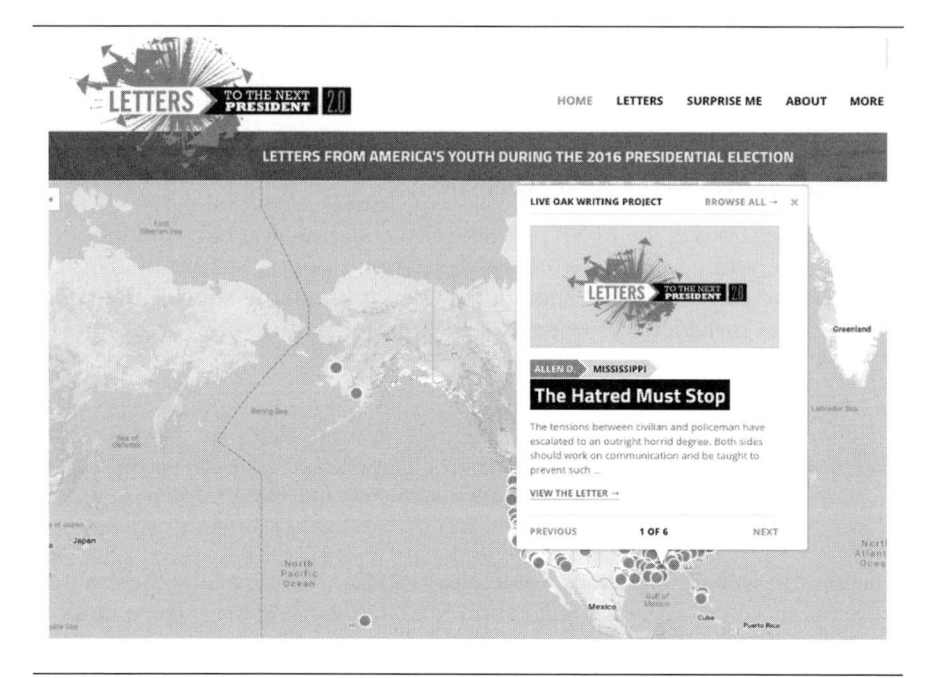

*Figure 1.* **A screenshot of letters2president.org.**

1. What topics did students choose to write to the next U.S. president about and how do these topics vary according to school demographics?
2. Through close textual analysis of a smaller set of letters, how did students write civically? In particular, how did students use evidence, emotional appeals, ethical appeals, logic, and personal experiences in arguing for legislative change?

As part of a reflection on the recent election, this study intentionally centered on the voices that are often positioned as the most disengaged in civic education literature; our textual analysis of student writing focused particularly on class sets of writing in schools that serve a majority of students of color and/or a majority eligible for free or reduced-price lunch (FRL). Furthermore, we also purposefully focused on schools located in states considered "swing states"[1] during the election, so that we might later study how potentially contrasting media messages were reflected in student writing; this analysis, however, is not a focus of this article.

## Data Analysis

We analyzed the data in two phases. First, we engaged in a quantitative analysis of the national data set. Specifically, we wanted to know, for the 11,035 student writers, what issues were most salient and how did these

*Youth Civic Writing*

Table 1
**Breakdown of Letters**

|  | Number of Letters | Percentage of Total |
|---|---|---|
| School designation | | |
| Letters from public school classes | 1,0153 | 92 |
| Private schools | 728 | 6.6 |
| Other organizations/programs | 154 | 1.4 |
| Title I eligible (schoolwide) | 4,285 | 39 |
| 50% or more students eligible for free or reduced-price lunch | 1,998 | 18 |
| 50% or more non-White students | 4,021 | 36 |
| Geographic markers | | |
| South | 2,464 | 22.3 |
| Midwest | 3,334 | 30.2 |
| West | 3,250 | 29.4 |
| Northeast | 1,987 | 18 |
| City | 3,648 | 33 |
| Suburban | 5,218 | 47 |
| Town | 613 | 6 |
| Rural | 1,508 | 14 |
| Total number of letters | 11,035 | 100 |

vary across regional, socioeconomic, and other factors across school sites. Again recognizing the need to understand the kinds of civic issues that interest youth of color as well as students of lower economic status, our analysis focused on differences between these communities others. Second, we engaged in qualitative analysis of five diverse schools in different regions of the country. As we detail below, we present a broad picture of youth civic writing through this approach: exploring both *what* interests students civically and *how* youth may have engaged in civic writing practices.

*Phase 1: Quantitative Analysis of Letter Topics*

In preparation for the first phase analyses, we expanded and reorganized the data set to understand the demographics of the sites that the letters came from, merging the previously noted data on the letters with school-specific data from the National Center for Education Statistics (NCES). By linking each letter to school information via the NCES ID number of each school, we were able to perform descriptive analyses to break down broad socioeconomic patterns within the corpus based on characteristics of the school population (no demographic data were collected at the individual student level). A key summary of the letters in this data set are shared in Table 1 and represented in Figure 2.

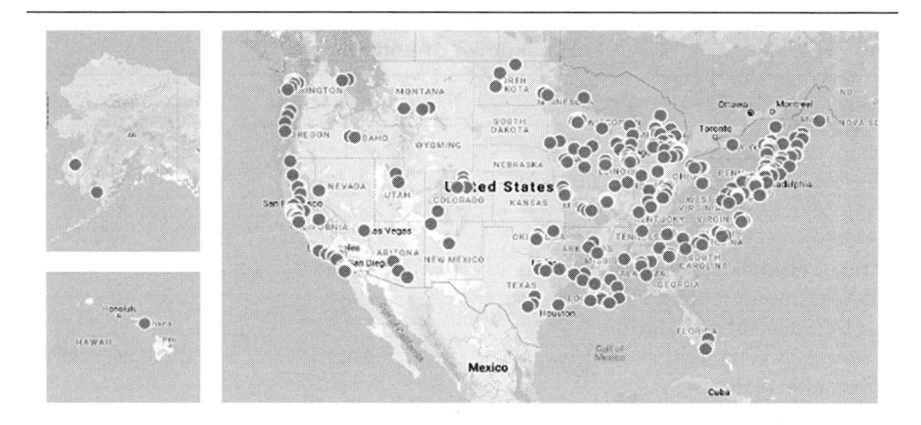

*Figure 2.* **Map of participating sites.**

To be able to make sense of the many topics students wrote about, we consolidated the thematic "tags" that students had applied to their letters. The submission system required students to manually enter all "tags" (relevant topic keywords) for their letters, entering up to five tags per letter; the system did not provide a menu list of common issues for students to choose from. Due to this open-entry system, students generated a total of 1,636 different issue tags, including a large proportion of redundant and related tags (e.g., tags included "animal lives matter," "animal life matters," "animal life," "animal rights," "animal lives," "animal cruelty," "animal treatment," and many others that suggested similar themes). We developed a scheme to code the tags and consolidate them into 69 distinct, nonoverlapping topic categories, clustering tags that were thematically related and/or redundant in order to allow us to feasibly conduct analyses based on broader topics (see Table 2).

Luis's letter at the beginning of this article, for example, included five different original issue tags: "racism," "inequality," "all lives matter," "police brutality," and "gun issues." In our consolidated categories, these tags were part of the "Race/Ethnicity," "Equality & Fairness," "Police," and "Guns" categories. While consolidating removes some of the nuance from the student-generated tags, we could not have conducted our analyses with the 1,636 original tags. Additionally, some tags remained uncategorized and often shed little light on the content of the letter (e.g., tags such as "problem," "issue," "U.S. issues"). However, more than 98.5% of letters had tags that were categorized in at least one of our 69 topic areas. Some of the remaining 1.5% had no tags included by the writer.

We used SPSS statistical analysis software to analyze the letter topics and school site data using chi-square tests to determine whether associations

*Table 2*
**Topic Issues**

| | | |
|---|---|---|
| Abortion & Reproductive Issues | Future, Change, & Reform | Protest & Free Speech |
| Animal Rights | Global Issues | Race/Ethnicity |
| Arts | Guns | Refugees |
| Black Lives Matter | Health | Religion |
| Bullying | Homelessness & Housing | Safety (General)/Security |
| Children & Teens | Human Rights & Human Trafficking | School/School-Related |
| Climate/Climate Change | Immigration | School Costs |
| College/Higher Ed | International/Foreign Affairs | School Hours |
| Corruption | Labor & Wages | School Homework & Grading |
| Death | Law/Criminal Justice | Science/Nuclear |
| Disability | LGBTQ | Sex, Sexuality, & Pregnancy |
| Discrimination & Prejudice | Love | Sexual Violence |
| Driving/Transportation | Media & Technology | Sports |
| Drugs, Alcohol, & Tobacco | Medical Issues | States |
| Economy | Mental Health | Suicide & Self-Harm |
| Education | Money | Terrorism |
| Education: Testing | Muslims | Unity & Diversity |
| Elections & Voting | Personal Traits & Values | Violence |
| Energy | Police | Veterans |
| Environment & Wildlife | Politics & Government | War/Peace |
| Equality & Fairness | Pollution/Garbage | Women/Gender |
| Family & Community | Population | Uncategorized |
| Food, Nutrition, & Hunger | Poverty | |
| Freedom/Institutional Control | | |

*Note.* LGBTQ = lesbian, gay, bisexual, transgender, and queer or questioning.

*Garcia et al.*

exist between letter topics that emerged and three binary school characteristics: (1) whether or not the school was eligible for Title I schoolwide, (2) whether or not more than 50% of students were eligible for FRL, and (3) whether the school had more than 50% students of color. Title I eligibility and FRL data are only available for public schools; however, 92% of letters were submitted from public schools, so we were able to analyze relationships based on these variables for the vast majority of letters.

*Phase 2: Qualitative Analysis Within Case Study Schools*

In the second phase, we selected a subset of five schools within the data set for a deeper analysis of letter content. The sites selected were identified as swing states during the 2016 presidential election and were located in different regions of the country. We selected schools that serve more than 50% students of color to shine light on the issues and concerns raised in these students' letters specifically, students whose civic engagement has often been doubted in the literature. The schools and courses that fit the selection criteria represent students of various grade levels (Grades 8–12), as well as varied contexts in which students were asked or given the opportunity to write. Selective courses such as Honors or AP (Advanced Placement) were excluded as were three sets of letters where all students in one class wrote on the same topic. All but one selected school are either Title I eligible schoolwide or serve a majority of students eligible for FRL (at the remaining school, 30% of students are eligible for FRL). The five schools represented in this study are detailed in Table 3.[2]

In this second, qualitative phase of the project, we performed content analysis (Schreier, 2012), to code *how* students constructed arguments within their letters. Specifically, we developed codes based on the modes of argumentation and persuasion frequently taught in schools; codes for letters focused on if they included logical, ethical, and empathetic and emotional appeals. We coded every letter in the case study selection to reveal how student writing manifested civic arguments. This set of letters also allowed us to code using a constant comparative analysis to develop general thematic categories (Strauss & Corbin, 1994) in order to look at youth conceptualizations of civics, their understanding of a president's responsibility and power, and the ways in which students approach civic ideas in their writing (Schreier, 2012). Emergent themes were documented in analytic memos (Hammersley & Atkinson, 1995), with a subset of letters coded by three coders to allow for interrater reliability analyses (Saldanña, 2009).

*Argumentation.* Given the political landscape of the 2016 election, in which facts, morality, and human empathy were all under active contestation, we wanted to see how students engaged in forms of argumentation. With this in mind, letters were initially coded for argumentation based on

Table 3
Case Study Schools

| State, City/Town | School Level | Urbanicity | % Students Eligible for FRL | % Students of Color | Title I eligible Schoolwide | n | Teacher-Generated Description |
|---|---|---|---|---|---|---|---|
| Florida, North Miami Beach | High school | Suburb | 62 | 73 | Yes | 13 | U.S. government (no grade specified) |
| Michigan, Flint | High school | Suburb | 60 | 74 | No | 46 | English (Grade 12) |
| Nevada, Las Vegas | High school | City | 30 | 53 | No | 27 | Creative writing (Grade 9) and American Literature (Grade 11) |
| North Carolina, Kannapolis | Middle school | Suburb | 100 | 65 | Yes | 31 | English (Grade 8) |
| Ohio, Cleveland | High school | City | 0[a] | 60 | Yes | 21 | "Choose to participate" (no grade specified) |

*Note.* FRL = free or reduced-price lunch.

[a] National Center for Education Statistics data indicate that this school is eligible for Title I schoolwide, but that 0% of students are eligible for FRL. It is possible that FRL eligibility was not recorded for this school.

*Garcia et al.*

Table 4
**Argumentation Codes**

| Code | Definition | Example |
|---|---|---|
| Appeal to logic | The letter makes an argument through the use of logical reasoning, including the presentation of a claim supported by cause and effect reasoning, or the presentation of specific evidence in support of a claim. A letter that makes an argument through logical appeal likely includes use of statements that follow the form of "if . . ., then . . ." and/or linking words such as "because." | "Homelessness is a major issue in america right now the percentage of homelessness is rising. (According to The state of homelessness in america) 'On a single night in Jan 2015 514,708 people were experiencing homelessness, in 2014 2 million people were poor and in households were doubled up with family and friends.' The homeless people are struggling to support their families with buying medications, clothes, a house and food." (Ariel H.) |
| Appeal to ethics | The letter conveys that action must be taken or that something is wrong due to a moral or value-based appeal. This appeal relies on ideas of what is right, fair, or should be true or universally available. | "I Believe that every kid *should* live a safe and healthy life." (Juan B.) |
| Appeal to empathy | The letter includes a direct appeal to perspective taking, walking in a different set of shoes, or to a personal perspective. Often this includes words like, "Imagine . . ." or "How would you feel if . . .?" | "Innocent people are losing their lives because of the death penalty; *how would you feel* if a loved one of yours was executed for no reason?" (Elizabeth K.) |

an a priori set of codes derived from the three Aristotelian modes of persuasion: logos, ethos, and pathos (loosely aligning with facts, morality, and empathy) a framework for argument that is often taught in middle or high school English courses (Burke, 1966). After an initial phase of coding, the codebook was refined and argumentation was operationalized around the codes: appeal to logic (originally logos), appeal to ethics (originally ethos), and appeal to empathy (originally pathos). Full definitions and an example for each are provided in Table 4.

*What counts as evidence.* Considering evidence as a core component of argumentation, and with similar concerns about the political landscape as articulated above, three different forms of evidence were coded: personal experience, citation, and unsourced data. Personal experience included any story or anecdotal connection to the author's life, including the experience of a close friend or family member. Letters were coded as containing citation if a direct quote or idea was attributed to a speaker, which at times included song lyrics or a component of a politician's speech, or if facts were directly sourced, referencing the website, newspaper, or other (usually online) source of that information.[3] When discrete facts were shared but not cited or sourced, the letter was coded as including "unsourced data."[4] Definitions and examples provided in Table 5.

**Study Limitations**

By nature, the LTNP project sample, while diverse in terms of geography, school profiles, and topics addressed, is not necessarily representative of students or of schools in the United States. While we are able to share analyses from a large sample of 11,035 letters, this is not a random or nationally representative sampling of youth civic identity or beliefs. However, within this corpus of data we have attempted to focus intentionally on students who are historically underserved and underrepresented—and most overlooked within civic education literature—by sampling schools fitting that profile in the second phase of the study.

Another limitation is that the findings from the first phase of this study are based on the tags that students selected and that we categorized in Table 1. The original tags for some letters may not have been representative of the content of the letters that students wrote and our process of consolidating may not fully convey the dimensionality of all letters. Additionally, our analysis is based on demographic information about the school sites as a whole, but we do not have demographic data on individual student writers, nor the specific pedagogies and examples teachers used when assigning the project.

Finally, this article provides an overview of how students across all schools in this study participated in this predominantly school-based, civic writing activity. We recognize that there are myriad dimensions of this work that we cannot cover in a single article. Further work builds on the findings here focused on the impact of this work on teacher identities (Garcia & Gargroetzi, 2019), and breakdown of student writing on topics such as immigration (A. Levinson & Garcia, 2019) In this article, by looking broadly across *all* the letters and closely at a purposeful sample of letters, we seek to begin to draw on the knowledge, demands, and uncertainties voiced by thousands of youth on the precipice of Trump's presidential victory.

*Garcia et al.*

*Table 5*
**Evidence Codes**

| Code | Definition | Example |
| --- | --- | --- |
| Personal experience | The letter includes references to or commentary on personal experience. This includes direct personal experience as well as experiences of friends and family. These references may directly support an argument made within the letter or provide context. | "As a 17 year old that has been through it, realize it or not, I say something has to be done to wipe discrimination off the face of the Earth. Whenever I go to a store to buy something, certain people look at me like I'm going to do something wrong when I didn't have any intentions of doing anything" (T'Onia M.) |
| Citation | The letter includes specific information or a quotation that is credited or sourced. This includes reference to a website as well as the citation of individuals such as public figures or artists as in the case of the citation of lyrics or a reference to the argument or works of an historical figure, even without direct quotation. | "According to the U.S. Bureau of Justice Statistics an African American male born in 2001 has a 32% chance of going to jail in his lifetime, while a white male only has a 6% chance." (Jonathan P.) |
| Unsourced data | The letter includes specific information or details that are not cited. In the case of a works cited section included at the end, but not indication of linking information to source within the letter, the letter was coded as including unsourced data. | "Miami-dade county public schools is the third biggest school district in the country, however it doesn't rank amongst the best academic districts in the country" (David L.) |

# Findings

Below, we explore two aspects of this significant corpus of student writing: (1) the topics students wrote about, including how letter topics related to socioeconomic characteristics of school sites among the 92% of letters from public schools and (2) the ways in which students wrote civically about these topics in a set of 138 letters from five case study schools.

*Youth Civic Writing*

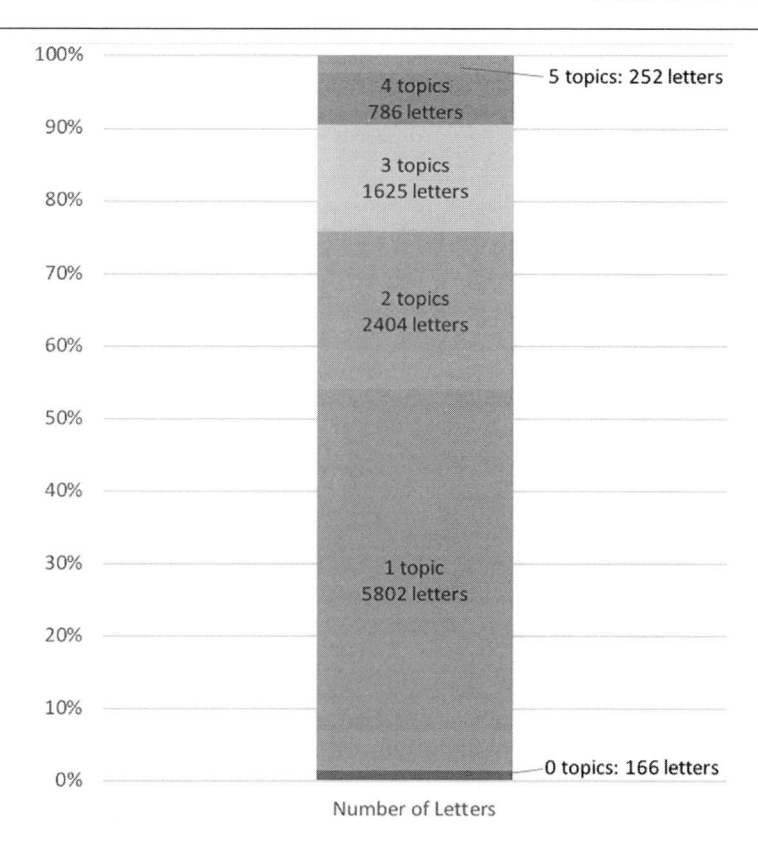

*Figure 3.* **Distribution of topics per letter.**

### What Topics Did Students Write About?

Looking broadly at the corpus of 11,035 letters reveals that students tagged their letters with a wide range of issues. There was no single topic that overshadowed all others, and some letters covered several diverse topics. The median number of topics per letter is 1, but the mean number of topics was 1.82 with a standard deviation of 1.088; thus while over half of letters' original tags (5,802 letters) fell into only one topic, 22% of letters fell into two topics, and 26% of letters fell under three or more topics (see Figure 3 below). Some letters were categorized with related topics (e.g., "environment" and "climate change"), while others addressed more disparate ones. This diversity demonstrates, as one might expect, that the issues at the forefront of students' minds ranged widely—later in this section, we address some of the ways that topics varied across different settings.

*Garcia et al.*

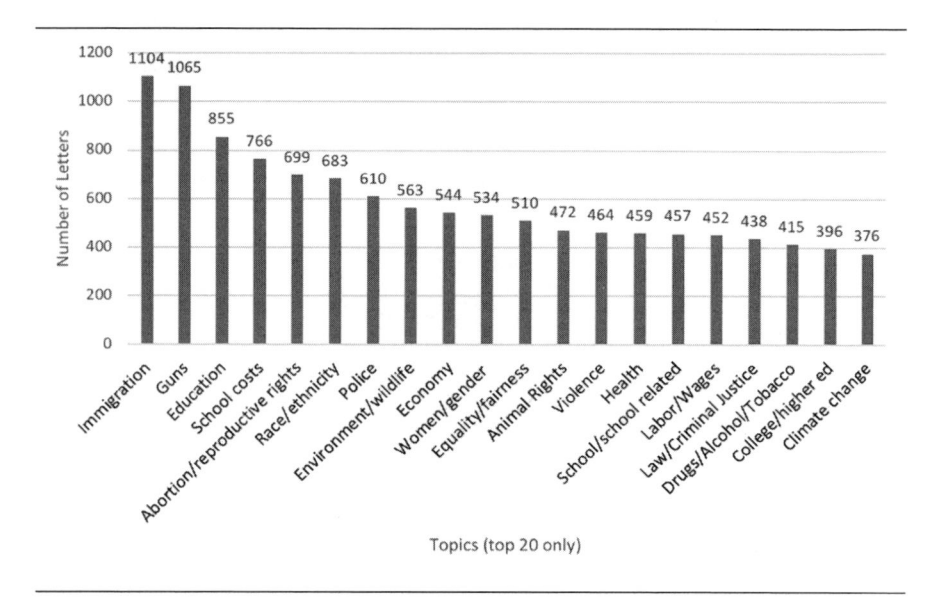

[Bar chart showing number of letters by topic:
Immigration 1104, Guns 1065, Education 855, School costs 766, Abortion/reproductive rights 699, Race/ethnicity 683, Police 610, Environment/wildlife 563, Economy 544, Women/gender 534, Equality/fairness 510, Animal Rights 472, Violence 464, Health 459, School/school related 457, Labor/Wages 452, Law/Criminal Justice 438, Drugs/Alcohol/Tobacco 415, College/higher ed 396, Climate change 376]

Topics (top 20 only)

*Figure 4.* **Top 20 topic categories.**

Of the 69 condensed topics we generated based on student tags, some were more prevalent than others, and these "top topics" were similarly diverse—including guns, race and ethnicity, the environment, health, school-related issues, LGBTQ (lesbian, gay, bisexual, transgender, and queer or questioning) issues, the economy, and more. Figure 4 shows the 20 most prevalent topics (of the total 69).

Immigration was a prevalent topic on the minds of letter writers—nearly 10% applied tags in the immigration topic category, which included "immigration," "border patrol," "illegal immigration," "deportation," and others. As part of the analysis, we sought to identify classes in which teachers may have assigned a topic for student letters and found three class groups in which all letters focused on immigration (166 letters).[5] Immigration was the only topic that appeared to be assigned to some classes. Even when we put aside these 166 letters, 941 remaining letters addressed the topic of immigration, more than any other topic except for guns.

Guns (constructed of issue tags such as "assault weapons," "gun control," "gun issues," "gun laws," "second amendment," "school shootings," and "shootings") was also a top topic, representing 9.65% of all letters. While high school students have gained greater visibility in the struggle for firearms regulation since activist efforts emerged from the school shooting in Parkland, Florida, in February 2018, it is notable that more than 1 year prior, guns were at the forefront of students' concerns. Close to 10% of all letters in this corpus were tagged with a gun-related issue.

Education and school-related issues are also—not surprisingly—a key concern for teens. The topic "School Costs" alone ranked highly (applied in 766 letters), suggesting students' anxiety over the expense of college tuition. In addition to the "School Costs" topic category and the "School/School-Related" category (457 letters) that aggregated more general tags such as "school," "school climate," "school education," "classrooms," "school funding," and "teachers," there were additional categories including "School Hours" (68 letters) and "School Homework & Grading" (98 letters) that cluster more specific concerns. The "College/Higher Ed" topic included 396 letters. Finally, the less specific "Education" topic category was present in 855 letters. Together these different education-related topic categories, a total of 1,963 letters (17.7% of all letters) addressed at least one topic related to education or schooling.

Race and ethnicity, women/gender, LGBTQ, discrimination and prejudice, and equality and fairness were all prevalent topics, indicating that youth wrote about issues pertaining to specific communities including identity and social justice, as well as broad patterns of inequity. "Economy" and "Labor & Wages" counted among the most prevalent topics as well. "Environment & Wildlife," and "Climate Change" were present in 563 and 376 of the letters, respectively, while "Animal Rights" occurred in 472 letters. Additional topics in the top 20 include "Health," "Drugs, Alcohol, & Tobacco," and "Violence."

Though we cannot discern the specific nature of letter arguments from the issues and topics alone—a tag of "Police" does not reveal the nature of the writer's message regarding policing—relationships between topics (e.g., which topics were often co-occurring in the same letters) suggest some trends. For example, the topics "Black Lives Matter," "Race/Ethnicity," and "Police" co-occur frequently in the same letters—more than 25% of "Police" letters were also in the "Black Lives Matter" and/or "Race/Ethnicity" topics—suggesting that a cluster of letters likely address racism in policing and recent tragedies of race-related police violence. With regard to education-related issues, "College/Higher Ed" and "School Costs" were also among the most co-occurring topics—67.5% of letters tagged in the "College/Higher Ed" topic were also tagged in the School costs topic—suggesting that letters addressing higher education often focused on its financial costs. Links between "Labor/Wages" and "Women/Gender," "Discrimination/prejudice" and "Race/Ethnicity" suggest that concern about inequities facing women and people of nondominant backgrounds, specifically with regard to jobs and earnings, also formed a cluster. Future qualitative research will explore issue-specific letters qualitatively to understand patterns in students' arguments, beliefs, and calls to action (A. Levinson & Garcia, 2019; Zummo, Gargroetzi, & Garcia, 2019).

Although topics students chose to address in their letters overlap to some extent with top issues among young Americans in other research, there are also marked differences. For example, respondents to the Harvard Public Opinion Project (Harvard University Institute of Politics, 2016), ages 18 to 29 years, listed the economy and terrorism as top concerns in fall 2016. While the economy was the ninth most prevalent issue in the LTNP, terrorism was not in the top 20 topics and occurred in only 342 letters, just over 3%. In the Harvard poll, "reducing inequality," "uniting the country," and "dealing with immigration" were the third, fourth, and fifth most important items that 18 to 29 year olds prioritized, respectively, and these resonate more closely with the issues that stand out among the student letters. Issues such as health, climate change, education, and gun control—prominent topics for our student writers—were only listed as priorities by 1% to 2% of the youth poll respondents. With differences in ages, prompts, and possibly demographics, we cannot project what might account for the differentiation in the issues that teens addressed in their letters and those that young adults prioritized in the poll. Existing survey reports also have not published analyses that compare socioeconomic status (SES), racial groups, or regional groups. A clear message from our reviews of existing data is that further research investigating teens' civic lives is needed.[6]

### How Did Letter Topics Vary According to School Demographics?

Building on the frameworks of civic "gaps," our analysis explored how letter topics varied across different contexts around the country. Among the 92% of letters submitted from public schools (10,152 letters), we identified significant associations between the topics addressed in student letters and demographic characteristics of the school population, among which we focused on three:

1. whether schools were eligible for Title I schoolwide;
2. whether schools had a majority (more than 50%) of students eligible for FRL; and
3. whether schools had a majority (more than 50%) students of color.

These analyses were performed with the letters from public schools; for the remaining 8% of letters from private schools or other institutions, we do not have access to the same site data.

Chi-square tests revealed that among the 10,152 letters from public school students, 43 of the 69 letter topics were associated significantly with at least one demographic characteristic of the school site student populations (Tables 6 and 7). Of these, 31 topics were associated with one or both socioeconomic factors (schoolwide Title I eligibility or a majority of students eligible for FRL), while 32 topics had associations with schools that serve more than 50% students of color or that serve more than 50% White

*Table 6*
**Topics Significantly Associated With School Socioeconomic Indicators**

| Topics | School Eligible for Title I Schoolwide (%) | School Not Eligible for Title I Schoolwide (%) | More Than 50% Students Eligible for FRL (%) | Less Than 50% of Students Eligible for FRL (%) |
|---|---|---|---|---|
| Topics associated with lower socioeconomic status schools | | | | |
| Black Lives Matter | 2.8 | 2.7 | 4.4** | 2.3 |
| Children & Teens | 2.6 | 2.6 | 3.4* | 2.4 |
| Discrimination & Prejudice | 3.1 | 3.3 | 4.2** | 3.1 |
| Family & Community | 1.7 | 1.7 | 2.7** | 1.5 |
| Homelessness & Housing | 2.4 | 2.7 | 3.3* | 2.3 |
| **Immigration** | 14.1** | 7.6 | 16.6** | 9.0 |
| **Police** | 6.2* | 5.2 | 9.2** | 4.7 |
| **Race/Ethnicity** | 6.5* | 5.6 | 8.2** | 5.8 |
| Religion | 0.6* | 0.3 | 0.6 | 0.4 |
| Sex, Sexuality, & Pregnancy | 1.2 | 1.0 | 1.7* | 0.9 |
| Sexual Violence | 1.4 | 1.4 | 2.1** | 1.2 |
| **Violence** | 5.3** | 3.4 | 7.2** | 3.4 |
| Topics associated with higher socioeconomic status schools | | | | |
| Abortion & Reproductive Issues | 6.3 | 6.2 | 5.3 | 6.5* |
| Climate Change | 3.5 | 3.2 | 2.5 | 3.5* |
| **College/Higher Ed** | 2.6 | 4.7** | 2.7 | 3.9** |
| **Drugs, Alcohol, & Tobacco** | 2.8 | 4.6** | 3.0 | 3.9* |
| **Education: Testing** | 0.7 | 1.4** | 0.7 | 1.1* |
| **Energy** | 1.2 | 2.2** | 0.7 | 2.0** |
| Environment & Wildlife | 5.0 | 5.2 | 4.1 | 5.4* |
| Food, Nutrition, & Hunger | 1.5 | 2.7* | 1.8 | 2.2 |
| Guns | 9.4 | 9.2 | 7.5 | 9.8** |
| Health | 3.1 | 5.0* | 3.0 | 4.4** |
| **Labor & Wages** | 3.2 | 4.4** | 2.9 | 4.3* |
| **Law/Criminal Justice** | 3.4 | 4.2* | 4.3 | 3.9 |
| **LGBTQ** | 2.8* | 3.7* | 2.9 | 3.7** |
| Media & Technology | 1.0 | 1.7* | 1.3 | 1.5 |
| Money | 1.4 | 1.9* | 1.4 | 1.8 |
| Pollution/Garbage | 3.0 | 2.9 | 2.2 | 3.1* |
| **School Costs** | 5.9 | 8.0** | 5.3 | 7.4* |
| School Homework & Grading | 0.7 | 1.2* | 0.8% | 1.0 |
| **School Hours** | 0.4 | 0.9** | 0.3% | 0.8** |
| School/School-Related | 3.3 | 4.7** | 3.4% | 4.2 |
| Terrorism | 3.5 | 2.9 | 1.9 | 3.5** |
| Veterans | 0.6 | 0.8 | 0.4 | 0.8* |
| War/Peace | 1.6 | 2.2* | 1.6 | 2.0 |
| **Women/Gender** | 4.1 | 4.9* | 3.3 | 4.8** |

*Note.* LGBTQ = lesbian, gay, bisexual, transgender, and queer or questioning. Boldface indicates significant relationship for both variables.
*$p$ < .05. **$p$ < .01.

### Table 7
### Topics Significantly Associated With School Racial Majority

| Topic | Majority Students of Color (%) | Majority White Students (%) |
|---|---|---|
| Topics associated with majority students of color | | |
| **Black Lives Matter** | 3.9** | 2.0 |
| Bullying | 1.7* | 1.2 |
| Corruption | 0.3* | 0.1 |
| **Discrimination & Prejudice** | 4.1** | 2.9 |
| **Immigration** | 15.3** | 7.1 |
| Muslims | 0.5** | 0.2 |
| **Police** | 8.1** | 4.1 |
| **Race/Ethnicity** | 8.2** | 5.0 |
| **Sexual Violence** | 1.7* | 1.3 |
| Topics associated with majority White students | | |
| **Abortion & Reproductive Issues** | 5.2 | 6.9** |
| Animal rights | 3.5 | 4.7** |
| Driving/Transportation | 0.4 | 0.7* |
| **Energy** | 1.4 | 1.9** |
| **Environment & Wildlife** | 4.7 | 5.5* |
| **Food, Nutrition, & Hunger** | 1.6 | 2.3** |
| **Guns** | 7.7 | 10.7** |
| **Health** | 3.3 | 4.6** |
| **Labor & Wages** | 3.3 | 4.5** |
| Personal Traits & Values | 0.7 | 1.1* |
| Politics & Government | 2.5 | 3.2* |
| **Pollution/Garbage** | 2.5 | 3.1* |
| Protest & Free Speech | 0.3 | 0.6* |
| Refugees | 0.7 | 1.1* |
| **School Costs** | 6.3 | 7.4* |
| **School Homework & Grading** | 0.5 | 1.1** |
| **School Hours** | 0.4 | 0.8* |
| Sports | 0.5 | 0.9** |
| **Terrorism** | 2.4 | 3.6** |
| Unity & Diversity | 0.2 | 0.5* |
| **Veterans** | 0.4 | 0.9** |
| **Women/Gender** | 3.5 | 5.2* |

*Note.* Boldface indicates topics that were also significantly related to one or both socioeconomic status indicators.
*p < .05. **p < .01.

students.[7] The relationships in Tables 6 and 7 are all significant at the 95% level. Although some topics were associated with racial majority and not with socioeconomic indicators, and vice versa, there were no topics

*Youth Civic Writing*

associated with both lower SES and White majority, nor with both higher SES and majority students of color. We stress that socioeconomics and race are not to be conflated. There was, however, covariance in this sample that may contribute to some topics converging on both racial and socioeconomic groups (e.g., immigration was more likely written about in schools serving a majority students of color, as well as among schools serving a majority lower income students), and this is difficult to parse out as many of the schools with one of these characteristics also possessed the other. Finally, 26 letter topics had no relationship to either school SES or racial indicators. These included "Education," "Equality & Fairness," "Economy," "Freedom/Institutional Control," "International/Foreign Affairs," and "Mental Health," among others. The following sections focus in on key issues where student letter topic focus varied according to the aforementioned demographics.

*Race and Discrimination*

What do these results tell us? First, topics of race and discrimination, including "Race/Ethnicity," and "Discrimination & Prejudice," were prevalent among letter writers overall, but were significantly more likely to be written about in letters from schools serving a majority of students of color and also significantly more likely (although slightly less so) to come from schools serving a majority lower income students (>50% receiving FRL). The topic "Black Lives Matter" also occurred more frequently among both of these groups of schools. Madisyn, a student from a school serving a majority of lower-income students and also a majority of students of color (largely Latinx and African American) in Oklahoma, who tagged her letter only with "Race" (categorized under "Race/Ethnicity") writes,

> *Our communities are plagued by death. Mothers and fathers are scared to send their children out because of the fear that they might be killed by doing something as simple as walking home or to a friend's house. 2016 has caused a lot of pain in the hearts of friends and families of the black community. Lives have been lost and in most cases no full justice is actually served. All lives will not matter until black lives matter, too.*

> *When you are scared or need help in an emergency, who do you call? The local police department is what most people would say, but what do you do if the police could care less about your life and judge you by your skin color, then who would call? That is a question most people have no answer to. We are fighting for equality and justice. We want to be able to walk down the streets and go about our day stress free. Lives have been taken for the simplest things: books, CDs, cigars, car trouble, and for things you would think you would be safe. These are real life examples of reasons people lives have been taken within the past two years. People mistake this movement as a violent movement. It is meant to cause awareness but people sometimes portray it as a way to start problems, creating fear for everyone involved.*

*Garcia et al.*

> We want justice. We want these murderers in jail. In most of these cases the people are not punished at all. Cops have been given so many chances this year. The system has not given us any justice; they get leave with pay and get to enjoy their lives while we grieve in pain. Recently a Tulsa police officer shot and killed a man, and she was let off on bond. Mostly every case is ignored, and even when it is "handled" things like that happen. My point is, black lives do matter, meaning all lives matter, but until black lives are equally treated we will continue to stand up for our people. Justice is all we want.

Madisyn's letter speaks to issues broader and deeper than the letter tag denotes. Although Madisyn applied only one tag of "Race" and did not tag her letter with "Police," "Violence," or anything else, her letter speaks to students' deep and related concerns around discrimination, violence, and specifically the role of police. The "Police" topic, which Madisyn's letter was not categorized in but clearly addresses, was also associated with higher poverty schools and was nearly twice as likely to be present in letters from schools with a majority students of color. Although our analyses here cannot discern the nature of students' arguments about police, the co-occurrence with race- and discrimination-related topics indicates that at least a cluster of police-focused letters address racial profiling among police and police brutality targeted at people of color. Given that about half of letter writers chose tags in only one topic area, it is likely that other letters like Madisyn's touched on more topics than may be obvious from the tags students applied.

The more frequent occurrence of the "Race/Ethnicity" topic and "Discrimination & Prejudice" topic among letters from schools serving a majority students of color corroborates similar disparities reported in polling data of Americans ages 18 to 29 years (Harvard University Institute of Politics, 2016), which suggests the majority of young people of color felt they were "under attack." The poll study also found that 62% of young Americans thought race relations would worsen if Trump were elected president, as compared with only 22% if Clinton were the winner. The prevalence of race and ethnicity as well as discrimination and prejudice across all student letters but particularly among letters from schools serving a majority students of color suggests that young people in communities of color in particular feel that the threats of racial and ethnic discrimination are affecting them in pronounced ways and that the president can take action for change.

*Violence and Guns*

Violence and sexual violence are also more prevalent topics for students from schools serving lower income communities and communities of color. One conjecture as to why this pattern occurs is that youth living in lower income communities may feel more viscerally the threat of violence,

*Youth Civic Writing*

particularly if living in areas that are underserved with regard to safety and social services, which is often the case with both lower income communities and communities populated largely by people of color. "Violence" was among the top 20 topics in the sample overall; "Sexual Violence" was present in a smaller group of 157 letters.

The "Guns" topic category, which might have been expected to follow a similar trend as "Violence," was more prevalent among lower poverty schools and schools with majority White students. To parse out the specific issues and stances students take in their letters and better understand why "Violence" may be more commonly addressed by students from schools with more lower income students and students of color, while, for example, "Guns" is more commonly addressed by students from other schools, further qualitative work is needed. However, among the tags included in "Guns" were "gun issues," "second amendment," "gun control," and so on, indicating topics related to firearms rights, legislation, and restrictions, whereas the "Violence" topic included tags such as "violence," "abuse," "murder," and "killing," suggesting a focus on actual violent acts rather than on the weapons used. Thus, while students from higher income communities and students from largely White schools may be more concerned about firearm issues, students in schools serving largely students of color as well as serving largely lower income students were more likely to express concern about violent acts and abuse.

*Education Costs*

Surprising findings emerged around the topics of "College/Higher Ed" and "School Costs." National education policy in recent years has heavily emphasized the importance of college and college readiness for all students. While we might have expected that students from schools serving the *less* affluent would be more concerned about challenges of accessing college and paying tuition, results showed an opposite trend. One potential explanation is that these data reflect the deep inequality in students' college opportunities and expectations. Higher SES students are more assured of attending college and thus are more likely to choose related issues to write about. Another reason students from less privileged backgrounds were more likely to write about topics other than college could be that there are other issues that they are *more* worried about, that they may experience on a day-to-day level in their communities, and that they are more motivated to call to the attention of the future president—however important college may be to them. As higher education is often seen as a means of climbing a social and economic ladder in the United States, findings around the themes of these letters highlight that perhaps the students that might most benefit from advocating for government action around school costs are also those that do not have the luxury to focus their attention on issues of higher education. The

"School Costs" topic was also more prevalent in schools with a majority White students, but the relationship with regard to school racial makeup is less strong than with the SES indicators.

*Immigration*

The topic of immigration had some of the strongest associations, both with socioeconomic characteristics of the student body as well as with racial makeup of the school. "Immigration" was a topic in more than 15% of letters from schools with a majority students of color, the topic being more than twice as likely to appear in letters from those schools than from schools with a majority of White students. Taking a deeper look at school demographic data for racial majority groups, while 7.1% of White majority schools' letters were categorized under immigration, 22.6% of letters from Latinx majority schools were categorized under immigration (immigration themed letters constituted 8.9% of letters from Asian majority schools, 12.8% of letters from Black majority schools, and 13.5% of letters from schools with "other" as the majority). All these patterns raise many questions, including what concerns youth most deeply regarding immigration? What are students' stances on the issue? Is immigration more commonly addressed among schools with a majority of students of color, and among students at majority Latinx schools, due to students' personal proximity to the issue, and if not what accounts for the relationships? These topics are addressed in a related analysis focused specifically on this topic (A. Levinson & Garcia, 2019). What is evident here is that immigration is an important issue among teens from a diverse array of schools but particularly for students from lower SES communities and communities of color. Investigating these patterns further could help understand what specific aspects of immigration concern students and what divisions might exist between students in different communities across the country.

With regard to all the topics associated with SES and/or racial indicators, differences also likely reflect the types of events, discussions, and debates that are prevalent among schools and/or communities that share characteristics, and that contribute to shaping young people's civic thought.

## Qualitative Analysis From Five Focal Sites

In the second phase of this research, we engaged in close textual analysis of 138 letters from five focal schools (see Table 3) to understand *how* students write civically and in particular how students use evidence, emotional appeals, logic, and personal experiences in arguing for legislative change. As described in the Methods section, we chose five school sites that were socioeconomically diverse, served high percentages of students of color, and were located in different regions of the country.

Students from the five school sites wrote on a wide range of topics, touching on 49 of the 69 total topics identified in the full set of 11,035 letters. Throughout the qualitative findings, we see that schools and classrooms are places where local concerns and norms of writing can be shared, but can also vary greatly. The two most prevalent topics from the nationwide letter set, immigration and guns, are written about at two and four of the schools, respectively, but come after other topics in frequency. Looking at the five sets of letters cumulatively, the most frequent topics are "Race/Ethnicity" (17), "Police" (16), "Equality & Fairness" (14), "Discrimination & Prejudice" (12), and "Education" (11). Given the demographic selection criteria for the schools included in this qualitative analysis, concerns about issues of racial and ethnic equality or discrimination and police relations are aligned with patterns described above. However, when each set of letters is viewed individually, different clusters of student concern emerge. For example, at one school, the most frequent topic was "Sexual Violence" (10 letters), a topic that was not tagged in letters from any of the other case schools. At another one of the five schools, LGBTQ was one of the most frequent topics (5 letters), but it was only mentioned in one letter from the other four case schools combined. Table 8 provides a summary of topics and frequency by school.

**How Did Students Write?**

*Forms of Argumentation and Use of Evidence*

In their letters, students developed compelling arguments employing varied argumentation forms that included appeals to logic, to empathy, and to ethical standards. They provided stories of their own experiences, cited data from internet news sources, cited songs or public figures, and provided facts or statistics that were sometimes left unsourced. Consider the letter below from North Carolina eighth grader, Adeline S., which we use to illustrate our qualitative findings in the remainder of this article.

> *Dear Mr. President,*
>
> *I am sure that you are in knowledge of the problems that are going on in our society. In fact, you're probably trying to do something about it at this moment. I just wanted to bring this one specific issue to your attention. Everyday people get upset because of the discrimination that people face on a day to day basis. Racism is a huge problem that the United States has been facing ever since probably when the first settlers arrived at the Americas. But over the years it has evolutionized to race and not just religion anymore. Apparently to other people, if you are from a different color, race or country, you are just bound to be discriminated and that's when the whole issue of racism starts. Racism is causing a lot of problems and it needs to stop before anything else happens.*

Table 8
Summary of Topics and Frequency by School

| Michigan (n = 46) | Ohio (n = 21) | North Carolina (n = 31) | Nevada (n = 27) | Florida (n = 13) |
|---|---|---|---|---|
| Sexual Violence: 10 | Discrimination & Prejudice: 4 | Race/Ethnicity: 6 | Education: 6 | Economy: 3 |
| Police: 6 | Equality & Fairness: 4 | Discrimination & Prejudice: 4 | Equality & Fairness: 6 | Climate/Climate Change: 2 |
| Race/Ethnicity: 6 | Health: 4 | Police: 4 | LGBTQ: 5 | Drugs, Alcohol, & Tobacco: 2 |
| Violence: 5 | Uncategorized: 4 | Food, Nutrition, & Hunger: 3 | Uncategorized: 3 | Children & Teens: 1 |
| College/Higher Ed: 4 | Labor & Wages: 3 | Immigration: 3 | Abortion & Reproductive Issues: 3 | Education: Testing: 1 |
| Discrimination & Prejudice: 4 | Children & Teens: 2 | Politics & Government: 3 | Police: 3 | Elections & Voting: 1 |
| Homelessness & Housing: 4 | Education: 2 | Terrorism: 3 | Race/Ethnicity: 3 | Environment & Wildlife: 1 |
| Black Lives Matter: 3 | Mental Health: 2 | College/Higher Ed: 2 | War/Peace: 3 | Food, Nutrition, & Hunger: 1 |
| Equality & Fairness: 3 | Police: 2 | Guns:– 2 | Animal Rights: 2 | Future, Change, & Reform: 1 |
| Family & Community: 3 | Politics & Government: 2 | Safety (General)/Security: 2 | Corruption: 2 | Guns: 1 |
| Health: 3 | Race/Ethnicity: 2 | School/School-Related: 2 | Freedom/Institutional Control: 2 | Health: 1 |
| Law/Criminal Justice: 3 | Abortion & Reproductive Issues: 1 | Violence: 2 | Politics & Government: 2 | Immigration: 1 |
| Women/Gender: 3 | Disability: 1 | Abortion & Reproductive Issues: 1 | Sex, Sexuality, & Pregnancy: 2 | Police: 1 |
| Abortion & Reproductive Issues: 2 | Drugs, Alcohol, & Tobacco: 1 | Black Lives Matter: 1 | Violence: 2 | Politics & Government: 1 |
| Education: 2 | Family & Community: 1 | Children & Teens: 1 | Women/Gender: 2 | School Homework & Grading: 1 |
| School Costs: 2 | Freedom/Institutional Control: 1 | Education: 1 | Black Lives Matter: 1 | Science/Nuclear: 1 |
| Bullying: 1 | Global issues: 1 | Environment & Wildlife: 1 | Bullying: 1 | |
| Children & Teens: 1 | Guns: 1 | Equality & Fairness: 1 | Climate/Climate Change: 1 | |
| Drugs, Alcohol, & Tobacco: 1 | LGBTQ: 1 | Homelessness & Housing: 1 | Education: Testing: 1 | |
| Education: Testing: 1 | School Costs: 1 | Labor & Wages: 1 | Family & Community: 1 | |
| Guns: 1 | Suicide & Self-Harm: 1 | Poverty: 1 | Global Issues: 1 | |
| Labor & Wages: 1 | War/Peace: 1 | School Costs: 1 | Labor & Wages: 1 | |
| Mental Health: 1 | Women/Gender: 1 | School Hours: 1 | Mental Health: 1 | |
| Politics & Government: 1 | | Sports: 1 | Personal Traits & Values: 1 | |
| Pollution/Garbage: 1 | | Suicide & Self-Harm: 1 | Poverty: 1 | |
| Sex, Sexuality, & Pregnancy: 1 | | War/Peace: 1 | Refugees: 1 | |
| Sports: 1 | | | Religion: 1 | |
| Uncategorized: 1 | | | Safety (General)/Security: 1 | |
| Unity & Diversity: 1 | | | School/School-Related: 1 | |

Note. LGBTQ = lesbian, gay, bisexual, transgender, and queer or questioning.

*Youth Civic Writing*

> *I personally have experienced racism a lot of times. For example, My sister and I were walking inside Wal-Mart and an elderly white lady kept looking at us funny. While we were walking, we were talking in Spanish and the lady comes up to us and says, "You're not in Mexico anymore so stop." My sister and I were in shock because we did not know how to respond. It makes me mad to just know the fact that people actually think that they are more superior than others just because they were born in the United States. From my knowledge, you're an immigrant unless you descend from the Native Americans, the Aztecs or the Mayans and the Vikings.*
>
> *According to alternet.com and Huffington Post, the system defends itself and not the public. And that is true. Racism has started so many things like police brutality all over the U.S, violent protest, athletes to kneel down during the National Anthem and so many more things. Several African Americans have been killed because of police being racist and shooting for no reason. How do you think that makes them feel? Mexicans get called out every day for being "rapist", "drug dealers", and "criminals." How do you think that makes them feel? The government isn't doing anything to put a stop to this, they're only helping themselves but not us.*
>
> *Racism is a huge issue that needs to be put a stop too. I am sure that you're busy with other things too but, try to do something about it. We build this Nation for everyone, not just one race. Every race, religion, country is full of people who have helped build the United States into what it is today. We should all be treated equally, just like this country was built to do.*
>
> *Adeline S.*
>
> *Work Cited:*
>
> *By Steven Rosenfeld/AlterNet. "8 Horrible Truths About Police Brutality and Racism in America Laid Bare by Ferguson." Alternet. N.p., 26 Nov. 2014. Web. 26 Oct. 2016.*
>
> *Almendrala, Anna. "Be Wary of Studies That Deny Racial Bias in Police Shootings." HuffingtonPost.com. N.p., 27 Sept. 2016. Web. 26 Oct. 2016.*

Adeline tagged her letter with the terms "racial injustice," "race," "discrimination," and "police brutality." Within our large-scale analysis, her letter was categorized under the topics "Race/Ethnicity," "Discrimination & Prejudice," and "Police," three of the most common issues within both the full set and the case study set of letters. In her letter, Adeline appeals to both ethical standards and to the empathy of the reader. Ethically, she refers to the foundational principles of the United States as she understands them:

> We build this Nation for everyone, not just one race. Every race, religion, country is full of people who have helped build the United

States into what it is today. We should all be treated equally, just like this country was built to do.

She appeals to empathy in her demand that the reader consider, "How do you think that makes them feel?" when discussing police violence and racism toward Mexicans. She illustrates her argument both with evidence from personal experience as well as by backing up her own opinions and experiences with references to claims made in digital online news sources, and she even includes full reference information in a "Works Cited" section at the end (this was not the case with other letters in Adeline's class).

Adeline, like other students, combines multiple forms of argumentation and evidence to present a letter expressing civic concern to an incoming president. The three forms of argumentation identified and coded across the set of letters from five schools were appeals to logic, appeals to ethics, and appeals to empathy. Within the case study letters, the most common form of argumentation used was an appeal to logic. Almost three quarters of the letters (71.74%) used logic to form an argument, building on explanations that followed from claims and evidence, employing cause and effect, or relying on if-then-because statements. Nearly two thirds of letters (61.59%) employed ethical appeals, referencing moral standards or what is "right." One quarter of the letters (25.36%) made appeals to empathy, asking the reader how they would feel in the shoes of another. Furthermore, these statistics reveal that students approached their civic writing tasks from multiple rhetorical angles (see Table 9). Like Adeline's, more than half of the letters (53.62%) made use of multiple forms of argumentation, and 13 letters used all three (9.42%). Six letters (4.35%) were coded that made use of none of the three forms of argumentation. These letters included letters that were largely informational or personal without broader claims, and one that was a semiabstract poem.

Differences between schools are marked, with students in Ohio and Nevada using logical arguments in fewer than 50% of their letters and students at the other three schools (Michigan, North Carolina, and Florida), using logical appeals in more than 80% of their published letters. Interestingly, the school-based differences in the use of logic in argumentation are not consistent across ethical or empathic appeals. The focal school in Michigan joins those in Ohio and Nevada in using ethical appeals in more than 65% of their letters, while the North Carolina and Florida sites used ethical appeals in fewer than 35% of their letters. Appeals to empathy were used in more than 50% of the letters only in Florida, and less than 40% at the other four schools. These patterns suggest both the power of classroom and teacher-specific norms for letter writing at the same time as revealing the diversity of approaches even within one local classroom setting.

### Table 9
**Frequency (%) of Use of Different Forms of Argumentation Across Sets of Letters**

| Characteristic | Full Set (n = 138) | Michigan High School (n = 46) | Ohio High School (n = 21) | North Carolina Middle School (n = 31) | Nevada High School (n = 27) | Florida High School (n = 13) |
|---|---|---|---|---|---|---|
| Appeal to logic | 71.74 | 82.61 | 42.86 | 87.10 | 44.44 | 100.00 |
| Appeal to ethics | 61.59 | 69.57 | 85.71 | 32.25 | 66.67 | 23.08 |
| Appeal to empathy | 25.36 | 23.91 | 38.10 | 16.13 | 29.63 | 53.85 |
| No logical, ethical, or empathic appeal | 4.35 | 0.00 | 4.76 | 6.45 | 11.11 | 0.00 |
| Two or more (logic, ethics, empathy) | 53.62 | 65.22 | 61.90 | 35.49 | 44.44 | 61.54 |
| All three (logic, ethics, empathy) | 9.42 | 10.87 | 9.52 | 6.45 | 7.41 | 15.38 |

*Citing Evidence*

At the same time that students were writing these letters, online bots were part of a substantial disinformation campaign to sow uncertainty and mistrust during the election season (Hindman & Barash, 2018). As students developed particular rhetorical arguments, we questioned to what extent evidence was utilized—and perhaps taught—across the sites in this study.

Like Adeline's letter, most letters in the case study set included some form of evidence. More than 80% included evidence as operationalized in the form of (1) personal experience, (2) a direct quotation or cited information (citation), or (3) a reference to specific facts that went unsourced (unsourced data). A summary of frequencies can be found in Table 10. Direct citation was the most frequent form of evidence followed by unsourced data and then personal experience, with the range relatively narrow at just over 10 percentage points. Use of personal experience ranged from 25% to 45% of letters across the five sets of letters. Almost 30% of letters used more than one form of evidence. Yet, less than 5% of letters have all three forms. Importantly, the differences in kinds of evidence used did not appear random.

Similar to the school-based differences in argumentation, students engaged with evidence differently across the five schools. For example, the Ohio and Florida sites both have much lower frequencies of citation as compared with Michigan and North Carolina. At the same time, students at the Florida school in particular referenced specific facts or data in their letters, oftentimes doing so without including a source (unsourced data) at a greater frequency than students at any other school. Students at the Nevada school in contrast used citation at a frequency below the average,

Garcia et al.

Table 10
**Frequency (%) of Use of Different Forms of Evidence Across Sets of Letters**

| Characteristic | Full Set (n = 138) | Michigan High School (n = 46) | Ohio High School (n = 21) | North Carolina Middle School (n = 31) | Nevada High School (n = 27) | Florida High School (n = 13) |
|---|---|---|---|---|---|---|
| Citation | 43.48 | 65.21 | 9.52 | 58.06 | 29.63 | 15.38 |
| Unsourced data | 38.41 | 47.83 | 38.10 | 35.49 | 18.52 | 53.85 |
| Personal experience | 33.33 | 26.09 | 42.86 | 29.03 | 44.44 | 30.77 |
| At least one form of evidence | 82.61 | 95.65 | 66.67 | 90.31 | 74.07 | 61.54 |
| Two or more | 28.26 | 39.13 | 23.81 | 25.81 | 18.52 | 23.08 |
| All three | 4.35 | 43.48 | 0.00 | 6.45 | 0.00 | 15.38 |

and they used unsourced data with the lowest frequency across the five schools but used personal experience with the highest frequency as compared with the other schools. Students at the Michigan site used personal experience with the lowest frequency across the five schools, but citation at the highest and unsourced data at the second highest frequency.

These contrasts between schools in both forms of argumentation and types of evidence used suggests different emphases in instruction with regard to what makes a convincing argument as well as what counts as evidence. These findings indicate that teacher instruction in areas such as defending and supporting an argument could play a substantial role in how student civic writing practices are developed likely reflecting teachers' varied instructional support of student participatory readiness (Allen, 2016) and instructional expectations across grade levels, courses, and states. Likewise, the kinds of evidence emphasized in particular schools may point to what educators "count" as evidence. For example, the use of citation and unsourced data signals the notion of external validity and credibility. The use of personal experience to support an argument or illuminate an issue gestures to a different epistemological orientation to what counts, focusing on lived, embodied, or experiential knowledge as well as the writer's relational power for communication and convincing a reader—a core component of the participatory readiness described by Allen (2016).

In recognizing the different ideological stances implicit in the kinds of evidence letter writers utilize, looking across uses of evidence and kinds of arguments, too, reveals useful patterns. The co-occurrence of forms of argumentation with forms of evidence (Table 11) suggests subtle relationships between the use of appeals to ethics and appeals to empathy with

*Table 11*
**Code Co-occurrence: Frequency (%) Within of Total Set of Letters (Frequency [%]Within Letters of That Argumentation Form)**

|  | Logical Appeal ($n = 99$) | Ethical Appeal ($n = 85$) | Empathic Appeal ($n = 35$) |
|---|---|---|---|
| Personal experience | 19.57 (27.27) | 21.74 (35.26) | 10.87 (42.86) |
| Citation | 34.47 (48.48) | 26.09 (42.35) | 11.59 (45.71) |
| Unsourced data | 29.71 (41.41) | 26.09 (35.29) | 12.32 (48.57) |

personal experience. Personal experience was most likely to be used within letters that made empathic appeals, and then ethical appeals, and least likely to be used in letters making logical appeals. This association suggests one version of a more relational approach to verbal empowerment and participation (Allen, 2016). While one might expect that citation and the use of data, even unsourced data, would be most closely associated with argumentation based on appeals to logic, citation was similarly likely to be used across all three forms of argumentation. Unsourced data was most likely in letters making empathic appeals followed closely by logical and then ethical appeals. While Adeline's letter fits these gentle trends in that it combines ethical and empathic argumentation with personal experience and citation, the variation between students in their uses of different forms of argumentation and evidence is more pronounced than any single pattern.

Particularly considering the role of false information—later defined as "fake news"—during the 2016 election, how students engage with evidence and for what topics has been a vastly underexplored aspect in youth civic education. Research about and after the election illustrates the vast difficulties youth have faced in evaluating credible sources online (boyd, 2018; Stanford History Education Group, 2016). Yet Adeline's letter suggests the possibility of flexible weaving together of varied forms of evidence and argumentation to communicate with power and conviction about a topic of concern.

## Discussion

### Interrogating Definitions of Civic Engagement

Looking at the thousands of letters submitted, students articulated complex civic arguments that demanded action on a broad range of topics they found personally resonant. These youth authors demand to be recognized as engaged civic actors—even within the contexts of school-based writing activities, challenging the notion of "gaps" in civic empowerment (M. Levinson, 2012) and in opportunity (Kahne & Middaugh, 2008) for youth in the United States today. Importantly, the distribution of topics that students wrote about differed substantially based on schoolwide socioeconomic factors.

We are struck that specific topics were more likely to be written about in higher and lower SES schools; what do the civic ideals and concerns of young people *mean* when topics such as college, health, and drugs/alcohol are significantly associated with higher SES schools and topics such as immigration and Black Lives Matter are associated with students of color and higher poverty schools? Whereas some youth could take for granted their safety and legal status in the country and author letters about higher education and climate change, many letters from lower SES schools suggest that core issues of immediate safety must be at the heart of civic dialogue and instruction today. These voices are ever present, demanding action, and—rightfully—mistrustful of government forces that have caused harm within communities.

These differences in what topics students write about point to important considerations around differentiation, cultivating "verbal empowerment" (Allen, 2016, p. 40), and the kinds of educational disparities students encounter across the country. The analysis of these letters demonstrates that voices of students like Luis, Madisyn, and Adeline echo the national issues that are mobilizing youth activism outside of schools, and in so doing challenges existing definitions of civics and the assumed *gaps* that separate civic learning in U.S. schools today. Furthermore, as every letter demonstrates a form of participatory readiness (e.g., Allen, 2016), the variation in topics illustrates stark differences in what counts within the civic domain of U.S. youth. For example, existing articulations of a civic empowerment gap acknowledge that youth of color may be "mistrustful and cynical" of political processes (M. Levinson, 2012, p. 37). However, these themes of cynicism and distrust are grounded in the very topics that were more likely to be tagged in letters from urban areas and schools serving lower SES students such as discrimination, Black Lives Matter, immigration, police, and violence. In recognizing the civic value of mistrust and it leading to movements like Black Lives Matter, these letters suggest that what has historically counted as topics of civic engagement and learning may exclude the valid feelings of cynicism of those most vulnerable within civic society.

The kind of cynicism and mistrust that M. Levinson (2012) ascribes to historically marginalized youth properly captures the specific locus of activated participatory readiness in these letters and during the lead up to the 2016 election. For instance, the amplification of resistance to police violence in communities of color can be read as the sociocultural construction of civic voices that have been stifled in traditional definitions of civic education, participation, and engagement. Likewise, considering that during the months that these letters were published, then candidate Trump sought to "build a wall" between the United States and Mexico and that Black Lives Matter continued to organize for racial justice, many of these letters operated in parallel with local and national organizing. These are not students writing in a bubble as their schooling experiences are interlinked with the political events happening in the "real" world; these students wrote alongside and in solidarity with ongoing civil rights movements.

Placing youth writing within this national context, these letters speak to new demands on political structures in the United States and challenge a status quo that historically disempowers youth of color and working-class individuals. For example, Black Lives Matter, as an organizing movement, describes its efforts as "working for a world where Black lives are no longer systematically targeted for demise" (Black Lives Matter, 2018). Likewise, a large portion of the letters tagged as violence and as police point to a distrust in government actors acting to uphold the safety and well-being of students, their families, and their communities. These are letters that ascribe mistrust and, perhaps, cynicism toward the institutional mechanisms in which they purportedly believe. While the letters in this study challenge contemporary definitions of civic learning, we are unsurprised that students of diverse backgrounds voiced diverse civic concerns within a project like this. Reflecting the wide array of thought and opinion that is the bedrock of a historically American embrace of difference, these letters course with eagerness to improve and to seek justice across the many faucets through which flow opportunity today.

### Teaching for Participatory Readiness

Based on the findings in this study, classrooms varied substantially with regard to the kinds of topics students took up as well as the particular approaches to writing about these topics. Considering the substantive differences that emerge around argumentation in the qualitative portion of this study, these differences are likely due to variations in instructional practice. Teachers, we can infer, played a central role in how students decided on and articulated particular arguments in this participatory civic project. And while we emphasize implications for educators here, we want to restate that our analysis does not evaluate the effectiveness or the quality of the writing students produced. Furthermore, we write this acknowledging that the Common Core State Standards have made substantive shifts to national writing practices in the years leading up to this study; particularly relevant here is shifts to argumentative forms of writing that rely on students' using evidence in their writing. These themes emphasize both standards-based approaches to teaching for participatory readiness and ways that educational policy frame what kinds of data are valued in school-based civic writing. How students are taught what forms of expression are valued in classrooms is a necessary area for future exploration.

Though there are emerging resources and data for evaluating student civic writing (e.g., NWP, 2017), the variation in writing approaches we analyzed in this study suggests that civic literacy instruction must focus on authentic writing for clearly articulated communities. As Adeline shares personal insights to articulate a nuanced argument about racial discrimination, her words are rooted in a particular time, place, and lived cultural experience. And because we can see substantial variation in classrooms supporting particular forms of argumentation for the public audience that can peruse these letters online, this study highlights a need to conceptualize how teachers teach for engagement with authentic audiences. Considering the possibilities for voicing powerful civic thought, these approaches need to guide students to see themselves as actors "in a living history and potential agents of transformation" (Ayers & Ayers, 2011, p. 6).

At the same time that these implications encourage local, authentic forms of writing, we are also aware that these suggestions arise during a time in which the spaces for dialogue and empowerment that embody contemporary participatory readiness require teacher courage. We know that some teachers are fearful of engaging in political discourse, particularly around controversial topics (Ayers & Ayers, 2011; Hess & McAvoy, 2015; Swalwell & Schweber, 2016). This study's findings imply a growing necessity for teachers to take on the mantle for supporting youth participatory readiness that is deftly enacted outside of schools daily. Considering that the letters in this study were written prior to an ascribed "Trump effect" (Costello, 2016), teachers must recognize and accommodate existing student civic agency to further foster participatory readiness through school-based spaces. These were not civically dormant students suddenly activated after the election; these letters highlight the capacity for empowered civic learning that mirrors what is more frequently seen outside of classrooms (e.g., Sawchuk, 2019).

## Listening to and Centering Youth Voice as Pedagogy

Recognizing that there are wide differences between the kinds of civic learning opportunities provided to students in schools in the United States today, this study builds from the understanding that *every* student engages in civic thought. Though our qualitative analysis emphasizes how differently students' civic writing may be from one school to another, the simple fact is that every student engaged in civic thought as demonstrated by all these letters having a self-identified set of tags and a call to action. Ultimately, these letters highlight the role of teachers in shaping how youth approach and engage in civic dialogue.

Central to the lessons of these letters is the need to *listen* to youth. When it comes to civic beliefs, concerns, and hopes, young people have a lot to say. Not only do young people have a lot to say, but they also voice their civic beliefs in myriad ways. When provided a space and platform for voicing civic thought, students articulated complex statements about their needs, hopes, and fears during a particularly caustic political moment. As much as this study's analysis focused on what students wrote and how they constructed their arguments, we are wary of losing sight of these letters as intentional statements made by students demanding to be heard. Not merely an act of recognizing youth voice, the civic participation evidenced in these letters speaks to Kirshner's (2015) reminder that "youth and societal institutions are strengthened when young people, particularly those most disadvantaged by education inequity, turn their critical gaze to education systems and participate in efforts to improve them" (p. 4).

As youth wrote about topics that were personally meaningful to them, we see the findings from this study emphasizing instruction as a conduit through which youth voice shapes and—ultimately—guides the "societal institutions" that Kirshner speaks to (p. 4). A pedagogy that centers *listening* requires youth-driven topics of inquiry and places student expertise alongside researched evidence, as explored in the qualitative analysis of letters in this study. Such instruction, too, can engage a "youth lens" (Petrone, Sarigianides, & Lewis, 2015) that leverages student writing to consider media "representations of adolescence" within policies and decision-making structures (p. 508). Additionally, as we've inferred that these letters are substantially shaped by teachers' instructional practices, we must question *how* educators gain practice at recentering classrooms on student voice. Fretting about youth civic engagement and students' lack of preparation for a media landscape bombarded by fake news largely ignores the fact that student civic identities are substantially shaped by schools and teachers. Our exploration of writing in just five sites within this study reveals that youth in areas that civic literature frequently sees as disengaged are highly vocal around civic issues that surround them; more important, the patterns of what kinds of evidence they utilize and what kinds of arguments they make suggest that teachers have guided how student civic voice is articulated, with what resources, and in acknowledgment of what other communities and movements.

Finally, many of the letters written by the students in this project were written as part of a required class activity. Even within this context, these letters often elucidated moments of student expertise and personal experience. Tacit knowledge played a consistent role in how and why students made their arguments. Considering the role of personal experience in many letters, the limitations of traditional civic boundaries, and the sociocultural signposts that shaped the corpus of letters, youth voice is guided and shaped along historically predictable lines; the expansiveness of where civic thought might tread is often hewn in by the boundaries of socioeconomic markers.

## Conclusion

Nearly 80 years before the 2016 election, alluding to a very different political and economic crisis roiling the United States, John Dewey (1939/1988) argued that solutions would only be found through "inventive effort and creative activity" (p. 225). Specifically, Dewey was noting that "for a long period we acted as if our democracy were something that perpetuated itself automatically" (p. 225). Just as Dewey argued—eight decades ago—that democracy must stretch to the new dimensions of a changing America, so too must we recognize that the definitions of civics that buttress youth learning must transmute in the modern day.

Based on reading these letters as indicators of participatory readiness, we find it difficult to reconcile the thousands of youth voices actively engaged in the LTNP project with the popular media narrative of youth civic complacency. Across every letter we analyzed, students identified topics of civic concern, presented calls for action, and often cited evidence to bolster their civic claims. At least for the participants whose teachers chose to submit to the LTNP project, every student voiced a civic concern. Even in this singular, school-based context, student civic imagination is vast.

Considering the thousands of students that identified and articulated civic issues, this study shines a light on the various ways that youth write civically within school contexts. Recognizing that civic topics range widely for different socioeconomic communities, the boundaries of civic participation extend beyond presupposed gaps; nearly every student in this study expressed a topic for civic action—both visceral issues of violence and aspirational needs like college tuition.

At the same time, we conclude this study recognizing that we end with more questions than answers. While we present a broader understanding of the topics that are salient to youth, teachers and teacher educators must sustain pedagogies that incorporate youth voice and participatory readiness. Likewise, the top issues that emerged point to powerful directions for understanding youth perspectives on broad national issues. Collectively, these powerful letters announce the civic demands of young people on the cusp of adulthood. Building from these voices, we must design new approaches to supporting the civic needs of students across the diverse contexts from which they are heard.

## ORCID iD

Antero Garcia https://orcid.org/0000-0002-8417-4723

## Notes

We would like to thank the Spencer Foundation for its support of this study. We would also like to thank AJ Alvaro, Elyse Eidman-Aadahl, Shadab Hussain, Caitlin Martin, Denise Sauerteig, KQED, and the National Writing Project for their assistance in conducting this study. Finally, we are grateful to Morgan Ames, Alicia Blum-Ross, Joseph Kahne, Nicole Mirra, Matt Rafalow, and three anonymous reviewers for their feedback on various drafts of this study.

[1]We identified the following as swing states: Florida, Iowa, Michigan, Nevada, New Hampshire, North Carolina, and Ohio. These seven states represent the consensus among political news sources as to which states were swing states in the 2016 presidential

election (www.politico.com, www.270towin.com, www.realclearpolitics.com); 2,691 (24%) of the letters came from swing states.

[2]Though the location and first name of students is publicly displayed on the LTNP site, we have removed the specific school names from these findings to obscure teacher and student identities as much as possible.

[3]Interestingly, at least a few letters cited other student's LTNP, either directly or in a Works Cited section.

[4]In the case where a "Works Cited" section was included at the end of the letter, but there was no indication of the source of those facts within the body of the letter, the letter was *not* coded as including citation, and it was also *not* coded as including unsourced data.

[5]Immigration appeared to be the only topic that appears to have been assigned to entire classes.

[6]In examining the topics, it is important to note that our categorization of students' issue tags affects this hierarchy. For example, 376 letters were tagged with issues that were categorized within the "climate change" topic, and 563 letters whose issue tags were categorized as "environment/wildlife," making these the 8th and 19th most prevalent topics, respectively. While there is some co-occurrence (letters that carried tags in each of these categories), had there been a single "Environment" topic rather than two distinct topics, it would have contained more letters and ranked higher on the list of topics written about. Similarly, "Education" and "School/School-Related" would combine into a larger set, and so forth.

[7]"Lower SES schools" refers to schools that are eligible for schoolwide Title I support and/or where 50% of students or more are eligible for FRL. "Higher SES schools" are schools not eligible for Title I schoolwide and/or where less than 50% of students are eligible for FRL.

## References

Allen, D. S. (2016). *Education and equality*. Chicago, IL: University of Chicago Press.

Atkins, R., & Hart, D. (2003). Neighborhoods, adults, and the development of civic identity in urban youth. *Applied Developmental Science, 7*, 156–164.

Ayers, R., & Ayers, W. (2011). *Teaching the taboo: Courage and imagination in the classroom*. New York, NY: Teachers College Press.

Bakhtin, M. (1986). *Speech genres and other late essays*. Austin: University of Texas Press.

Barron, B., Gomez, K., Pinkard, N., & Martin, C. K. (2014). *The Digital Youth Network: Cultivating digital media citizenship in urban communities*. Cambridge: MIT Press.

Black Lives Matter. (2018). About. *Black Lives Matter*. Retrieved from https://blacklivesmatter.com/about/

boyd, d. (2018). *SXSW EDU keynote: What hath we wrought?* Retrieved from https://www.youtube.com/watch?v=0I7FVyQCjNg

Boyte, H. (2003). A different kind of politics: John Dewey and the meaning of citizenship in the 21st century. *The Good Society, 12*(2), 1–15.

Burke, K. (1966). *Language as symbolic action: Essays on life, literature, and method*. Berkeley: University of California Press.

Carnegie Corporation of New York & Center for Information and Research on Civic Learning and Engagement. (2003). *The civic mission of schools*. New York, NY: Author.

Cohen, C., Kahne, J., Bowyer, B., Middaugh, E., & Rogowski, J. (2012). *Participatory politics: New media and youth political action*. Irvine, CA: DML Research Hub. Retrieved from https://ypp.dmlcentral.net/sites/default/files/publications/Participatory_Politics_Report.pdf

Costello, M. B. (2016). *The Trump effect: The impact of the presidential election on our nation's schools*. Southern Poverty Law Center. Retrieved from https://www.splcenter.org/sites/default/files/splc_the_trump_effect.pdf

Damon, W. (2001). To not fade away: Restoring civil identity among the young. In D. Ravitch & J. Vitteriti (Eds.), *Making good citizens: Education and civil society* (pp. 122–141). New Haven, CT: Yale University Press.

Dewey, J. (1988). Creative democracy: The task before us. In J. A. Boydston (Ed.), *John Dewey: The later works, 1925-1953* (Vol. 14, pp. 224–230). Carbondale: Southern Illinois University Press. (Original work published 1939)

Flanagan, C., & Levine, P. (2010). Civic engagement and the transition to adulthood. *The Future of Children, 20*, 159–179.

Garcia, A. (2017). *Good reception: Teens, teachers, and mobile media in a Los Angeles high school*. Cambridge: MIT Press.

Garcia, A., & Gargroetzi, E. (2019). *Teachers' civic beliefs in practice: Exploring civic writing pedagogy during the Letters to the Next President project*. Manuscript in preparation.

Garcia, A., & Mirra, N. (2019). "Signifying nothing": Identifying conceptions of youth civic identity in the English Language Arts Common Core State Standards and the National Assessment of Educational Progress' Reading Framework. *Berkeley Review of Education, 8*, 195–223.

Gonchar, M. (2018, October 4). Wasted ballots? A lesson exploring why more young people don't vote, and what students can do about it. *The New York Times*. Retrieved from https://www.nytimes.com/2018/10/04/learning/lesson-plans/wasted-ballots-a-lesson-exploring-why-more-young-people-dont-vote-and-what-students-can-do-about-it.html

Gould, J. (2011). *Guardian of democracy: The civic mission of schools*. Philadelphia: Leonore Annenberg Institute for Civics of the Annenberg Public Policy Center, University of Pennsylvania.

Green, E. L., Benner, K., & Pear, R. (2018, October 21). "Transgender" could be defined out of existence under Trump administration. *The New York Times*. Retrieved from https://www.nytimes.com/2018/10/21/us/politics/transgender-trump-administration-sex-definition.html

Hale, J. N. (2016). *The freedom schools: Student activists in the Mississippi civil rights movement*. New York, NY: Columbia University Press.

Hammersley, M., & Atkinson, P. (1995). *Ethnography: Principles in practice* (2nd ed.). New York, NY: Routledge.

Harvard University Institute of Politics. (2016). *Survey of young Americans' attitudes toward politics and public service* (31st ed., Harvard Public Opinion Project). Retrieved from http://iop.harvard.edu/sites/default/files/content/docs/161025_Harvard%20IOP%20Fall%20Report_FINAL%5B1%5D.pdf

Hess, D., & McAvoy, P. (2015). *The political classroom: Ethics and evidence in democratic education*. New York, NY: Routledge.

Hindman, M., & Barash, V. (2018). Disinformation, "fake news" and influence campaigns on Twitter. *Knight Foundation*. Retrieved from https://kf-site-production.s3.amazonaws.com/media_elements/files/000/000/238/original/KF-DisinformationReport-final2.pdf

Hobbs, R. (2011). *Digital and media literacy: Connecting culture and classroom*. Thousand Oaks, CA: Corwin Press.

Ivanič, R. (1998). *Writing and identity: The discoursal construction of identity in academic writing*. Amsterdam, Netherlands: Benjamins.

Jenkins, H., Shresthova, S., Gamber-Thompson, L., Kligler-Vilenchik, N., & Zimmerman, A. M. (2016). *By any media necessary: The new youth activism*. New York, NY: New York University Press.

Kahne, J., & Middaugh, E. (2008). *Democracy for some: The civic opportunity gap in high school* (Circle Working Paper No. 59). College Park, MD: Center for Information & Research on Civic Learning and Engagement.

King, M. L., Jr. (1994). *Letter from the Birmingham jail*. San Francisco, CA: Harper San Francisco.

Kirshner, B. (2015). *Youth activism in an era of education inequality*. New York, NY: New York University Press.

Knight Foundation. (2018). *American views: Trust, media and democracy*. Retrieved from https://knightfoundation.org/reports/american-views-trust-media-and-democracy

Laskas, J. M. (2017, January 17). To Obama with love, and hate, and desperation. *The New York Times*. Retrieved from https://www.nytimes.com/2017/01/17/magazine/what-americans-wrote-to-obama.html

Levinson, A., & Garcia, A. (2019). *Immigration as a focus in youth letters to the president: Arguments and argumentation in student writing in a polarized debate*. Manuscript in preparation.

Levinson, M. (2010). The civic empowerment gap: Defining the problem and locating solutions. In L. Sherrod, J. Torney-Purta, & C. A. Flanagan (Eds.), *Handbook of research on civic engagement* (pp. 331–361). Hoboken, NJ: Wiley.

Levinson, M. (2012). *No citizen left behind*. Cambridge, MA: Harvard University Press.

Literat, I., Kligler-Vilenchik, N., Brough, M., & Blum-Ross, A. (2018). Analyzing youth digital participation: Aims, actors, contexts and interests. *Information Society*, *34*, 261–273.

McDonnell, L. M. (2000). Defining democratic purposes. In L. McDonnell, P. M. Timpane, & R. W. Benjamin (Eds.), *Rediscovering the democratic purposes of education* (pp. 1–18). Lawrence: University Press of Kansas.

McLuhan, M. (1962). *The Gutenberg galaxy: The making of typographic man*. Toronto, Ontario, Canada: University of Toronto Press.

Mirra, N. (2018). *Educating for empathy: Literacy learning and civic engagement*. New York, NY: Teachers College Press.

Mirra, N., & Garcia, A. (2017). Re-imagining civic participation: Youth interrogating and innovating in the multimodal public sphere. *Review of Research in Education*, *41*, 136–158.

Monaghan, J., & Saul, W. (1987). The reader and the scribe, the thinker: A critical look at the history of American reading and writing instruction. In T. S. Popkewitz (Ed.), *The formation of the school subjects: The struggle for creating an American institution* (pp. 85–122). London, England: Falmer Press.

National Assessment Governing Board. (2014). *Civics framework for the 2014 National Assessment Educational Program*. Washington, DC: U.S. Department of Education.

National Writing Project. (2017). Civically engaged writing analysis continuum. *National Writing Project*. Retrieved from https://cewac.nwp.org/

Nie, N. H., Junn, J., & Stehlik-Barry, K. (1996). *Education and democratic citizenship in America*. Chicago, IL: University of Chicago Press.

Payne, K. A., & Journell, W. (2019). "We have those kinds of conversations here . . .": Addressing contentious politics with elementary students. *Teaching and Teacher Education*, *79*, 73–82.

Perman, M. (2001). *Struggle for mastery: Disfranchisement in the South, 1888-1908*. Chapel Hill: University of North Carolina Press.

Petrone, R., Sarigianides, S. T., & Lewis, M. A. (2015). The youth lens: Analyzing adolescence/ts in literary texts. *Journal of Literacy Research, 46,* 506–533.

Saldaña, J. (2009). *The coding manual for qualitative researchers.* Thousand Oaks, CA: Sage.

Sawchuk, S. (2019, July 31). Schools teach civics. Do they model it? *Education Week.* Retrieved from https://www.edweek.org/ew/articles/2019/05/08/schools-teach-civics-do-they-model-it.html

Schreier, M. (2012). *Qualitative content analysis in practice.* Thousand Oaks, CA: Sage.

Schwartz, S. E. O., Chan, C. S., Rhodes, J. E., & Scales, P. C. (2013). Community developmental assets and positive youth development: The role of natural mentors. *Research in Human Development, 10,* 141–162.

Southern Poverty Law Center. (2016). *Ten days after: Harassment and intimidation in the aftermath of the election.* Montgomery, AL: Author. Retrieved from https://www.splcenter.org/sites/default/files/com_hate_incidents_report_final.pdf

Stanford History Education Group. (2016). *Evaluating information: The cornerstone of civic online reasoning.* Retrieved from https://stacks.stanford.edu/file/druid:fv751yt5934/SHEG%20Evaluating%20Information%20Online.pdf

Strauss, A., & Corbin, J. (1994). Grounded theory methodology. In N. Denzin & Y. Lincoln (Eds.), *Handbook of qualitative research* (pp. 273–285). Thousand Oaks, CA: Sage.

Swalwell, K., & Schweber, S. (2016). Teaching through turmoil: Social studies teachers and local controversial current events. *Theory & Research in Social Education, 44,* 283–315.

Tripodo, A., & Pondiscio, R. (2017). Seizing the civic education moment. *Educational Leadership, 75,* 20–25.

Tyack, D. B., & Cuban, L. (1995). *Tinkering toward utopia.* Cambridge, MA: Harvard University Press.

Watts, R., & Flanagan, C. (2007). Pushing the envelope on civic engagement: A developmental and liberation psychology perspective. *Journal of Community Psychology, 35,* 779–792.

Westheimer, J., & Kahne, J. (2004). What kind of citizen? The politics of educating for democracy. *American Educational Research Journal, 41,* 237–269.

Wineburg, S. S. (2018). *Why learn history (When it's already on your phone).* Chicago, IL: University of Chicago Press.

Wray-Lake, L., & Hart, D. (2012). Growing social inequalities in youth civic engagement? Evidence from the National Election Study. *Political Science & Politics, 45,* 456–461.

Zummo, L., Gargroetzi, E., & Garcia, A. (2019). *Youth voice on climate change: Using factor analysis to understand the intersection of science, politics, and emotion.* Manuscript in preparation.

Manuscript received October 30, 2018
Final revision received July 20, 2019
Accepted July 22, 2019

# Fostering Democratic and Social-Emotional Learning in Action Civics Programming: Factors That Shape Students' Learning From *Project Soapbox*

Molly W. Andolina
Hilary G. Conklin
DePaul University

*This research examines the factors that shape high school students' experiences with an action civics program—Project Soapbox—that fosters democratic and social-emotional learning. Drawing on pre- and postsurveys with 204 students, classroom observations, teacher interviews, student work samples, and student focus group interviews, the study illuminates how specific features of the curriculum and its implementation are linked to its promising outcomes. Our findings indicate that the curriculum's emphases and structure, along with instructional decisions and context, play key roles in influencing student outcomes. Project Soapbox's power lies in its alignment with many well-established civic education best practices and in its intentional linkage with key social-emotional learning practices, many of which are newly recognized as having particular civic import.*

KEYWORDS: action civics, civic education, democratic education, high schools, social-emotional learning

---

MOLLY W. ANDOLINA is an associate professor of political science at DePaul University. Her field of expertise includes public opinion and youth political engagement. She has published work on the political activism of millennials, college student activism, Americans' attitudes toward gay rights, the challenges of survey research measurement, and public opinion on the Monica Lewinsky scandal.

HILARY G. CONKLIN is an associate professor of education at DePaul University, 2247 North Halsted, # 320, Chicago, IL 60614-3624; e-mail: *hconkli1@depaul.edu*. Her research interests are broadly rooted in the need to provide equitable, intellectually rich, authentic learning opportunities for all students. Her current scholarship explores the design of teacher preparation experiences, the impact of these experiences on teachers' practices and their students' learning, and youth learning from civic education.

Civic education, broadly conceived, is widely viewed as an essential part of the K–12 education curriculum by educators and the public alike. This is evidenced by the fact that almost every state has a civics requirement, and it is further supported by surveys where over 90% of high school students report taking at least one civics course (Center for Information and Research on Civic Learning and Engagement [CIRCLE], 2013). Yet the widespread incidence of civics education conceals a growing concern among researchers and practitioners about the *quality* of most civics programming. Comprehensive studies of civic education programs indicate that the inadequate civics instruction currently available to most students is at least partially to blame for the low levels of civic engagement among youth (CIRCLE, 2013; Gould, Jamieson, Levine, McConnell, & Smith, 2011; Levine & Kawashima-Ginsberg, 2015). Indeed, while there is general agreement on what constitutes best practice in civic education—practice that centers around teaching young people the skills of civic participation and the orientations of lifelong civic engagement, thus moving beyond the political knowledge that tends to be the focus of most curricular efforts (c.f. Campbell, 2008; Niemi & Junn, 1998; Syvertsen, Flanagan, & Stout, 2007; Torney-Purta, 2002; Youniss, 2011)—such quality instruction is not widely available. Perhaps even more disturbing, when quality civics education exists, the best curricular and cocurricular programs are disproportionately available to students who have higher socioeconomic status, are enrolled in more challenging academic programs, and are most likely to go to college (Kahne & Middaugh, 2008; Levinson, 2012). The differential civic opportunities available to students in high schools across America contribute to the differential participation rates among young adults.

## High-Quality Civic Education Practices and Outcomes

Despite these inequities, there is general agreement on what high-quality practices involve, as well as compelling evidence of their promising impact when they are implemented. The best curricular practices engage students in discussions of current events, create a classroom climate that promotes the open exchange of ideas, include teacher encouragement of independent thinking and expression of opinions, provide opportunities for service learning and participation in simulations, and promote programs that allow students to select issues that are relevant to their own lives (Gibson & Levine, 2003; Campbell, 2008; Kahne & Middaugh, 2008; Kahne & Sporte, 2008; Syvertsen et al., 2007; Torney-Purta, 2002; Youniss, 2011). Studies designed to evaluate the impact of civic education best practices document an array of positive outcomes, including gains in factual knowledge, increases in anticipated civic engagement, the development of skills of democratic deliberation, and more attention to political news (Kahne, Crow, & Lee, 2013; Longo, Drury, & Battistoni, 2006; McDevitt & Kiosis, 2006).

Moreover, scholars have found that the impact is not limited to students who are participating in civics programming but can spill over into positive influences on parents (in increased attention to the news and a greater propensity to turn out and vote) (McDevitt & Kiosis, 2004, 2006), and it can last years after the program has ended (McDevitt & Kiosis, 2006; Torney-Purta & Amadeo, 2003). And while these best practices are disproportionately available to White, wealthier students in the most selective classes in high-achieving schools, research also shows that disadvantaged youth who have opportunities to engage in high-quality civics education are more participatory as adults (Wilkenfeld, 2009).

*Action Civics*

Many of the best practices in civic education are common in programs that come under the framework of action civics. Proponents of action civics education contend that one does not learn how to be a citizen by studying processes, watching adults, or reading texts but, rather, by actively taking part in the work of citizenship (Gingold, 2013; National Action Civics Collaborative [NACC], 2010; Warren, 2019). Action civics programming requires students to engage with authentic issues in their communities: Students identify the issues of importance to them and their communities and then are provided with guidance, skills instruction, and opportunities that enable them to "do civics and *behave as citizens*" (Levinson, 2012, p. 32).

Action civics integrates aspects of the strength-based approaches of Positive Youth Development (e.g., Benson, Scales, Hamilton, & Sesma, 2006), the problem-solving and relationship skills of social-emotional learning (SEL), the community orientation and real-world contexts of service learning, the collective power of youth organizing, and knowledge of the political systems of traditional civic education, with the goal of empowering youth in the most marginalized communities (Gingold, 2013). In contrast to civic education practices that focus exclusively on civic knowledge, or service-learning projects in which students participate as volunteers in community organizations, action civics curricula are grounded in four guiding principles: action, particularly collective; youth voice, knowledge, and expertise; youth agency; and reflection (Gingold, 2013; NACC, 2010). Drawing on these common principles, action civics theory posits that when youth voice and expertise are valued, and young people have authentic opportunities for expression, engagement, and reflection, then powerful civic learning can occur, thereby narrowing the civic empowerment gap and strengthening our democracy (e.g., Gingold, 2013; NACC, 2010). While specific action civics programming varies, generally, students progress through six common steps: examining their community, identifying issues of importance to them, conducting research on the issues, developing a strategy for action, taking action, and reflecting on the process (Gingold, 2013).

Although research on action civics is still emerging, there is a growing repertoire of studies of single programs that establish a link between action civics curricula and a host of promising outcomes, including civic skills such as public speaking and community mapping, social capital, political efficacy, and content knowledge. The research—which includes case studies on a schoolwide initiative in Massachusetts (Berman, 2004), the Building Civic Bridges program (LeCompte & Blevins, 2015), the iEngage summer civics institute (Blevins, LeCompte, & Wells, 2016), Project 540 (Battistoni, 2004), the We The People curriculum (Walling, 2007), the Constitutional Rights Foundation's City Works Initiative (Kahne, Chi, & Middaugh, 2006), and the Student Voices program (Feldman, Pasek, Romer, & Jamieson, 2007; Syvertsen et al., 2009)—identifies positive outcomes associated with key action civics components such as an emphasis on student voice and the creation of open classrooms where students discuss and debate current events and are encouraged to speak their minds. In addition, in many of these programs, which are implemented with diverse populations across the United States, students engage with civic leaders and the broader community, often as part of a service-learning opportunity. In our study of the impact of the action civics curriculum *Project Soapbox* on participating high school students (Andolina & Conklin, 2018), students reported gains in rhetorical proficiency, increased confidence for public speaking, and heightened willingness and desire to become involved in political action. In addition, the student participants in our study noted the impact of listening to their peers' speeches: They expressed a greater sense of connection to other students, a deeper understanding of their peers' and their own experiences, and an enhanced appreciation for perspectives other than their own (Andolina & Conklin, 2018). Thus, there is a growing consensus of research linking action civics curricula to positive outcomes.

Yet while scholars have established outcomes associated with various action civics curricula and their features, there has been a paucity of research that has explored variations in implementation or context (for an exception, see Ballard, Cohen, & Littenberg-Tobias, 2016) or the specific factors that shape the curricula's outcomes. Civic education practices are subject to state requirements, district and school support, as well as teacher implementation effects (Kahne & Middaugh, 2008; Niemi & Junn 1998; Westheimer & Kahne, 2004). As Ballard et al. (2016) explain, civic education programs are not "homogenous interventions" (p. 378). Further, much of the research that has been conducted on action civics programs has relied primarily on survey measures. Thus, we know less about the processes by which action civics programs create their positive outcomes and the specific aspects of program implementation that may shape differential outcomes across varied contexts.

## Social-Emotional Learning and Relational Skills for Democratic Citizenship

Alongside the growth in attention to action civics has been a growing recognition that listening and other social and emotional capacities are vital for both academic development and civic engagement (cf. Cramer & Toff, 2017; Levine, 2013; Weissberg, Durlak, Domitrovich, & Gullotta, 2015). Scholars have emphasized the relational dimensions of citizenship as central to solving the problems facing us as democratic societies, arguing that we should aim to increase interpersonal practices such as listening, particularly to those different from ourselves, in order to improve trust, develop community, build empathy, and foster equity (Allen, 2004; Cramer & Toff, 2017; Dobson, 2012; Levine, 2013). Democratic theorists as well as experts on social-emotional development suggest that attentive listening engenders empathy, allows for vulnerability, builds relationships, and develops a sense of connection among individuals—democratic orientations that lead, in turn, to broader outcomes such as building trust and bridging political rifts (Allen, 2004; Cramer & Toff, 2017; Levine, 2013; Weissberg et al., 2015). The development of trusting social relationships among teachers and students contributes to youths' sense of belonging, their affective connection to the broader society, their development of a public identity, and their inclination to act in the interest of the common good (Flanagan, Stoppa, Syvertsen, & Stout, 2010).

The burgeoning interest in the relational citizenship skills engendered and associated with democratic listening parallels the growing emphasis on cultivating SEL skills. The SEL domains of social awareness and relationship skills include the abilities to empathize, feel compassion, and listen actively (Weissberg et al., 2015), and these are competencies that are well aligned with the developmental needs of adolescence (Williamson, Modecki, & Guerra, 2015). Well-implemented SEL programs—which necessitate teacher practices that offer strong emotional support and opportunities for student voice and autonomy—have demonstrated not only improved academic outcomes but also greater empathy and stronger peer and adult relationships (Weissberg et al., 2015).

While SEL competencies sometimes focus on individual or interpersonal skills, as noted above, many of these competencies are also vital to the development of social trust, civic identity, and democratic orientations (Allen, 2004; Cramer & Toff, 2017; Flanagan et al., 2010; Levine, 2013). Given that a growing number of states have implemented SEL standards (Collaborative for Academic, Social, and Emotional Learning, 2019; Gingold, 2013), and schools are thus increasingly called upon to teach SEL skills, the cultivation of SEL that is civically oriented offers numerous potential benefits. Thus, bringing the worlds of civic education and SEL together provides an opportunity to sharpen our understanding of the relational skills that are essential to both.

In sum, there is growing interest in the promising practices and outcomes of action civics—and this interest aligns with the growing interest in the relational dimensions of citizenship and SEL. However, we have limited empirical insight into the process by which specific elements of action civics programs produce their positive outcomes or the ways in which context and implementation contribute to the positive outcomes that research has documented. Further, with limited exceptions (cf. Barr et al., 2015), there is little research that explores the social-emotional dimensions of civic education programming.

## Research Question

In this article, we examine an action civics program for high school students—*Project Soapbox*—that fosters democratic learning and SEL. Our prior research established that students' participation in *Project Soapbox* not only increased their confidence in their rhetorical skills and expectations for future political engagement but also cultivated their sense of empathy for others' experiences and their feelings of connectedness to others (Andolina & Conklin, 2018). Building on these findings, here we seek to understand how specific features of the curriculum and its implementation are linked to its promising outcomes. The research question we explore is

- What factors shape students' experiences with and learning from an action civics curriculum?

## *Project Soapbox* in Chicago

*Project Soapbox*, started by the Chicago-based nonprofit, nonpartisan Mikva Challenge, is a public speaking curriculum comprising five detailed lessons that are designed to be useable as a stand-alone, weeklong curriculum for approximately hourlong class periods. The *Soapbox* curriculum includes reproducible handouts, rubrics, and suggested resources and is available for a nominal fee on Mikva's website (see https://secure.mikva challenge.org/project-soapbox). Mikva notes that all of its programs are "grounded in the principles of Action Civics" and are designed to "provide youth with authentic and transformative democratic experiences," "develop agency and future commitment to civic action," and "provide youth with skills and knowledge to be effective citizens"—all key programming goals (Mikva Challenge, 2019).

In the curriculum, students choose a community issue of importance to them around which they will develop a speech. To prepare them for this task, students learn about the structure of good speeches; analyze sample speeches; learn how to use different forms of evidence to support arguments, how to grab audiences' attention, and to use other rhetorical devices; outline and write rough drafts of their own speeches; learn tools for effective delivery of a speech; and practice delivering these speeches with their peers. Finally, students deliver their finished speech to their classroom of peers, along with outside adult judges from the community. These adults are recruited through Mikva's networks and include businesspeople, lawyers, public officials, parents, clergy, or other city residents. While the role of these adults varies by community, because *Project Soapbox* in Chicago is structured as a competition, adult judges there complete rubrics to evaluate students' speech structure, content, and delivery.

When students deliver their speeches, the curriculum encourages teachers to establish clear expectations among students that they listen to each speech without interruption, complete peer feedback forms for one another, and give "wild applause" after each speech is completed. The top speakers from individual schools advance to a citywide competition, which is also judged by community members. Given that the curriculum features opportunities for open exchange of ideas, the development of public speaking skills, and the role of authentic youth voice, the curriculum and goals of *Project Soapbox* are well aligned with the best practices of civic education (Campbell, 2008; Niemi & Junn, 1998; Syvertsen et al., 2007; Torney-Purta, 2002; Youniss, 2011).

The curriculum is typically implemented in social studies or English language arts classrooms. A teacher's decision to use *Project Soapbox* is voluntary, and most teachers incorporate it as part of their regular school-day curriculum. Some teachers implement *Soapbox* as a stand-alone curriculum, while others use it as part of a larger district civics curriculum or as part of Mikva's Issues to Action curriculum, both of which include components that link students' *Soapbox* speeches with broader public engagement. To support teachers' implementation of the program, Mikva provides professional development opportunities—in person, online, or both. As the curriculum guide notes, while the lessons can be taught in the span of a week, many teachers take longer in order to allow students to develop their speeches further—often between 2 and 3 weeks. While *Project Soapbox* is now implemented in many cities across the country at different times of the year, in Chicago, it is typically implemented in the early part of the school year—usually in September, October, or November—often timed to allow students to connect their speech topics to upcoming elections. The citywide competition occurs in mid- to late November.

## Research Method

To study the factors that shaped students' learning from and experiences with *Project Soapbox*, in the fall of 2015, we recruited a sample of 19 classrooms that included nine teachers (six social studies, three language arts) who were implementing *Project Soapbox* at nine different Chicago public high schools. All the participating teachers were incorporating *Project Soapbox* as part of the regular school-day curriculum. We collaborated closely with Mikva to recruit our sample: When teachers signed up to implement *Project Soapbox* at the beginning of the school year ($N = 40$ teachers), Mikva sent them an email informing them about our research, encouraging them to participate if they wished, and providing a link to our consent form and further information if they were interested. From the teachers who completed this consent form and agreed to participate ($N = 14$), we selected those teachers who were starting their curriculum in October ($N = 11$), allowing the lead time to distribute student assent/consent and parental permission forms, return to the classroom to gather those forms, and administer an initial survey (see below) prior to students beginning the curriculum. Because of scheduling constraints, we were unable to include two teachers who had agreed to participate, leaving us with our sample of nine teachers in nine schools.

The nine schools in our sample served either majority Hispanic or majority African American student populations, and all but one of the schools in our sample had 90% or more low-income students. The school sample included the range of types of high schools found within Chicago Public Schools (CPS)—neighborhood, charter, magnet, small, military, and alternative—and served student populations ranging in size from 124 to 2,927 students. The schools also encompassed the range of school rating levels that exist within CPS: Of the five possible performance levels, the schools in our sample included three "below average" (Level 2), two "average" (Level 2+), one "high" (Level 1), and three "highest" (Level 1+).

We administered surveys to students before and after they took part in the curriculum to examine change over time. While 232 students completed the initial survey, 204 completed both pre- and postsurveys. Most measures in the pre- and postsurveys were either replicated exactly or modified slightly from previous studies, including the Chicago Consortium on School Research's (2013) "Five Essentials" survey, the California Civic Index (Kahne, Middaugh, & Schutjer-Mance, 2005), and the Civic Engagement Questionnaire (Zukin, Keeter, Andolina, Jenkins, & Delli-Carpini, 2006). Factor analyses conducted on all scales revealed single factors, and when tested for reliability, all scales posted solid to strong alpha scores (ranging from .64 to .928). We also created survey items corresponding to the Common Core speaking and listening standards, given that the *Project Soapbox* curriculum is aligned with these standards. Our postsurvey also included an additional set of measures that Mikva Challenge regularly uses to evaluate *Project Soapbox* (program evaluation measures). These measures included Likert-scale questions asking students for their level of agreement with statements in response to their participation in the program, such as "I feel more confident," "I feel I am a better public speaker," and "I plan to speak up on issues that are important to me in the future" (see Table 1). This portion

## Table 1
**Project Soapbox Program Evaluation Measures**

| Indicator | Response Categories | List of Items |
|---|---|---|
| | | How much do you agree or disagree with each statement about how you feel after preparing and giving a speech? As a result of the activities and competition I participated in, |
| More confident | | I feel more confident. |
| My ideas were heard | | I feel like my ideas were heard by my peers. |
| Better public speaker | *Strongly disagree,* | I feel I am a better public speaker. |
| Expert on my topic | *Disagree, Agree,* | I feel like I am an expert on my topic. |
| Less nervous to speak | *Strongly agree* | I feel less nervous to speak up in front of a group. |
| Plan to speak up on issues | | I plan to speak up on issues that are important to me in the future. |
| Plan to work to make a difference | | I plan to work to make a difference on the issue I spoke about. |
| I want to do this again | | This is something I want to do again. |

of the postsurvey also included open-ended questions such as "What do you think you learned as a result of participating in *Project Soapbox*?" Finally, we incorporated a series of questions designed to capture the ways in which different factors might impact student learning. In addition to demographic information, we assessed student political socialization at home, and we asked the students if they had ever given a speech before, how often they practiced their speech, their level of engagement in the classroom, and their assessment of school and class climate. While some measures were curriculum specific (e.g., the number of times the students practiced), others were adapted from previous studies of youth civic engagement (e.g., Flanagan, Syvertsen, & Stout, 2007). A complete list of the variables and scales is provided in Tables 2 and 3.

The student sample closely mirrored the demographics of CPS: Our sample was 41% African American, 48% Hispanic, and predominantly low income. Of the students who completed the pre- and postsurveys, 55% were female, and the majority (76%) were between 17 and 19 years old, with the remainder (24%) aged 14 to 16 years.

To complement the student survey data, we selected five classrooms—each in a different school—in which to collect qualitative data on teachers' practices and students' engagement with the curriculum. Our aim in selecting these five

Table 2
**Project Soapbox Independent Variables in Ordinary Least Squares Regression**

| Indicator | Type | Response Categories | List of Items in Measure/Question if Single Item |
|---|---|---|---|
| Year in school | Single item | *9th, 10th, 11th, 12th* | Grade: |
| Given a speech before | Single item | *Yes, more than once; Yes, once; No, never* | Until today's *Project Soapbox* competition, had you ever given a speech in front of a group of people before? |
| Number of times practiced | Single item | | How many times did you practice your speech? |
| Family political socialization | Scale; pre only; $\alpha = .64$ | *Never, Occasionally, Sometimes, Regularly* | How often would you say each of the following occurs, if at all? I talk to my parents/guardians about politics and current events. My parents/guardians volunteer in our community. My parents/guardians vote in elections. My parents/guardians are involved in local politics (school board/city council). |
| Classroom climate | Scale; pre only; $\alpha = .91$ | *Strongly disagree, Disagree, Agree, Strongly agree* | Please indicate how much you agree or disagree with each statement *about this class*: In this class, students have a voice in what happens. In this class, students can disagree with a teacher, if they are respectful. In this class, students can disagree with each other, if they are respectful. In this class, students are encouraged to express opinions. |
| Mother's education | Single item | *Less than a high school degree, High school graduate or GED certificate, Some college, College degree, Postgraduate or professional degree* | What is your mother's highest education level? |
| How much do you care about the topic? | Single item | *Not at all, Not very much, Somewhat, A lot* | How much do you care about this topic? |

Table 3

**Project Soapbox Dependent Variables in Ordinary Least Squares Regression (for Survey Items With Pre/Post Measures)**

| Indicator | Reliability (Alpha Score): Pre/Post | Response Categories | List of Items in Measure |
|---|---|---|---|
| Class engagement | α = .824/.771 | *Strongly disagree, Disagree, Agree, Strongly agree* | Please indicate how much you agree or disagree with each statement *about this class:*<br>I usually look forward to this class.<br>I work hard to do my best in this class.<br>The topics we are studying are interesting. |
| Civic competencies | α = .877/.900 | *I definitely can't; I probably can't; Maybe; I probably can; I definitely can* | If you found out about a problem in your community that you wanted to do something about (e.g., illegal drugs were being sold near a school or high levels of lead were discovered in the local drinking water), how well do you think you would be able to do each of the following?<br>Create a plan to address the problem.<br>Get other people together to care about the problem.<br>Express your views in front of a group of people.<br>Identify individuals or groups who could help you with the problem.<br>Call someone on the phone you had never met before to get his or her help with the problem.<br>Contact an elected official about the problem.<br>Post something on a social media website like Facebook or Twitter to inform people about the problem. |
| Teacher rating | α = .910/.917 | *Strongly disagree, Disagree, Agree, Strongly agree* | Please indicate how much you agree or disagree with each statement *about your teacher in this class:*<br>My teacher often connects what I am learning to life outside the classroom.<br>My teacher encourages students to share their ideas about things we are studying in class.<br>My teacher often requires me to explain my answers.<br>My teacher encourages us to consider different solutions or points of view.<br>My teacher wants us to become better thinkers, not just memorize things. |

*(continued)*

## Table 3 (continued)

| Indicator | Reliability (Alpha Score): Pre/Post | Response Categories | List of Items in Measure |
|---|---|---|---|
| Listening skills | α = .877/.867 | *Not at all confident, Not very confident, Somewhat confident, Very confident* | How confident are you in your ability to do each of the following *when listening to someone speak?* <br> Determine a speaker's point of view on an issue. <br> Evaluate how well a speaker uses evidence. <br> Evaluate how well a speaker uses rhetoric, such as word choice, points of emphasis, and tone. |
| Presentation skills | α = .928/.913 | *Not at all confident, Not very confident, Somewhat confident, Very confident* | How confident are you in your ability to do each of the following in *presenting information to others?* <br> Explain a problem clearly to an audience. <br> Provide different kinds of evidence to explain the importance of an issue. <br> Present a well-organized, easy to follow speech. <br> Grab an audience's attention through the opening and closing of a speech. <br> Deliver a speech effectively through clear body language, and effective pacing and volume. |
| Intended activism after high school | α = .769/.817 | *Not at all likely, Not very likely, Somewhat likely, Very likely* | When you think about life *after high school*, how likely is it that you would do each of the following? <br> Contact or visit someone in government. <br> Vote in an election. <br> Express your opinion on an issue by contacting a newspaper, or radio or TV talk show. <br> Try to raise awareness about an issue by posting an article on a social media site like Facebook or Twitter. |

*(continued)*

Table 3 (continued)

| Indicator | Reliability (Alpha Score): Pre/Post | Response Categories | List of Items in Measure |
|---|---|---|---|
| Future activism* | α = .853/.890 | *Strongly disagree, Disagree, Agree, Strongly agree* | We are also interested in what you think about your future activities. How much do you agree with each of the following? Being actively involved in community issues is my responsibility. In the next 3 years, I expect to work on at least one community problem that involves a government agency. I have good ideas for programs and projects to help solve problems in my community. In the next 3 years, I expect to be involved in improving my community. |

*Adapted from the California Civic Index.

classrooms was to explore dimensions of variation in implementation of the curriculum, drawing from the pool of teachers who agreed to more in-depth study. Thus, we selected English language arts ($N = 1$) and social studies ($N = 4$) classrooms to better understand how the curriculum was implemented in these differing subject matter contexts, teachers who were both experienced ($N = 4$) and novice ($N = 1$) at implementing the curriculum, and teachers using *Project Soapbox* in different curricular contexts—as part of a larger 11th/12th-grade civics curriculum, a 9th-grade criminal law curriculum, an 11th/12th-grade sociology course, an 11th-grade social science course, and, in the case of the English language arts classroom, a junior/senior AP Language and Composition course, alongside a civics classroom in a partner social studies classroom. We also chose classrooms that represented differing student demographic makeups and differing school types and ratings, as described above. The teachers in these classrooms spent between 8 and 15 class periods on *Project Soapbox*; three of the teachers followed the curriculum quite closely, making only small modifications (e.g., showing additional video speech samples), while the other two teachers used the majority of the curriculum but made more significant modifications (e.g., creating different rubrics for the speeches).

In these selected classrooms, we observed and recorded detailed notes on 3 days of implementation of the curriculum, including at least 1 day in each classroom when students delivered their finished speeches, to see how the teachers employed the curricular materials, how students engaged with the curriculum, and students' performances when delivering their speeches. We then interviewed these five teachers to better understand their goals for using the curriculum and their pedagogical decision making around the materials; these interviews each lasted between 40 and 50 minutes and were audio-recorded. We collected students' written speeches to examine their topics and evaluate their use of rhetorical elements emphasized in the curriculum, and we conducted student focus groups in four of the selected classrooms to gain further insight into students' experiences with the curriculum and the teachers' practices. The student focus groups included between 3 and 9 students, depending on how many students from each school consented to participate and were available at the times their teacher designated (e.g., lunchtime or after school); these interviews lasted between 35 and 60 minutes. Finally, we observed the citywide competition, recorded observations about the competition and process and examined video recordings of the 10 student finalists' speeches.

### Data Analysis

In order to examine the factors that shaped students' experiences with and learning from this action civics curriculum, we began with an analysis of the student surveys, focusing first on the quantitative measures. We employed ordinary least squares regression to determine which factors were most instrumental in predicting key outcomes among the participants. One set of dependent variables was taken from the postsurvey and created by Mikva to measure students' self-assessments of the impact of the curriculum (program evaluation measures), as described above. In addition, we created a scale to measure students' evaluations of their rhetorical skills and a scale to measure students' intended political action in the future. These additional scales were based on questions that were asked of students both before and after their participation in *Project Soapbox*, which allowed us to measure change. When using variables that had both pre- and postsurvey measures, we used students' scores on the premeasure as a control variable for the postscore (as suggested by Molnar, Smith, & Zahorik, 1998; Singer & Andrade, 1997). Our independent variables included self-reported gender, race, ethnicity, socioeconomic status (measured by mother's education), political socialization at home (measured by parental role modeling and political discussion), classroom climate (measured by students' responses to presurvey questions), the number of times students practiced their speeches, past history of speech giving, how much students reported that they cared about their topic, and the student's year in school.

Our qualitative analysis included the students' open-ended survey responses and the student focus group interviews, classroom observations, and teacher interviews. We categorized the students' open-ended survey responses into themes, including the broad categories under which the students' speech topics fell. A large percentage of students elected to speak about topics like gun and gang violence in the city of Chicago, police brutality and misconduct, and domestic violence. Students also discussed issues such as college tuition costs, school start times, the importance of education for voting, unemployment, and gentrification. Thus, we created speech topic categories such as domestic violence, community violence, racism/discrimination, and education. Similarly, we grouped students' perceptions of what they had learned from the curriculum into themes (e.g., learned speaking skills, research skills, etc.).

We transcribed all the teacher interviews and student focus groups and then coded all the transcripts and classroom observation notes both inductively and deductively using categories aligned with the research question and action civics theory. For example, we initially developed codes to correspond with the factors we expected to shape students' learning, (e.g., teacher, students' connection to their topic, etc.); then, after reading through the transcripts and open-ended student survey responses, we developed additional codes that captured the themes that emerged from the data about students' learning (e.g., the structure of the curriculum). After coding these data, we wrote analytical memos (Glesne & Peshkin, 1992) that identified patterns and themes about the factors shaping students' learning from the curriculum and selected quotes from the various data sources that most clearly represented each of the themes.

# Findings

In the following, we build on our previous findings about what students said they had learned from the curriculum (Andolina & Conklin, 2018) to focus on the factors that shaped the positive outcomes students reported from their participation. First, we provide an overview of the factors we explored through our quantitative analysis of the student surveys and discuss those factors that were most salient to student outcomes. Then, we discuss the qualitative data to highlight both those factors that reinforce the quantitative student survey findings and those that emerged from a close examination of the open-ended survey responses, classroom observations, teacher interviews, student speeches, and student focus groups.

**Factors That Shaped Curricular Outcomes: Student Survey Findings**

Our first analysis of the relative influence of key factors employed the program evaluation variables—items like "I feel more confident" and "I feel I am a better public speaker"—of students' self-reported gains as the dependent variables. For context, most of the students indicated that they had been positively affected by their participation, with the majority indicating in the exit survey that they were "less nervous about speaking" and that they felt their "ideas were heard." In evaluating what factors were most important for predicting these outcomes, we regressed each program evaluation variable on the independent variables detailed above.

The results (see Table 4) indicate the influence of the political environment of the home (where parents talk about politics and they volunteer and vote), the number of times students practiced their speech, and the classroom climate (how students felt about sharing their opinions and disagreeing with others in the classroom) on students' perceived outcomes. For example, students who came from homes where parents discussed politics and modeled civic participation (positive political socialization) were more likely to say that after participation, they felt like an expert on their topic. Similarly, the more students practiced their speeches, the greater their confidence in their speaking skills and their sense that their voice had been heard. And if students felt that teachers respected and encouraged their opinions (classroom climate), then they were more likely to say they planned to speak up on these issues again in the future. It should be noted, however, that the impact of these variables on the various program evaluation measures is uneven. As Table 4 indicates, while political socialization, speech practice, and classroom climate were often key predictors of various outcomes, none of the three was consistently significant across all eight program evaluation equations.

## Table 4
### Ordinary Least Squares Regressions: Predicting the Program Evaluation Measures Among *Project Soapbox* Participants

| | More Confident | My Ideas Were Heard | Better Public Speaker | Expert on My Topic | Less Nervous to Speak | Plan to Speak up on Issues | Plan to Work to Make a Difference | I Want to Do This Again |
|---|---|---|---|---|---|---|---|---|
| Constant | −.66(.61) — | .24 (.60) — | −.23 (.76) — | −.86 (.77) — | −1.16 (.78) — | .11 (.70) — | .11 (.65) — | −1.42 (.73)* — |
| Year in school | .12 (.81) *.09* | −.02 (.09) *.10* | .03 (.11) *.02* | .07 (.11) *.04* | .09 (.11) *.05* | −.02 (.10) *−.01* | −.14 (.10) *−.10* | .06 (.10) *.04* |
| Given a speech before | .01 (.09) *.00* | .13 (.09) *.09* | −.05 (.18) *−.03* | .18 (.11) *.11* | .04 (.11) *.02* | .04 (.10) *.03* | −.01 (.10) *−.01* | .03 (.10) *.02* |
| Number of times practiced speech | .42 (.07) *.37\*\** | .22 (.07) *.21\*\** | .37 (.09) *.29\*\** | .09 (.09) *.07* | .22 (.09) *.17\** | .17 (.08) *.15\** | .24 (.07) *.21\*\** | .27 (.08) *.21\*\** |
| Family political socialization | .04 (.03) *.09* | .04 (.03) *.10* | .04 (.03) *.09* | .08 (.03) *.18\*\** | .01 (.03) *.03* | .06 (.03) *.16\** | .07 (.03) *.17\** | .07 (.03) *.16\** |
| Class climate | .05 (.03) *.11* | .07 (.03) *.19\*\** | .04 (.03) *.09* | .04 (.04) *.08* | .08 (.04) *.17\** | .06 (.03) *.14\** | .01 (.03) *.03* | .06 (.03) *.13* |
| Mother's education | −.04 (.06) *−.04* | −.04 (.06) *−.05* | −.02 (.07) *−.01* | −.00 (.07) *−.00* | .10 (.08) *.10* | .03 (.07) *.03* | .03 (.06) *.04* | .13 (.07) *.13* |
| How much cared about the topic | .53(.11) *.30\*\** | .48 (.11) *.29\*\** | .48 (.14) *.24\*\** | .64 (.14) *.32\*\** | .58 (.14) *.28\*\** | .46 (.13) *.25\*\** | .70 (.12) *.39\*\** | .60 (.13) *.30\*\** |
| Adjusted $R^2$ | .34** | .25** | .20** | .17** | .18** | .17** | .27** | .26** |

*Note.* The cells contain *B* values (unstandardized coefficients) and standard errors (in parentheses) in the first row and *B* values (standardized coefficients, in italics) in the second row.
*$p \leq .05$. **$p \leq .01$.

While these influences are noteworthy, the one factor that was both more influential than the others and most consistent throughout every dependent variable was how much a student cared about a topic. How much students cared about a topic was significantly (and strongly) related to how highly they ranked the impact of the curriculum on all of the program evaluation measures—their confidence (and their decreased nervousness), their desire to want to make a difference, their intention to speak up about the issue in the future, their sense of being an expert, their feeling that they had been heard, and their enthusiasm for participating in *Project Soapbox* again.

Similarly, we found that how much students cared about their topic was a significant predictor of a variety of student self-assessments of their civic and literacy skills after the conclusion of the curriculum. As reported earlier, students made modest but significant gains in their assessment of their post–high school civic engagement, as well as some of their rhetorical skills (Andolina & Conklin, 2018). To determine what best accounts for the observed change, we regressed the postsurvey scale on all the independent variables included in the first analysis, and we added the presurvey measure as a control variable. Again, how much students cared about their topic was a significant and substantial predictor of their postsurvey assessment of how engaged they were with the class (e.g., how much they look forward to the class, work hard, and find the topics interesting) (see Table 5). Similarly, how much students cared about their topic was strongly related to their assessment of their civic competency to address a community problem, their rating of their teacher (e.g., in connecting learning to life outside the classroom, encouraging critical thinking), and their confidence in their academic listening skills. There were three instances in which caring about one's topic did not reach the level of statistical significance: postsurvey assessments of rhetorical skills and two different measures of anticipated political engagement. In these cases, presurvey orientations, as well as how often students practiced their speech and, in the case of political intentions, family political socialization, held sway.

While not uniformly consistent, it is important to note that a student's assessment of the classroom climate showed up as a significant predictor for three of the program evaluation outcomes and the postsurvey scores for students' assessment of their classroom engagement and civic competencies. The questions designed to measure classroom climate were included in the presurvey only and indicated that prior to the implementation of the curriculum, many of the students in the sample felt that their teachers had created environments that supported student voice, allowed them to respectfully disagree with one another and their teachers, and encouraged them to express their own opinions. As illustrated in Figure 1, an overwhelming majority of students in the presurvey agreed that they were given a voice in the classroom, that their teacher allowed them to respectfully

## Table 5
**Ordinary Least Squares Regressions: Predicting Postsurvey Outcomes Among *Project Soapbox* Participants**

| | Class Engagement Scale | Civic Competencies Scale | Teacher Rating | Listening Skills | Presentation Skills | Intended Activism After High School | Future Activism |
|---|---|---|---|---|---|---|---|
| Constant | -.64 (.83) | 2.71 (2.32) | 1.75 (1.44) | 4.23 (1.07) | 3.81 (1.64) | 1.97 (1.52) | 1.89 (1.35) |
| | — | — | — | — | — | — | — |
| Year in school | -.00 (.12) | -.28 (.32) | .12 (.20) | -.31 (.15) | .11 (.23) | -.37 (.21) | -.09 (.20) |
| | -.00 | -.05 | .03 | -.12* | .03 | -.10 | -.03 |
| Given a speech before | .15 (.11) | -.90 (.33) | .21 (.20) | -.14 (.14) | -.42 (.22) | .09 (.22) | -.01 (.20) |
| | .06 | -.14** | .05 | -.06 | -.10 | .02 | -.00 |
| Number of times practiced speech | .22 (.09) | .78 (.26) | .03 (.16) | .15 (.12) | .48 (.18) | .62 (.17) | .42 (.15) |
| | .18* | .16** | .01 | .08 | .15** | .20** | .15** |
| Family political socialization | .02 (.03) | .14 (.10) | .11 (.06) | .09 (.04) | .06 (.07) | .22 (.07) | .15 (.06) |
| | .03 | .08 | .09 | .13* | .05 | .21** | .15* |
| Classroom climate | .21 (.04) | .21 (.11) | .04 (.03) | .00 (.05) | .11 (.07) | .05 (.07) | -.01 (.06) |
| | .30** | .11* | .10 | .00 | .09 | .04 | -.01 |
| Mother's education | .05 (.08) | .13 (.22) | .13 (.10) | -.15 (.10) | .10 (.15) | -.10 (.14) | -.17 (.13) |
| | .03 | .03 | .12 | -.10 | .04 | -.04 | -.07 |
| How much cared about the topic | .82 (.15) | .85 (.44) | .74 (.26) | .38 (.19) | .49 (.30) | .35 (.29) | .26 (.25) |
| | .27** | .10* | .15** | .12* | .09 | .07 | .06 |
| DV in presurvey | .34 (.05) | .45 (.05) | .53 (.09) | .46 (.06) | .45 (.05) | .46 (.06) | .60 (.06) |
| | .38** | .53** | .55** | .50** | .53** | .45** | .61** |
| Adjusted $R^2$ | .61** | .56** | .52** | .40** | .49** | .44** | .27** |

*Note.* The cells contain *B* values (unstandardized coefficients) and standard errors (in parentheses) in the first row and β values (standardized coefficients, in italics) in the second row. DV = dependent variable.
*$p \leq .05$. **$p \leq .01$.

## In this class... (pre survey)

[Bar chart showing percentages for Agree and Strongly Agree across four categories:
- Students have a voice: Agree 50, Strongly Agree 29
- Students can respectfully disagree with teacher: Agree 47, Strongly Agree 41
- Students can respectfully disagree with each other: Agree 50, Strongly Agree 44
- Students are encouraged to express opinions: Agree 46, Strongly Agree 44]

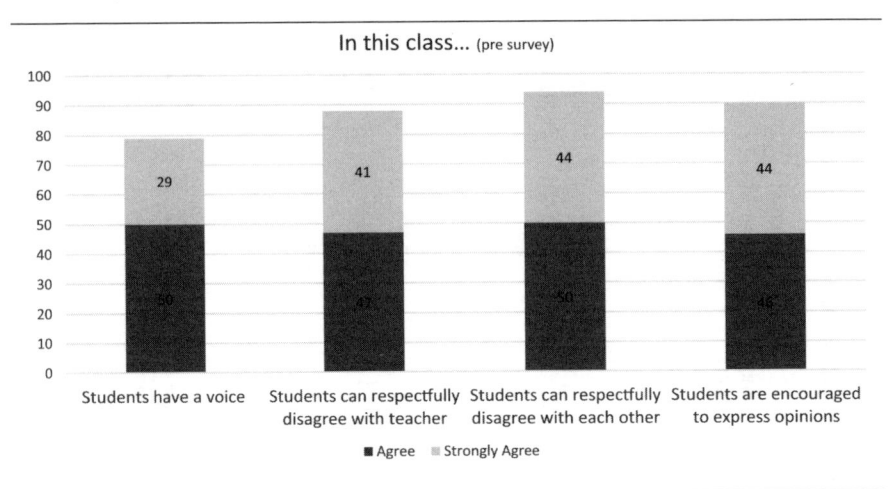

*Figure 1.* **Classroom climate.**

disagree with him or her and with one another, and that their teacher encouraged them to express their opinions. Clearly, across our entire sample, the majority of students were experiencing key elements of civic education best practices prior to engaging with the curriculum.

However, not all classrooms were equivalent in the implementation of a positive classroom climate. For example, as illustrated in Figure 2, in Teacher A's classroom, not only did an overwhelming number of students agree that they had a voice in the classroom, but fully half of them strongly agreed with this assessment. Other teachers received higher overall scores for their classroom environment, yet the magnitude of student assessment was weaker, as illustrated by the data from Teacher B, where 96% of the students agreed that they were given a voice but only 17% strongly agreed with this description. And other teachers, such as Teacher C, were not as successful, with only 42% describing the classroom as a place where students have a voice in what happens. As the multivariate analysis described above illustrates, student assessment of classroom climate was a significant predictor for gains in students' class engagement and, importantly, increases in how students rated their civic competencies.

In sum, the quantitative findings from the student survey data reveal that some factors outside the classroom—such as students living in homes where parents discussed politics and modeled civic participation—shaped students' experiences with and learning from the *Project Soapbox* curriculum. However, many of the influences on students' learning—such as how much a student cared about his or her speech topic and the student's assessment of the classroom climate—related to the structure of the curriculum

## In this class, students have a voice... (pre survey)

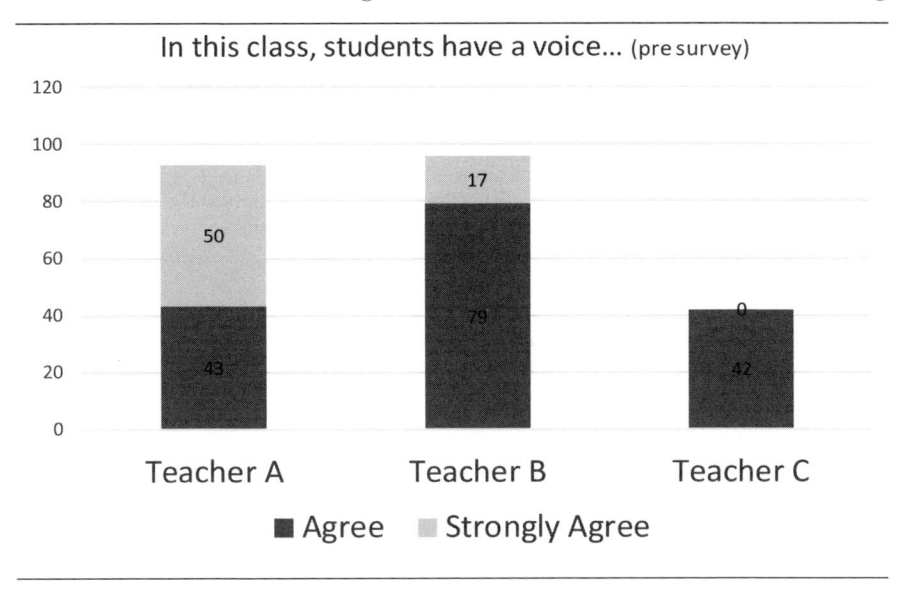

Bar chart showing stacked bars of Agree and Strongly Agree responses by teacher:
- Teacher A: Agree 43, Strongly Agree 50
- Teacher B: Agree 79, Strongly Agree 17
- Teacher C: Agree 42, Strongly Agree 0

*Figure 2.* **Classroom climate by teacher.**

itself as well as the instructional context in which the curriculum was implemented. We now turn to a discussion of the qualitative data, which provide a deeper and more textured understanding of the ways in which these and other curricular and instructional factors influenced students' learning from *Project Soapbox*.

### Features of the Curriculum That Shaped Student Outcomes

Our observations of the five focal classrooms, interviews with the teachers and students, analysis of the student speeches, and examination of the *Project Soapbox* curricular materials suggest that much of the power of the curriculum lies in the elements it includes, the way its creators carefully scaffold students' speech development, and how students' speech delivery is structured. In the first four of the five lessons, the curriculum provides teachers video links to a range of sample speeches, reproducible handouts that enable students to analyze the quality of sample speeches, and additional organizers to lead students from initial speech brainstorming to drafts in which they "spice up" their speeches with rhetorical devices to create polished products. Our classroom observations and teacher interviews illustrated that the teachers relied on these materials and found them very valuable: We observed the teachers showing students many of the curriculum's recommended speech examples and using the curriculum handouts to have students analyze these sample speeches and craft their own. As Ms. Bowman[1] explained, one of the curriculum's strengths was its structure:

> Walking through how to create a speech . . . the very brief rough draft, and then the ways in which you could add to it with the grabbers and the closers and the rhetorical devices, and making sure that you have the call to action, and making sure that you have explained the assets of the community that they already have . . . [for students] . . . it kind of alleviated some stress that they had of being like . . . "I don't even know where to start."

The curricular scaffolding, then, provided helpful tools for teachers to assist their students in crafting speeches.

Students' speeches, in turn, revealed the elements emphasized in the curriculum: Students incorporated repetition and imagery, attention grabbers, logical and emotional appeals, and calls to action. One student, for example, spoke with emotion about the need for immigration reform, explaining,

> My mom was 8 months pregnant when she came to the States. . . . She was in the desert . . . crossed the river. . . . It was so dangerous. I didn't understand . . . but she wanted me to get an education. We are hurt, humiliated, tired.

Other students cited statistics alongside personal stories, such as one who wrote her speech about domestic violence:

> A woman is beaten every 9 seconds. Two million injuries and 1,300 deaths are caused each year as a result of domestic violence. . . . My sister is a domestic survivor; she was 6 months pregnant when she was abused by the father of her children [sic].[2]

Speeches like these indicate that the students were using the suggestions featured in the curriculum to create powerful speeches.

Students' increased confidence in their rhetorical and presentation skills, along with the speeches they produced, are indicators that the teachers' use of the scaffolded curriculum enabled students to develop valuable speech-writing and delivery skills. While the general sequence and structure of the curriculum appeared to be key factors shaping the teachers' practices and students' learning, two related elements of the curriculum stood out as particularly influential: students being able to choose their own topic and the curriculum's emphasis on the use of emotional appeals.

## Students' Choice of Speech Topics

As the quantitative analysis highlighted, how much students cared about their speech topic was a significant and powerful predictor of positive outcomes from the curriculum; the fact that students care about their topics, in turn, is directly connected to the curriculum's emphasis on having students choose their own topic. The first lesson includes a brainstorming handout for students that asks them to first think of what they are proud of in their school/community/city/society and what they wish they could change; the handout then asks them to "name an issue that is very important to you and explain why it is important to you." Because of this curricular requirement, students' *Soapbox* speeches centered on topics that were closely connected to their daily experiences and about which they cared a great deal.

Both the quantitative and the qualitative data indicated the salience of choosing a topic for students in the *Soapbox* experience. From the survey, we learned that three quarters of the students—that is, most of them—reported caring a lot about the topics they had chosen to speak about. And in response to an open-ended survey question that asked, "What was the best part about preparing and giving a speech?" about one in six students replied, "Choosing the topic." When asked in the focus group interviews what it was like to participate in *Project Soapbox*, many students focused on their opportunity to choose a topic that had personal meaning. Students said things like "I liked that I got the chance to speak about something that I care about" and "I really liked picking the topic and then breaking it down and really doing a lot of research on it."

The opportunity for high school students in these schools to choose a community issue of importance to them appeared to be a crucial factor in their engagement with the curriculum, thereby facilitating their speech writing and delivery skills. Many students explained that their connection to the topic motivated them and made them willing to develop and deliver their speech—something that they might have otherwise not been interested in or willing to do. For example, in the focus group interviews, students made comments like the following:

> I thought it was really fun, because I'm not really much of a public speaker, so just because it's difficult for me. But yeah, it made it a lot easier that I know the subject that I wanted to talk about and express how I felt about it.

> I wrote about killing, and I added my cousin's death up in there. So that connection—it really meant something to me. That's how come I was passionate about it and willing to share with everyone else, because it was something that I felt very strong about.

The teachers, too, indicated that the *Soapbox* curriculum was motivational to their students. Mr. Cahill spoke of one student who decided to write about bullying, based on the student's own experiences: "I was pleased particularly with one student who has given me trouble all year long in terms of motivation and being on task in class. . . . He won the classroom and schoolwide [*Soapbox* competition] and [is] going on to citywide." Ms. Bowman indicated that choosing a topic helped students connect more deeply with what they do care about. She said,

> It forced them to actually consider things in their life that they do give a crap about. . . . And a lot of times they're not asked about that. . . . The fact that the majority of the kids really got up and . . . gave their speeches showed that they took this project seriously.

Thus, being able to choose a topic that was meaningful to them spurred many students to participate willingly in the activities of *Project Soapbox* and enabled them to develop new confidence, skills, and understandings through the process.

### Incorporation of Emotional Appeals

Students' ability to choose a topic dovetailed with their incorporation of emotional appeals to persuade their audiences—something that the curriculum encouraged and appears connected to the curriculum's impact in fostering connection and empathy. The curriculum, which emphasizes the rhetorical skill of knowing one's audience, suggests that personal stories enable a speaker to appeal emotionally to an audience. Given students' close connections to their topics, many students integrated such personal stories, often with very moving effects. Student speeches included statements such as "I was bullied for years," "My cousin got killed," and "We shouldn't have to be afraid of being deported." Some students delivered speeches through tears, such as one who spoke about domestic violence and described her aunt not feeling safe, until "one day, she just stopped calling." Another student began to cry as she spoke about her father: "You came, you left, you left my life a mess."

In our classroom observations, the emotional impact of many students' speeches was palpable. In some cases, audience members voiced this impact, such as one student who, after hearing a classmate's speech about a friend who was bullied and died, said, "I feel kind of shitty because I myself have been a bystander. I could possibly change a life . . . so thank you for telling me that story." When we observed the citywide competition and audience members were given the opportunity to share reflections from the speeches, many students and adults in the room responded with great emotion. After hearing a speech on domestic violence in which a student shared that "I'm here with no father and no mother. . . . They are both in the same cemetery," several parents in the room as well as other students were moved to tears. Thus, the curriculum's emphasis on the use of personal stories and other emotional appeals seemed to cultivate audience members' sense of connection along with their appreciation of the experiences of others.[3]

## Features of the Speech Presentations

While the elements described above appeared salient in shaping the development of students' speeches, the data suggest that the way the curriculum structures the speech presentations themselves is also crucial for fostering students' confidence, developing their sense of agency, and cultivating empathy and connection through attentive listening. All students are expected to deliver their speeches in front of one another as part of the classroom competition, and the *Soapbox* curriculum encourages teachers to establish an authentic audience—including adult judges such as parents, other school staff members, and community members—that is highly supportive of all student speech givers. The curriculum directs teachers to explain that "all speeches should receive wild applause when they are completed," meaning that "everyone cheers loudly and enthusiastically." The curriculum also recommends that students complete peer feedback forms as they listen to one another; these forms are provided in the curriculum and ask students to focus on the content, delivery, and effectiveness of their peers' speeches.

Although we discuss below how teachers varied in their implementation of "wild applause" and peer feedback, for those who adhered to it, the curricular emphasis on creating supportive, attentive audiences appeared to foster students' confidence and sense of agency. Several teachers discussed the effect of this support for their students. Mr. Gilroy spoke about a new student from the Dominican Republic:

> She's incredibly nervous. . . . she's learning not only about her topic, but she's gaining the confidence. . . . but she's going to have so much support from her peers when she goes up there because we have this norm that you clap wildly for everyone . . . and then just really be energetic and supportive of one another. . . . I think that they get a lot of criticism . . . from a lot of the adults in their life. And just to be wildly applauding them for being them and for the work they're doing is great.

Similarly, Ms. Bowman described how "heartwarming" it was to see students' "respect and appreciation for each other." These teachers' comments mirror our observations; in many of the classrooms and at the citywide competition, students offered one another enthusiastic support.

Further, in all of the classrooms we observed, students appeared very engaged with and focused on their peers' speeches. Students corroborated this observation by noting the power of having others hear their stories, an idea echoed by many students in the focus group interviews. One student explained,

> At first I had a big weight on my heart. It was like when I spoke to everybody how I feel, I felt kind of better, because I felt like

everybody was listening to me. I really didn't have no one to really talk to me to really get it off my chest. So I felt more open and relieved.

Another student noted, "I learned how to speak out my opinion, and how, in different ways, people may hear me. About my point of view about my community and how I may change it." Because the students were focused on one another's words, student speakers felt "heard."

In addition to feeling heard by having the opportunity to present their speeches publicly, many students also indicated a developing sense of agency by virtue of delivering their speeches to audiences that included their peers and adults from the community. In the focus group interviews, students said things like "I learned that as an individual we have a voice and a voice that could potentially give us power in the future." One student who participated in the citywide competition commented,

> There was [an adult] lady behind me, and it pretty much made me know that I came for a reason. She basically told me that—because I picked abortion—that my speech had got through to her. She said she had never seen abortion in that light.

Similarly, Ms. Bowman explained how a fellow teacher in her school was impacted by hearing one of her students' speeches on the problematic nature of the term *Black on Black crime*, noting, "The teacher across the hall, he's a young Black teacher, [said], 'I use that term all the time. . . . After that speech, I will never use it again. . . . I had never thought of it like that.'"

Students could see that giving their speeches had the potential to influence others. Teachers also noted the empowerment for students that came from speaking out in front of their peers. Ms. Bowman explained that her students were "really, really proud of themselves"; many had been very nervous and doubted their abilities to deliver the speech, and "then they went up and you couldn't tell that they were nervous at all." Mr. Gilroy pointed out the particular value that speaking out publicly had for marginalized students:

> I think there are general benefits that are evident when you have a large group of disproportionately Black and Brown students who are disproportionately nonvocal leading up to 16 and 17 years old. . . . There are some situations in which children are still raised in families where the expectation is that children are seen, not heard. And this . . . should begin to turn that on its head.

Other teachers echoed these sentiments, such as Ms. Vogel who spoke about a very quiet, shy student:

> She got up and gave a really, extremely personal, and I think really well-done speech on immigration. . . . her family has a really dramatic

immigration story. She's a DREAMer, and this is an issue that's really close to her heart. It was just amazing to see her do it. . . . I think it's empowering to them in different ways.

These comments indicate that speaking publicly through *Soapbox* gives students a sense of agency, which may be especially empowering for those students who have been traditionally disenfranchised.

A key aspect of students feeling heard and feeling that their words had impact was their sense that their audiences were actively listening. In the focus group interviews, students made comments like the following:

When people was giving their speeches, I seen people shaking their head and just showing . . . that they're listening and that they got something from that *Soapbox* speech, basically. . . . It probably did change somebody's life or improve their image of how they looked at stuff.

Another student explained, "By other people clapping for you, you saw that you were able to get the message across of what the problem you were dealing with was."

Indeed, part of what appears to be powerful about *Project Soapbox* is that the curriculum puts students in a position to hear, learn from, and connect with one another. Unlike the structure of classroom discussion, in which participants often listen to one another in order to develop a response, the structure of *Soapbox* speech presentations encourages participants to listen solely to hear. Mr. Gilroy spoke about this unique aspect of the curriculum shaping his choice to use it: "It's the first time for many of the young people to be able to structurally listen to their peers, and for adults to listen to their students about the issues that matter to them. . . . There's a structure for listening." Because the format of the speech presentations encourages listening, students have the opportunity to hear new perspectives and develop a deeper understanding of one another and the issues important to their peers.

Furthermore, because the *Soapbox* speech presentations encourage listening, several teachers noted that the curriculum actually serves to foster a sense community—both among students and with the teacher. Ms. Vogel explained that one of her reasons for using *Soapbox* is that it's "good for building classroom community." Other teachers commented that the curriculum not only helps students learn about and connect with one another but it also helps them as teachers learn about their students. Mr. Cahill explained how *Soapbox* informs his AP class content: 'Through *Soapbox*, I find out what they care about, and it often . . . directs my course for the rest of the year." He went on to explain that being able to learn about his students through the *Soapbox* speeches had a profound effect on his view of teaching more broadly:

> [*Project Soapbox*] changed my teaching. It made me realize that if you don't listen to students, it's one of the main reasons they don't listen to you. . . . Students have knowledge, and that knowledge can impact change, and it's unique and important. . . . When they're listened to, they feel taken seriously, and they take things seriously.

The way in which the *Soapbox* curriculum structures the opportunity for students to speak publicly about issues they care about—and listen to one another—produces powerful outcomes for them, their peers, and the adults who hear their speeches.

## Instructional Choices and Context

While the curriculum, as written, provides the opportunities described above for students to experience powerful learning, the qualitative data revealed that the teachers' varied instructional choices and contextual factors shaped students' experiences. Not surprisingly, the teachers we interviewed had differing goals in their teaching, generally, and also for using *Project Soapbox*, which led them to emphasize different aspects of the curriculum. Some teachers were most interested in fostering public speaking skills, some particularly valued the research skills cultivated, while others prioritized the listening to and learning about one another that come from participating in the curriculum. Ms. Vogel, for example, noted the match between *Soapbox* and her law class, explaining, "One of the things I'm really enforcing this year is everybody has to talk. . . . Your voice has value. You have something to contribute." Similarly, Ms. Estrada focused on the cultivation of public speaking and research skills in her social science course:

> I want them to feel more confident speaking up so that their professors [in college] will know them. . . . I think public speaking skills are just important if you ever want to advocate for yourself . . . but . . . my primary interest is research . . . and I like the idea of strengthening an argument with evidence.

For these teachers, *Soapbox* offered practice in particular skills that were relevant to their broader curricula. Yet, as referenced earlier, teachers like Mr. Gilroy use *Soapbox* intentionally because it cultivates listening among students as well as in adults, while Ms. Bowman and Mr. Cahill similarly value learning about their students through the speeches.

The teachers' differential goals and emphases, in turn, showed up in the differences in their instructional practices and in the strength of each classroom climate—findings that are consistent with the quantitative data that revealed distinctions in classroom climates. In some cases, the teachers had intentionally invested significant time in cultivating a classroom community in which students developed trust and respect for one another. Ms. Bowman explained that, leading up to *Project Soapbox*, she had engaged

## Fostering Democratic and Social-Emotional Learning

students in many community-building activities in which they practiced "diplomatic skills . . . being able to see someone's else . . . point of view and appreciating others' viewpoints," which she believed helped them feel less nervous when they spoke in front of their peers. Because of these exercises, she observed an increase in students' respect for one another:

> I think a big contribution to that has been because of the amount of activities that we do where you have to talk to each other and you have to listen to each other. . . . And they're also appreciating each other and respecting each other more, which is just making everything better.

Similarly, Mr. Cahill, after observing that some of his students were primarily focused on the competitive aspect of *Project Soapbox*, led his students through a series of compassion meditation exercises in which he asked them to

> [imagine] people further outside our circle and the things they go through and how their needs might be met . . . to get thinking about the emotional reasons of why we're doing this. . . . It fed into the idea of considering your audience too, of having that step back to think empathetically about someone you really disagree with in order to figure out how to better argue with them.

In both of these classrooms, the teachers actively cultivated a sense of concern for others among their students. These were also the classrooms in which, based on our observations and our focus group interviews, students appeared the most energized by participation in *Project Soapbox*.

Similarly, the teachers we observed made varied instructional decisions on how closely to follow the curriculum—with some following it closely, some omitting parts, and others enhancing it—and these decisions too influenced students' experiences. For example, Mr. Cahill asked his students to "throw praise" on each other to make the speech competition feel "celebratory of [students'] voices," Ms. Vogel implemented the "wild applause" after each speech, and Ms. Bowman designated a student for each speech to give feedback and a compliment. Yet in Ms. Estrada's class, there was silence after each student's speech. While Ms. Estrada did require her students to provide written evaluations for two of their peers' speeches, the silence after each speech created a much more subdued atmosphere than in the more celebratory classes. Particularly in cases where students' speeches were deeply personal and emotional, the absence of applause or verbal affirmation left an air of emotional uncertainty. Likewise, in our observation of the semifinal round of the citywide competition, in one classroom, the use of polite applause instead of wild applause appeared to reduce the sense of community within this space.

Finally, some teachers' instructional choices on how to adapt the curriculum were shaped by their curricular context—another factor that influenced students' experiences with *Project Soapbox*. Of our five focal teachers, Mr. Cahill was the only English Language Arts teacher; because he was implementing *Soapbox* in American literature and AP Language and Composition courses, he emphasized the craft of writing for a specific audience and modified the *Soapbox* rubric to align more closely with his course goal of "focusing on the use of pathos, logos, and ethos in arguments . . . specifically tailored to a specific audience." According to him, the focus on a specific audience helped his students craft their emotional appeals more powerfully.

In some of the classrooms, the broader curricular context in which *Soapbox* was embedded also appeared to reinforce its impact. Ms. Bowman's course, for example, used a districtwide curriculum focused on democratic participation. The first unit was about power, participation, and democracy, and *Soapbox* was its culmination; the curriculum then went on to a unit on public policy. In Mr. Cahill's course, although his was an English class, the school in which he taught was part of a network of "democracy schools," which meant that the students, as he explained, are "already very familiar with *Soapbox* and the idea of using your voice to make change, being politically active, the avenues through which they can have civic discourse and civic engagement."

In sum, the qualitative data reveal how both the structure of the *Project Soapbox* curriculum and the teachers' instructional decisions played key roles in influencing student outcomes. The scaffolded lessons encouraged and provided students the tools to write speeches that include rhetorical devices like attention-grabbing openers, statistics-based evidence, emotional appeals, and calls to action. The emphasis on student choice of topics appears to have engaged and motivated students and enabled them to craft and deliver personally relevant speeches with powerful emotional appeal. The focus on student choice also provided teachers an opportunity to learn about their students, which allowed them to better understand their students as people and, in some cases, influenced their instruction. Meanwhile, the curriculum's attention to supportive audiences through the use of wild applause and focused listening provided a supportive environment for students to share vulnerable stories of personal import, which built their confidence, sense of agency, and feelings of connection with one another. Finally, this research reveals the ways in which students' experiences with action civics programming such as *Project Soapbox* may differ depending on the learning goals and instructional choices of various teachers.

## Limitations

While we deliberately employed a mixed-methods research approach to gain the benefits of both quantitative and qualitative data, our qualitative sample may not be representative of the range of opinions in the full sample. We were not able to observe every classroom or every instructor, so we cannot establish the incidence of key implementation processes. Further, given that our student sample included predominantly low-income students and students of color, we do not know how representative the experiences of our sample with this curriculum are of other student samples. Indeed, it would be valuable to know the extent to which the same factors are salient in shaping the learning of students who are more demographically diverse than the students in our sample (e.g., a sample that includes greater socioeconomic or racial/ethnic diversity).

In addition, we did not include our measure of classroom climate in the postsurvey, so we cannot document how *Project Soapbox* may have contributed to increases in students' assessments here. Initially, we believed that this would be a static condition, established prior to implementation. Our qualitative data reveal otherwise. Further, while social-emotional factors emerged throughout findings, our instruments did little to specifically probe these relational dimensions. Subsequent research would do well to design instruments that are more focused on assessing these relational dimensions and their impact.

## Discussion

The growing consensus that varied action civics curricula are consistently associated with positive student learning outcomes has allowed us to turn greater attention to the factors that shape these positive outcomes and the ways in which context and implementation may influence students' experiences with such curricula. Our quantitative analysis, which is able to control for many individual variables, such as gender, socioeconomic status, and family political socialization, provides additional support for the impact of many of the best practices in civic education on positive outcomes such as gains in anticipated political engagement. For example, the data from this study add empirical evidence to the notion that student autonomy, or the practice of allowing students to choose topics of importance to them, is directly related to their outcomes (e.g., Kahne & Middaugh, 2008; LeCompte & Blevins, 2015), a finding that holds up here when various predispositions are taken into account.

More important, however, our qualitative data allow us to go beyond the quantitative analysis to consider key elements of this action civics curriculum and its implementation that are not easily captured by survey questions. The data from this study provide insight into the process by which student learning from *Project Soapbox* occurs and illustrate the differential experiences that result from the ways in which a common curriculum may be shaped by instructor implementation. Our analysis of the curriculum and our comparison of the various ways in which the teachers implemented their lessons provide a fuller understanding of the power of action civics and what factors may boost or undermine the general student outcomes documented by the empirical evidence.

Many of our findings about the importance of various features of the *Soapbox* curriculum are consistent with theories of motivation, as well as the features of action civics and high-quality civic education more broadly. The centrality of student choice of speech topics in the *Project Soapbox* curriculum and the importance of its impact are not surprising given the well-established understanding that providing opportunities for choice and self-direction supports students' need for autonomy and spurs intrinsic motivation (Ryan & Deci, 2000). Prior research has established that such choices must be related to students' personal interests, values, and goals; optimally challenging; and given in a warm, empathic, and accepting context (Katz & Assor, 2007). Framed this way, *Project Soapbox*'s success lies in the way in which it blends action civics' emphasis on youth voice, expertise, and issue identification (Gingold, 2013) with the motivation that is enabled by the positive environment provided by the wild applause (and, in some cases, explicit praise) of peers.

Further, *Project Soapbox*'s carefully scaffolded curriculum appears to fulfill many elements of action civics' broader programming goal that "adults scaffold opportunities for students to launch youth-driven civic projects through a multi-step process" (NACC, 2010). Although the brief *Project Soapbox* curriculum does not lead students through all six steps of the action civics framework as comprehensively as other programming does, students who participate in the curriculum are practicing the habits and orientations of citizenship within a social context (McIntosh & Youniss, 2010; Torney-Purta, Amadeo, & Andolina, 2010). They are examining their community to identify issues of importance to them, conducting research on these issues, and articulating strategies for action through their speeches and calls to action. And while some may contend that *Soapbox* participants are not taking action in their communities, the act of voicing their ideas publicly among their peers at school as well as other adults is in itself a public action. Flanagan et al. (2010) argued that schools are mini polities—public spaces where we engage with one another about the choices we are making about the type of society we live in. In giving these speeches and making their views public, students are "enter(ing) in the political realm" (McIntosh & Youniss, 2010, p. 26). Thus, students themselves (and their classrooms and schools) constitute a community of public, competing ideas, opinions, and values. In addition, the explicit inclusion of adults from the community (not just peers) to listen to students' speeches sends a powerful message about the value of youth voice. By participating in the carefully scaffolded *Project Soapbox* curriculum, students are practicing democracy within their classrooms.

In addition, research has documented that speaking out on issues that matter in the presence of trusted others fosters strength, connecting people to one another and providing the foundation for civic engagement. Connections to one another, as Flanagan (2003) has argued, have an inherently civic component because "the ties that bind young people to the polity are based on participating in local community groups where they feel respected and where their voice is taken seriously" (p. 257). Thus, students' willingness to express their—often deeply personal—views publicly, among trusted adults and peers, is a component of acting and developing politically. Indeed, when implemented under optimal conditions, our data suggest, *Soapbox* may have the potential to build the kind of "public" that Flanagan et al. (2010) describe: Youth experience trusting relationships and are allowed to express their opinions in open classroom climates, and these trusting relationships enable them to develop a sense of belonging that leads to the development of collective, civic identities oriented toward the common good.

Finally, this research helps us understand the process by which an action civics curriculum can foster the listening and relational capacities that scholars and the public alike increasingly recognize as vital to SEL and democracy. *Project Soapbox*'s curricular elements position students to listen supportively and actively to those different from themselves, and in doing so, they offer the potential to increase trust, empathy, and equity—the democratic orientations, social awareness, and relationship skills that are increasingly prized and necessary in our world (Allen, 2004; Cramer & Toff, 2017; Levine, 2013; Weissberg et al., 2015). At the same time, teachers can amplify these outcomes through deliberate community-building exercises, overt and audible enthusiasm and support by students for one another, and other experiences that reinforce the democratic orientations emphasized in the curriculum.

Thus, *Project Soapbox* is powerful in part because it is well aligned with many of the well-established best practices in civic education. Our analysis contributes to the growing understanding of action civics curricula by identifying the key factors that shape student experiences with the curriculum and the differential impact created by the various ways the program is implemented. However, *Project Soapbox* is also powerful because the curriculum is intentionally linked to key SEL practices, many of which are newly recognized as having particular civic import (Levine & Kawashima-Ginsberg, 2017). When action civics programming is embedded with SEL practices, the potential payoff is even greater. As McKay Bryson and Warren (2018) argue, "Social and emotional learning and action civics are not just compatible, they are necessary and interdependent academic complements." With today's hyperpartisan politics, the crippling deadlock of political institutions, and the increasing violence in political rhetoric, civic education that addresses both key civic skills and important SEL competencies is critical.

## Implications

This research provides evidence for clearly identifiable practices in the context of action civics learning that teachers can adopt that will contribute to the learning outcomes of their students not only in terms of the key skills they need to be successful in school but also to provide them with the critical skills that will allow all students, regardless of their status in society, to participate fully in our democracy. For example, the findings of this study suggest the importance of providing verbal feedback and enthusiastic snaps and claps for speakers; creating supportive, attentive classroom environments; and—especially notable in our era of increasingly standardized curricula—giving students the opportunity to choose their own topics to explore. In addition, the findings indicate that educators would be well served to spend time on community building prior to implementation of the curriculum.

While research on action civics programs is growing (e.g., Blevins et al., 2016), scholarly work on the various factors that make such programs successful remains in its early stages. This study uses quantitative data to deepen our understanding of the key influences that are associated with student learning across two different domains—civic engagement and rhetorical skills. At the same time, the qualitative data from this study illuminate those factors that shape students' learning. And, perhaps most important, this research draws directly upon the voices and insights of those students who are traditionally left out of high-quality civic engagement opportunities, thus providing evidence of the efficacy of the curriculum for populations that are in most need of the instruction.

### Notes

We are extremely grateful to those teachers and students who were willing to participate in this study. We thank Kara Gonnerman and Claire Kalinowski for their invaluable research assistance for this study.

This research has been generously supported by grants from the Spencer Foundation, the Brinson Foundation, and DePaul University.

[1] All the teachers' names are pseudonyms.

[2] All the student comments, both written and verbal, are included here exactly as the students expressed them.

[3] Because some students' speeches include stories of personal trauma, the most recent version of the *Project Soapbox* curriculum has been updated to include guidelines and recommendations for teachers on trauma-informed practice.

### References

Allen, D. (2004). *Talking to strangers: Anxieties of citizenship since Brown v. Board of Education*. Chicago, IL: University of Chicago Press.

Andolina, M. W., & Conklin, H. G. (2018). Speaking with confidence and listening with empathy: The impact of *Project Soapbox* on high school students. *Theory & Research in Social Education*, 46, 374–409. doi:10.1080/00933104.2018.1435324

Ballard, P., Cohen, A., & Littenberg-Tobias, J. (2016). Action civics for promoting civic development: Main effects of program participation and differences by project characteristics. *American Journal of Community Psychology*, 58, 377–390. doi:10.1002/ajcp.12103.

Barr, D. J., Boulay, B., Selman, R. L., McCormick, R., Lowenstein, E., Gamse, B., . . . Leonard, M. B. (2015). A randomized controlled trial of professional development for interdisciplinary civic education: Impacts on humanities teachers and their students. *Teachers College Record, 117*, 1–52.

Battistoni, R. (2004). Student-powered solutions. *Principal Leadership, 5*, 22–24.

Benson, P. L., Scales, P. C., Hamilton, S. F., & Sesma, A. (2006). Positive youth development: Theory, research and applications. In W. Damon & R. Lerner (Eds.), *Handbook of child psychology* (pp. 894–941). New York, NY: Wiley.

Berman, S. H. (2004). Teaching civics: A call to action. *Principal Leadership, 5*(1), 16–20.

Blevins, B., LeCompte, K., & Wells, S. (2016). Innovations in civic education: Developing civic agency through action civics. *Theory and Research in Social Education, 44*, 344–384. doi:10.1080/00933104.2016.1203853

Campbell, D. (2008). Voice in the classroom: How an open classroom climate fosters political engagement among adolescents. *Political Behavior, 30*, 437–454.

Center for Information and Research on Civic Learning and Engagement. (2013). *All together now: Collaboration and innovation for youth engagement* (The report of the Commission on Youth Voting and Civic Knowledge). Retrieved from http://www.civicyouth.org/wp-content/uploads/2013/09/CIRCLE-youthvoting-individualPages.pdf

Chicago Consortium on School Research. (2013). *Measures and item statistics in 2012 surveys*. Retrieved from https://consortium.uchicago.edu/sites/default/files/publications/2012studentsurveymeasurestatistics.pdf

Collaborative for Academic, Social, and Emotional Learning. (2019). *State scan scorecard project*. Retrieved from https://casel.org/state-scan-scorecard-project-2/

Cramer, K. J., & Toff, B. (2017). The fact of experience: Rethinking political knowledge and civic competence. *Perspective on Politics, 15*(3): 754–770. doi: 10.1017/S1537592717000949

Dobson, A. (2012). Listening: The new democratic deficit. *Political Studies, 60*, 843–859. doi:10.1111/j.1467-9248.2012.00944.x

Feldman, L., Pasek, J., Romer, D., & Jamieson, K. H. (2007). Identifying best practices in civic education: Lessons from the student voices program. *American Journal of Education, 114*, 75–99. doi:10.1086/520692

Flanagan, C. (2003). Developmental roots of political engagement. *PS: Political Science & Politics, 36*, 257–261. doi:10.1017/S104909650300218X

Flanagan, C., Stoppa, T., Syvertsen, A. K., & Stout, M. (2010). Schools and social trust. In L. Sherrod, J. Torney-Purta, & C. Flanagan (Eds.), *Handbook of research on civic engagement in youth* (pp. 307–329). Hoboken, NJ: Wiley.

Flanagan, C., Syvertsen, A., & Stout, M. (2007, May). *Civic measurement models: Tapping adolescents' civic engagement* (CIRCLE Working Paper No. 55). Retrieved from https://civicyouth.org/PopUps/WorkingPapers/WP55Flannagan.pdf

Gibson, C., & Levine, P. (2003). *The civic mission of schools*. New York, NY/College Park, MD: Carnegie Corporation of New York/Center for Information and Research on Civic Learning and Engagement.

Gingold, J. (2013, August). *Building an evidence-based practice of action civics: The current state of assessments and recommendations for the future* (CIRCLE Working Paper No. 78). Retrieved from the https://civicyouth.org/wp-content/uploads/2013/08/WP_78_Gingold.pdf

Glesne, C., & Peshkin, A. (1992). *Becoming qualitative researchers: An introduction*. White Plains, NY: Longman.

Gould, J., Jamieson, K. H., Levine, P., McConnell, T., & Smith, D. B. (Eds.). (2011). *Guardian of democracy: The civic mission of schools.* Philadelphia, PA: Leonore Annenberg Institute for Civics of the Annenberg Public Policy Center at the University of Pennsylvania. Retrieved from https://civicyouth.org/wp-content/uploads/2011/09/GuardianofDemocracy.pdf

Kahne, J., Chi, B., & Middaugh, E. (2006). Building social capital for civic and political engagement: The potential of high school civics courses. *Canadian Journal of Education, 29,* 387-409.

Kahne, J., Crow, D., & Lee, N. J. (2013). Different pedagogy, different politics: High school learning opportunities and youth political engagement. *Political Psychology, 34,* 419–441. doi:10.1111/j.1467-9221.2012.00936.x

Kahne, J., & Middaugh, E. (2008, February). *Democracy for some: The civic opportunity gap in high school* (CIRCLE Working Paper No. 59). Retrieved from https://pdfs.semanticscholar.org/1788/690ef884cb9d737323324d6b190c3d-d2a42b.pdf?_ga=2.233674396.309243143.1564219535-1205025470.1564219535

Kahne, J., Middaugh, E., & Schutjer-Mance, K. (2005). *California Civic Index* [Monograph]. New York, NY/Los Angeles, CA: Carnegie Corporation/Annenberg Foundation.

Kahne, J., & Sporte, S. (2008). Developing citizens: The impact of civic learning opportunities on students' commitment to civic participation. *American Education Research Journal, 45,* 738–766. doi:10.3102/0002831208316951

Katz, I., & Assor, A. (2007). When choice motivates and when it does not. *Educational Psychology Review, 19,* 429–442. doi:10.1007/s10648-006-9027-y

LeCompte, K., & Blevins, B. (2015). Building civic bridges: Community-centered action civics. *Social Studies, 106,* 209–217. doi:10.1080/00377996.2015.1059792

Levine, P. (2013). *We are the ones we have been waiting for: The promise of civic renewal in America.* New York, NY: Oxford University Press.

Levine, P., & Kawashima-Ginsberg, K. (2017, September 21). *The republic is (still) at risk—and civics is part of the solution* (A briefing paper for the Democracy at a Crossroads National Summit) [Monograph]. Retrieved from https://www.civxnow.org/static/media/SummitWhitePaper.fc2a3bb5.pdf

Levine, P. L., & Kawashima-Ginsberg, K. (2015). *Civic education and deeper learning* (A commissioned paper for Jobs for Future Deeper Learning Paper Series) [Monograph]. Washington, DC: Jobs for Future.

Levinson, M. (2012). *No Citizen Left Behind.* Cambridge, MA: Harvard University Press.

Longo, N. V., Drury, C., & Battistoni, R. M. (2006). Catalyzing political engagement: Lessons for civic educators from the voices of students. *Journal of Political Science Education, 2,* 313–329. doi:10.1080/15512160600840483

McDevitt, M., & Kiosis, S. (2004, September). *Education for deliberative democracy: The long-term influence of KidsVoting USA* (CIRCLE Working Paper No. 22). Retrieved from https://civicyouth.org/PopUps/WorkingPapers/WP22McDevitt.pdf

McDevitt, M., & Kiosis, S. (2006, August). *Experiments in political socialization: KidsVoting USA as a model for civic education reform* (CIRCLE Working Paper No. 49). Retrieved from https://pdfs.semanticscholar.org/eddd/4d59fa0e-d26e7d0cd92d4b1bad4c6ffc5b8e.pdf?_ga=2.59216687.309243143.1564219535-1205025470.1564219535

McIntosh, H., & Youniss, J. (2010). Toward a political theory of political socialization of youth. In L. Sherrod, J. Torney-Purta, & C. Flanagan (Eds.), *Handbook of research on civic engagement in youth* (pp. 23–41). Hoboken, NJ: Wiley.

McKay Bryson, M., & Warren, S. (2018, October 15). Civics is social and emotional. *Education Week*. Retrieved from https://blogs.edweek.org/edweek/learning_social_emotional/2018/10/civics_is_social_and_emotional.html

Mikva Challenge. (2019). *Action civics model*. Retrieved from https://www.mikvachallenge.org/programs/the-action-civics-model/

Molnar, A., Smith, P., & Zahorik, J. (1998, December 1). *1997–98 Evaluation results of the Student Achievement Guarantee in Education (SAGE) program* [Monograph]. Retrieved from https://nepc.colorado.edu/publication/1997-98-evaluation-results-student-achievement-guarantee-education-sage-program

National Action Civics Collaborative. (2010). *Action civics: A declaration for rejuvenating our democratic traditions*. Retrieved from http://actioncivicscollaborative.org/about-us/action-civics-declaration/

Niemi, R., & Junn, J. (1998). *Civic education*. New Haven, CT: Yale University Press.

Ryan, R., & Deci, E. (2000). Self-determination theory and the facilitation of intrinsic motivation, social development, and well-being. *American Psychologist, 55*, 68–78.

Singer, J. M., & Andrade, D. F. (1997). Regression models for the analysis of pretest/posttest data. *Biometrics, 53*, 729–735.

Syvertsen, A. K., Flanagan, C. A., & Stout, M. D. (2007). *Best practices in civic education: Changes in students' civic outcomes* (CIRCLE Working Paper No. 57). Retrieved from https://civicyouth.org/PopUps/WorkingPapers/WP57Flanagan.pdf

Syvertsen, A. K., Stout, M. D., Flanagan, C. A., Mitra, D. L., Oliver, M. B., & Sundar, S. S. (2009). Using elections as teachable moments: A randomized evaluation of the Student Voices civic education program. *American Journal of Education, 116*, 33–67. doi:10.1086/605100

Torney-Purta, J. (2002). The school's role in developing civic engagement: A study of adolescents in twenty-eight countries. *Applied Developmental Science, 6*, 203–212. doi:10.1207/S1532480XADS0604_7

Torney-Purta, J., & Amadeo, J. (2003). A cross-national analysis of political and civic involvement among adolescents. *PS: Political Science & Politics, 36*, 269–274. doi:10.1017/S1049096503002208

Torney-Purta, J., Amadeo, J., & Andolina, M. (2010). A conceptual framework and multi-method approach for research on political socialization and civic engagement. In L. Sherrod, J. Torney-Purta, & C. Flanagan (Eds.), *Handbook of research on civic engagement in youth* (497–524). Hoboken, NJ: Wiley.

Walling, D. R. (2007). The return of civic education. *Phi Delta Kappan, 89*, 285–289.

Warren, S. (2019). *Generation citizen: The power of youth in our politics*. Berkeley, CA: Counterpoint Press.

Weissberg, R. P., Durlak, J. A., Domitrovich, C. E., & Gullotta, T. P. (2015). Social and emotional learning: Past, present, and future. In J. Durlak, C. E. Domitrovich, R. P. Weissberg, & T. P. Gullotta (Eds.), *Handbook of social and emotional learning: Research and practice* (pp. 3–19). New York, NY: Guilford Press.

Westheimer, J., & Kahne, J. (2004). What kind of citizen? The politics of educating for democracy. *American Educational Research Journal, 41*, 237-269.

Wilkenfeld, B. (2009, May). *Does context matter? How the family, peer, school, and neighborhood contexts relate to adolescents' civic engagement* (CIRCLE Working Paper No. 64). Retrieved from https://civicyouth.org/PopUps/WorkingPapers/WP64Wilkenfeld.pdf

Williamson, A. A., Modecki, K. L., & Guerra, N. G. (2015). SEL programs in high school. In J. Durlak, C. E. Domitrovich, R. P. Weissberg, & T. P. Gullotta

(Eds.), *Handbook of social and emotional learning: Research and practice* (pp. 181–196). New York, NY: Guilford Press.

Youniss, J. (2011). Civic education: What schools can do to encourage civic identity and action. *Applied Developmental Science, 15*(2), 98–103. doi:10.1080/10888691.2011.560814

Zukin, C., Keeter, S., Andolina, M., Jenkins, K., & Delli-Carpini, M. (2006). *A new engagement? Political participation, civic life and the changing American citizen*. New York, NY: Oxford University Press.

Manuscript received November 2, 2018
Final revision received July 17, 2019
Accepted July 22, 2019

# Dynamics of Reflective Assessment and Knowledge Building for Academically Low-Achieving Students

Yuqin Yang
*Central China Normal University*
Jan van Aalst
Carol K. K. Chan
*University of Hong Kong*

*This study investigates designs for developing knowledge building (KB) and higher order competencies among academically low-achieving students. Thirty-seven low-achieving students from a ninth-grade visual arts course in Hong Kong participated. The design involved principle-based KB pedagogy, with students writing on Knowledge Forum® (KF), enriched by analytics-supported reflective assessment. Analysis of the discourse on KF showed that the low achievers were able to engage in productive discourse, with evidence of metacognitive, collaborative, and epistemic inquiry. Analysis illustrates how the design supported student engagement, including (1) reflective inquiry and social metacognition; (2) reflective meta- and epistemic talk; (3) evidence-based reflection for collective growth; and (4) reflection embedded in community ethos. Implications of reflective assessment for supporting low achievers for inquiry learning and KB are discussed.*

---

YUQIN YANG is an associate professor of the learning sciences in the School of Educational Information Technology, Hubei Key Laboratory of Educational Informationization, Central China Normal University, No. 152 Luoyu Road, Wuhan, Hubei, China; e-mail: *yangyuqin@mail.ccnu.edu.cn*. Her research interests include pedagogy and assessment of knowledge building, learning analytics, metacognition, and collaborative learning.

JAN VAN AALST is an associate professor of the learning sciences in the Faculty of Education, University of Hong Kong, Pokfulam, Hong Kong SAR, China; e-mail: *vanaalst@hku.hk*. His research interests include pedagogy and assessment of knowledge building and computer-supported collaborative learning.

CAROL K. K. CHAN is a professor of the learning sciences and psychology in the Faculty of Education, University of Hong Kong, Pokfulam, Hong Kong SAR, China; e-mail: *ckkchan@hku.hk*. Her research interests include knowledge building, dialogic education, and teacher professional development.

KEYWORDS: reflective assessment, metacognition, collaborative inquiry, technology-enhanced learning, academic low achievers

Twenty first-century education calls for the development in students of high-order competencies such as metacognition, collaboration, agency, and creativity (Bereiter, 2002; National Research Council, 2000; Trilling & Fadel, 2009). All students, regardless of socioeconomic status and academic background, need equitable access to opportunities to develop these competencies. In particular, successful learning experiences that focus on high-order competencies are critically important for low-achieving students. These can not only help students improve their academic performance and thus narrow the achievement gap, they can also create a cycle of continuous improvement (Becker & Luthar, 2002; Snell & Lefstein, 2018). Lower expectations of students and associated instructional approaches that are geared toward lower order skills do not provide low-achieving students the necessary opportunities to improve their academic performance (Becker & Luthar, 2002). Addressing the needs of all learners was identified in the recently published *How People Learn II* report (National Academy of Science, Engineering and Medicine, 2018) as an area needing substantial research.

Collaborative and inquiry-based instructional approaches that emphasize higher order competencies have many benefits for learners, such as deep understanding, higher order competencies, and self-efficacy (Chan, 2013). Positive collaborative inquiry engagement requires metacognitive skills, that is, goal setting, monitoring, and reflection (Brown, 1997; Järvelä et al., 2015; National Research Council, 2000); quality social interaction (Barron, 2003; Kaendler, Wiedmann, Rummel, & Spada, 2015; Stahl, 2006); and epistemic dispositions (Barzilai & Chinn, 2018). Low achievers have various difficulties and fewer opportunity to develop competencies in these areas. This creates a vicious cycle. Engaging them in successful collaborative inquiry and providing access to educational opportunity is not only an important educational endeavor but also a great challenge for educators, where *engagement* refers to "a goal-directed state of active and focused involvement in a learning activity" (D'Mello, Dieterle, & Duckworth, 2017, p. 106).

Although the literature includes many intervention studies on low achievers' task engagement (Baxter, Woodward, & Olson, 2001; Dietrichson, Bøg, Filges, & Jørgensen, 2017; Han, Capraro, & Capraro, 2014), many focus on educational achievement and not high-order thinking and collaborative inquiry. Research on collaborative inquiry and scaffolding in the learning sciences has flourished in the past two decades, but there is limited research concerning students from different academic tracks (Raes, Schellens, & De Wever, 2014). Our informal analysis of research published in the two flagship journals of the International Society of the Learning Sciences in the past 5 years suggests few studies have specifically investigated underprivileged populations, although most classrooms have included students with low academic achievement. This also applies to research on knowledge building, the educational model we consider in this article. Furthermore, when research has compared differential effects on high and low achievers, there are few explanatory frameworks (Han et al., 2014; So, Seah, & Toh-Heng, 2010). How academic low achievers and at-risk learners can be scaffolded to engage in high-level collaborative inquiry and knowledge building, supported by technology, are important questions that remain to be investigated.

The term *knowledge building* (KB), as used in this article, refers to the educational model developed by Scardamalia and Bereiter since the 1990s (Scardamalia & Bereiter, 2006, 2014). The primary aim of KB is to introduce students to the practices by which the state of knowledge in a community is advanced. As an educational model, it goes beyond understanding the core concepts of a domain to understanding the nature of knowledge in that domain and how knowledge is created. Collective effort toward this community-level goal, student agency, metacognition, the improvability and social nature of knowledge and knowing, all are essential features. KB discourse takes place in Knowledge Forum® (KF), a computer-supported collaborative environment (Scardamalia & Bereiter, 2014). In KB classrooms, students generate questions and co-construct explanations, using both online and offline discourse to pursue progressively deeper understanding collectively.

In this study, we designed a KB environment enriched by reflective assessment supported by analytic tools to engage low achievers in KB inquiry. Reflective assessment refers to how students take on collective agency to set learning goals, monitor personal and community progress, use feedback to identify knowledge gaps, and examine how to improve their ongoing learning addressing broader problems (Lei & Chan, 2018; Yang, van Aalst, Chan, & Tian, 2016). Earlier research on reflective assessment has examined self and peer assessments focusing on individual progress (White & Frederiksen, 1998). In KB, reflective assessment is a collaborative process in a community; not everyone needs to be metacognitive at the same pace but collectively students can pursue shared metacognition and agency for community and personal advances.

The online discourse in KB occurs in KF, which can be augmented by assessment tools in KF that students use to reflect on their KB progress. We have developed one tool, the Knowledge Connections Analyzer (KCA) and an accompanying framework, which collects information from KF relevant to a few basic questions about the KB process (van Aalst, Chan, Tian, Teplovs, Chan, & Wan, 2012). Research using the KCA tools with cohorts of low achievers show that they were able to use the tool, make sense of the data it provided, and use these data to improve their KB (Yang, 2019; Yang et al., 2016). However, these preliminary studies did not investigate the *classroom dynamics* and knowledge practices, including how students engaged in *reflective assessment*, the changes that occurred in the classroom, design implications for developing agency and metacognition, or the principles needed to support low achievers in interventions. The present study aims to provide a framework addressing student difficulties by examining the principles, design, and dynamics of KB for engaging low-achieving students in KB and inquiry-learning. To our knowledge, it is one of the first studies on KB to specifically examine work by a *cohort of low-achieving students*, and in which tools for reflecting on discourse are used *by the students*. KB provides only a context here; the bigger challenge is to promote low-achieving students' collaboration, reflection, and epistemology for productive inquiry more generally.

## Literature Review and Framework

### Higher Order Competencies and Difficulties for Low-Achieving Students

Low-achieving students, who often are from low socioeconomic and diverse ethnic backgrounds (Dietrichson et al., 2017; Slavin, Lake, Davis, & Madden, 2011), enter schools with fewer cognitive, metacognitive, and social skills; and with limited epistemic dispositions needed for educational achievement (Dietrichson et al., 2017) and productive inquiry-based learning (White & Frederickson, 1998; Tsai & Shen, 2009). Many have low motivation and efficacy (Becker & Luthar, 2002) and have more difficulty developing higher order competencies than higher achieving students. Helping low achievers gain successful learning experience in collaborative inquiry can benefit them greatly (Raes et al., 2014; White & Frederiksen, 1998).

Metacognitive skills—such as planning, monitoring, and reflection—are crucial for developing various capabilities for collaborative inquiry (White & Frederiksen, 1998; Zohar & Dori, 2003), particularly for academic low achievers (Yang et al., 2016). Skilled readers are more aware of the purpose of reading than poor readers—setting goals, allocating time, using strategies, and monitoring their comprehension (Wong, 1987); similarly, low achievers in math learning seldom appropriately select, monitor, or adapt strategies (Montague, Enders, & Dietz, 2011). Learners benefit most from collaborative inquiry when adept at metacognitive monitoring, reflection, and regulating (Azevedo, 2005). However low achievers often have cognitive, metacognitive, and collaborative inquiry skill difficulties (Yang et al., 2016). They often focus on low-level strategies (e.g., searching without a goal) and are less able to deploy/develop key metacognitive skills (Brown & Campione, 1994).

Collaborative and discursive skills are essential for productive collaborative inquiry. Students who collaborate well often show sound question-explanation exchanges, argumentation and uptake of ideas, and rise-above by synthesizing diverse ideas (van Aalst & Chan, 2007; Zhang, Scardamalia, Lamon, Messina, & Reeve, 2007). Low achievers are generally unfamiliar with group skills such as articulating viewpoints, listening to others, building on others' ideas, and thinking together, which are important collaborative inquiry competencies. Primarily, low achievers and at-risk students lack a developed cultural sense of what collaboration is about, and opportunity to engage in productive and collaborative talk (Duschl & Osborne, 2002). Due to a lack of social support and opportunities in school and nonschool settings, they have limited experience communicating and expressing ideas, examining the validity of ideas, and developing discursive practices.

*Epistemic disposition* is important in collaborative inquiry that involves students' epistemic understanding about what knowledge is and how it develops as well as engagement in epistemic goals and processes (Greene, Sandoval, & Bråten, 2016). Research has shown students' epistemic understanding and dispositions influence student achievement, thinking and problem solving (Greene et al., 2016). Primarily, immature learners have less developed views of what knowledge involves, seeing it as linear and static, rather than evolving and extendable, and they lack epistemic goals and purposes (Barzilai & Chinn, 2018). With less developed epistemic goals and dispositions, low-achieving students often believe there are "certain" and definitive answers and thus less likely to exert efforts inquiring thus hampering their higher order and collaborative inquiry.

With lower teacher expectations (Zohar, Degani, & Vaaknin, 2001), subsequent inequitable exposure to learning opportunities, and negative appraisal messages, many low achievers feel powerless over their own learning potential, which impedes the development of self-efficacy (Becker & Luthar, 2002; Zohar & Dori, 2003). While the literature has revealed these difficulties, we do not use a deficit model viewing inadequate skills as something inherent to low-achieving students. Rather, in their educational histories, these students have lacked opportunities to engage productively. Research has shown that when low achievers are provided appropriate instruction, they can engage in higher order thinking (Zohar & Dori, 2003). Accordingly, efforts to improve students' academic achievement and develop their higher order competencies require corresponding higher teacher expectations and support.

## Educational Interventions for Diverse Learners and Low Achievers

Research has examined low achievers' *educational achievement* and the instruction they receive, emphasizing task engagement and peer-assisted learning (Baker, Gersten, & Lee, 2002; Baxter et al., 2001; Hawkins, Doueck, & Lishner, 1988). Research on diverse learners and low–socioeconomic status students similarly highlights cooperative learning (Dietrichson et al., 2017; Slavin et al., 2011), student agency, and the importance of being heard (Wallace & Chhuon, 2014). Research in STEM (science, technology, engineering, and mathematics) education using problem-based learning shows student engagement yields positive results, comparing high and low achievers (Han et al., 2014). Barton and Tan (2010) suggested low-income urban youth could actively appropriate project activities and tools to challenge their traditional classroom roles. Direct instruction involving explicit teaching of principles (Baker et al., 2002; Kroesbergen, van Luit, & Mass, 2004), formative assessment that provides students data (Baker et al., 2002), and progress monitoring (Dietrichson et al., 2017) all are important strategies. Effective intervention strategies for low achievers include task engagement (Barton & Tan, 2010; Han et al., 2014), peer-assisted and cooperative learning (Dietrichson et al., 2017; Slavin et al., 2011), direct instruction (Baker et al., 2002), and formative assessment for progress monitoring (Baker et al., 2002; Dietrichson et al., 2017). These studies also show low achievement is less a psychological trait than an *artifact* of the learning context and history. While some progress has been made, interventions are needed to help low achievers develop the high-order competencies fundamental to continuous development (Becker & Luthar, 2002; Snell & Lefstein, 2018).

From a learning-sciences perspective, historically, cognitive and learning scientists have examined the importance of designing for low achievers' higher order competencies, focusing on metacognition, and social support. Seminal research on reciprocal teaching involved students with learning difficulties taking increased *cognitive responsibility* for teaching their peers key strategies for understanding, supported by social context (Palincsar & Brown, 1984). Reciprocal teaching, later extended to include fostering communities of learners in low–socioeconomic status contexts (Brown, 1997; Brown & Campione, 1994), has focused on students scaffolding each other with multiple zones of proximal development with a learn-to-learn *community* ethos. White and Frederiksen (1998) developed reflective assessment to promote metacognition in scientific inquiry. Zohar and Dori (2003) found low-achieving students significantly progressed using authentic problems and hands-on activities. In one of the few studies using technology-supported environment, Raes et al. (2014) found low achievers benefited more than their inquiry science counterparts from using a web-based environment when phenomena are made concrete and visible using visualization. Research suggests low achievers can achieve higher order learning goals, given appropriate supports and scaffoldings. Thus far, few studies have examined collaborative and epistemic inquiry among low achievers. Epistemic inquiry refers to endeavors to build knowledge together.

### Knowledge Building as a Principle-Based Approach

KB is an educational model focusing on students' collective responsibility for idea improvement and community growth (Bereiter, 2002; Scardamalia & Bereiter 2006, 2014) supported by technology. KB aims to bring to schools the creative processes in knowledge communities—school-aged students can be cultivated to work like a community of scientists, contributing and extending frontiers of knowledge. In KB classrooms, students inquire and pursue problems using classroom and KF discussion. They post questions, build on ideas, construct explanations, and direct further inquiry to deepen and synthesize knowledge.

Students' KB inquiry is supported by KF, a computer-supported collaborative discourse environment designed to support communal idea improvement (see Figure 1). KF includes collaborative workspaces (views) with a graphical interface, where students can post questions and ideas for collective idea improvement. Students write notes using metacognitive scaffolds (e.g., "I need to understand," "My theory") and can synthesize the development of ideas by linking their notes to a synthesis note or using the rise-above function. KF includes assessment and analytics tools to track student progress (Scardamalia & Bereiter, 2016). Three decades of design-based research in KB indicates that students can engage in advanced KB practice with positive effects on learning outcomes (Chan, 2013; Chen & Hong, 2016). However, most studies involve regular students with mixed backgrounds; few focus on cohorts of primarily low achievers.

KB is a *principle-based*, open-ended model that highlights students' collective efforts for idea improvement. A principle-based approach "defines core values and principles, leaving to teachers . . . discretionary judgment . . . making adaptive classroom decisions to accommodate their different contexts and possibilities" (Zhang, Hong, Scardamalia, Teo, & Morley, 2011, p. 263).

A system of 12 KB principles, formulated by Scardamalia (2002) guide KB pedagogical design and research. Several principles summarized below illuminate metacognitive, social, and epistemic competencies, which are important for working with low-achieving students:

*Epistemic Agency*

High-level agency with students taking initiative, negotiating the fit between own and others' ideas, and taking charge of high-level inquiry (e. g., goal setting, monitoring, and evaluation) is emphasized in KB. Low achievers manifest limited agency and metacognition, await teacher direction, and lack task motivation. In KB settings, students perform high-level knowledge work with related goals, motivations, evaluations, and long-range planning normally left to teachers. Students are encouraged to have agency and think about what they know and need to know. While this is often difficult for low achievers, they are supported in a social and community context, where they can see their ideas through others and compare them; they can see their own thinking through others' lens, thus encouraging metacognition.

*Collective Responsibility for Community Knowledge*

Contributing ideas to the community is as or more prized than individual performance. Communities working together contribute valuable and

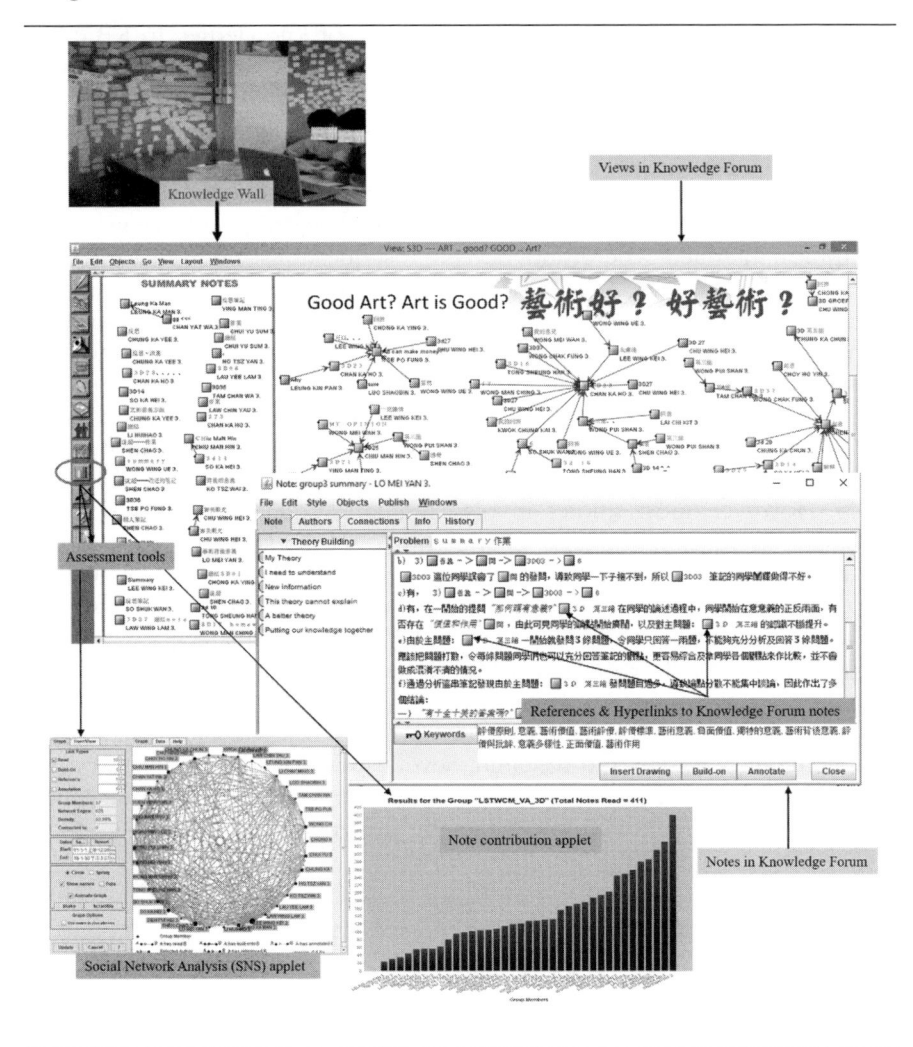

*Figure 1.* **Knowledge forum and features and knowledge building wall.**

*Note.* (1) Knowledge building wall (top) with students posting ideas on note-cards mounted on classroom wall; (2) knowledge forum view (middle) as collaborative work space for note writing; each square icon is a computer note with lines as links between them; reference notes include hyperlinks to other notes; (3) assessment tools for number of notes contributed (bottom right) and interactions among students (bottom left).

diverse ideas; knowledge advances cannot be done by individuals, and KB provides low achievers opportunities to advance together. Low achievers face competition problems, as schools often compare individual performance. Diversity and individual differences are often considered obstacles, and educational approaches like differentiation have been used to address low achievers' problems. Community knowledge emphasizes that everyone, regardless of accomplishment and background, can add value in a KB community; diverse ideas and learners are community assets that contribute to the progress of ideas.

*Improvable Ideas*

All ideas are improvable and continuous and collective efforts can improve ideas' quality, coherence, and utility. Low achievers generally lack epistemic understanding of knowledge and inquiry, believing knowledge to be static, based on fixed, external standards. In KB communities, learning focuses on progress, not a fixed end-product. Inquiry can be increasingly deepened and diverse learners can improve ideas together, thus supporting new ways of viewing knowledge, and helping low achievers develop more mature epistemic dispositions.

*Concurrent, Embedded, and Transformative Assessment*

Assessment is an integral component of KB and adds an inquiry component to the community's work and progress, leading to new actions that enhance both. Embedded and transformative assessment helps students actualize and develop their metacognitive skills, which is key to low achievers' inquiry and KB process (more details in the next section).

Although KB, as an open-ended community approach, is intended for all learners, students with academic difficulties may still face problems due to a lack of strategic moves, communication, and epistemic dispositions in technology-enhanced inquiry environments. When working on KB, students need to post ideas, build on and work with multiple KF posts, which can be challenging for low achievers lacking in agency and collaborative skills. With ideas distributed across individual postings over time, students easily get into short or fragmented discussions lacking in conceptual progress and knowledge integration (Yang et al., 2016; Zhang et al., 2018). Students, particularly low achievers lacking in collaborative and metacognitive skills, need additional designs and tools to scaffold their engagement in collective monitoring of and reflection on their online discourse.

## KB Enhanced by Reflective Assessment

Fundamentally, we highlight the importance of reflection, first emphasized in *How People Learn* report (National Research Council, 2000) as a key strategy for promoting students learning, understanding, and KB. It is widely accepted that inquiry alone is inadequate; students also need to reflect on their inquiry (Sandoval, 2005). In the seminal study on reflective assessment, White and Frederiksen (1998) provided students scientific inquiry criteria for peer assessment to help them improve their metacognition and found below-average students gained more in physics knowledge than above-average students. Encouraging students to think back scaffolds them to become metacognitive and realize what they are doing and what to do next, which is particularly important for low achievers.

In KB, we define reflective assessment as students taking active roles in identifying personal and community knowledge gaps and examining how to move forward, personally and as a community (Lei & Chan, 2018; Scardamalia, 2002; Yang et al., 2016). Compared with formative assessment, which focuses on closing the gap between current and desired performance (Taras, 2009), reflective assessment focuses more on cultivating student agency for continuing inquiry.

While earlier studies on reflective assessment have involved peer assessment (Toth, Suthers, & Lesgold, 2002; White & Frederiksen, 1998), we use reflective assessment as a *collaborative* KB community process. Not every community member develops at the same pace, but students can scaffold each other's metacognitive development through modeling and collective work. Students who have difficulty engaging in metacognition, monitoring their personal progress, or enacting individual metacognition can be sustained in a community, as reflective assessment in KB takes on richer dimensions. In a community, students can engage in metacognitive activities through monitoring and reflecting on group progress (Hmelo-Silver & Barrows, 2008; van Aalst & Chan, 2007; Yang et al., 2016), asking questions and explaining, and scaffolding one another's metacognition through shared agency; such pooled intelligence is crucial for low achievers.

There is increased research on the use of learning analytics in supporting student inquiry in the learning sciences (Wise & Schwarz, 2017). KB research also encompasses learning analytics on KF, particularly students using analytic tools for collective agency (Zhang et al., 2018). Reflective assessment, premised on the KB principle of concurrent and embedded assessment supports student inquiry when it is *embedded* in their KB work; *concurrent* with the use of evidence-supported tools to help them visualize and understand where they are heading; and *transformative* as they reflect on their inquiry and change their KB processes (Scardamalia, 2002). Reflective assessment helps students to reflect on what individuals and their community know and what need to be improved—it involves goal setting, monitoring, rise-above, and meta-discourse (meta-talk), which is critical for productive KB discourse (Scardamalia & Bereiter, 2006).

## Rationale, Framework, and Design of the Study

This study examines reflective assessment supported by technology in a KB classroom to help academic low achievers develop higher order competencies and productive KB.

*Knowledge Building and Reflective Assessment for Low Achievers*

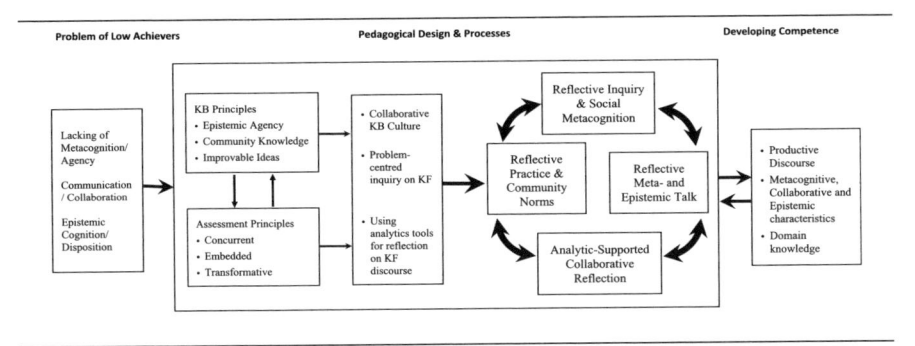

*Figure 2.* **A design framework of reflective assessment for academic low achievers in knowledge building (KB).**
*Note.* KF = Knowledge Forum.

    This reflective-assessment design uses a variety of tasks, scaffolds, and tools to support students to *reflect and assess their classroom and KF discussion* and to support their complex inquiry and productive discourse creation.

    Reflective assessment can benefit all students, but is particularly important for low achievers, due to their assumed difficulties. First, it promotes low achievers' engagement in *metacognitive* processes—planning, monitoring, reflecting on their online discourse, deploying and developing metacognition and agency—through *principle-guided reflection*, aided by analytics tools and evidence. Second, it can help low achievers develop *collaboration* through awareness of *principle-based norms emphasizing collective effort and community examples*. Third, by asking students to reflect on and improve ideas for continuing pursuit (not just fixed answers), it can help low achievers to develop productive *epistemic dispositions* through cultivating reflective and *collaborative culture for idea improvement*.

    We propose a framework using KB with reflective assessment, supported by analytics for transformative learning (Figure 2). Primarily, we first identify academically low-achieving students' common difficulties regarding metacognition and agency, collaboration, and epistemic dispositions. This framework emphasizes key KB principles (*epistemic agency, community knowledge*, and *improvable ideas*) that target metacognition, collaboration, and epistemic dispositions. These key principles are further enhanced by the reflective assessment principle that is concurrent (with analytics-tool feedback), embedded (intertwined with the main inquiry), and transformative (student agency for changing their learning). We designed key phases and different tasks/scaffolds to enact the principles (see Table 2), but these tasks can vary depending on contextual situations. Principle-based design, rather than highly scripted instruction (Zhang et al., 2018) is important in KB pedagogy (Scardamalia, 2002) and most useful for adaptation to the emergent needs of academic low achievers.

*Yang et al.*

We also conjecture how the design as a system will support student reflection including (1) *Reflective inquiry and social metacognition* help scaffold students to reflect on different inquiry tasks; ideas are made public that help them see theirs and others' points of views in collective work. (2) *Reflection through meta- and epistemic talk* encourage students to reflect on KF work by discussing examples of good discourse to help them develop an explicit and metacognitive understanding of higher order inquiry. (3) *Analytics-supported collaborative reflection* for collective growth help to visualize KF inquiry that support students to engage in collaborative reflection. (4) *Reflective practice and community norms*, in which students see reflection as a community norm that becomes their part of thinking and habit of mind. The framework also examines the learning outcomes and competencies reflected in KF discourse and domain knowledge.

The design framework features KB principles, tasks, and processes aligned with three assertions. First, reflective assessment has been shown to promote *metacognition* (White & Frederiksen, 1998); we now enhance that in a social context. Collective inquiry and reflection are intertwined, supported by different scaffolds, including KB wall, portfolio, meta-talk, and visualization from analytics tools (Raes et al., 2014; discussed later). Students will reflect on questions they have (goal setting), what they have/have not learnt (monitoring), and new inquiry (planning); individual low achievers may not be metacognitive but can develop social metacognition supported by tool-based visualization (Raes et al., 2014). Second, reflection is done in small groups and community, as students interact, they have opportunities to develop social and collaborative competencies; KB talk takes place throughout and the KCA data necessitates students' discussing how they make meaning and synthesize ideas; the use of tool can help widen the dialogic and reflective space (Wegerif, 2007). Third, reflective assessment emphasizes not just reaching correct answers but also deepening understanding and idea improvement (Scardamalia, 2002). As students engage in collaborative reflection supported by tools, they can see multiple perspectives and can begin to realize that knowledge is not fixed and certain. Concurrent feedback (no need to wait for teacher appraisal) can help students develop epistemic dispositions needed for higher level inquiry.

The current study and proposed framework are a systematic program premised on the KB model and reflective assessment. In previous research, regular high-school students taking geography and chemistry courses created *electronic KF portfolios* in which they reflected on and assessed their online KF discourse and accomplishments, and areas and questions yet to be considered. Reflective assessment helped students to develop explanatory and productive KB discourse (Lee, Chan, & van Aalst, 2006; van Aalst & Chan, 2007). More recently, we designed reflective assessment enhanced with *analytic tools* (the KCA) to help a wider range of students to reflect on their inquiry and discourse (Yang, 2019; Yang et al., 2016). While these studies have shown positive results, the integrated system of classroom processes and dynamics of reflective assessment has not been investigated to unveil how reflective-assessment design in a KB environment can support academic low achievers' higher order inquiry, an area much needed for equity in current education.

### Research Goal and Questions

The goal of this study was to examine how academic low achievers can be supported to engage in higher order inquiry and KB in technology-supported environments. We developed a pedagogical design using KB with reflective assessment supported by analytic tools and examined the processes and dynamics of reflective assessment. KB pedagogy involves students pursuing inquiry into self-generated problems using offline and KF discussion, developing theory, and building knowledge. To support academic low achievers, we enriched KB using reflective assessment, with students engaging in reflective tasks and dialogic talks supported by analytics while collectively reflecting on their classroom inquiry and KF discourse to chart their knowledge advance.

We first investigated if and how low achievers could engage in KB demonstrating productive discourse with metacognitive, social, and epistemic characteristics. We argue that, if students gradually take on higher level agency supported by reflective assessment, they may show productive KF discourse moves with collaborative interaction, metacognitive competence in identifying gaps, regulating group inquiry, and epistemic orientation reflected in synthesis, conceptualization, and sustained pursuit of inquiry. We expected to see low achievers could engage in productive online discourse illustrating these productive discourse moves with changes over time. Second, we analyzed how academic low achievers engaged in reflective assessment and how reflective-assessment designs supported their metacognitive, social, and epistemic growth. We employed qualitative analysis using multiple rich data sources to examine how students enact reflective assessment in KB contexts, identifying key themes to illuminate processes. Table 1 shows our research focus, research questions, and proposed analyses.

## Method

### Research Context and Participants

Hong Kong secondary schools are classified into three bands, based on students' public examination results. This study was conducted at a Band-3 school, with students performing at or below the 10th percentile of the student population at admission. The sample was a class of 37 Grade 9 students taking a Visual Arts course. Participants were typical of low-achieving students. They were taught in Chinese and had no previous KB experience.

Table 1
**Summary of Research Focus, Research Questions, and Analysis**

| Focus | Research Questions (RQ) | Analyses |
|---|---|---|
| 1. Student engagement in knowledge building (KB): Knowledge Forum (KF) online writing and development with metacognitive, collaborative, and epistemic characteristics; collective inquiry and idea improvement | RQ1a: To what extent did academic low achievers engage in productive online discourse in KB illustrating metacognitive, collaborative, and epistemic inquiry discourse moves, and did those moves improve over time? | • Analyses of productive discourse and higher order competencies in inquiry threads using discourse moves as unit of analysis; discourse moves reflecting metacognitive (regulation), collaborative (explanation), and epistemic (synthesis) characteristics<br>• Comparison of discourse moves over time<br>• Analysis of collective growth in inquiry threads |
| Student KF discourse and domain understanding | RQ1b: Could individual students gain new knowledge through KB augmented by reflective assessment? and how was their online KB/KF engagement related to domain understanding? | • Pre- and posttest comparisons on domain knowledge measured by open-ended tests and exam results<br>• Correlation analysis of KF participation, KF discourse processes, and domain knowledge |
| 2. Classroom processes and dynamics of reflective assessment in a KB classroom influencing students' growth | RQ2: How did low achievers engage in reflective assessment for collective inquiry and KB? Specifically, how did the design support collaborative reflection and promote metacognitive, collaborative, and epistemic growth? | • Qualitative analysis of students' reflective assessment using multiple data sources including classroom videos, field notes, audio-video recordings of student collaborative reflection activities, student artifacts including group concept-maps, weekly written reflections, pencil-and-paper portfolio notes, student and teacher interviews, and KF and Knowledge Connections Analyzer (KCA) data. |

The teacher was an experienced visual-arts teacher who had employed KB-based teaching for approximately 8 years.

### Pedagogical Design: Knowledge Building and Reflective Assessment

Students carried out a 5-month inquiry into the topics of *art* and *art evaluation*, with weekly lessons including key questions such as "What is art?" and "How is art appreciated?" Course content was flexibly arranged in response to students' emergent inquiries. Student work comprised whole-class discussions, small-group collaboration, individual and collaborative note writing, and reflection (online and offline). Building on previous KB studies (Chan, 2011; Yang et al., 2016), the development of collaborative KB culture, problem-centered collective inquiry, deepening inquiry, and assessment were emphasized. This study includes new designs of reflective assessment integrated throughout supported by analytic tools. Aligned with the design framework (Figure 1), Table 2 details the pedagogical design of analytics-supported reflective assessment for academic low achievers in KB.

**Phase 1: Developing a collaborative KB culture and reflective inquiry (Weeks 1-9).** To increase students' motivation and enhance their inquiry, collaborative, and reflection skills, small- and whole-class discussions were organized to create *an error-free culture* of open discourse, sharing, inquiring, negotiation, and learning and to develop their epistemic approach to knowledge and inquiry. The teacher engaged low achievers in discourse integrated with manipulating objects—for example, constructing three-dimensional objects from wires, explaining how selected pictures represented art, and visiting nearby villages to observe historical artifacts as arts objects. Based on these experiences, the class constructed a *knowledge-building wall* (KB wall, Figure 1) by attaching index cards with ideas and questions to the classroom wall for their peers' review, questions, and reflection. Exemplar KB wall questions were identified to scaffold students to pose good inquiry questions. To develop reflective skills, students wrote reflections after each lesson, as well as pencil-and-paper summary notes (portfolio) that involved selecting at least six exemplar notes from the KB wall and writing a *reflective* statement to explain their theory (e.g., why a drawing of a rubbish dump can be art) and show why and how these notes supported their theory.

**Phase 2 (Weeks 10-12): Engaging in collective problem-centered inquiry on KF and reflective talks.** Following their KB wall inquiry, students worked in four- to six-student groups to formulate questions, for example, what is art, and how is art evaluated. After whole-class discussion, students selected the most interesting questions for further inquiry on KF (Figure 1). Students started their KF inquiry by posting questions and ideas, then built on others' work to address the problems. KF affordances include visuals and co-authored notes that ameliorated their writing difficulties. Online and offline discourse were intertwined, and regular KB talks provided students opportunities to reflect on their discourse—reviewing what ideas had been discussed, what progress had been made, and the nature of good discourse. Through KF inquiry and

Table 2

**Pedagogical Design of Reflective Assessment in Knowledge Building (KB)**

Goals: Students Develop Productive Inquiry and Higher Order Competencies in KB

| KB Principles | Pedagogical Phases | Designs for Reflective Assessment |
|---|---|---|
| Epistemic Agency<br>Community Knowledge<br>Improvable Ideas<br>Embedded, Concurrent, and Transformative Assessment | 1. Developing a collaborative KB classroom culture and reflective inquiry | • Inquiry tasks and reflection on inquiry<br>• Authentic tasks (e.g., constructing three-dimensional objects using wires) and problems for developing question-explanation and discursive skills<br>• KB wall with ideas made visible and public for questions, inquiry, and reflection<br>• Reflective journals for tracking one's own understanding—metacognitive growth<br>• Collaborative reflection on and assessment of public ideas on KB wall using portfolios—knowledge extendable and epistemic dispositions |
| | 2. Engaging in collective problem-centered inquiry on KF and reflective talks | • KB inquiry supported by KF<br>• Frequent and dialogic talks to help students "reflect" on their KB inquiry and KF writing—how they were writing and progressing, and nature of good KF inquiry and discourse |
| | 3. Using analytic-tools for reflection on KF discourse | • Using KF accompanying tools "applets" to motivate students to write and read more notes (quantity of notes)<br>• Using the KCA and prompt sheets to help students to engage in productive reflective assessment.<br>  ✓ Are we a community that collaborate? helped low achievers reflect and focus on *collaboration and community knowledge*<br>  ✓ Are we putting knowledge together? helped low achievers become more *metacognitively aware of synthesis and rise-above* (meta-discourse) of KF inquiry<br>  ✓ How do our ideas develop over time? helped low achievers reflect on *conceptual progress and idea connections of their ideas*<br>• Evidence-based reflection as collective responsibility to promote an inquiry-oriented sharing culture for the development of epistemic dispositions |

*Note.* KF = Knowledge Forum; KCA = Knowledge Connections Analyzer.

*Knowledge Building and Reflective Assessment for Low Achievers*

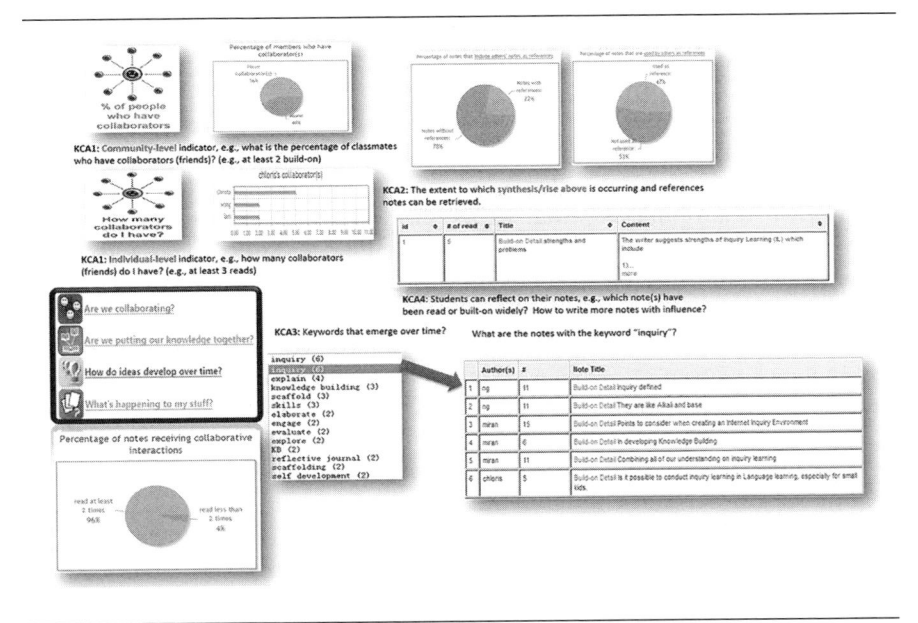

*Figure 3.* **Knowledge Connections Analyzer (KCA): Interface, questions, and output.**

reflection, students developed the sense that ideas were *improvable* and that the role of the community was to support progress collectively.

Phase 3 (Weeks 13–21): Using analytic tools (KCA) for reflection on KF discourse. Students used analytic tools to help them reflect collectively on their KF writing. After several weeks of KF writing, the teacher introduced the KF assessment applet tools, which show KF participation indices (used in other KB classrooms), and the KCA, an analytic tool to help students reflect on their collaborative inquiry on KF (Figure 3), set goals, monitor progress, and plan for further inquiry (see van Aalst et al., 2012; Yang et al., 2016).

Aligned with the KB emphasis on student agency (Scardamalia, 2002), students were provided opportunities to use the KCA, followed by *reflective discussion on the visualized data*, to help them review what they have done on KF, analyze problems, and set goals for future work.

The KCA tools includes a framework with four intuitive questions that allow young students, including low achievers, to reflect on their KB work from different angles (van Aalst et al., 2012). The KCA set of questions include (1) "Are we a community that collaborates?" that taps into the notion of community knowledge (collaboration)—the extent to which all members contribute and collaborate when writing on KF; (2) "Are we putting knowledge together?" that addresses the notion of *synthesis* and *rise-above* and *synthesis*—the extent to which the class community synthesized individual ideas and makes "rise-above" contributions, which provide a higher level of conceptualization of them; (3) "How do our ideas develop over time?" that touches on improvable ideas, conceptual progress, and idea connections—the extent to which class members take agency collectively to improve their ideas and discourse; and finally (4) "What is happening to my own contributions?" that helps show the impact of students' work in the community, and how different notes influence others' ideas and development over time. After selecting one of these questions, students could choose and vary its parameters and the KCA output would show data on what students had done (e.g., how many friends we have?).

In the classroom, students worked together using the KCA in whole-class situations enriched with after-class small-group discussion. Initially, the teacher demonstrated KCA and explained the need to review and reflect on their KF inquiry. Typically, the teacher introduced one KCA question each lesson and discussed why the visualization and data from each question were important for KB inquiry. The teacher also demonstrated and discussed productive ways of interpreting the KCA data; students tried the KCA in dyads and groups, using metacognitive questions (e.g., What have we found? Why run this analysis? How would we plan to improve our inquiry?; see Appendix) supported by reflective talks on KCA findings. Classwork was enriched with small-groups learning after class, six groups included, one at a time, so students could learn more about KCA and interact with closer guidance using the KCA. With the collaborative reflection opportunities using the KCA, students progressively became more aware of what they were doing on KF and made plans to improve their KF writing; they also put together their ideas in collaborative concept-maps, wrote rise-above notes, and posted them on KF. In sum, online and offline discourse are intertwined, sustained with reflection, as students wrote on KF using analytic tools to help them assess and reflect collectively on their KF writing and KB inquiry.

## Data Sources

### Classroom Observations, Videotaping, and Student Artifacts

We observed and kept a record of classroom events, capturing both student and teacher activities through field notes and photographs of all lessons, and video recordings of most lessons; seventeen 50-minute videos were collected in total (850 minutes). We collected all artifacts, including students' weekly written reflections that recorded what they had learned about art and design and questions they had—specifically they wrote about what they had and had not known and what they would like to inquire in the future. Students wrote their ideas mounted on post-it cards on the KB wall, and for individual pencil-and-paper portfolios, they identified and selected the important ideas from themselves, their peers, and the class community from the KB wall. We also collected collaborative concept maps that recorded the key points of group and KF discussions and indicated the changing understanding of domain knowledge. In total, 80 concept maps produced by 8 student groups were collected.

### Video Recordings of Reflective-Assessment Sessions

We video-recorded students' reflective-assessment activities while on KF, in-class and after-class group sessions, including their interpretations of and reflections on the data, and plans for their KF discourse. Detailed field notes were included, and interpretations discussed with the teacher. Six in-class reflective-assessment sessions (whole-class) of 30 to 50 minutes and six after-class reflective-assessment sessions of 60 to 90 minutes were video-recorded. In each after-class session, student groups (2 to 6 volunteer students in each) ran the KCA and reflected on their collaboration, knowledge synthesis, and idea improvement using the KCA data.

### Student and Teacher Interviews

Semistructured interviews were used to examine students' reflective-assessment experiences. We interviewed groups of two to five students, either before or immediately after class, and also after they used the KCA. Most interviews were informal and lasted 20 to 30 minutes. The interview questions tapped into their reflective experience, for example, "What did you write on KF after the last KCA analysis?" The teacher's reflection was collected regularly with systematic interviews over different periods, each lasting for 30 minutes, to capture teacher's design and understanding of how KB and reflective assessment was enacted.

### Domain Understanding

To examine students' knowledge gains on their inquiry topics, pre-and posttests designed by the teacher were administered at the beginning and the end of the course. The pretest questions were What do you know about the topics of "What is art?" and "How is art appreciated?" The posttest questions were What have you learnt about the topics of "What is art?" and "How is art appreciated?" Students were given about 15 minutes to complete the test on both occasions.

### KF Participation and the KCA Data

The Analytic Toolkit (Burtis, 1998) was used to collect information about the number of notes written and read, and the percentage of notes linked to each other from log data. Data from 400 KF notes were collected and analyzed. The Analytic Toolkit data have been used widely in published studies (e.g., Lai & Law, 2006; Lee et al., 2006; So et al., 2010). Using the KCA, the researcher further retrieved quantitative data on student collaboration in terms of reading notes, building on notes, and synthesizing notes. We used the KCA data to indicate the extent to which students were collaborating with others, synthesizing ideas, and writing rise-above notes.

## Data Analysis

### Analyzing KF Writing

Analysis was conducted to examine students' engagement in productive discourse for KB (i.e., How students made knowledge advance together) and cognitive, metacognitive, collaborative, and epistemic characteristics of discourse. The unit of analysis was an "inquiry thread," a sequence of KF notes contributed by different community members to address a problem (e.g., "How to appreciate art pieces") illuminating students' collective pursuit of knowledge (Zhang et al., 2007). We analyzed all 400 computer notes; 17 inquiry threads were identified for analysis. A second researcher independently placed 40% of the notes into inquiry threads, leading to an intercoder reliability of .80 (Cohen's kappa). Within each inquiry thread, we coded students' KF notes using different categories illuminating cognitive, metacognitive, collaborative, and epistemic characteristics adapted from our coding framework (Table 3; Yang et al., 2016). Different categories reflect students' developing competences including ideation (cognitive), regulation (metacognition), synthesis (collaborative-epistemic) reflecting rise-above and higher level conceptualization. Two raters independently coded notes from three inquiry threads ($n$ = 120, 30%), with an interrater reliabilities of .78 for *questions*, .78 for *ideas*, and .77 for *community* (Cohen's kappas).

### Synthesis/Rise-Above KF Notes

Students also wrote group synthesis notes before and after KCA use. Students' synthesis/rise-above KF notes (a meta-level note that consisted of hyperlink to other notes) were analyzed with a 5-point coding scheme, modified from an earlier study (Lei & Chan, 2018). These synthesis/rise-above notes were examined to provide evidence of students' ability to engage in higher level collaborative and epistemic inquiry. These notes varied from listing notes and copying information from others' notes with no explanations to meta-conversation using a "we" perspective to reflect on discourse goals, identification of gaps, and investigation of what the discourse was about. Two raters independently scored all the synthesis notes, leading to an interrater reliability of .72 (Cohen's kappa).

### Assessment of Domain Understanding

Students' responses were rated based on degrees of understanding and whether a clear and coherent explanation were provided using a 4-point

Table 3
**Coding Scheme for Examining Discourse Moves in Inquiry Threads in Knowledge Forum**

| Code | Definition/Defining Features |
|---|---|
| Questions | |
| Fact-seeking | Questions on definition of the terms or concepts, or seeking factual information |
| Explanation-seeking | Questions seeking open-ended responses with elaborative explanations |
| Metacognitive | Questions prompting metacognitive monitoring, reflecting on and regulation of inquiry process and/or individual or joint understanding, referring to group dynamics, monitoring, regulatory learning, and clarification-seeking questions |
| Idea | |
| New idea | Concept/idea proposed not previously introduced |
| Simple claim | Opinion stated without elaboration or justification |
| Explanation | Inferences supported with reasons, examples and evidence |
| Metacognitive statement | Statements and explanation toward monitoring, reflecting or regulating individual or collective understanding and inquiry-related process |
| Community | |
| Lending support | Inquiry suggestions with related expert resources for further inquiry |
| Deepening inquiry | Commenting and developing peers' ideas; expressing alternative ideas |
| Regulating inquiry | Monitoring and/or repairing question-explanation exchange process by asking questions or requesting explanations |
| Synthesizing and rise-above | Rise-above notes; summarizing the group's understanding or a string/cluster of notes and attempting to achieve new insights |

scale from 1 to 4 (see Supplementary Materials available in the online version of the journal). Two raters independently scored all of the data; the interrater reliability was .78 (Cohen's kappa).

*Thematic/Narrative Analysis*

We employed qualitative analyses using thick data to understand how low achievers engaged in KB and reflective assessment. We first browsed the videos and transcripts to develop an overall sense of the reflective-assessment process, followed by identifying "digestible" chunks of the videos: with major episodes of reflective assessment. These video segments were contextualized and linked to develop a story line. We analyzed classroom data as well as analytic-based reflective assessment and other sources of data (e.g., interviews, artifacts, classroom observations). We identified and selected important classroom events guided by principles of KB and assessment outlined in framework. For example, during each phase of the pedagogical design, we first identified the classroom events that best illustrated the practices of KB and reflective assessment and examined how they support the development of capabilities necessary for engaging in collective KB. Constant comparison of these different episodes, narratives, and critical events mapping to different instructional phases bring about the key emerging themes.

# Results

## Research Question 1: Students' Productive Discourse and Change and Relations With Domain Knowledge

*Productive and Sustained Inquiry in KF Discourse*

*Inquiry thread analysis.* This analysis examined the entire inquiry threads to show how students engaged in distributed work and sustained inquiry. Figure 4 illustrates how low achievers could engage in distributed work and sustained inquiry. No student dominated the process, and many threads (e.g., #1, #3, #4, #7, #8, and #13) involved most students as authors, demonstrating their ability to collaborate and sustain inquiry. Most inquiry threads lasted more than 7 weeks, indicating different students showed sustained interest pursuing inquiry into these topics, suggesting students' developing epistemic-oriented dispositions.

Qualitative analysis of threads illustrates how low achievers collectively pursued KB and engaged in progressive problem solving (e.g., #1, #3, #4, #7, #8, and #9). In these threads, students proposed interesting problems and explanations, monitored and regulated their inquiry by asking relevant questions and seeking clarification, addressed problems at increasing depth, and

## Knowledge Building and Reflective Assessment for Low Achievers

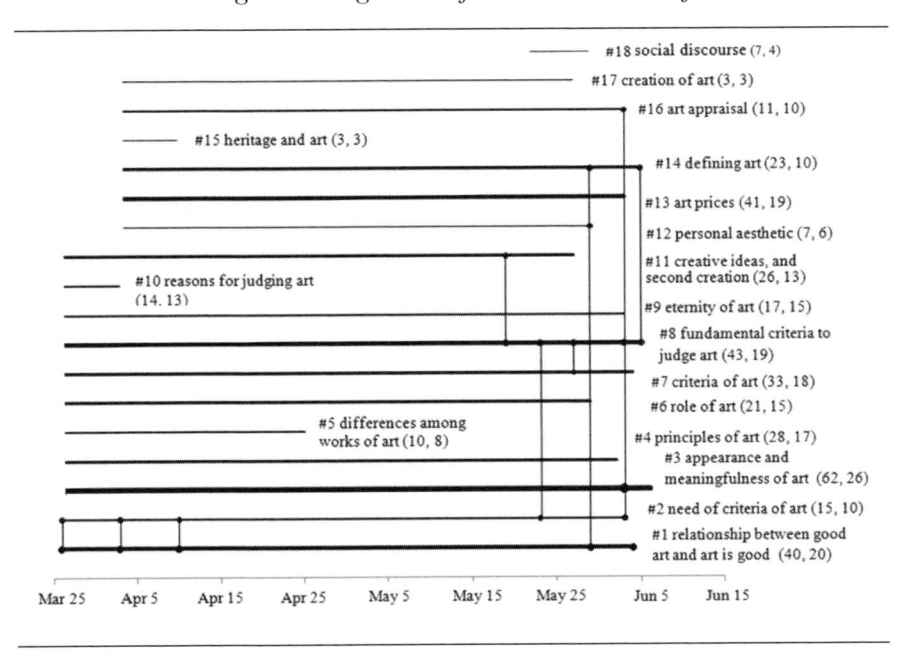

*Figure 4.* **KF inquiry threads with distributed work for collective inquiry.**
*Note.* Each thread identified with problem, number of notes, and authors in parentheses respectively; dotted lines as bridging notes for inquiry threads.

produced higher levels of conceptualization. For example, in the art evaluation thread (Thread #8), students initially asked how to judge whether a piece of art were successful, leading them to the understanding that the judging practice was influenced rather than determined by a personal aesthetic vision. This spawned further inquiry problems and statements of what students knew, and students generated summaries of what they discussed to identify problems for further inquiry; for example, "What do you mean by . . .? Are they contradictory?" (Student [S2]) "Why not talk about the meanings behind the art instead of personal aesthetic vision, seemingly an endless conversation?" (S21) "Let me summarize the above selected notes . . . successful art should be meaningful, which means thought-provoking," (S7) and "Idea improvement [scaffold]. . . . Both the appearance of and meanings behind art determine the critical and defining qualities of successful art" (S15).

*Analysis of discourse types.* We further examined the extent of productive KB discourse by coding the 17 inquiry threads (van Aalst, 2009) to distinguish among increasingly advanced KB discourse patterns: (1) knowledge sharing, mere accumulation of information; (2) knowledge construction

1263

## Table 4
### Characteristics of Discourse: Frequency in Questioning, Ideas and Community in Inquiry Threads

| Inquiry Threads | Questions - Factual Questions | Questions - Explanatory Questions | Questions - Metacognitive Questions | Ideas - New Ideas | Ideas - Simple Claim | Ideas - Elaborated Explanation | Ideas - Metacognitive Statement | Community - Lending Support | Community - Deepening Inquiry | Community - Regulating Inquiry | Community - Synthesis/Rise-Above |
|---|---|---|---|---|---|---|---|---|---|---|---|
| #1 | 1 | 3 | 9 | 4 | 9 | 8 | 4 | 2 | 20 | 9 | 4 |
| #2 | 0 | 1 | 5 | 2 | 3 | 1 | 2 | 1 | 6 | 5 | 2 |
| #3 | 0 | 1 | 11 | 1 | 17 | 18 | 11 | 1 | 34 | 13 | 8 |
| #4 | 1 | 0 | 4 | 5 | 7 | 4 | 4 | 1 | 16 | 6 | 2 |
| #5 | 1 | 0 | 3 | 0 | 4 | 1 | 1 | 1 | 2 | 6 | 0 |
| #6 | 0 | 3 | 1 | 4 | 4 | 2 | 4 | 1 | 11 | 4 | 2 |
| #7 | 0 | 1 | 4 | 2 | 11 | 10 | 4 | 0 | 24 | 5 | 3 |
| #8 | 0 | 2 | 4 | 4 | 8 | 6 | 16 | 0 | 17 | 6 | 16 |
| #9 | 0 | 1 | 2 | 0 | 7 | 6 | 1 | 0 | 13 | 2 | 1 |
| #10 | 1 | 0 | 2 | 0 | 6 | 3 | 0 | 0 | 5 | 2 | 0 |
| #11 | 2 | 0 | 3 | 5 | 5 | 10 | 1 | 3 | 18 | 3 | 1 |
| #12 | 0 | 0 | 0 | 1 | 2 | 3 | 1 | 0 | 5 | 0 | 1 |
| #13 | 2 | 2 | 2 | 0 | 14 | 20 | 1 | 1 | 33 | 2 | 1 |
| #14 | 2 | 0 | 3 | 0 | 7 | 5 | 5 | 0 | 9 | 2 | 5 |
| #15 | 1 | 0 | 0 | 0 | 2 | 0 | 0 | 0 | 2 | 0 | 0 |
| #16 | 1 | 0 | 0 | 0 | 6 | 3 | 1 | 0 | 9 | 0 | 1 |
| #17 | 0 | 0 | 1 | 0 | 1 | 1 | 0 | 0 | 1 | 1 | 0 |
| Total | 12 | 13 | 53 | 27 | 111 | 97 | 48 | 11 | 222 | 66 | 40 |
| M | 0.71 | 0.76 | 3.12 | 1.59 | 6.53 | 5.71 | 2.82 | 0.65 | 13.06 | 3.88 | 2.35 |
| SD | 0.45 | 0.77 | 2.76 | 1.50 | 3.60 | 4.50 | 5.05 | 0.60 | 8.83 | 3.23 | 4.77 |

*Notes.* SD = standard deviation. There were 12 *bridging notes*, which belonged to more than one inquiry thread. Therefore, the total frequency in each column, representing the net frequency of all inquiry threads, should be ≤ the sum of the numbers in all inquiry threads.

*Knowledge Building and Reflective Assessment for Low Achievers*

(questions, explanations, and co-construction of ideas); and (3) KB/creation discourse (progressive problem solving, rise-above meta-discourse and community advances). Of the 17 inquiry threads, 4 were classified as knowledge sharing, 7 as knowledge construction, and 6 as KB. This is a relatively positive result, compared to previous studies on knowledge quality of KB threads among regular students (Fu, van Aalst, & Chan, 2016; van Aalst, 2009).

*Discourse Characteristics: Discourse Moves and Change Over Time*

*Analysis of discourse moves.* KF writing was coded to examine student engagement in productive discourse moves (questions, ideation, community) illuminating metacognitive, collaborative, and epistemic characteristics (see Table 3 and Supplementary Materials available in the online version of the journal). Table 4 shows students wrote more notes with new ideas and collaborative explanations (124) than simple claims (111), suggesting collaboration; KF progress was reviewed and shown in metacognitive questions (53 notes), and metacognitive/discursive statements (48 notes). Analysis of community perspective showed 66 notes as regulating inquiry and 40 as synthesizing notes. These results suggest students were involved in regulating their group/community inquiry, aligned with social metacognition and agency; they also generated synthesis and rise-above notes (with high-level conceptualization) that reflect progress in collaborative-epistemic work; putting sustained efforts to help the community synthesize, rise above and advance knowledge (see examples, Figure 5).

*Changes in discourse characteristics over time.* To examine change over time, the KF notes in each inquiry thread were sequenced based on when they were last modified, then equally distributed into three stages (Stages 1, 2, 3; Zhang et al., 2007), and each stage analyzed. Fourteen large inquiry threads (10 or more notes) were analyzed, and several smaller threads excluded. Analysis of higher level discourse moves—metacognitive questions and statements and synthesis notes—was conducted. Table 5 indicates students mostly contributed questions and ideas during Stage 1; during Stages 2 and 3 they deepened their Stage 1 inquiry questions, generated metacognitive questions and statements to regulate their inquiries, and wrote synthesis and rise-above notes that reflected the class' collective pursuit of knowledge advancement. Overall, the results suggest low achievers were increasingly able to use sophisticated metacognitive, collaborative, and epistemic discourse moves over time as they engaged in productive KB.

*Knowledge Gains and Relations Between KB and Domain Knowledge*

*Change in domain knowledge.* Students' individual knowledge gains were examined, and significant differences obtained between pretest ($M =$

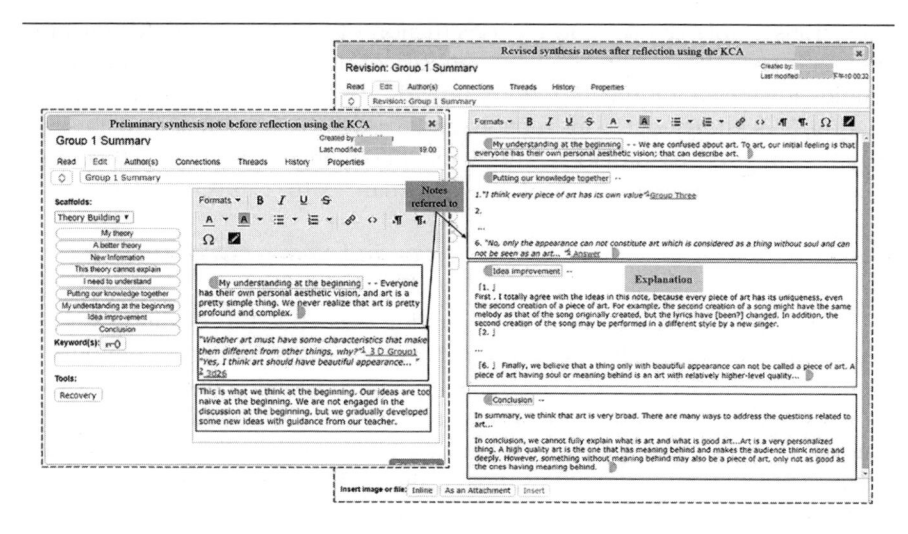

*Figure 5.* **Comparison of group synthesis notes written before and after KCA reflection.**

Note. KCA = Knowledge Connections Analyzer; KF = Knowledge Forum. The two synthesis notes from Group 1 were extracted from KF and translated into English.

1.81, $SD$ = 0.60) and posttest scores ($M$ = 3.08, $SD$ = 0.80), $t$ (36) = −12.71, $p$ < .01. We also examined students' knowledge gains based on examination results. Paired sample $t$ test indicated significant differences in examination scores before ($M$ = 56.17, $SD$ = 13.50) and after ($M$ = 67.92, $SD$ = 11.18) the program, $t$ (36) = −8.48, $p$ < .01. While no control class was available and the results must be interpreted with caution, preliminary evidence was obtained indicating benefits to learning outcomes.

*Relationships between online discourse and individual knowledge gains.* Table 6 shows Pearson correlation coefficients between KF participation (e.g., notes written from log files), discourse moves, domain understanding, and examination scores. Notes written, reflecting cognitive contributions, was significantly correlated with domain understanding ($r$ = .43) and examination scores ($r$ = .68). KF metacognitive statements and revision were significantly correlated with domain understanding ($rs$ = .52 and .45), and examination scores ($rs$ = .65 and .45). The "references" in syntheses notes and explanations were significantly correlated with domain understanding ($rs$ = .52 and .33) and examination scores ($rs$ = .51 and .53). These results suggest students were more likely to understand domain knowledge better when involved in active contribution, collaboration, and

### Table 5
**Increased Frequency of Metacognitive, Collaborative, and Epistemic-Oriented Discourse Moves**

|  | Stage 1 | Stage 2 | Stage 3 |
|---|---|---|---|
| General questioning | 11 | 2 | 1 |
| General ideation | 26 | 45 | 26 |
| Metacognition | 16 | 31 | 62 |
|    Metacognitive questions | 15 | 20 | 19 |
|    Metacognitive statements | 1 | 11 | 43 |
| Synthesis and rise-above inquiry | 0 | 7 | 39 |

### Table 6
**Correlation Analysis of KF Activities (Written, Revision, References) and KF Discourse (Explanation, Metacognition, Synthesis) With Domain Understanding and Examination Scores**

|  | 1 | 2 | 3 | 4 | 5 | 6 | 7 |
|---|---|---|---|---|---|---|---|
| 1. No. of KF notes written | — | | | | | | |
| 2. No. of KF notes revision | .36* | — | | | | | |
| 3. No. of KF references (hyperlinks) | .53** | .70** | — | | | | |
| 4. No. of explanations | .85** | .20 | .46** | — | | | |
| 5. No. of metacognitive statements | .68** | .59** | .51** | .41* | — | | |
| 6. Level of synthesis notes | .36* | .48** | .68** | .29 | .43** | — | |
| 7. Domain knowledge understanding | .43** | .45** | .52** | .33* | .52** | .56** | |
| 8. Examination results | .68** | .45** | .51** | .53** | .65** | .55** | .76** |

*Note.* KF = Knowledge Forum.
$*p < .05. **p < .01.$

metacognitive and epistemic-oriented processes during their online discourse.

Taken together, the analyses show that these academic low achievers were involved in productive KB discourse illustrating distributed and sustained inquiry and engaged in metacognitive, collaborative, and epistemic discourse moves with changes over time. As in other KB studies (Lee et al., 2006), low achievers' discourse engagement was related to domain understanding. As well, both KF discourse *quality* (KB threads; Fu et al., 2016) and KF participation *quantity* (e.g., number of notes read/written, use of scaffolds) were comparable to results from published studies using regular cohorts (e.g., Lai & Law, 2006; Lee et al., 2006). Low achievers not only could improve with the interventions, they could, given appropriate scaffolds, engage in KB inquiry at levels similar to regular students'.

## Research Question 2: Student Engagement in Reflective Assessment, Processes, and Dynamics

We examined classroom processes and dynamics to investigate how students engaged in reflective assessment and how the pedagogical design supported their development, in terms of metacognition, collaboration, and epistemic dispositions. Four interrelated themes on reflective assessment mapping to instructional phases were examined: (1) reflection on inquiry tasks and social metacognition, (2) reflection on KF inquiry and meta- and epistemic talk, (3) analytics-supported reflection for collective growth, and (4) reflection as social practice and community norms.

### Reflection on Inquiry Tasks and Social Metacognition

From the start, principle-based inquiry and reflective tasks were designed to support students' engagement, agency, and metacognition situated in a social/community context. Students began by experiencing a community culture of inquiry, metacognition, and collaboration. They visited a village; created (in groups) objects to illustrate what art is; collaboratively constructed concept maps; and asked questions of each other. One goal of these activities was to help low achievers engage in communication—that is, "to talk with others . . . to listen to others" (Teacher interview, May 20). Another was to help low achievers develop inquiry dispositions and metacognition in *setting goals* through formulating questions. As noted in the teacher interview, "For students with low achievement, it is helpful to start with something authentic, such as wired objects and field visits; they can be motivated and self-directed to some extent to ask more questions and work together to create." The teacher helped students develop inquiry and reflective skills using scaffolds for ideas to place on the KB wall (Figure 1): "In your writing of your ideas, you can play with these coloured cards and . . . openers: "My idea," "I need to understand," "My explanation," and "A better theory" (Teacher instruction on KB wall). Students were scaffold to become metacognitively more aware of inquiry processes, as these ideas and scaffolds were *visually* displayed for inquiry and reflection. Students put forth ideas on what they knew and wondered about, such as "How can the rubber duck at Kowloon West be art?" They reflected on what they needed to know, generated questions on note-cards placed on KB wall, connected cards with strings, and collaboratively responded to each other (Figure 1). A primary goal was to help students develop metacognitive skills for asking deepening questions and collaborative inquiry, and the epistemic disposition that inquiry is open-ended. The KB wall note cards made it possible for students to track their development and *reflect on* what they thought in the context of other classmates' ideas. Ideas on KB walls are improvable and can be extended through collective efforts. In the interview, the teacher elaborated,

*Knowledge Building and Reflective Assessment for Low Achievers*

> My students have poor communication and thinking skills, but can benefit from scaffolds, such as making objects, using the KB wall where they physically work on arranging the notes . . . with that . . . even low achievers can understand the public nature of nature of discussion . . . the visual display is the class's shared ideas—it is physically present and visually displayed in our room. (Teacher interview)

The teacher was alluding to the *epistemic* aspects linking the public nature of discussion to KB. Students also engaged in continuous reflective assessment through writing learning journals after each lesson. Writing learning journals based on their experience was a regular feature that helped students not only engage in continuous monitoring and gradually internalize metacognitive skills but also to take increasing responsibility for their own learning and develop personal efficacy. As S9 commented, "Writing these diaries seems helpful though I was kind of unwilling to at the beginning. . . . Keep thinking and questioning . . . I have a sense of achievement when I [go back] and read all my writing at the end . . . I really make it" (Informal interview, February 13).

Students found it difficult to be metacognitive about their own work, but could do so together (socially) via viewing own and others' ideas on KB wall and carrying out collective reflection tasks such as paper portfolios, wherein students would select several notes (their own and their classmates' ideas) from the KB wall, and then *reflect* on how these ideas (theirs and their classmates') were useful or relevant. The portfolios helped them set learning goals (identify the original question), track what was discussed (monitor), consider what was missing (identify gaps), and reflect on new learning and questions (set plans and goals).

In summary, reflective-assessment tasks were embedded with inquiry activities to transform students' learning. These different reflective tasks enabled students to develop metacognitive and collaborative competence, and the epistemic view that knowledge is extendable. Metacognition is socially developed, undergirded by principles of agency (e.g., question asking), community knowledge (working together), and improvable ideas (new directions) that set the stage to transform students' knowledge and competencies.

*Reflection on KF Inquiry Using Meta- and Epistemic Talk*

When first introduced to KF, students seemed to lack motivation and did not know how to write quality notes, ask good questions, and build on others' notes. Frequent and opportunistic reflective talks were conducted to help students reflect on their KF work and understand the criteria for and nature and standards of good inquiry and discussion. For example,

> Teacher (T) Ok . . . "What is good inquiry question" . . . What are the elements of good inquiry questions?
> S13 Have points! [content]

T Yes. Have points! . . . (microphone passed to another group)
S8 They can help other classmates to think more.
T Yes, question that makes others to think deeply. What do you mean about this kind of question? . . . (one student raised hand, and the teacher passed microphone to the student)
S33 Meaningful and constructive
S19 Have room to discuss . . . flexible . . .
S8 New questions can come from the question
T Could you elaborate a little bit more?
S8 For example, I put forth a question [on KF], then another question follows from that . . .
T Yes, that mean the questions can give birth to other questions, and make you keep thinking, right?
S17 New . . . Newsy . . . does not repeat what other classmates have already said . . .
S25 It is an unusual idea.
S11 Um . . . How to be unusual?
S25 Um . . . Unexpected
S14 Use questions to address questions
T Good, use questions to address questions. Any more ideas? . . . Just now, you contributed several good [ideas]. Good questions and discussion is open, newsy . . . and provoking, and make others think more questions, OK? Wow, you are really great. So, let us go and *create such a question* or issue on KF around "Good art . . . Art is good." OK? (from classroom videos, March 27).

This example illustrates how the teacher helped students reflect on the quality of their KF writing by focusing on generating questions. Through this reflective talk, students were the ones to develop the criteria for good inquiry questions in their KF writing. Low achievers were also not good at responding and building on others' notes. Here is another example of how KB talk helped them reflect and develop an epistemic understanding of discourse.

T . . . another thing . . . to consider is whether students are responding to others? Um we call that *build on* . . . *But why respond to others* . . .
S16 I think responding to others can support the arguments . . . or [help to] ask deeper questions.
T Yes . . . you mean supporting others' arguments, or ask deeper questions to help him/her clarify their arguments . . . Any other ideas?
S3 Um, share our own opinions.
T Sharing your own ideas . . . *Why is responding to others about sharing your own ideas*. What is the connection? . . . same or different? . . . How does it work?
S20 First, sharing our own ideas; then deepening . . . and making our own ideas clear.
S6 Why is [responding] about making our own ideas clear?
S17 Um, first, we present our own ideas; then . . . we need to deepen the inquiry . . . and clarify our own ideas when we think these ideas are not so good [Remark: S17 was setting goal of deepening and reflecting on gaps]

## Knowledge Building and Reflective Assessment for Low Achievers

T Yes, well done. She speaks out one of the dance steps . . . that is voicing our own opinions, then deepening and clarifying. This is the dance steps from this group. This may be Cha-Cha. Is there a second Cha-Cha? (S8 raised hand and the teacher passed the microphone to her.)

S8 *I think we can see the problems of others' ideas when responding to others. But some of their points can also help us change our own ideas.* Then we add other ideas to enrich our own explanation and even rise above our own ideas.

T Good, thanks. Can you hear What S8 said? Cannot? (Microphone passed to S8 again) Can you speak aloud again? Make it clearer?

S8 What I mean . . . *responding to others' notes can help us deepen our own ideas, because we may never think about [what we think] ourselves . . . um . . . we can even make a summary and refer to what others have written. [metacognition in social context]*

This example illustrates how students engaged in reflective talks to become more aware of the nature of productive discourse—they noted that they can see their own views better through the lens of others. Reflective and epistemic talks about KF discourse helped scaffold students toward more productive discourse for ongoing work—they used their understanding to inform their KF work. One student commented that one of the teacher's "most important" tasks was to "structure discussion [by asking us to evaluate and reflect on] [what] notes are good and why . . . [and] how to produce good notes" (S12, informal interview immediately after the KB talk).

Reflective KB talks were used opportunistically, undergirded by principles of community and improvable ideas as the need arose—for example, students built on others and made deeper inquiries. KB classroom talk often focused on content and idea development; however, emphasis here was placed on the epistemic nature of inquiry and discourse. These reflective meta-talks helped low achievers become metacognitively aware of the discourse process, develop epistemic criteria for good discourse, and were encouraged to use these understanding from collective reflection in their continued KF inquiry.

### Reflection Supported by Analytics for Collective Growth

KB inquiry on KF involving multiple posts, and the need to create community discourse with synthesis and conceptualization can be complex for low achievers. Initially, the teacher employed KF's integrated assessment applets to show students graphs on how much they had written on KF both collectively and individually (Figure 1). These visualization of their KF participation provided concurrent feedback that could help them *reflect on their progress* and is generally motivating (Student comment: "When I see how much we have written, I felt I can do better and I will write more")

While these initial works using applet tools seemed to motivate students to write more notes, nevertheless, students were not doing collaborative

KB—that is, synthesizing, creating, and advancing knowledge. They presented ideas with limited explanations or articulating an isolated understanding of concepts (e.g., the appearance of art). The KCA was then introduced to help low achievers engage in deeper discourse, reflect, and become more *metacognitively aware* of their online work, as discussed in the Method section. Students could run the KCA questions on their KF writing—KCA, supported by reflective prompts (e.g., What do you want to find out?), helped students engage in reflection collaboratively with peers, while developing metacognitive competence. Here we include an example of KCA reflective assessment elicited by the second question (*Are we putting our knowledge together?*). A classroom discussion ensued, with the teacher working with students to interpret and reflect on the data, and scaffolding students toward the KB goal of synthesis/rising above as a collective responsibility.

>  T. . . 19 notes have references. [Notes with hyperlinks to other students' KF notes for collective work] Only 8% of the notes have references; 92% do not. . . . Now, have a look at the notes in the table (*reading the KCA output*). What does this mean? Do you know how we could make it better? [Using analytics and evidence to support inquiry]
>  S11 We didn't use each other's notes to write our own notes. . . We are not used to summarizing (*synthesizing*) what we have discussed. [students reflect on their problems . . . focus on "we" as community]
>  S16 I think we need to write more notes with references
>  S12 We can use each other's notes (*as references*) to support our own notes.
>  S8 Write more synthesis notes and incorporate as many references as we can [planning ahead] . . .
>  T. . . OK, let's have a look at the synthesis notes and see whether they really synthesize the ideas they refer to . . . Let's look at this note (*written by Group One*) . . . Are you happy with the quality of the note? . . . What would you do to improve it?
>  S8 I think we just listed the notes to which we referred, without explaining why we incorporated them . . . [noting problems]
>  S4 Our conclusion is kind of problematic: it's so general. It seems unrelated to the reference notes . . . [identifying issues]
>  T. . . When preparing synthesis notes, you could propose and explain your arguments . . . with a short introduction . . . you could explain how the reference notes illustrate your arguments, and what topics need further inquiry . . . (Video-recording of reflective-assessment session, May 15)

These excerpts suggest how students engaged in KB reflection with a focus on the collective, raising issues and suggestion plans. Students appreciated these analytic-supported reflections and became more aware of the process. For example, S15 commented, "the use of [KCA] and discussion was quite helpful . . . we learned how to collaboratively reflect on and analyze the assessment data *and . . . how to analyze. . . improve our notes*" (informal interview after a KB talk, May 23).

## Knowledge Building and Reflective Assessment for Low Achievers

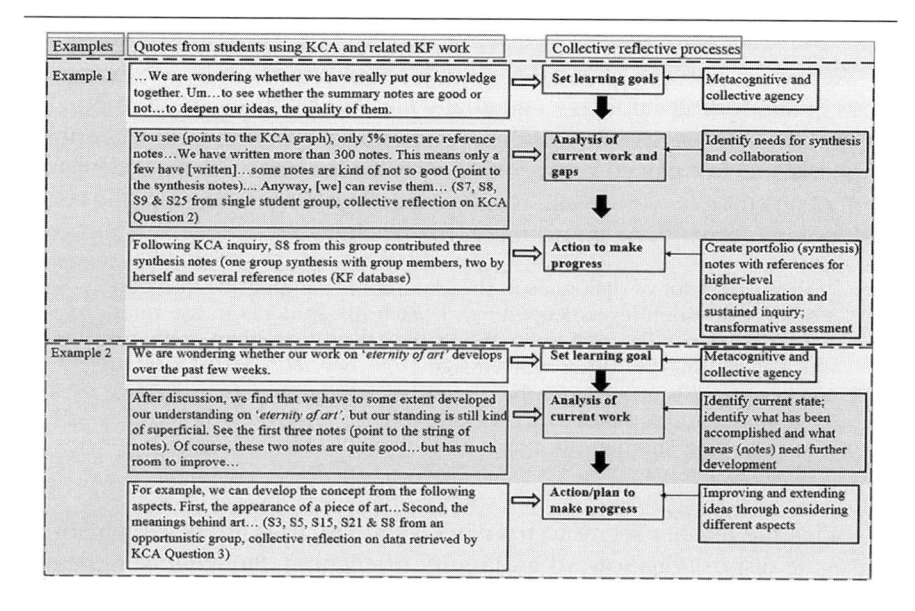

| Examples | Quotes from students using KCA and related KF work | | Collective reflective processes |
|---|---|---|---|
| Example 1 | ...We are wondering whether we have really put our knowledge together. Um...to see whether the summary notes are good or not...to deepen our ideas, the quality of them. | ⇨ Set learning goals | Metacognitive and collective agency |
| | You see (points to the KCA graph), only 5% notes are reference notes...We have written more than 300 notes. This means only a few have [written]...some notes are kind of not so good (point to the synthesis notes).... Anyway, [we] can revise them... (S7, S8, S9 & S25 from single student group, collective reflection on KCA Question 2) | ⇨ Analysis of current work and gaps | Identify needs for synthesis and collaboration |
| | Following KCA inquiry, S8 from this group contributed three synthesis notes (one group synthesis with group members, two by herself and several reference notes (KF database) | ⇨ Action to make progress | Create portfolio (synthesis) notes with references for higher-level conceptualization and sustained inquiry; transformative assessment |
| Example 2 | We are wondering whether our work on 'eternity of art' develops over the past few weeks. | ⇨ Set learning goal | Metacognitive and collective agency |
| | After discussion, we find that we have to some extent developed our understanding on 'eternity of art', but our standing is still kind of superficial. See the first three notes (point to the string of notes). Of course, these two notes are quite good...but has much room to improve... | ⇨ Analysis of current work | Identify current state; identify what has been accomplished and what areas (notes) need further development |
| | For example, we can develop the concept from the following aspects. First, the appearance of a piece of art...Second, the meanings behind art... (S3, S5, S15, S21 & S8 from an opportunistic group, collective reflection on data retrieved by KCA Question 3) | ⇨ Action/plan to make progress | Improving and extending ideas through considering different aspects |

*Figure 6.* **Reflective assessment using Knowledge Connections Analyzer (KCA) (excerpts from group inquiry).**

Figure 6 shows two examples of student group inquiry using the KCA—setting learning goals, reviewing what they had done, analyzing gaps, and setting new plans—primarily framed by the notion of collective responsibility helping the group and community to improve on KF inquiry. With the support of KCA-aided reflection, these low achievers continued to engage in inquiry reaching 400 KF notes, and some produced synthesis notes that depicted change after KCA reflection (see Figure 5). Students not only made sense of the assessment data in thoughtful ways and created a shared understanding of the purpose of data use but were also equipped with critical cognitive and metacognitive skills such as questioning, explaining, and summarizing. Evidence-based reflection supported by tools could help low achievers to analyze gaps in their collective knowledge, take actions to address gaps, advance collective knowledge and take increased ownership. The reflective instances, framed by collective responsibility, promoted an inquiry-oriented sharing culture for the development of epistemic dispositions that led students to converse about the data productively and thoughtfully.

## Reflection Developed as Social Practice and Community Norms

Reflective assessment is a developing social practice with a community ethos of supporting students to gradually take up practices and is important for tackling problems of epistemic dispositions. In this KB classroom, emphasis was not placed on specific activities, but on the general development of an ethos of wondering, working together to find out, stepping back, and looking forward. As an example, the teacher said,

> There are a lot of dialogues in the classroom. I emphasize question-answer-question [ever deepening]. I teach my students to use questions to clarify the problems. When they face a problem, they will use questions . . . they will engage . . . *they will create a habit.* When there is a question, they know something new is happening . . . they dare to think, to argue with you . . . and when they argue, there will be another question, and they will just keep going on . . . (Teacher interview)

What the teacher seems to have been alluding to is the development of epistemic dispositions toward an inquiry orientation. Students were developing community norms and practices of reflecting on their own and others' work to make collective progress. In a KB talk on Q2 of the KCA, S11 commented on a revised KF synthesis note of another group, calling it "quite messy and fragmented, especially toward the end . . . so many ideas, but incoherent and not well integrated." The student also said, "I can learn from some other notes . . . how to synthesize ideas; some peer notes . . . provide useful perspectives . . . . I think [the above-mentioned note] could be improved incorporating more notes . . ."

The above example suggests that students were developing the practice and that they were able to articulate, assess, evaluate, and contribute to others' ideas for collective progress. There were no comparisons or criticisms of others' work, because collaborative reflection for idea improvement support class progress and transformative learning. Reflection and pursuit for improvement were gradually becoming classroom norms and practice in this community, sustaining its continuing growth.

Another aspect of the community developing reflective and inquiry practices was its cumulative *and progressive nature*. New skills, competences, and epistemic dispositions were gradually and progressively developed, supported by reflection that mediated students' new activities and competence. Reflective inquiry, opportunistic KB talk, and analytics-supported reflection were integral to the process. When students engaged in inquiry such as question posing on KB wall, they were engaged in reflection and idea development, which helped them progressively develop reflective and collaborative skills in their KF writing; KB meta-talk and analytics-supported reflection helped them further to develop the KB process of productive discourse.

With reflection as developing skills and social practices, students also seemed capable of reflecting on and *interweaving* different parts of their learning experience. Working on a KCA-supported assessment, S8 commented,

> We realized the problems with our note through this discussion . . . we were also motivated to improve the note. I think, for example, that we should have *followed the format* we used to prepare the KF wall summary portfolio note to improve it. . . . We should have explained why we use particular notes as references and commented on their strengths and weaknesses. (Interview after KB talks, May 15).

In this example, S8 was referring back to how their start-of-the-year KF wall and portfolio experience helped them tackle the new task of producing quality synthesis notes. The student also demonstrated metacognitive competences and seemed able to identify problems, set improvement goals, and identify strategies by drawing from his peers' collective learning: "We realized the problem," "We should have followed the format." The emphasis on reflective assessment may have helped students realize the need to integrate and consolidate their developing competences. Primarily, students used their newly gained skills and interests to engage in new activities that allowed them to progressively advance ideas in the communal space.

Collaborative reflection is a community norm. Their increased metacognitive and epistemic understanding helped them engage in the KB process by collectively advancing ideas and helping improve each other's notes. Combined with skills such as inquiry, collaboration and metacognition, students gradually exhibited increased engagement in KB, supported by reflective assessment using the KCA, resulting in their production of rich KB discourse.

Taken together, these four themes could also be mapped onto the pedagogy design phases, as they were interrelated and progressive. Students first engaged in reflective inquiry with guided reflection to develop metacognition in a social context, including setting learning goals. These inquiry and reflective tasks were developed further through opportunistic KB talks, and students used analytic-supported assessment to develop metacognitive goal setting, planning, and reflection skills. In the KB community, competencies and skills were cumulative and integrative, developing into social practices supported by community ethos and norms.

## Discussion

This study examined how reflective-assessment design in a KB environment supported academic low achievers' collaborative inquiry and development of higher order competencies and investigated sociocognitive dynamics associated with that development. Figure 2 provides a summary, including problems of low achievers, pedagogical designs, sociocognitive processes, and learning outcomes. In the following, we first discuss the evidence regarding the research questions, and then provide an explanatory account of sociocognitive dynamics, and finally outline the educational implications for developing higher order inquiry and productive KB for low achievers.

## Effects of KB Reflective Assessment on Productive Discourse and Higher Order Competencies

Analysis of the KF inquiry threads (Figure 4) indicated that many students contributed to them; KF work is distributed, and most threads developed over many weeks. These results are important for *equity*, because they suggest not only some students can benefit—many students were involved and collectively pursued sustained inquiry. The analysis suggested these low achievers gradually shifted from viewing online KF writing as question-answer short exchanges toward developing epistemic dispositions for open-ended inquiry. Analysis of KF discourse showed that, despite their difficulties and low prior achievement, students engaged in productive KB discourse, including collaborative explanations, metacognitive and regulatory processes, and epistemic-oriented inquiry using synthesis and emerging questions. The coding of online discourse moves illustrated students' developing metacognitive, collaborative, and epistemic competence (Table 4). Students were increasingly engaged and motivated to take responsibility for the advancement of collective knowledge (Table 5).

Although it was not possible to include a control group, previous KB research provides a solid precedent for using changes over time to measure student development; for example, Zhang et al. (2007) assessed changes in students' personal ideas across three stages to investigate discourse development and KB. Comparisons with published KB studies indicate these students participated actively and produced KB discourse similar to that of regular cohorts (Lee et al., 2006; van Aalst, 2009; Lai & Law, 2006). With reflective-assessment design, KB activities not only affected low achievers' higher order inquiry, including their metacognition, inquiry, and epistemic dispositions but also their domain knowledge.

KF analysis findings from this study are consistent with our earlier research using analytics-supported reflective assessment with different classes (Yang, 2019; Yang et al., 2016). In the present article, we provide a clearer analysis of productive KB examining online discourse illustrating students' collective responsibility, using discourse coding, thread content, and summary notes. KF analysis generally focuses on knowledge advance (e.g., Zhang et al. 2007); our analyses also suggest the discourse moves reflect students' higher order competencies in metacognition, collaboration, and epistemic dispositions in community context, for example, students wrote synthesis note as they became more epistemic in their approach to knowledge. As well, such high-quality discourse is generally absent from other online discussion environments or even KB classrooms lacking strong design. Without a clear focus on the *collective* supported by reflective assessment, students may focus on individual work, resulting in fragmented writing commonly found in online contexts (Hew & Cheung, 2008).

The findings are consistent with KB research on the role of reflection (van Aalst & Chan, 2007) and designed intervention on productive KB discourse (Zhang et al., 2018). While it was expected KB interventions would bring about improvement, this is one of the few studies to focus on a whole cohort of low achievers. Using detailed analysis of online discourse and other qualitative data, these low achievers were shown to engage in productive KB discourse and complex inquiry comparable to other KB cohorts. Different excerpts and data provide corroborating evidence illustrating their increased metacognitive competence (e.g., Figure 6 KCA examples setting goals, analyzing gaps), collaborative discourse (e.g., constructive comments for classmates' progress in KF writing), and epistemic dispositions (e.g., discussion of what inquiry involves). As shown in research on academic interventions (e.g., Dietrichson et al., 2017) and higher order thinking (Zohar & Dori, 2003), the problems of low achievers is not a deficit model but lack of engagement opportunities. Our findings add to the literature that, given appropriate designs and scaffolds in the learning environment, academic low achievers can engage in complex collaborative inquiry and KB.

**Dynamics of KB Reflective Assessment Scaffolding Academic Low Achievers**

Qualitative analyses illustrated how academic low achievers engaged in reflective assessment for KB in alignment with the framework (Figure 2). The analyses show how low achievers initially had difficulties writing/discussing questions, and how the designs and dynamics of reflective assessment— including (1) reflective inquiry and social metacognition, (2) reflective meta- and epistemic talk, (3) evidence-based reflection for collective growth, and (4) reflective practice embedded in classroom system and community ethos helped them to gradually take up collective agency and engage in productive inquiry. Teacher interviews suggested that authenticity of design and pedagogical intentions align with assumptions of KB and reflective assessment. In the following, we discuss how the design features and processes work coherently as a system of KB practice, supporting low achievers in inquiry-based learning and KB. While this study on reflective assessment was in a KB classroom, its themes are applicable to other classrooms.

*Developing Reflective Inquiry and Social Metacognition*

Metacognitive reflection and agency are pivotal to learning and inquiry, but often lacking among low achievers. Reflective assessment builds on seminal research with low achievers and at-risk students developing metacognitive reading in social contexts (Palincsar & Brown, 1984) and reflective inquiry in science (White & Frederiksen, 1998). Similar to reciprocal teaching asking children with learning difficulties to take up *increased responsibility*, KB enhanced with reflective assessment emphasizes collective responsibility for low achievers—reflection is social, and students support each other in a KB community.

This study shows how low achievers had many opportunities to work on authentic and reflective tasks in a community to develop a sense of *agency* and collective responsibility. KB inquiry activities were enriched using learning journals and paper-and-pencil portfolio. Students wrote notes asking questions and setting learning goals; they monitored their progress through comparing their own ideas with those of others (e.g., KB wall), and evaluated their own and the class' progress. Thinking about thinking is not easy for low achievers; however, when working as a community wherein ideas are improvable (Scardamalia, 2002), they can ask questions of an audience, see what others think, learn from others' examples and strategies, and become more aware of what they know and what they need to know through interacting with others.

Using metacognition to teach low achievers has been a key theme (Brown, 1997; White & Frederiksen, 1998; Zohar & Peled, 2008), and we emphasize developing shared cognition and metacognition in a KB community. Primarily, through reflective assessment in a community, students can use others' lenses to sharpen their understanding of their own thinking and develop goal setting, monitoring, and planning collectively in a social context. Not everyone becomes metacognitive at the same pace; however, they can scaffold each other, follow up on others' examples, use others' ideas to sharpen their own, and become metacognitive collectively, as reflected in their KF discourse and classroom inquiry.

### Reflective Meta- and Epistemic Talk

Students with low achievement commonly have difficulties in communicating, collaborating, and engaging in discursive talk; as such, educational approaches for low achievers typically involve peer interaction (Han et al., 2014; Hawkins et al., 1988; Slavin et al., 2011). Analysis shows the classroom teacher regularly used talks to help students articulate and reflect on their understanding. Low-achieving students were continually involved in dialogue in dyads, groups, and classroom communities, with different students building on others' ideas, reflecting on experience, and the teacher being just another discussion member—dialogue was ongoing.

This study also extends the notion of classroom and KB talk, adding metacognitive and epistemic dimensions for low achievers. Most inquiry-based classroom and KB talks are about content and improving ideas (Zhang et al., 2007). This proposed framework added another layer and helped low achievers develop an understanding of the epistemic nature of discussion. Similar to the idea of thinking about thinking, reflective talk involves *talking about their talk* to enhance collaborative competence. Low achievers lacking collaborative skills may have difficulties in inquiry-based environments. In reflective talks, students engaged in discussing their dialogue/discourse on KF; initially, low achievers might not know what to write and, through such reflective talk, could use examples from peers to construct criteria and standards. Assessing others' responses required them to comprehend and to process, while productive talk helped them to articulate. Primarily, reflective and opportunistic talks can help low achievers become metacognitively aware of their own actions and gradually develop the epistemic disposition viewing knowledge as extendable. The notion of explicitly teaching principles has been examined in academic interventions for low achievers (Baker et al., 2002); our study shows students can work together to construct these principles through reflective and epistemic talk in a community context.

*Evidence-Based Reflection for Collective Growth*

Reflective assessment, as a design for academic low achievers, is also important from the perspective of formative assessment—providing data to students on how they are performing (Baker et al., 2002) and progress monitoring (Dietrichson et al., 2017) in academic interventions. Analytics-supported reflective assessment emphasizes giving students agency and having them reflect on their own data and progress. Raes et al. (2014) discussed the role of visualization of data and phenomenon in facilitating low achievers' inquiry in computer-supported environments.

We developed these ideas of progress monitoring, linked to the principle of concurrent, embedded, and transformative assessment (Scardamalia, 2002). We designed learning analytics tools and used KCA's intuitive questions to engage academic low achievers. Visualization of KF data provided *through* technology and analytics—applets, scaffolds on prompt sheets, and the KCA output—facilitated collective reflection for inquiry and provided specific areas for them to discuss. Running the KCA and different tools, students could see their and their classmates' contributions on KF for timely feedback (concurrent assessment), then use that evidence to improve their KF writing (transformative assessment). Premised on the *community* principle, students were not just working for their own progress but for the progress of the class community ("*We* found out the problems. *We* can try out . . ."). As students assessed their own work—supported by evidence and discussed goals, problems, and plans—they gradually took on metacognitive/regulatory competencies and transformed their understanding. Reflective assessment using analytics is consistent with the notion of formative assessment and feedback (Dietrichson et al., 2017), but goes beyond fixed goals, with students

*Yang et al.*

continually pursuing deepening inquiry using data and evidence to guide their collaborative reflection and collective growth—the ideas that there are no specific answers and things are extendable through collective efforts can motivate academic low achievers.

*Reflection Embedded in Classroom System and Community Ethos*

Reflective assessment builds on the idea of fostering communities of learners (Brown, 1997), emphasizing reflective practice as a community norm. The analysis suggests reflection and epistemic dispositions for diverse learners and low achievers need to be developed *progressively*, supported by a community ethos. The strength of the classroom community emerged from the multiple and intertwined ways in which students' discourse and knowledge development was supported by the sociocultural contexts, online environment, and reflective assessment.

Learning in this KB classroom involved both individual and collective goals, supported in different ways, including the following: principles governing classroom behavior, such as KB's collaborative culture and norms; the allocation of roles for collective responsibility in advancing discourse and reflection; the design of tools, including KF/KCA; and opportunistic meta-talks and prompt sheets. The goals and intentions of this classroom were unlike those in more traditional ones—emphasis was on developing a collaborative KB culture for collective advancement, using KF and the KCA to make the processes tangible and achievable by providing a record of community inquiry. KB's community ethos was important for reflective assessment, as students were not judging their personal attainment, but how the whole class progressed.

Our study demonstrates that low achievers—albeit with initial difficulties and tensions—could gradually move toward high-level inquiry and discourse, in part because KB's collaborative ethos encouraged their agency and belief in improvable ideas to sustain their continual engagement. The teacher noted KB is an "error-free" safe environment for his students. Research has shown the important role of cooperative learning in academic interventions for low achievers (Slavin et al., 2011); reflective assessment in KB also draws on students as resources but focuses on the *collective*. For developing higher order inquiry, KB culture goes beyond fixed goals to afford students the openness to pursue and create new goals. Low-achieving students can work together toward collective accomplishments and community responsibility, a new kind of identity emphasizing collective rather than individual achievement.

These explanatory themes are progressive and cyclical and address low achievers' metacognitive, social, and epistemic difficulties. The importance of reflection and self-peer assessments has always been emphasized in designing learning environments. This study emphasizes collaborative reflection—students reflecting on and assessing their own inquiry and discourse collaboratively. The progressive and interrelated themes of reflective inquiry, reflective talks, evidence-based reflection, and reflective practice are premised on KB community principles, and can be applicable in different learning environments.

This study was conducted in a Hong Kong classroom, where learning sciences research is still a recent phenomenon. Chinese students, particularly those with low academic abilities, are usually regarded as passive learners (Chan, 2011); this study reveals the possibility of nurturing a community of knowledge builders among such low-achieving Chinese students. The study suggests the general applicability of KB theories and designs and documents the adaptive expertise of the teacher who implemented the KB approach in a specific classroom context. While students tend to be more compliant in Asian settings, the results cannot be explained by compliance behavior. KB is different from behavioral approaches in which students merely follow teacher instructions and perform practice. The change involves metacognitive work, sustained collaborative inquiry, and epistemic work and discourse—all normally difficult for low achievers. This intervention was grounded in KB theory, and its technology-supported assessment design enriched the teacher's adaptive expertise. While its findings are specific to this Asian classroom, its analysis, processes, and implications are applicable for designing KB and higher order inquiry in other classrooms in different cultural contexts.

**Limitation and Implications for Future Research**

While this study is limited—paradoxically—by the teacher's competence and high-level adaptive expertise, this is also its strength. With the paucity of research on designing collaborative and epistemic work for low achievers, in-depth case study is important. This study has provided rich data and key themes that may be translated into other reflective-assessment designs to be further tested. It is important to examine the extent to which the designs implemented depended on this specific teacher, and whether they can be used effectively by other teachers working in different KB classroom contexts. There may also be questions regarding the study's focus on local circumstances, and on whether the KB activities and analytic-supported reflective assessment designs should be separated to test KCA's causal mechanisms. Nevertheless, this study builds on a research tradition in KB studies; the notion of a case study is to illuminate designs, processes, and dynamics, even in a best-scenario case, examining what is possible and what it says about theories and design of KB and assessment. We have offered explanatory processes and design principles; in the tradition of design-based research, these designs and processes need to be continually tested in other classroom and educational settings.

In examining higher order inquiry and 21st-century education competencies, it would be useful to examine whether the observed changes are *transferable*. What changes in students' understandings about art might emerge months after the conclusion of the KB intervention? To what extent do KB and reflective assessment prepare students for future learning or KB in other domains? Transfer is an important issue in the learning sciences (Lobato, 2008), and responses to these questions will advance this research.

This study included data on domain knowledge and pretest-posttest change. As there was no comparison group, it is difficult to ascertain causal change to the pedagogical design. It would be useful to test the reflective assessment design in future studies incorporating different classes in different contexts and to investigate more closely how the design and dynamics influence students' higher order competencies, KB, and domain knowledge. As well, this study examined and identified ideas such as social metacognition, meta-talk, and collaborative reflection that are important for theory and design in learning sciences. Our findings provide some characterization, but they need to be unpacked further through iterative studies and analyses.

## Conclusions

This study has addressed the important problem of providing equity and access to diverse learners to develop higher order collaborative and epistemic inquiry for 21st-century education. While there have been many discussions on meeting the needs of all students (e.g., National Academy of Science, Engineering and Medicine, 2018, *How People Learn II*), this is one of the few studies to document how a cohort of low-achieving students can engage in productive KB and develop high-order competencies, including metacognition, collaboration, and epistemic disposition. Contemporary educational interventions generally focus on educational achievement for low achievers (Baker et al., 2002; Dietrichson et al., 2017); this study directly examined how low achievers can succeed in collaborative inquiry and KB, addressing the need to provide access for all learners through innovative learning designs.

This study contributes to research on scaffolding inquiry-based learning among low achievers by proposing an explanatory framework. We started with the difficulties facing low achievers and proposed a design premised on KB principles enriched with reflective assessment for promoting their metacognitive, collaborative, and epistemic growth. The integrated design and themes—including reflective inquiry and social metacognition, reflective meta- and epistemic talk for understanding, evidence-based reflection for collective growth, and reflection embedded in classroom system and community ethos—provide explanatory themes with implications for designing learning environments to support collaborative inquiry among academic low achievers in different classroom contexts.

*Knowledge Building and Reflective Assessment for Low Achievers*

This study builds on current research by enriching conceptualization for scaffolding diverse learners using the KB reflective-assessment perspective. Educational research has examined support for low achievers using principles of task engagement, authentic problems, and peer interaction/cooperative learning. This study builds on the literature emphasizing metacognition/agency, community, and improvable ideas for low achievers. While low achievers are passive learners, these principles paradoxically would encourage them to take collective responsibility for productive inquiry. The community perspective is important—cognitive and learning-sciences research has examined metacognition in reflective assessment (White & Frederiksen, 1998), and we extend that to collaborative reflective assessment and social metacognition supported in a community context. Metacognition and reflection are difficult, and the novelty of this study is to use learning analytics tools to provide visualization of students' ideas to support their metacognition and reflection and to widen the dialogic and reflective space for low achievers' KB. Progress monitoring and feedback are discussed in the academic intervention literature (Dietrichson et al., 2017); we extend this with the use of analytics tools placed in the hands of students to promote their agency in directing their own inquiry. We have developed the design and demonstrated one of the first studies helping low achievers to use analytics-supported tools to reflect on their inquiry in complex technology environment. A key contribution of this study is that it is not only a proof-of-concept study but also provides an explanatory framework supporting low achievers in complex inquiry using KB reflective assessment that can be examined and tested in other inquiry-based technology environments.

This study also has implications for KB research. This is one of the few KB studies to include deep analyses of *both* online and offline discourse; such analysis may enrich theory and analysis of KB with design implications for scaffolding inquiry for low achievers. This study also contributes to KB assessment through unpacking the principles of embedded, concurrent, and transformative assessment, and points to continuing work in this direction. This implementation of KB in a different cultural context is also beneficial for examining the application and robustness of the theory and opens up new design possibilities.

This study also has classroom-level implications for facilitating academic low achievers in engaging in higher order competencies, collaborative inquiry, and KB. Several key principles are discussed based on the findings. First, teachers can enhance metacognition in a social context by supporting students to reflect, inquire, and use others' examples and different lenses to reflect on their own models (e.g., One student noted she understood what she thought when responding to others). They can scaffold students to enhance both their personal and social metacognition by asking them what they earlier thought, what they discussed, and how that changed their understanding. Second, a dialogic approach is needed to engage low achievers in productive dialogues and reflective meta-talk (e.g., teacher said my classroom is full of student talks). Classroom talks are often about content, but they can be extended to help low achievers develop explicit understandings of what makes good inquiry and discourse and reflect on their dialogue activity. Third, reflective assessment gives cognitive responsibility to the students and enables assessment for transformative purposes; even low achievers can assess their own and others' learning and collaboration, increasing their agency. Teachers can provide meaningful concurrent feedback using different tools, prompting students to reflect and explain; technology can be placed in students' hands to afford them more agency in assessing their work. Finally, developing a collaborative KB classroom ethos is important for supporting reflective culture and practices. KB involves students working as a community of learners and adding value to the community, regardless of their level of competence. It focuses not on individual achievements, but on collective efforts and progress; the classroom norm is students working together, reflecting together, and developing epistemic dispositions to inquire and improve collectively.

This study of KB and reflective assessment has theoretical and educational implications for examining and designing classroom-level learning environments intended to develop higher order thinking, inquiry, and collaboration among academic low achievers. While it has examined KB, this study has theoretical value and design implications that offer insights into the relationships among assessment, collaborative inquiry, and instructional practice; the potential affordances of KB for students with low achievement; and the nature and dynamics of reflective assessment.

# Appendix

## Prompts for Reflection Using Knowledge Connection Analyzer (KCA)

Date_____ We have written Knowledge Forum (KF) notes this week– Yes/No

**Our analysis**:
Question 1: What have we done with the Knowledge Connections Analyzer and what are the results?

**Our goal**:
Question 2: Why did we do this analysis?

**Our understanding and discovery:**
Question 3: What problems have we discovered? What have we found out from the analysis?

**Our wonderment/We don't understand:**
Question 4: What are some questions we have? What is something we don't quite understand?

**Our plan:**
Question 5: Keep on or improve our present work on Knowledge Forum? If we try to improve, how would we plan to do it?

**Other questions and comments**

## Notes

Supplemental material is available for this article in the online version of the journal.
This research was partly supported by a grant to the first author from Ministry of Education of the People's Republic of China (Grant No. 18YJC880107), and a grant to the second and third authors from the University Grants Committee, Research Grant Council of Hong Kong (Grant No. 752508H).

## References

Azevedo, R. (2005). Using hypermedia as a metacognitive tool for enhancing student learning? The role of self-regulated learning. *Educational Psychologist, 40,* 199–209.

Baker, S., Gersten, R., & Lee, D. S. (2002). A synthesis of empirical research on teaching mathematics to low-achieving students. *Elementary School Journal, 103,* 51–73.

Barron, B. (2003). When smart groups fail. *Journal of the Learning Sciences, 12,* 307–359.

Barton, A. C., & Tan, E. (2010). We be burnin'! Agency, identity, and science learning. *Journal of the Learning Sciences, 19,* 187–229.

Barzilai, S., & Chinn, C. (2018). On the goals of epistemic education: Promoting apt epistemic performance. *Journal of the Learning Sciences, 27,* 353–389.

Baxter, J. A., Woodward, J., & Olson, D. (2001). Effects of reform-based mathematics instruction on low achievers in five third-grade classrooms. *Elementary School Journal, 101,* 529–547.

Becker, B. E., & Luthar, S. (2002). Social-emotional factors affecting achievement outcomes among disadvantaged students: Closing the achievement gap. *Educational Psychologist, 37*, 197–214.

Bereiter, C. (2002). *Education and mind in the knowledge age.* Mahwah, NJ: Lawrence Erlbaum.

Brown, A. (1997). Transforming schools into communities of thinking and learning about serious matters. *American Psychologist, 52*, 399–413.

Brown, A. L., & Campione, J. C. (1994). Guided discovery in a community of learners. In K. McGilly (Ed.), *Classroom lessons: Integrating cognitive theory and classroom practice* (pp. 229–270). Cambridge: MIT Press.

Chan, C. K. K. (2011). Bridging research and practice: Implementing and sustaining knowledge building in Hong Kong classrooms. *International Journal of Computer-Supported Collaborative Learning, 6*, 147–186.

Chan, C. K. K. (2013). Collaborative knowledge building: Towards a knowledge-creation perspective. In C. E. Hmelo-Silver, C. A. Chinn, C. K. K. Chan & A. O'Donnell (Eds.), *The International handbook of collaborative learning* (pp. 437–461). New York: Routledge.

Chen, B., & Hong, H. Y. (2016). Schools as knowledge-building organizations: Thirty years of design research. *Educational Psychologist, 51*, 266–288.

Dietrichson, J., Bøg, M., Filges, T., & Jørgensen, A. (2017). Academic interventions for elementary and middle school students with low socioeconomic status: A systematic review and meta-analysis. *Review of Educational Research, 87*, 243–282.

D'Mello, S., Dieterle, E., & Duckworth, A. (2017). Advanced, Analytic, Automated (AAA) measurement of engagement during learning. *Educational Psychologist, 52*, 104–123.

Duschl, R. A., & Osborne, J. (2002). Supporting and promoting argumentation discourse in science education. *Studies in Science Education, 38*, 39–72.

Fu, E. L. F., van Aalst, J., & Chan, C. K. K. (2016). Toward a classification of discourse patterns in asynchronous online discussions. *International Journal of Computer-Supported Collaborative Learning, 11*, 441–478.

Greene, J. A., Sandoval, W. A., & Bråten, I. (2016). *Handbook of epistemic cognition.* New York, NY: Routledge.

Han, S., Capraro, R., & Capraro, M. M. (2014). How science, technology, engineering, and mathematics (STEM) project-based learning (PBL) affects high, middle, and low achievers differently: The impact of student factors on achievement. *International Journal of Science and Mathematics Education, 13*, 1089–1113.

Hawkins, J. D., Doueck, H. J., & Lishner, D. M. (1988). Changing teaching practices in mainstream classrooms to improve bonding and behaviour of low achievers. *American Educational Research Journal, 25*, 31–50.

Hew, K. F., & Cheung, W. S. (2008). Attracting student participation in asynchronous online discussions: A case study of peer facilitation. *Computers & Education, 51*, 1111–1124.

Hmelo-Silver, C. E., & Barrows, H. S. (2008). Facilitating collaborative knowledge building. *Cognition and Instruction, 26*, 48–94.

Järvelä, S., Kirschner, P. A., Panadero, E., Malmberg, J., Phielix, C., Jaspers, J., . . . Järvenoja, H. (2015). Enhancing socially shared regulation in collaborative learning groups: Designing for CSCL regulation tools. *Educational Technology Research and Development, 63*, 125–142.

Kaendler, C., Wiedmann, M., Rummel, N., & Spada, H. (2015). Teacher competencies for the implementation of collaborative learning in the classroom: A framework and research review. *Educational Psychology Review, 27*, 505–536.

Kroesbergen, E. H., van Luit, J. E. H., & Mass, C. J. M. (2004). Effectiveness of explicit and constructivist mathematics instruction for low-achieving students in the Netherlands. *Elementary School Journal, 104*, 233–251.

Lai, M., & Law, N. (2006). Peer scaffolding of knowledge building through collaborative groups with differential learning experiences. *Journal of Educational Computing Research, 35*, 123–144.

Lee, E. Y., Chan, C. K. K., & van Aalst, J. (2006). Students assessing their own collaborative knowledge building. *International Journal of Computer-Supported Collaborative Learning, 1*, 277–307.

Lei, C., & Chan, C. K. K. (2018). Developing meta-discourse through reflective assessment in knowledge building environments. *Computers & Education, 126*, 153–169.

Lobato, J. (2008). Research methods for alternative approaches to transfer: Implications for design experiments. In A. E. Kelly, R. A. Lesh, & J. Y. Baek (Eds.), *Handbook of design research methods in education: Innovations in science, technology, engineering, and mathematics learning and teaching* (pp. 167–194). New York, NY: Routledge.

Montague, M., Enders, C., & Dietz, S. (2011). Effects of cognitive strategy instruction on math problem solving of middle school students with learning disabilities. *Learning Disability Quarterly, 34*, 262–272.

National Academy of Science, Engineering and Medicine. (2018). *How people learn II: Learners, contexts, and cultures*. Washington, DC: National Academies Press. doi:10.17226/24783

National Research Council. (2000). *How people learn: Brain, mind, experience, and school* (Expanded ed.). Washington, DC: National Academies Press.

Palincsar, A. S., & Brown, A. (1984). Reciprocal teaching of comprehension-fostering and comprehension-monitoring activities. *Cognition and Instruction, 1*, 117–175.

Raes, A., Schellens, T., & De Wever, B. (2014). Web-based collaborative inquiry to bridge gaps in secondary science education. *Journal of the Learning Sciences, 23*, 316–347.

Sandoval, W. A. (2005). Understanding students' practical epistemologies and their influence on learning through inquiry. *Science Education, 89*, 634–656.

Scardamalia, M. (2002). Collective cognitive responsibility for the advancement of knowledge. In B. Smith (Ed.), *Liberal education in a knowledge society* (pp. 67–98). Chicago, IL: Open Court.

Scardamalia, M., & Bereiter, C. (2006). Knowledge building: Theory, pedagogy, and technology. In R. K. Sawyer (Ed.), *The Cambridge handbook of the learning sciences* (pp. 97–115). New York, NY: Cambridge University Press.

Scardamalia, M., & Bereiter, C. (2014). Knowledge building and knowledge creation: Theory, pedagogy, and technology. In R. K. Sawyer (Ed.), *The Cambridge handbook of the learning sciences* (2nd ed., pp. 397–417). New York, NY: Cambridge University Press.

Slavin, R. E., Lake, C., Davis, S., & Madden, N. (2011). Effective programs for struggling readers: A best-evidence synthesis. *Educational Research Review, 6*, 1–26.

Snell, J., & Lefstein, A. (2018). "Low ability," participation, and identity in dialogic pedagogy. *American Educational Research Journal, 55*, 40–78.

So, H. J., Seah, L. H., & Toh-Heng, H. L. (2010). Designing collaborative knowledge building environments accessible to all learners: Impacts and design challenges. *Computers & Education, 54*, 479–490.

Stahl, G. (2006). *Group cognition: Computer support for building collaborative knowledge*. Cambridge: MIT Press.

Taras, M. (2009). Summative assessment: The missing link for formative assessment. *Journal of Further and Higher Education, 33*, 57–69.

Toth, E. E., Suthers, D. D., & Lesgold, A. M. (2002). "Mapping to know": The effects of representational guidance and reflective assessment on scientific inquiry. *Science Education, 86*, 264–286.

Trilling, B., & Fadel, C. (2009). *21st century skills: Learning for life in our times.* San Francisco, CA: Jossey-Bass.

Tsai, C. W., & Shen, P. D. (2009). Applying web-based self-regulated learning and problem-based learning with initiation to involve low-achieving students in learning. *Computers in Human Behavior, 25*(6), 1189–1194.

van Aalst, J. (2009). Distinguishing knowledge sharing, construction, and creation discourses. *International Journal of Computer-Supported Collaborative Learning, 4*, 259–288.

van Aalst, J., & Chan, C. K. K. (2007). Student-directed assessment of knowledge building using electronic portfolios. *Journal of the Learning Sciences, 16*, 175–220.

van Aalst, J., Chan, C., Tian, S. W., Teplovs, C., Chan, Y. Y., & Wan, W.-S. (2012). The Knowledge Connections Analyzer. In J. van Aalst, K. Thompson, M. J. Jacobson, & P. Reimann (Eds.), *The future of learning: Proceedings of the 10th international conference of the learning sciences (ICLS 2012)*–Volume 2, short papers, symposia, and abstracts (pp. 361–365). Sydney, Australia: ISLS.

Wallace, T. L., & Chhuon, V. (2014). Proximal processes in urban classrooms: Engagement and disaffection in urban youth of color. *American Educational Research Journal, 51*, 937–973.

Wegerif, R. (2007). *Dialogic education and technology: Expanding the space of learning.* New York, NY: Springer.

White, B., & Frederiksen, J. (1998). Inquiry, modelling, and metacognition: Making science accessible to all students. *Cognition and Instruction, 16*, 3–118.

Wise, A. F., & Schwarz, B. B. (2017). Visions of CSCL: eight provocations for the future of the field. *International Journal of Computer-Supported Collaborative Learning, 12*, 423–467.

Wong, B. Y. (1987). How do the results of metacognitive research impact on the learning disabled individual? *Learning Disability Quarterly, 10*, 189–195.

Yang, Y. (2019). Reflective assessment for epistemic agency of academically low-achieving students. *Journal of Computer Assisted Learning.* Advance online publication. doi:10.1111/jcal.12343

Yang, Y., van Aalst, J., Chan, C. K. K., & Tian, W. (2016). Reflective assessment in knowledge building by students with low academic achievement. *International Journal of Computer-Supported Collaborative Learning, 11*, 281–311.

Zhang, J., Hong, H.-Y., Scardamalia, M., Teo, C., & Morley, E. (2011). Sustaining knowledge building as a principle-based innovation at an elementary school. *Journal of the Learning Sciences, 20*, 262–307.

Zhang, J., Scardamalia, M., Lamon, M., Messina, R., & Reeve, R. (2007). Socio-cognitive dynamics of knowledge building in the work of 9-and 10-year-olds. *Educational Technology Research and Development, 55*, 117–145.

Zhang, J., Tao, D., Chen, M. H., Sun, Y., Judson, D., & Naqvi, S. (2018). Co-organizing the collective journey of inquiry with idea thread mapper. *Journal of the Learning Sciences, 27*, 390–430.

Zohar, A., & Dori, Y. J. (2003). Higher order thinking skills and low-achieving students: Are they mutually exclusive? *Journal of the Learning Sciences, 12*, 145–181.

Zohar, A., & Peled, B. (2008). The effects of explicit teaching of metastrategic knowledge on low and high-achieving students. *Learning and Instruction, 18*, 337–353.

Zohar, A., Degani, A., & Vaaknin, E. (2001). Teachers' beliefs about low-achieving students and higher order thinking. *Teaching and Teacher Education, 17*, 469–485.

Manuscript received January 29, 2017
Final revision received June 30, 2019
Accepted July 23, 2019

# Differences at the Extremes? Gender, National Contexts, and Math Performance in Latin America

Ran Liu
Andrea Alvarado-Urbina
Emily Hannum
*University of Pennsylvania*

*Studies of gender disparities in STEM (science, technology, engineering, and mathematics) performance have generally focused on average differences. However, the extremes could also be important because disparities at the top may shape stratification in access to STEM careers, while disparities at the bottom can shape stratification in dropout. This article investigates determinants of gender disparities in math across the performance distribution in Latin American countries, where there is a persistent boys' advantage in STEM performance. Findings reveal disparate national patterns in gender gaps across the performance distribution. Furthermore, while certain*

---

RAN LIU is an incoming assistant professor (starting January 2020) at the Department of Educational Policy Studies, School of Education, University of Wisconsin-Madison, 235 Education Building 1000 Bascom Mall Madison, WI 53706; e-mail: *ranliu42@gmail.com*. Her research interests include gender, STEM education, and quantitative methods. Her recent publications appear in *Social Forces, Comparative Education Review,* and *Comparative Education.*

ANDREA ALVARADO-URBINA is a doctoral student in sociology at the University of Pennsylvania. Her research interests are education and migration in developing countries, with a focus on Latin America. She is currently working on a project on multicultural ethnic identities in the incorporation of immigrant students in the north of Chile.

EMILY HANNUM is a professor of sociology and education at the University of Pennsylvania. Her research interests are poverty and child welfare, social stratification, and sociology of education. Current projects include studies of childhood poverty and inequality in China and the impact of large-scale school consolidations on educational attainment in China. Recent publications include "Education in East Asian Societies: Postwar Expansion and the Evolution of Inequality" (2019, *Annual Review of Sociology,* with Hiroshi Ishida, Hyunjoon Park, and Tony Tam) and "Home, School, and Community Deprivations: A Multi-Context Approach to Childhood Poverty in China" (2019, *Journal of Contemporary China,* with Weiwei Hu and Albert Park).

*national characteristics are linked to gender gaps at the low- and middle-ranges of the performance distribution, female representation in education is the only characteristic associated with a reduced gender gap at the top level.*

KEYWORDS: STEM education, gender, Latin America

## Introduction

The phenomenon of gender differences in STEM (science, technology, engineering, and mathematics) performance is a continuing concern, as it relates to the underrepresentation of women at the highest levels of STEM (Else-Quest, Hyde, & Linn, 2010). Some scholars have argued that there may be greater variability in performance among boys than girls (Feingold, 1992; Hedges & Nowell, 1995; Hyde, Lindberg, Linn, Ellis, & Williams, 2008), which suggests the importance of considering patterns of gender disparity at the extremes of the performance distribution. The extremes of the distribution could also be important because disparities at high performance levels may shape stratification in access to high-level STEM education and careers (Fan, Chen, Matsumoto, & Fan, 1997; Xie & Shauman, 2003), while disparities at the bottom can shape stratification in grade repetition and dropout (Janosz, LeBlanc, Boulerice, & Tremblay, 1997; Jimerson, Egeland, Sroufe, & Carlson, 2000). However, cross-national studies of gender disparities in student STEM performance have generally focused on average differences.

There are other limitations in the comparative literature on gender gaps in STEM performance. Analyses of the national sources of variation in gender gaps in educational performance have tended to focus on national economic development and national gender equality indicators in various societal domains—education, economics, politics, and cultural norms (Else-Quest et al., 2010; Guiso, Monte, Sapienza, & Zingales, 2008; Penner, 2008; Riegle-Crumb, 2005). National education system characteristics, such as degree of privatization, standardization, or stratification (branching or tracking), while studied extensively in the context of socioeconomic stratification (Park, 2008; Van de Werfhorst & Mijs, 2010), have not been considered as routinely in studies of gender disparities in math performance, with few exceptions (Ayalon & Livneh, 2013). In addition, very little comparative research has analyzed factors shaping gender differences in math performance across Latin America.

This article addresses these limitations. We investigate determinants of gender disparities in math performance across the performance distribution in Latin American countries, where there is a persistent boys' advantage in math performance in most countries despite a female-favoring gender gap in educational attainment (United Nations Educational Scientific and Cultural Organization [UNESCO], 2018). Using the Third Regional Comparative and Explanatory Study data from 15 Latin American countries, we address two questions: First, does the gender difference in math performance vary across the performance distribution? In particular, is there greater male variability? Second, are gender differences across the math performance distribution associated with national-level factors, including economic development, gender equality regimes, and education system characteristics?

In addressing these questions, this article begins to rectify a significant regional imbalance in empirical work on gender and STEM performance and will illuminate the question of whether there is a need to consider separately inequalities at high and low performance levels. The remainder of this article reviews comparative and Latin America-specific literature in English and Spanish relevant to gender disparities in math performance, introduces the data and methods, presents results, and discusses implications.

## Framework

### Gender Gaps at the Extremes

Previous studies in the United States have shown that gender differences in mean mathematic performance are very small and sometimes favor girls, depending on the sample, measure, and educational stage. For example, a recent meta-analysis of U.S. studies found no significant gender difference in elementary and middle school, but small gender gaps in complex problem solving favoring male students in high school and college (Lindberg, Hyde, Petersen, & Linn, 2010). Similarly, using National Assessment of Educational Progress data from 1990 to 2015, researchers find that male students have a negligible advantage in math in fourth grade and no advantage in eighth grade; a meaningful advantage only emerges in high school (Fahle & Reardon, 2018). However, a new study using the U.S. state accountability test data from third- to eighth-grade students finds that although there is no overall gender achievement gap in math, there are considerable variations across school districts. Math gaps tend to favor male students more in socio-economically advantaged school districts as well as in districts with larger gender differences in adult socioeconomic status (Reardon, Fahle, Kalogrides, Podolsky, & Zárate, 2019). This district variation suggests that gender gaps in math performance may be linked to broader social and structural contexts.

International studies also show considerable ambiguity: a gender gap in math performance persists in some countries, while not in others. Using data from the 2003 Trends in International Mathematics and Science Study (TIMSS, which surveys fourth- and eigth-grade students) and the Programme for International Student Assessment (PISA; which surveys 15-year-old students), Else-Quest et al. (2010) find that although over 60% of the countries show a negligible to small gender difference in math performance, this gender gap varies greatly across countries. The gender effect size, measured by the difference between boys' and girls' means divided by the pooled within-gender standard deviation, varies from −0.42 in Bahrain to 0.40 in Tunisia.

One potentially significant limitation in the existing literature is its focus on mean differences in math performance, rather than differences at the extremes of performance. Attention to the extremes is important, because gender differences at the extreme ends of performance are often more substantial than gender difference at the means (Baye & Monseur, 2016). Moreover, disparities at high performance levels may shape stratification in access to high-level STEM education and careers (Fan et al., 1997; Xie & Shauman, 2003), while disparities at the bottom can shape stratification in grade repetition and dropout (Janosz et al., 1997; Jimerson et al., 2000). For these reasons, an exclusive focus on average differences could elide socially significant disparities in math performance. An important exception to this characterization is Penner (2008), who examines extreme math performance through logistic regression and quantile regression models. Penner's sample of 22 countries, however, are mostly Western developed countries.

Furthermore, there is an ongoing debate about the "greater male variability hypothesis": The notion that independent of mean differences, male students have a greater variance than female students in math ability and therefore are more likely to be at both the top and the bottom of the distribution of math performance (Lindberg et al., 2010). This hypothesis is sometimes used to explain the underrepresentation of women in scientific research fields, given that if women had smaller variability in math performance, they would be underrepresented in the top of the distribution. Thus, a test of the greater male variability hypothesis could provide insights into the origin of the excess of male students at the top levels of math performance and math-intensive careers (Hedges & Friedman, 1993; Lindberg et al., 2010).

Certain evidence is consistent with the greater male variability hypothesis. The variance of male students' math performance is larger compared to female students' in various data sets from the United States and other countries (Feingold, 1992; Hedges & Nowell, 1995; Hyde et al., 2008). A meta-analysis of 242 studies show that the overall variance ratio (VR) between male and female students' variance in math performance is 1.08, indicating a slightly greater male variability (Lindberg et al., 2010). But importantly, there is also contrary evidence that suggests smaller male variability in some national and international data sets. A cross-national study using the PISA data set, for example, shows that in Germany, Lithuania, and the Netherlands, women have greater variability than men in math performance scores (Penner, 2008). Lindberg et al. (2010) analyze large U.S. adolescent data sets covering the past 20 years and find that the male-female VR ranges from 0.88 to 1.34. These findings cast doubt on the universal applicability of the male variability hypothesis and suggest that features of national context may be linked to whether there are gender gaps in performance at the extremes (Hyde & Mertz, 2009; Lindberg et al., 2010; Penner, 2008).

## Comparative Perspectives on Gender Differences in Math Performance

The significant national variation suggests that patterns of gender difference in math performance may be shaped by macro-level structures. What macro structures might be tied to gender disparities in math performance? One classic line of thinking sometimes referred to as the modernization hypothesis implies that gender disparities recede with national economic development, as modern competitive pressures increase and egalitarian values become institutionalized (Baker & LeTendre, 2005; Inglehart & Norris, 2003). However, some studies find little effect of national economic development on gender gaps in math performance (Guiso et al., 2008) or even reveal larger gender gaps in math attitudes in more affluent countries (Charles, Harr, Cech, & Hendley, 2014; Sikora & Pokropek, 2012). Scholars have thus suggested that in countries with existential security and culture favoring individual self-expression, students' instrumental concerns with lucrative careers would decrease, and pursuit of more personally expressive and gendered careers would increase, leading to larger gender gaps in math attitudes and, by possible extension, math performance (Charles et al., 2014).

Beyond theories about the role of economic development in driving gender disparities in educational performance, scholars have turned to what might be called the national gender equity context. Structural factors associated with more and less gender egalitarian societies may shape gender differences in math performance through two mechanisms: by creating incentive structures through promoting female representation in education, the labor market, and politics and by attaching gendered values to different academic subjects and careers through gender norms and stereotypes (Penner, 2008). Using the PISA data and the Global Gender Gap Index (GGGI) developed by the World Economic Forum, Guiso et al. (2008) examined 40 countries and reported a smaller gender gap in math performance in countries with higher overall gender equity. Narrower gender gaps in math were also found in countries with higher gender equality in politics (Else-Quest et al., 2010; Guiso et al., 2008; Penner, 2008; Riegle-Crumb, 2005), school enrollment (Else-Quest et al., 2010), labor participation (Baker & Jones, 1993; Guiso et al., 2008; OECD, 2015), and research jobs (Else-Quest et al., 2010).

On the other hand, some studies have found counterintuitive results that gender equality at the national level has no effect on or even exacerbates gender inequality in math performance. For example, using the same measurement as Guiso et al. (2008), Fryer and Levitt (2010) find no link between the GGGI and the gender gap in math performance. They argue that their different finding is due to the inclusion of countries in the Middle East, where, despite high levels of gender inequality, there is little or no gender gap in math performance. In addition, Riegle-Crumb (2005) found no association between gender equality in the labor force participation rate and gender gaps in math performance. Penner (2008) even found a negative association between gender equality in labor force participation and the gender gap in math performance with a sample of Western developed countries. Penner suggests a possible explanation: in his sample, countries with greater female labor force participation also tend to have higher degrees of occupational gender segregation. When it comes to gendered cultural values, Penner (2008) and Riegle-Crumb both find that gender ideologies concerning the importance of home and children for women at the national level were not associated with gender gaps in math performance.

### National Education System Characteristics

An important limitation in previous research is the scant attention to national education system characteristics that may shape gender disparities (Ayalon & Livneh, 2013). One such feature is the level of standardization, and especially the autonomy of schools on what and how they teach (Ayalon & Livneh, 2013; Park, 2008; Van de Werfhorst & Mijs, 2010). Among the few scholars who have examined the interaction between gender and standardization of education systems, Ayalon and Livneh (2013) find that standardization of curriculum helps reduce advantages of boys over girls in math performance. Tsui (2007) also argues that Chinese students achieve higher gender parity in math performance compared to their U.S. counterparts because of the rigorous and standardized national mathematics curriculum.

A second feature of education systems is stratification, which is sometimes called differentiation, branching, streaming, or tracking. Stratification usually refers to the extent to which education systems have differential curricula and tracks based on students' performance and aspirations at the secondary level (Han, 2016). Sikora and Pokropek (2012) find that higher levels of stratification are associated with a lower chance of expecting a career in computer science or engineering for girls but not for boys. Han (2016) also finds a positive association between the level of stratification and gender gaps in STEM occupational expectations. Current evidence suggests a greater gender gap in STEM aspirations and expectations in more stratified education systems, but implications of stratification for math performance itself are not established.

A third feature that might be significant is scope of privatization. Ceron (2016) found that in Latin American countries, achievement inequality by family background is greater in countries with higher levels of privatization of the education system. Consistent with this insight, Torche (2005) found that inequality increased for cohorts who received education during and after the privatization of education system in Chile. Accordingly, in Chile, the association between schools' aggregate family socioeconomic status and students' test scores is much greater for private-voucher schools than for public schools, which results in pronounced socioeconomic stratification (Mizala & Torche, 2012). The effect of privatization on gender differences remains underexplored.

We have noted limitations in the existing literature: a lack of attention to extreme performance and a dearth of attention toward national characteristics that might shape gender gaps across the distribution. A further point that might be viewed as a limitation is the geographic coverage of existing evidence: very few studies have analyzed factors shaping gender differences in math performance in Latin America in comparative perspective. We next provide a brief overview of the context of gender and education in Latin America.

### Gender and Education in Latin America

Latin America has achieved significant expansions of education coverage, access, and progression in most countries in recent decades, such that by 2012 the region's literacy rate reached an average of 93.3%, compared to 88.9% in 2000 (UNESCO, 2014). However, there is considerable heterogeneity in advances within the region, and there are some equity issues within countries associated with class and location of residence. There are also remaining concerns regarding quality of education, as indicated by performance inequalities in the Third Regional Comparative and Explanatory Study (Tercer Estudio Regional Comparativo y Explicativo, hereafter TERCE). TERCE results show that 61% of third graders and 70% of sixth graders are in reading performance Levels I and II (the two lowest out of four levels), while 71% of third graders and 83% of sixth graders are in the lowest two levels for math performance (UNESCO, 2015, pp. 7–8).

Overall, gender stratification in education in many countries in Latin America encompasses a girls' advantage in general educational attainment but a persistent boys' advantage in STEM performance. A policy paper for the UNESCO global education monitoring report indicates a significant access advantage for women in Latin America and the Caribbean: for every 100 women, 96 men completed primary, 94 completed lower secondary, and 91 completed upper secondary education, while only 83 were attending some form of postsecondary education (UNESCO, 2018). A study of the impact of the 1980s economic crisis on inequality of educational opportunity in four Latin American countries for birth cohorts 1940 to 1975 finds a growing female advantage in educational attainment across cohorts, in line with trends in industrialized countries (Torche, 2010). Another study of educational stratification of adolescents growing up during the 1980s, 1990s, and 2000s in Latin America shows that girls have higher probabilities of school enrollment in all years and countries studied (Marteleto, Gelber, Hubert, & Salinas, 2012). Gender parity indices for 2010 and 2013 show gender parity in access to primary education and indicate a slight advantage for women in secondary education (UNESCO, 2014).

However, evidence also suggests that in the majority of countries in the region, boys show a fairly consistent advantage in math and science. A UNESCO report on Latin America and the Caribbean concludes, "It is clear that girls in the region (with the exception of Cuba and the Dominican Republic) consistently achieve on average lower results in scientific subjects than the male students" (UNESCO, 2014, p. 98). Other studies have highlighted the unevenness in gender patterns across Latin American countries. Analyses of the 2007 TIMSS data show significant gender disparities favoring men in math scores in El Salvador and Colombia, while the 2006 PISA results show similar findings for Argentina, Brazil, Chile, Colombia, and Uruguay (Valverde & Näslund-Hadley, 2011). On the other hand, while seconding UNESCO's observation that women outperform men in reading and men outperform women in math and science, a report by the Inter-American Development Bank contrasts evidence from different studies and finds that in some countries, this difference is not significant (Valverde & Näslund-Hadley, 2011). The 2009 Caribbean Certificate of Secondary Education (Certificado Caribeño de Educación Secundaria) entry exam even shows that women fared better than men in math and science in some English-speaking Caribbean countries.

Gender stratification patterns in education also vary across the performance distribution in Latin American countries. Abadía and Bernal (2017) analyze math, sciences, reading, and global performance by gender for Colombian 11th graders as reported by the 2014 entry exam to secondary education (SABER 11). The authors find a significant gender gap in math and science favoring boys that widens toward the top of the distribution. Another study analyzes the entry exam of the Universidade Federal de Pernambuco, the major university in the Northeast of Brazil. Results indicate a male advantage in all three subjects and greater variation among boys than girls. Quantile regression results further indicate that the math gender gap varies across the distribution, and the male advantage is smaller at the tail (Guimaraes & Sampaio, 2008).

Last, most studies conclude that observable individual, family, and school characteristics only partially explain gender gaps in test scores across the region (Abadía, 2017, p. 15; Abadía & Bernal, 2017, p. 27). Unobserved factors contributing to the gap may include the broader national context of economic development, norms about women's roles in society, and educational system features. However, few studies in Latin American contexts have directly assessed the role of national contexts. Utilizing TERCE data and focusing on country-level characteristics, this study fills in the gap in the previous literature.

## Hypotheses

Drawing on the comparative literature on gender and math performance and on evidence drawn from prior studies in Latin America, we first test whether there is greater male variability in students' math performance among Latin American countries. We then investigate whether national-level factors, including economic development, gender equality regime, and education system characteristics, are associated with gender differences across the distribution.

> *Hypothesis 1:* There is a greater variability in math performance among boys than among girls.

The greater male variability hypothesis states that boys tend to have a greater variance in math performance. This means that the distribution of math performance among boys is flatter compared to that among girls, thus boys are not only more likely to be at the top but also more likely to be at the bottom at the distribution. In addition, boys should be more advantaged at the top and more disadvantaged at the bottom of the distribution; this means that top-performing boys should have better performance compared to top-performing girls, while boys at the bottom should have worse performance compared to girls at the bottom.

Next, to evaluate national characteristics associated with modernization theories about gender gaps, we test the following hypotheses:

> *Hypothesis 2a:* A higher level of national economic development is associated with smaller gender gaps in math performance across the distribution.

Since narrower gender gaps in math have been found in countries with higher gender equality (Guiso et al., 2008; Penner, 2008; Riegle-Crumb, 2005), we also test the following hypothesis:

> *Hypothesis 2b*: A higher level of national gender equality and lower level of gender segregation in occupations are associated with smaller gender gaps in math performance across the distribution.

Finally, drawing on the studies about education system characteristics, especially Ayalon and Livneh (2013) and Tsui (2007) on standardization, Sikora and Pokropek (2012) and Han (2016) on stratification, and Ceron (2016) on privatization, we test the following hypotheses:

> *Hypothesis 3a:* A higher level of standardization of the curriculum is associated with smaller gender gaps in math performance across the distribution.
> *Hypothesis 3b:* A higher level of stratification of the education system is associated with larger gender gaps in math performance across the distribution.

*Hypothesis 3c:* A higher level of privatization of the education system is associated with larger gender gaps in math performance across the distribution.

## Methodology

### The TERCE Data Set

TERCE is a cross-national study of learning and achievement in Latin American countries administered in 2013 by the UNESCO; 15 countries participated in the study (Argentina, Brazil, Chile, Colombia, Costa Rica, Dominican Republic, Ecuador, Guatemala, Honduras, Mexico, Nicaragua, Panama, Paraguay, Peru, and Uruguay). TERCE evaluated third- and sixth-grader performance in reading, science, writing, and mathematics. In this study, we focus on gender differences in math performance among students in the sixth grade. The final combined sample includes 57,476 students in 15 countries.

TERCE presents student performance results in two different ways: the *test score* and the *performance level*. First, TERCE provides five scores called plausible values from a distribution with a regional mean of 700 and standard deviation of 100 points. Second, TERCE classifies students into four achievement levels based on their test scores. The fourth level represents the highest achievement (UNESCO, 2016). In this study, we use all five plausible values of test scores in quantile regression models and performance levels in logistic regression models as the dependent variables.

### Country-Level Variables

Based on findings from previous studies, we include measures of three dimensions of country-level factors into our analysis: economic development, gender equality regimes, and education system characteristics.

First, to measure national level of economic development, we use gross national income (GNI) per capita in 2013 (The World Bank, 2017). Second, to measure national level of gender equality, we follow Guiso et al. (2008) and Fryer and Levitt (2010) by including four indices from the Global Gender Gap Report 2013 (The World Economic Forum, 2013):

> *Index of educational attainment (EDU):* derived from the female-to-male ratios of literacy rate, net primary enrollment rate, net secondary enrollment rate, and gross tertiary enrollment rate.

> *Index of economic participation and opportunity (ECON):* derived from the female-to-male ratios of labor force participation, wage for similar work, total earned income, number of legislators, senior officials and managers, and number of professional and technical workers.

*Index of health and survival (HS):* derived from sex ratios at birth and female-to-male ratios of healthy life expectancy.

*Index of political empowerment (PE):* derived from the female-to-male ratios of current seats in parliament, positions at ministerial level, and number of years of a female head of state over the male value in the past 50 years.

As Penner (2008) points out, gender equality in economic participation might be correlated with gender segregation in the labor market, which may confound the findings. Therefore, we include gender segregation in the analysis. To measure national level of gender segregation in the labor market, we develop two Duncan Segregation Indices (Duncan & Duncan, 1955) based on data provided by the International Labor Organization. *Segregation by skills* is calculated from numbers of male and female employees working at different occupational skill levels (low, medium, and high). *Segregation by industry* is calculated from numbers of male and female employees working in different industries (agriculture; manufacturing; mining, quarrying, and electricity, gas and water supply; construction; market services; public administration or community, social and other services).

Else-Quest and Hamilton (2018) point out that composite gender equality measures may mask important factors and processes within each domain, and individual domain-specific gender equality measures can be utilized to reveal specific mechanisms. Therefore, we also tested for the effects of selected domain-specific gender equality variables that are used to construct the composite measures. In the domain of economic participation, we include female-to-male ratios of labor participation rates, female-to-male ratios in professional and technical jobs, and an indicator of gender wage equality. In the domain of education attainment, we include female-to-male ratios in literacy rate, primary education enrollment rate, secondary education enrollment rate, and tertiary education enrollment rate. In the domain of health and survival, we include female-to-male ratio at birth and female-to-male ratio of life expectancy. In the domain of political empowerment, we include female-to-male ratio in parliament. These variables are also extracted from the Global Gender Gap Report 2013 (The World Economic Forum, 2013). Due to data limitations, several variables are missing in certain countries: the female-to-male ratio in professional and technical positions is missing in Guatemala; the female-to-male ratio in primary enrollment rate is missing in Costa Rica; and the female-to-male ratio in secondary enrollment rate is missing in Brazil, Costa Rica, and Honduras. We use the composite measures in our main analysis to maximize the sample size of countries. We present results from models using the domain-specific indicators in Appendix B and Appendix C and discuss them in the notes.

Finally, to measure national education system characteristics, we develop the following three variables:

## Gender, National Contexts, and Math Performance in Latin America

To measure *standardization (STA)*, we follow the definition and operationalization of Bol and Van de Werfhorst (2013) and construct a scale based on three questions from the principals' questionnaire in TERCE. These questions describe school autonomy in deciding textbooks, course contents, and which courses to offer. We perform a factor analysis on these three variables to create a standardized scale, and then aggregate this scale to the country level with student weights to indicate the level of standardization of the national education system.

To measure the level of stratification of education system, we use *vocational enrollment (VOP)*, or the proportion of students in vocational secondary education (number of students in vocational secondary education divided by total number of students in secondary education, regardless of age). This variable is derived from the World Development Indicators.

To measure *privatization (PRIV)*, we follow Ceron (2016) and use the weighted proportion of students in urban private schools in each country based on the TERCE data.

All national-level variables are standardized when included in the models. Table 1 shows all national-level variable values for the 15 countries included in the sample. For a correlation matrix between all country-level variables, see Appendix A.

### Models

Following Penner (2008), we use both logistic and quantile regression models to examine how gender differences vary across the distribution of math performance. The logistic models examine the likelihood of being at or above various performance levels. The quantile regression models examine the size of gender differences at different percentiles across the distribution. For example, results from logistic models using "being at Level III or above" as the dependent variable report gender differences in the likelihood of being at Level III or above; quantile regression models at the 90th percentile, on the other hand, report the gender differences in test scores between the 90th percentile of boys' and girls' distribution. In addition, ordinary least squares (OLS) regression models and ordered logistic models are also used for comparison. OLS models report difference in conditional means, while quantile regression models provide more information on conditional differences at specified percentiles. Ordered logistic models treat the dependent variable as ordinal and assume proportional odds across different levels, meaning that the relationship between each pair of outcome levels should be the same. Our analysis, however, shows that this is not the case.

Within each country, we apply logistic regression models in the standard form:

$$\ln\left(\frac{p_i}{1-p_i}\right) = \text{Female}_i \beta + \varepsilon_i,$$

Table 1
Standardized Country-Level Factors

| Country | GNI | EDU | ECON | HS | PE | SEGS | SEGI | PRIV | VOP | STA |
|---|---|---|---|---|---|---|---|---|---|---|
| Argentina | 1.478 | 0.368 | −0.649 | 0.797 | 0.934 | 1.840 | 0.729 | 0.678 | 0.204 | 2.351 |
| Brazil | 0.347 | 0.672 | 0.516 | 0.797 | −0.423 | 0.441 | −0.257 | −0.246 | −1.224 | −1.074 |
| Chile | 1.516 | 0.616 | −1.413 | 0.797 | −0.417 | −0.393 | −0.112 | 3.405 | 0.785 | −1.243 |
| Colombia | −0.151 | 0.304 | 1.750 | 0.670 | −0.246 | 1.795 | −0.490 | −0.351 | −0.820 | −1.659 |
| Costa Rica | 0.167 | 0.672 | −0.531 | −0.455 | 1.035 | −1.110 | −0.421 | −0.871 | 0.698 | −0.481 |
| Dominican Republic | −0.242 | −0.751 | 0.844 | −1.375 | −0.628 | 0.587 | 0.334 | −0.680 | −1.105 | 0.009 |
| Ecuador | −0.402 | 0.208 | −0.016 | −0.174 | 1.308 | −0.958 | −1.140 | 0.041 | 0.281 | 0.821 |
| Guatemala | −1.070 | −3.151 | −1.452 | 0.797 | −1.196 | 0.636 | 1.359 | −0.258 | 1.515 | 0.522 |
| Honduras | −1.528 | 0.576 | −0.348 | −0.072 | −0.551 | −0.909 | 2.183 | −0.636 | 1.539 | −0.183 |
| Mexico | 0.590 | −0.039 | −1.319 | 0.797 | 0.395 | −1.585 | −0.356 | −0.714 | 0.254 | −0.200 |
| Nicaragua | −1.506 | 0.640 | −0.077 | −0.174 | 2.336 | −0.764 | 0.731 | −0.021 | −1.454 | 1.433 |
| Panama | 0.962 | 0.336 | 1.510 | −0.302 | −0.126 | 0.401 | −0.292 | −0.346 | −0.042 | 0.557 |
| Paraguay | −0.897 | 0.097 | 0.174 | −0.174 | −0.898 | −0.178 | −1.984 | 0.133 | 0.133 | −0.279 |
| Peru | −0.349 | −0.959 | 0.027 | −2.730 | −0.442 | 0.175 | −0.481 | 0.464 | −1.438 | −0.516 |
| Uruguay | 1.087 | 0.408 | 0.986 | 0.797 | −1.082 | 0.022 | 0.197 | −0.187 | 0.674 | 0.143 |

*Source.* World Bank; World Economic Forum; International Labor Organization; UNESCO (United Nations Educational Scientific and Cultural Organization).

*Note.* GNI = gross national income per capita in 2013; EDU = index of gender equality in education; ECON = index of gender equality in economic participation and opportunities; HS = index of gender equality in health and survival; PE = index of gender equality in political empowerment; SEGS = segregation by skills; SEGI = segregation by industries; PRIV = privatization of education system; VOP = proportion of vocational secondary students in all secondary students; STA = standardization of education system. All variables are standardized within the 15 countries.

where $p_i$ is the probability of student $i$ achieving a certain level or above; Female$_i$ is a dummy variable that equals 1 if student $i$ is a girl, 0 if student $i$ is a boy; $\varepsilon_i$ is the error term. Similarly, we apply the quantile regression models in the standard form:

$$Y_i = X_i \beta + \varepsilon_i,$$

where $Y_i$ is the math score for student $i$, $X_i$ is a vector of independent variables, and $\varepsilon_i$ is the error term. Standard errors were calculated using a Huber-White sandwich estimator adapted for quantile regression.

Next, to measure how national-level factors are associated with gender differences across the distribution of math performance, we estimate logistic and quantile regression models with gender at the individual level and include the cross-level interaction term between gender and national-level indices. The interaction terms estimate the effect of each national-level factor on gender differences at the individual level. To facilitate interpretation, each country-level factor and its interaction with gender is included in a separate model. In addition, we include country fixed effects in all models to control for unobserved heterogeneities across countries. The logistic regression models take the following form:

$$\ln\left(\frac{p_{ij}}{1-p_{ij}}\right) = \text{Female}_i \beta_1 + \text{Female}_i \times \text{Country Variable}_j \beta_2 + \sum_{i=1}^{15} \sigma_j \text{Country}_j + \varepsilon_{ij},$$

where $p_{ij}$ is the probability for student $i$ in country $j$ to achieve a certain level or above, Country Variable$_j$ is the value of the national-level variable of focus in country $j$, and $\sum_{i=1}^{15} \sigma_j \text{Country}_j$ is the country fixed effect. Similarly, the quantile regression models take the following form:

$$Y_i = \text{Female}_i \beta_1 + \text{Female}_i \times \text{Country Variable}_j \beta_2 + \sum_{i=1}^{15} \sigma_j \text{Country}_j + \varepsilon_{ij}$$

where the outcome $Y_i$ is the math score for student $i$, and all the other terms are the same.[1] We describe our findings in the next section.

## Analysis

### Gender Differences in Mean and Variability of Math Performance

Basic descriptive statistics show a gender gap favoring male students in most countries, but the size of this gender gap varies greatly across countries. Table 2 presents these descriptive statistics for each country in the TERCE data set. Column 5 reports the mean differences between male and female students' math scores in each country. In most countries, except for Panama and Chile, there is a significant gender gap favoring male students. The size of this gender difference, however, varies across countries. Among

Table 2
**Descriptive Statistics and Gender Differences by Country**

| Country | N (1) | % Female (2) | Overall M (3) | Overall SD (4) | Mean Difference (Male − Female) (5) | Variance Ratio (Male/Female) (6) | Effect Size (7) | Significance Level in Mean Difference (8) |
|---|---|---|---|---|---|---|---|---|
| Argentina | 3,639 | 0.496 | 720.992 | 85.887 | 7.130 | 1.009 | 0.083 | * |
| Brazil | 2,983 | 0.513 | 721.347 | 85.540 | 13.808 | 0.978 | 0.161 | *** |
| Chile | 5,044 | 0.505 | 807.643 | 103.148 | 4.743 | 1.046 | 0.046 | |
| Colombia | 4,308 | 0.490 | 710.801 | 81.771 | 17.161 | 1.092 | 0.210 | *** |
| Costa Rica | 3,520 | 0.499 | 740.702 | 75.154 | 13.812 | 1.033 | 0.184 | *** |
| Dominican Republic | 3,661 | 0.502 | 624.560 | 49.818 | 6.750 | 1.043 | 0.135 | *** |
| Ecuador | 4,818 | 0.474 | 708.534 | 82.183 | 14.535 | 1.032 | 0.177 | *** |
| Guatemala | 4,056 | 0.481 | 677.378 | 65.641 | 19.889 | 1.091 | 0.303 | *** |
| Honduras | 3,880 | 0.493 | 682.299 | 74.467 | 6.198 | 1.036 | 0.083 | ** |
| Mexico | 3,618 | 0.488 | 780.175 | 101.527 | 13.423 | 1.053 | 0.132 | *** |
| Nicaragua | 3,726 | 0.532 | 648.755 | 55.816 | 14.742 | 1.014 | 0.264 | *** |
| Panama | 3,413 | 0.510 | 658.632 | 74.373 | −3.613 | 1.021 | −0.049 | |
| Paraguay | 3,222 | 0.495 | 654.462 | 81.041 | 8.595 | 1.036 | 0.106 | ** |
| Peru | 4,789 | 0.494 | 718.457 | 103.674 | 22.795 | 0.981 | 0.220 | *** |
| Uruguay | 2,799 | 0.502 | 765.847 | 98.112 | 5.876 | 1.025 | 0.060 | *** |
| Total | 57,476 | 0.497 | 709.764 | 97.375 | 12.100 | 2.565 | 0.121 | *** |

*Note.* For each country, column 1 reports the number of observations; column 2 reports the proportion of the sample that is female; column 3 reports the mean math performance score; column 4 reports the pooled standard deviation of mathematics performance; column 5 reports the gender difference (male minus female) in the mean of mathematics literacy; column 6 reports the gender variance ratio (VR, male variance divided by the female variance) of mathematics literacy; column 7 reports the effects size in gender difference (mean difference divided by pooled standard deviation); column 8 reports the significance level of mean difference.
* $p < .05$. ** $p < .01$. *** $p < .001$.

the countries with a significant gender gap, the mean difference in column 5 ranges from 5.876 (Uruguay) to 22.795 (Peru), and the effect size in column 7 (mean difference divided by standard deviation) ranges from 0.060 (Uruguay) to 0.303 (Guatemala). This variation suggests the importance of national and social contexts when it comes to mean differences.

Results also partly contradict the male variability hypothesis. Column 6 in Table 2 reports the VR, or the male variance divided by the female variance (Hyde & Mertz, 2009). In most countries, male students indeed have greater variance in math performance (VR > 1.0), but the size of the gender difference in variances is small and varies by country, ranging from VR = 1.009 in Argentina to VR = 1.092 in Colombia. Moreover, in Brazil and Peru, female students actually have greater variances than male students (VR < 1.0), which shows that the greater male variability hypothesis does not hold across the board and offers evidence of cross-country differences.

### Gender Representation Across the Distribution

In addition to positing that boys tend to have greater variance in math performance, the greater male variability hypothesis (Hypothesis 1) also implies that boys are more represented at both the top and bottom tails of the distribution. Results from our analysis, however, show contradicting evidence. While boys are indeed better represented at the top of the distribution, we find that girls are more represented at the bottom in many countries.

Table 3 shows results of an ordered logistic model predicting the level of math performance (column 1) and a series of logistic models predicting the logged odds of achieving Level I (column 2), Level II and above (column 3, which is a flipped version of column 2 and is presented for comparison with columns 4 and 5), Level III and above (column 4), and Level IV (column 5). Findings show that in about half of the countries (Argentina, Brazil, Colombia, Guatemala, Honduras, Nicaragua, and Peru), girls are more represented at the bottom level; in other countries, there is no significant gender difference in the odds of being at the bottom level. In Guatemala, for example, the results show that the odds of girls to be at Level I is 1.368 times of the odds of boys to be at the same level (column 2); girls only have 73.1% of the odds of boys to be at or above Level II (column 3), 54.3% of the odds of boys to be at or above Level III (column 4), and 22.8% of the odds of boys to be at Level IV (column 5). We can also roughly interpret odds ratios here as counts: for every boy at Level I, there are 1.368 girls; for every boy at Level II and above, there are 0.731 girls; for every boy at Level III and above, there are 0.543 girls; for every boy at Level IV, there are only 0.228 girls.

Columns 6 to 8 present the $p$ values from adjusted Wald tests that the coefficients at different levels are equivalent.[2] In Guatemala, all three pairs of coefficients are significantly different from each other, showing heterogeneous gender effects at different levels of math performance: boys tend to be

## Table 3
### Estimated Odds Ratios Associated With Being Female From Selected Ordered and Binary Logit Models of Achievement of Specified Levels of Math Performance

| Country | Ordered Logistic (1) | Level I (2) | ≥Level II (3) | ≥Level III (4) | Level IV (5) | ≥Level II/ ≥Level III Test (6) | ≥Levels II/ IV Test (7) | ≥Levels III/ IV Test (8) |
|---|---|---|---|---|---|---|---|---|
| Argentina | 0.833* (0.064) | 1.219* (0.107) | 0.821* (0.072) | 0.857 (0.082) | 0.832 (0.145) | .676 | .938 | .847 |
| Brazil | 0.767*** (0.070) | 1.296* (0.136) | 0.771* (0.081) | 0.748** (0.084) | 0.875 (0.151) | .798 | .435 | .317 |
| Chile | 0.929 (0.065) | 0.942 (0.094) | 1.061 (0.106) | 0.922 (0.070) | 0.812* (0.076) | .148 | .024** | .078 |
| Colombia | 0.678*** (0.076) | 1.409** (0.172) | 0.710** (0.087) | 0.577*** (0.084) | 0.858 (0.249) | .134 | .603 | .194 |
| Costa Rica | 0.783** (0.059) | 1.162 (0.100) | 0.861 (0.074) | 0.680*** (0.065) | 0.735 (0.143) | .024* | .420 | .633 |
| Dominican Republic | 0.794 (0.108) | 1.240 (0.168) | 0.806 (0.109) | 0.304*** (0.095) | 0.121* (0.127) | .001*** | .073 | .383 |
| Ecuador | 0.878 (0.077) | 1.066 (0.097) | 0.938 (0.085) | 0.715** (0.089) | 0.642* (0.122) | .013* | .053 | .547 |
| Guatemala | 0.702*** (0.048) | 1.368*** (0.096) | 0.731*** (0.051) | 0.543*** (0.079) | 0.228*** (0.092) | .038* | .004** | .013* |
| Honduras | 0.784** (0.069) | 1.273** (0.116) | 0.785*** (0.071) | 0.774 (0.111) | 0.790 (0.153) | .912 | .974 | .915 |
| Mexico | 0.827** (0.059) | 1.081 (0.109) | 0.925 (0.093) | 0.791** (0.061) | 0.716** (0.082) | .133 | .067 | .283 |
| Nicaragua | 0.807* (0.074) | 1.209* (0.112) | 0.827* (0.077) | 0.440*** (0.087) | 0.391 (0.205) | .002** | .148 | .803 |
| Panama | 1.134 (0.093) | 0.878 (0.073) | 1.139 (0.094) | 1.077 (0.176) | 0.678 (0.257) | .723 | .171 | .160 |
| Paraguay | 0.929 (0.089) | 1.067 (0.105) | 0.937 (0.092) | 0.842 (0.125) | 0.724 (0.334) | .464 | .576 | .726 |
| Peru | 0.677*** (0.056) | 1.548*** (0.140) | 0.646*** (0.058) | 0.732*** (0.068) | 0.721* (0.113) | .140 | .453 | .906 |
| Uruguay | 0.860 (0.140) | 1.355 (0.298) | 0.738 (0.163) | 0.929 (0.140) | 1.012 (0.190) | .147 | .167 | .555 |

*Note.* For each country, column 1 reports the odds ratio of the female variable of the ordered logistic regression predicting levels of math performance, columns 2–5 report the odds ratio of being female from a logistic regression model estimating the logged odds of being at math performance Level I (column 2), at or above Level II (column 3), at or above Level III (column 4), and at Level IV (column 5). Column 6 reports the *p* value from an adjusted Wald test that the coefficients at or above Level II and at or above Level III are equivalent, column 7 reports the *p* value from an adjusted Wald test that the coefficients at or above Level II and at Level IV are equivalent, and column 8 reports the *p* value from an adjusted Wald test that the coefficients at or above Level III and at Level IV are equivalent. Standard errors in parentheses.

*p < .05. **p < .01. ***p < .001.

better represented at higher levels. This pattern is also seen in other countries and regions: in Costa Rica, Dominican Republic, Ecuador, Guatemala and Nicaragua, boys are better represented at or above Level III compared to Level II; in Chile and Guatemala, boys are better represented at or above Level IV compared to Level II; in Guatemala, boys are better represented at Level IV compared to at or above Level III.

### Gender Gaps Across the Distribution

We next present weighted quantile-quantile plots by country to compare male and female students' math score distributions within selected percentiles (Figure 1). Results show considerable variation across the performance distribution and distinct patterns across countries. The quantile-quantile plot is a plot of the quantiles of one data set against the quantiles of another data set; here, we show the plots of the quantiles of male students' math scores against the quantiles of female students' math scores. The red line is the reference line defined as $y = x$. If the two data sets come from a population with the same distribution, the points should fall approximately along this reference line. The greater the departure from this reference line, the greater the possibility that the two data sets have come from populations with different distributions. The points above the reference line indicate that boys' math scores are higher than girls' math scores at a certain percentile. The greater the departure from the reference line, the larger the gender difference is at the given percentile.

The plots in Figure 1 show distinct patterns. In Paraguay, Uruguay, Panama, and Chile, there seem to be no significant gender differences across the distribution. In Colombia, Ecuador, Dominican Republic, Guatemala, and Honduras, there are larger gender differences in favor of boys at the higher extreme. In Argentina and Brazil, there are larger gender differences at the lower extreme. To further test the significance and magnitude of gender differences across the distribution, we employ quantile regression models.

Table 4 further shows results from quantile regression models at different cutoffs of the math score. These results further confirm that the greater male variability hypothesis does not universally hold. Column 1 in Table 4 reports results from an OLS regression as a reference; by comparing results from the OLS and quantile regression models, we can better examine whether there are heterogeneous gender effects across different percentiles of math performance.

Results in Table 4 show that across selected percentiles in all countries, whenever there is a significant gender gap, it is almost always in favor of boys. It is worth noticing that while the OLS coefficient in Chile shows no significant overall gender difference, the quantile regression results show that girls' scores are 13.517 points lower than boys at the 90th percentile. Even in countries with consistent male advantages across different

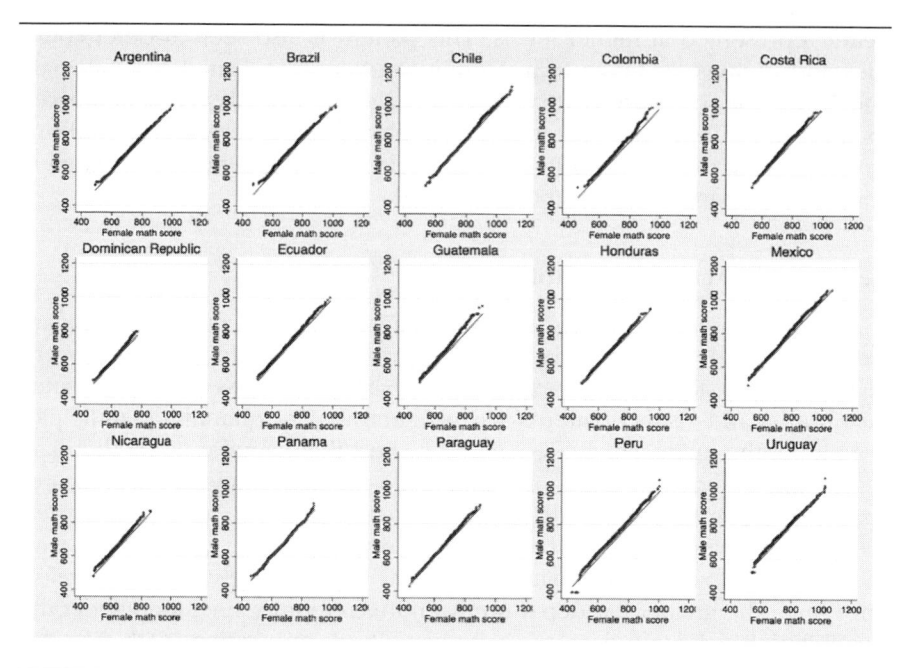

*Figure 1.* **Quantile-quantile plot comparing male and female students' math score distributions by country.**

*Note.* Figure 1 shows the quantile-quantile plot comparing male and female students' math score distributions. The quantile-quantile plot is a plot of the quantiles of one data set against the quantiles of another data set. The red line is the reference line defined as $y = x$. If the two data sets come from a population with the same distribution, the points should fall approximately along this reference line. The greater the departure from this reference line, the greater the possibility that the two data sets have come from populations with different distributions. In this data set, the points above the reference line indicate that boys' math scores are higher than girls' math scores at a certain percentile. The greater the departure from the reference line, the larger the gender difference is at the given percentile.

percentiles, the size of the gender gap varies at different positions of the distribution. For example, in Guatemala, boys score 14.052 points higher than girls at the 5th percentile, but this advantage increases to 17.418 points at the median, 27.518 points at the 90th percentile, and 32.015 points at the 95th percentile; similarly, in Colombia, boys score 13.111 points higher than girls at the 5th percentile, while this advantage increases to 19.813 points at the 75th percentile and 35.433 points at the 95th percentile.

Results in Table 4 extend the various patterns we see in Figure 1, suggesting distinct patterns across countries. For example, in Chile, there is a significant gender gap in favor of male students at a certain percentile, but no

## Table 4
## The Effect of Being Female on Math Score at Different Quantiles

| Countries | OLS (1) | 0.01 (2) | 0.05 (3) | 0.10 (4) | 0.25 (5) | 0.50 (6) | 0.75 (7) | 0.90 (8) | 0.95 (9) | 0.99 (10) |
|---|---|---|---|---|---|---|---|---|---|---|
| Argentina | −7.130* (3.391) | −0.756 (21.464) | −4.002 (9.911) | −5.663 (6.198) | −7.652 (4.557) | −6.945 (4.777) | −7.379 (4.831) | −8.862 (8.777) | −8.259 (9.329) | −12.277 (23.693) |
| Brazil | −13.808*** (3.479) | −19.179 (18.602) | −14.435 (9.275) | −15.338* (7.100) | −14.960** (5.133) | −13.789** (4.480) | −13.995** (4.937) | −13.081 (9.132) | −10.920 (9.027) | −8.158 (31.17) |
| Chile | −4.743 (3.480) | 5.883 (15.538) | 2.624 (6.553) | 0.387 (5.520) | −0.080 (4.376) | −2.463 (4.752) | −9.725 (5.590) | −13.517* (5.392) | −9.258 (8.869) | −8.739 (15.471) |
| Colombia | −17.161*** (2.785) | −12.728 (16.083) | −13.111* (5.430) | −13.982** (5.025) | −11.966** (4.302) | −13.133*** (3.676) | −19.813*** (4.972) | −27.959*** (5.872) | −35.433*** (8.012) | −46.083* (22.991) |
| Costa Rica | −13.812*** (3.638) | −3.846 (15.498) | −9.813 (5.340) | −8.440 (5.677) | −12.596* (5.882) | −15.361*** (4.181) | −16.725*** (4.629) | −12.741* (6.358) | −13.453 (9.052) | −17.239 (20.155) |
| Dominican Republic | −6.750** (2.135) | −10.514 (14.041) | −5.956 (5.235) | −6.439 (4.202) | −4.997 (2.744) | −5.517 (2.868) | −7.553* (3.827) | −8.182 (4.751) | −11.346 (6.693) | −17.897 (13.072) |
| Ecuador | −14.535*** (2.948) | −15.564 (17.68) | −10.484 (6.817) | −13.695*** (5.378) | −10.759*** (3.944) | −14.141*** (4.080) | −16.714*** (3.851) | −18.432** (6.525) | −19.506* (9.058) | −19.667 (20.88) |
| Guatemala | −19.889*** (2.997) | −14.906 (15.403) | −14.052* (5.683) | −15.655* (6.156) | −17.765*** (3.556) | −17.418*** (3.485) | −19.805*** (5.325) | −27.518*** (6.368) | −32.015*** (8.849) | −40.555* (16.52) |
| Honduras | −6.198* (2.714) | 1.957 (16.212) | −6.201 (6.656) | −4.648 (4.817) | −4.146 (3.611) | −4.722 (3.277) | −5.615 (4.731) | −13.048* (6.069) | −14.131 (8.276) | −12.081 (20.434) |
| Mexico | −13.423*** (3.982) | −13.373 (21.077) | −9.106 (8.360) | −7.744 (7.848) | −5.571 (6.245) | −11.476 (6.460) | −19.782** (6.902) | −21.782** (7.871) | −20.630* (9.448) | −14.266 (20.605) |
| Nicaragua | −14.742*** (2.203) | −17.995 (12.392) | −13.918** (5.303) | −14.515*** (4.270) | −13.596*** (2.912) | −13.700*** (3.019) | −16.410*** (3.845) | −15.205*** (4.512) | −15.776** (7.492) | −19.389 (22.675) |
| Panama | 3.613 (3.248) | 5.637 (14.011) | 6.008 (9.169) | 5.005 (5.642) | 2.733 (4.294) | 4.170 (3.926) | 5.598 (4.919) | 0.334 (6.946) | −0.953 (9.203) | −12.1 (24.536) |
| Paraguay | −8.595* (3.691) | −9.946 (18.12) | −3.794 (8.097) | −6.427 (6.153) | −5.833 (4.621) | −7.305 (4.448) | −10.878* (5.023) | −13.704 (7.829) | −11.992 (8.199) | −9.86 (16.538) |
| Peru | −22.795*** (3.241) | −29.865 (23.472) | −23.297*** (6.343) | −25.154*** (5.228) | −26.033*** (4.940) | −22.145*** (4.423) | −20.814*** (5.337) | −18.345** (6.320) | −22.131** (8.114) | −21.56 (20.422) |
| Uruguay | −5.876 (4.470) | −1.826 (23.482) | −5.310 (10.224) | −5.027 (7.352) | −3.303 (5.786) | −4.782 (5.730) | −6.185 (8.385) | −8.786 (7.583) | −14.294 (11.873) | −24.605 (25.904) |

*Note.* OLS = ordinary least squares. For each country, column 1 reports the effect of being female on math score from an OLS regression, and columns 2 to 8 report the effect of being female on math score from quantile regression models estimated at the 1st percentile (column 2), 5th percentile (column 3), 10th percentile (column 4), the 25th percentile (column 5), the 50th percentile (column 6), the 75th percentile (column 7), the 90th percentile (column 8), the 95th percentile (column 9), and the 99th percentile (column 10). Standard errors in parentheses.

*$p < .05$. **$p < .01$. ***$p < .001$.

## Table 5
### Odds Ratios of Cross-Level Interaction Terms Between Gender and National-Level Variables From Logistic Models Predicting Log Odds of Scoring Above Various Levels

|  | Ordered Logistic (1) | At or Above Level II (2) | At or Above Level III (3) | Level IV (4) |
|---|---|---|---|---|
| Female × GNI | 1.068* (0.031) | 1.079* (0.037) | 1.168*** (0.051) | 1.072 (0.076) |
| Female × EDU | 1.046 (0.030) | 1.037 (0.032) | 1.049 (0.051) | 1.221* (0.115) |
| Female × ECON | 0.958 (0.031) | 0.928* (0.036) | 0.926$^+$ (0.041) | 1.070 (0.075) |
| Female × HS | 1.039$^+$ (0.024) | 1.056* (0.027) | 1.022 (0.029) | 1.018 (0.044) |
| Female × PE | 1.070$^+$ (0.038) | 1.106** (0.042) | 1.060 (0.061) | 0.928 (0.094) |
| Female × SEGS | 0.970 (0.026) | 0.935* (0.030) | 0.965 (0.033) | 1.053 (0.058) |
| Female × SEGI | 1.000 (0.027) | 0.973 (0.028) | 1.044 (0.051) | 1.018 (0.099) |
| Female × PRIV | 1.022 (0.027) | 0.991 (0.038) | 1.044 (0.030) | 1.030 (0.037) |
| Female × VOP | 1.062$^+$ (0.036) | 1.097* (0.041) | 1.069 (0.053) | 0.935 (0.075) |
| Female × STA | 1.040$^+$ (0.025) | 1.041 (0.029) | 1.053 (0.035) | 0.964 (0.056) |

*Note.* Number of observations = 57,476. Standard errors in parentheses. Ten national-level variables are included in the models: GNI = gross national income per capita; EDU = index of gender equality in education; ECON = index of gender equality in economic participation and opportunities; HS = index of gender equality in health and survival; PE = index of gender equality in political empowerment; SEGS = segregation by skills; SEGI = segregation by industries; PRIV = privatization of education system; VOP = proportion of vocational secondary students in all secondary students; STA = standardization of education system. All national-level variables are standardized within the 15 countries.
$^+p < .1.$ $*p < .05.$ $**p < .01.$ $***p < .001.$

overall gender difference based on the OLS regression. In such cases, quantile regression helps capture the nuances that the OLS regression alone would miss. In Brazil, the gender difference is larger at the lower extreme of the distribution. Conversely, in Guatemala, Ecuador, Mexico and Colombia, the gender differences are larger at the higher extreme of the distribution. These different patterns of distribution suggest that national context could be crucial in influencing gender gaps in math performance.

### Estimating the Effect of National-Level Characteristics

Table 5 reports the results of logistic regression models predicting the logged odds of being at or above certain levels. To model how country-level factors are associated with gender differences in math performance at different levels, we include country-level variables in the logistic regression models and interact them with the individual-level dummy variable for being female. Country fixed effects are also included to control for unobserved country-level characteristics. The coefficients of the cross-level interaction terms show how national-level variables moderate gender gaps across different levels. A significantly positive coefficient of the interaction term means a higher value of the country-level variable is associated with a smaller gender gap in math performance.

The interaction effects in Table 5 show that, consistent with Hypothesis 2a, GNI per capita is generally positively associated with girls' likelihood of scoring at higher levels relative to boys. Other country-level characteristics, on the other hand, are mainly associated with gender differences in the likelihood of being at or above Level II (column 2). For example, partly consistent with Hypothesis 2b, in countries with higher gender equality in health and survival and political empowerment, the gender difference in the probability of scoring at or above Level II is smaller. Surprisingly, in countries with higher equality in economic participation and opportunities, this gender difference is larger. This is consistent with some of the previous studies (Penner, 2008). Penner suggests that when countries provide female students with more opportunities for economic participation, it is plausible that these opportunities are more likely to be in female-dominated nontechnical sectors. To test this, we also included two gender segregation indices into the model. Results show that a higher level of gender segregation in skills is associated with a lower likelihood for girls to be at or above Level II, which is partly consistent with Penner's assumption. On the other hand, the level of gender segregation in industries does not matter much for gender differences at any level.[3]

When it comes to education system characteristics, only the proportion of vocational secondary students (VOP) is found positively associated with girls' likelihood of scoring at or above Level II, which is inconsistent with Hypothesis 3b.

Quantile regression models present similar patterns. Table 6 reports results from quantile regression models of math scores estimated at different percentiles across the distribution. Similar to Table 5, the interaction terms show how national-level variables affect gender differences at selected percentiles. Results indicate that, partly consistent with Hypothesis 2a, GNI per capita is associated with decreases in gender gaps in general (column 1) and at the lower end of the distribution (columns 3 to 6). The effects of national gender equality regimes, on the other hand, vary across the distribution. Female representation in education is associated with reduced gender gaps in math performance at the lower middle (columns 5 and 6) and the top (columns 9 and 10) of the distribution; gender equality in economic participation is associated with reductions in the gender gap only at the 75th percentile (column 7); national gender equality in health and survival is associated with reductions in the gender gap in general but exhibits no particular effect at each percentile; national gender equality in political empowerment exhibits no significant effect. Furthermore, gender segregation in

Table 6
**Coefficients of Cross-Level Interaction Terms Between Gender and National Variables From Regression Models Predicting Math Scores**

| | OLS (1) | 0.01 (2) | 0.05 (3) | 0.10 (4) | 0.25 (5) | 0.50 (6) | 0.75 (7) | 0.90 (8) | 0.95 (9) | 0.99 (10) |
|---|---|---|---|---|---|---|---|---|---|---|
| Female × GNI | 3.198*** (0.856) | 4.768 (3.225) | 3.325* (1.427) | 3.702** (1.350) | 3.259** (1.042) | 2.883* (1.139) | 2.323 (1.658) | 2.774 (1.752) | 4.008[+] (2.211) | 4.478 (3.971) |
| Female × EDU | 3.224*** (0.807) | 2.844 (3.094) | 2.16 (1.519) | 2.685[+] (1.523) | 3.069*(1.037) | 2.532* (1.002) | 2.104 (1.400) | 3.165[+] (1.717) | 4.685* (2.222) | 6.487* (3.212) |
| Female × ECON | 1.102 (0.832) | −0.789 (3.512) | 0.553 (1.784) | 0.396 (1.306) | 1.245 (1.191) | 1.962[+] (1.111) | 2.544* (1.264) | 2.183 (1.610) | 0.528 (2.363) | −2.434 (4.385) |
| Female × HS | 1.861[+] (0.904) | 3.497 (2.686) | 2.139 (1.321) | 2.466[+] (1.364) | 2.448[+] (1.299) | 1.477 (1.189) | −0.168 (1.258) | −1.534 (1.358) | −0.655 (2.163) | −0.299 (3.833) |
| Female × PE | −0.73 (0.883) | −1.893 (2.725) | −1.019 (2.182) | −0.724 (1.601) | −0.899 (1.019) | −1.334 (0.994) | −1.359 (1.204) | 0.37 (1.514) | 1.333 (2.091) | 1.841 (4.458) |
| Female × SEGS | −0.145 (0.802) | −0.573 (4.127) | −0.023 (1.723) | −0.65 (1.360) | −0.696 (1.395) | 0.563 (1.172) | 0.913 (1.153) | −0.469 (1.665) | −1.736 (2.496) | −5.465 (4.021) |
| Female × EGI | 0.665 (0.825) | 1.383 (5.432) | −0.099 (2.105) | 0.383 (1.754) | 0.108 (1.037) | 0.612 (0.957) | 1.148 (1.393) | −0.05 (2.156) | −0.397 (2.198) | −1.728 (3.521) |
| Female × PRIV | 1.256[+] (0.723) | 2.538 (2.226) | 1.99 (1.352) | 1.611 (1.059) | 1.163 (0.901) | 1.177 (0.929) | 0.404 (1.183) | 0.367 (1.358) | 2.013 (2.182) | 3.134 (3.121) |
| Female × VOP | 2.051* (0.877) | 5.169* (2.594) | 2.508[+] (1.496) | 2.726* (1.375) | 2.221* (1.027) | 1.381 (1.140) | 0.876 (1.226) | −0.655 (1.733) | −0.112 (2.360) | 0.738 (3.565) |
| Female × STA | 0.916 (0.769) | −0.05 (4.250) | 0.479 (1.881) | 0.331 (1.544) | 0.097 (1.142) | 0.341 (1.092) | 1.272 (1.113) | 1.786 (1.747) | 2.335 (2.323) | 1.252 (4.003) |

*Note.* OLS = ordinary least squares. Number of observations = 57,476. Standard errors in parentheses. Column 1 reports the OLS regression results; each of columns 2 to 8 reports the results from a quantile regression model at a different percentile. Female is a dummy variable for being female. Ten national-level variables are included in the models: GNI = gross national income per capita; EDU = index of gender equality in education; ECON = index of gender equality in economic participation and opportunities; HS = index of gender equality in health and survival; PE = index of gender equality in political empowerment; SEGS = segregation by skills; SEGI = segregation by industries; PRIV = privatization of education system; VOP = proportion of vocational secondary students in all secondary students; STA = standardization of education system. All national-level variables are standardized within the 15 countries.
[+] $p < .1$. * $p < .05$. ** $p < .01$. *** $p < .001$.

skills and industries does not exhibit a significant effect on the gender difference in general or across the percentiles.[4]

When it comes to the education system characteristics, no significant effect emerges for the level of standardization or privatization of education systems. However, a higher proportion of vocational students is associated with a smaller gender gap at the lower to middle percentiles but not the top end of the distribution; this finding is inconsistent with Hypothesis 3b. We will discuss implications in the next section.

## Discussion and Conclusion

Using cross-national data from Latin American countries, this article examines whether the greater male variability hypothesis holds for math performance across different Latin American countries, and whether national-level factors are correlated with gender gaps across the performance distribution at the micro level. Three findings are particularly important. First, the greater male variability hypothesis does not hold across the board. Second, a higher level of stratification of the education system is associated with a smaller gender gap in math performance at the lower end, which contradicts our hypotheses. Third, although many national-level factors are associated with gender differences from the lower end to the middle of the distribution, the same is not true at the top. Below, we discuss the implications of each of these findings.

Regarding the first finding, in some Latin American countries, girls are more likely to fall in the bottom of the distribution. This finding shows that analysis of mean differences alone obfuscates critical nuances in the gender gap across the distribution. Furthermore, patterns of gender differences in mean and variance vary across Latin American countries. Although in general, there is a significant gender gap favoring male students, the size of the gender effect varies greatly across countries. Similarly, although male students in most countries tend to have a greater variability in math performance, in Brazil and Peru, female students have greater variances than male students. We further show that gender effects are not necessarily more pronounced at the extremes of the performance distribution: for example, in Costa Rica, the gender differences are actually smaller at both extremes of the distribution. Contrary to the greater male variability hypothesis, these complex findings show that there is not always greater representation of boys at both ends of the distribution.

Our findings do confirm a persistent gender gap favoring boys among the top performers in several countries. However, they also point out a problem that has not attracted as much attention: the vulnerability of girls at the bottom level. In about half of the countries, girls are more likely to fall into the bottom level of performance; even at the bottom percentiles of the distribution, boys still have a performance advantage in most countries. This finding is surprising. Low-performance is often associated with elevated risk of subsequent grade repetition and school dropout (Janosz et al., 1997; Jimerson et al., 2000), yet patterns of educational attainment across many Latin American countries are female-favorable. One possible explanation for this apparent contradiction points to expectations about girls' performance: if it is culturally assumed that girls "are not good at math," then the association between low performance in math and dropout for girls might be weaker than for boys, more so than in other regions. This speculation points to the need for further research on gender differences in the consequences of low performance.

Regarding the second main finding, our analysis shows that the association between education system characteristics and the gender gap in math performance is partly inconsistent with our hypotheses. We found no significant association between standardization or privatization and the gender gap. On the other hand, a higher proportion of vocational students is associated with a smaller gender gap at the lower performance range. One explanation of the stratification effect is that girls may be more likely to choose academic schools rather than vocational tracks than boys; as a quasi-experimental study using Finnish school data shows, in a comprehensive system where students are tracked into vocational and academic schools at age 15 to 16 years, girls are more likely to choose the academic track than boys (Pekkarinen, 2008). Considering that the TERCE data in our study focus on sixth-grade students in primary education, while the stratification measure is derived from secondary education data in each country, it is possible that girls in countries with a high proportion of vocational secondary students feel more motivated to achieve higher scores in order to successfully enter the academic track. Further research on stratification of education systems is needed to understand the actual mechanisms that affect gender gaps in math. More broadly, given reports that educational segregation has increased in Latin America during the past two decades, with lower income students concentrating in often underresourced public schools (Arcidiacono et al., 2014), attention to measures of system stratification may be particularly important to monitor in the future.

Finally, results show that country-level factors are more consistently linked to gender gaps at the low- and middle-parts of the performance distribution and less so among the top performers. For example, we find that higher GNI per capita is associated with smaller gender gaps in math performance at the lower to middle percentiles of the distribution. This finding is partly consistent with modernization theory, which predicts smaller gender differences in more developed countries. However, higher GNI per capita is not associated with a smaller gender gap at the top of the distribution. Similarly, findings also indicate that a higher level of stratification of the education system is associated with a smaller gender gap at the lower to middle percentiles, but not the top end of the distribution.

*Gender, National Contexts, and Math Performance in Latin America*

In fact, among country-level factors, only national context of female representation in education is associated with reductions of the gender gap in performance at the top of the distribution. This finding suggests that a female-favorable national education context may be important for creating an incentive structure for top-performing girls. If we consider that a national context of gender equality might be more directly linked to cultural phenomena than measures of economic development, the fact that it is the only statistically significant national-level factor in our models might be an indicator of how relevant the cultural dimension of gender inequality in education is in Latin America. This interpretation is consistent with previous findings that only about half of the gender performance gap can be explained by observable individual, family, and school characteristics in Latin American countries, and the larger social-cultural context may play a role in affecting gender differences (Abadía, 2017; Abadía & Bernal, 2017). Considering that students from the upper tail of the math performance distribution are more likely to enter STEM fields (Fan et al., 1997), more work is needed to illuminate sources of the gender gap among top performers, as one part of the process that generates female underrepresentation in STEM careers.

There are two caveats to these analyses. First, this study focuses on selected macro-level measures and does not fully explore the complexity in national-level contexts. Although we use country fixed effects to control for country-level characteristics, additional national-level forces such as migration and urbanization and within-country variation across socioeconomic and sociocultural groups may also be important for understanding patterns of gender difference. Second, with cross-sectional data, we are only able to investigate associations and cannot make causal claims. Future studies should consider using longitudinal data sets with lagged outcome variables to further identify causal influences.

Despite these limitations, this study provides important insights into the problem of gender disparity in STEM education in Latin America. It also provides implications for future policies and initiatives. First, while it is important to study gender differences at the mean and the top, it is equally crucial to identify and provide assistance to disadvantaged girls at the bottom of the performance distribution. Second, a greater share of students in vocational education does not necessarily result in a larger gender gap in STEM performance, as has been suggested elsewhere. Finally, while national gender equality indicators in various domains may reflect structural opportunities for women, the measure most closely tied to gender parity among the highest performers is female representation in education. The policy implications of this relationship are complicated in a region where girls' primary and secondary enrollment outstrip boys' in many countries.

## Appendix A
### Correlation Matrix Between Country-Level Variables

|      | GNI    | EDU    | ECON   | HS     | PE     | SEGS   | SEGI   | PRIV   | VOP   | STA   |
|------|--------|--------|--------|--------|--------|--------|--------|--------|-------|-------|
| GNI  | 1.000  |        |        |        |        |        |        |        |       |       |
| EDU  | 0.314  | 1.000  |        |        |        |        |        |        |       |       |
| ECON | 0.022  | 0.290  | 1.000  |        |        |        |        |        |       |       |
| HS   | 0.326  | 0.168  | −0.209 | 1.000  |        |        |        |        |       |       |
| PE   | −0.076 | 0.425  | −0.140 | −0.028 | 1.000  |        |        |        |       |       |
| SEGS | 0.250  | −0.222 | 0.427  | 0.105  | −0.299 | 1.000  |        |        |       |       |
| SEGI | −0.216 | −0.227 | −0.265 | 0.165  | −0.045 | 0.085  | 1.000  |        |       |       |
| PRIV | 0.428  | 0.128  | −0.363 | 0.145  | −0.053 | 0.082  | −0.103 | 1.000  |       |       |
| VOP  | 0.065  | −0.166 | −0.499 | 0.459  | −0.271 | −0.260 | 0.331  | 0.089  | 1.000 |       |
| STA  | −0.026 | −0.096 | −0.180 | 0.028  | 0.494  | 0.071  | 0.298  | −0.118 | 0.084 | 1.000 |

*Note.* GNI = gross national income per capita; EDU = index of gender equality in education; ECON = index of gender equality in economic participation and opportunities; HS = index of gender equality in health and survival; PE = index of gender equality in political empowerment; SEGS = segregation by skills; SEGI = segregation by industries; PRIV = privatization of education system; VOP = proportion of vocational secondary students in all secondary students; STA = standardization of education system.

## Appendix B
### Odds Ratios of Cross-Level Interaction Terms Between Gender and National-Level Domain-Specific Gender Equality Variables From Logistic Models Predicting Log Odds of Scoring Above Various Levels

|  | Ordered Logistic (1) | At or Above Level II (2) | At or Above Level III (3) | Level IV (4) |
|---|---|---|---|---|
| Female × labor | 0.960 (0.031) | 0.918* (0.034) | 0.969 (0.039) | 1.060 (0.064) |
| Female × wage | 1.015 (0.033) | 1.038 (0.042) | 0.945 (0.043) | 0.894+ (0.060) |
| Female × professional | 0.996 (0.032) | 0.963 (0.036) | 0.968 (0.040) | 1.082 (0.073) |
| Female × literacy | 1.037 (0.028) | 1.030 (0.030) | 1.011 (0.037) | 1.107+ (0.068) |
| Female × primary | 1.026 (0.037) | 1.037 (0.042) | 1.051 (0.062) | 0.997 (0.105) |
| Female × secondary | 1.038 (0.033) | 1.018 (0.035) | 1.031 (0.061) | 1.241+ (0.138) |
| Female × tertiary | 1.020 (0.040) | 0.973 (0.043) | 1.016 (0.051) | 1.100 (0.087) |
| Female × birth ratio | 1.069+ (0.037) | 1.073+ (0.045) | 1.127* (0.053) | 1.000 (0.065) |
| Female × life expectancy | 1.017 (0.028) | 1.019 (0.031) | 1.001 (0.036) | 1.058 (0.059) |
| Female × parliament | 1.042 (0.034) | 1.082* (0.041) | 1.048 (0.046) | 0.936 (0.065) |

*Note.* Total number of observations = 57,476. Standard errors in parentheses. Domain-specific gender equality variables are included in the models, and all national variables are standardized within the 15 countries. Labor = female-to-male ratio in labor participation rate; wage = gender wage equality score; professional = female-to-male ratio in share of professional and technical workers; literacy = female-to-male ratio in literacy rate; primary = female-to-male ratio in primary education enrollment rate; secondary = female-to-male ratio in secondary education enrollment rate; tertiary = female-to-male ratio in tertiary education enrollment rate; birth ratio = female-to-male ratio at birth; life expectancy = female-to-male ratio in life expectancy; parliament = female-to-male ratio in parliament. The variable professional is missing in Guatemala; primary is missing in Costa Rica; secondary is missing in Brazil, Costa Rica, and Honduras. In models involving these three variables, countries with missing values are dropped from the sample.
+$p < .1$. *$p < .05$. **$p < .01$. ***$p < .001$.

## Appendix C
### Coefficients of Cross-Level Interaction Terms Between Gender and National-Level Domain-Specific Gender Equality Variables From Quantile Regression Models Predicting Math Scores

|  | OLS (1) | 0.01 (2) | 0.05 (3) | 0.10 (4) | 0.25 (5) | 0.50 (6) | 0.75 (7) | 0.90 (8) | 0.95 (9) | 0.99 (10) |
|---|---|---|---|---|---|---|---|---|---|---|
| Female × labor | −1.697* (0.864) | −3.593 (3.926) | −2.167 (1.468) | −2.583+ (1.569) | −2.367* (1.026) | −1.589 (1.024) | −0.713 (1.456) | 0.362 (1.989) | −0.234 (2.760) | −0.394 (4.100) |
| Female × wage | 0.503 (0.892) | −0.5 (2.618) | 0.871 (1.552) | 0.75 (1.552) | 0.795 (1.216) | 1.001 (0.988) | 2.036 (1.456) | 1.613 (2.224) | 0.576 (2.934) | −1.17 (3.416) |
| Female × professional | 2.694** (0.856) | 0.635 (3.903) | 1.809 (2.244) | 1.811 (1.976) | 2.716+ (1.428) | 3.110** (1.072) | 3.420** (1.313) | 1.933 (1.722) | 1.351 (2.232) | −1.871 (3.983) |
| Female × literacy | 4.266*** (0.771) | 3.893 (2.877) | 3.556* (1.550) | 4.117* (1.624) | 4.587*** (1.082) | 3.950*** (0.887) | 3.453** (1.190) | 3.758* (1.553) | 5.009* (1.970) | 4.759 (3.308) |
| Female × primary | −0.771 (0.860) | −0.667 (4.630) | −1.221 (1.594) | −0.794 (1.384) | −0.84 (1.097) | −1.041 (1.119) | −0.996 (1.232) | −1.056 (1.613) | −0.085 (2.512) | 2.586 (3.688) |
| Female × secondary | 4.219*** (0.909) | 2.277 (4.114) | 2.445 (1.969) | 2.907 (2.036) | 3.647** (1.211) | 3.669*** (1.064) | 3.823* (1.301) | 5.002** (1.799) | 5.832* (2.506) | 5.031 (4.389) |
| Female × tertiary | 4.511*** (0.837) | 4.038 (4.476) | 3.877* (1.945) | 3.773* (1.677) | 3.850*** (1.057) | 4.220*** (1.084) | 5.359*** (1.202) | 5.710*** (1.716) | 6.255** (2.177) | 3.622 (3.808) |
| Female × birth ratio | 3.157*** (0.823) | 3.046 (3.309) | 3.028+ (1.551) | 2.945 (1.450) | 2.814* (1.224) | 2.725** (1.035) | 2.964** (1.107) | 3.676* (1.735) | 5.199** (1.909) | 5.784+ (3.392) |
| Female × life expectancy | 1.509 (0.963) | 2.766 (2.762) | 1.593 (1.462) | 1.721 (1.438) | 1.812 (1.450) | 1.282 (1.238) | −0.124 (1.435) | −1.71 (1.494) | −1.626 (2.407) | −1.974 (3.859) |

*Note.* Total number of observations = 57,476. Standard errors in parentheses. Domain-specific gender equality variables are included in the models, and all national variables are standardized within the 15 countries. Labor = female-to-male ratio in labor participation rate; wage = gender wage equality score; professional = female-to-male ratio in share of professional and technical workers; literacy = female-to-male ratio in literacy rate; primary = female-to-male ratio in primary education enrollment rate; secondary = female-to-male ratio in secondary education enrollment rate; tertiary = female-to-male ratio in tertiary education enrollment rate; birth ratio = female-to-male ratio at birth; life expectancy = female-to-male ratio in life expectancy; parliament = female-to-male ratio in parliament. The variable professional is missing in Guatemala; primary is missing in Costa Rica; secondary is missing in Costa Rica, and Honduras. In models involving these three variables, countries with missing values are dropped from the sample.

*Gender, National Contexts, and Math Performance in Latin America*

ORCID ID

Emily Hannum    https://orcid.org/0000-0003-2011-9984

Notes

The authors gratefully acknowledge support from the University of Pennsylvania Provost's Global Engagement Fund to the project, from the Judith and William Bollinger Graduate Fellowship to Ran Liu, and from the Advanced Human Capital Program of the National Commission for Scientific and Technological Research (CONICYT) of Chile (Folio. 72150110) to Andrea Alvarado-Urbina. This article has benefited from feedback received at presentations at the University of Pennsylvania Education and Inequality Works-in-Progress Workshop and the Comparative and International Education Society meeting (March 2018, Mexico City).

[1] Using dummy variables for fixed effects in nonlinear models with maximum likelihood estimation can produce bias when the number of clusters is large and number of cases within each cluster is small; however, when the number of clusters is small and the number of cases within each cluster is large, the bias will be minimized. Please refer to Allison (2009) for details.

[2] The test is done using a general ordered logistic (gologit) model, which helps to relax the proportional odds assumption of the conventional ordered logistic model. After running the general ordered logistic model, we test whether coefficients across different levels are equivalent. For details, please refer to Williams (2005).

[3] We further tested the effects of selected domain-specific gender equality measures using the same set of logistic regression models. Results are presented in Appendix B and are largely consistent with results in Table 5. The effect of female representation in education at the top level is reflected in the effects of female-to-male ratio in literacy rate and female-to-male ratio in secondary education enrollment rate; although the effects of these two specific indicators are only marginally significant, the magnitudes of effects are similar to that of the composite measure. Furthermore, the gendering effect of gender equality in economic participation is mainly reflected in the effect of female-to-male ratio in labor participation rates. The effect of gender equality in health and survival is reflected in the effect of female-to-male sex ratio at birth. The effect of gender equality in political empowerment is reflected in the effect of female-to-male ratio in the parliament.

[4] Using the same set of quantile regression models, we further tested the effects of selected domain-specific gender equality measures in education attainment, economic participation, and health and survival. Results are presented in Appendix C and are largely consistent with results in Table 6. The effect of female representation in education is reflected in the effects of female-to-male ratio in literacy rate and female-to-male ratio in secondary education enrollment rate. The effect of gender equality in economic participation is reflected in the effects of female-to-male ratio in labor participation rate and female-to-male ratio in professional and technical positions. The effect of gender equality in health and survival is reflected in the effect of female-to-male sex ratio at birth.

References

Abadía, L. K. (2017). Gender score gaps of Colombian students in the PISA test. *Vniversitas Economica, 17*(8), 1–24.
Abadía, L. K., & Bernal, G. (2017). A widening gap? A gender-based analysis of performance on the Colombian high school exit examination. *Revista de Economia Del Rosario, 20*(1), 5–31. doi:10.12804/revistas.urosario.edu.co/economia/a.6144
Allison, P. D. (2009). *Fixed effects regression models*. Thousand Oaks, CA: Sage.
Arcidiacono, M., Cruces, G., Gasparini, L., Jaume, D., Serio, M., & Vazquez, E. (2014). *La segregacion escolar publico-privada en America LATINA* [Public-private

school segregation in Latin America] (No. 195). Santiago, Chile: Publicación de las Naciones Unidas.

Ayalon, H., & Livneh, I. (2013). Educational standardization and gender differences in mathematics achievement: A comparative study. *Social Science Research*, *42*, 432–445. doi:10.1016/j.ssresearch.2012.10.001

Baker, D. P., & LeTendre, G. K. (2005). *National differences, global similarities: World culture and the future of schooling*. Palo Alto, CA: Stanford University Press.

Baker, D. P., & Jones, D. P. (1993). Creating gender equality: Cross-national gender stratification and mathematical performance. *Sociology of Education*, *66*, 91–103. doi:10.2307/2112795

Baye, A., & Monseur, C. (2016). Gender differences in variability and extreme scores in an international context. *Large-Scale Assessments in Education*, *4*(1). doi:10.1186/s40536-015-0015-x

Bol, T., & Van de Werfhorst, H. G. (2013). *The measurement of tracking, vocational orientation, and standardization of educational systems: A comparative approach* (GINI Discussion Paper No. 81, p. 63). Retrieved from http://archive.uva-aias.net/uploaded_files/publications/81-3-3-1.pdf

Ceron, F. I. (2016). Beyond school effects: The impact of privatization and standardization of school systems on achievement inequality in Latin America. *SSRN Electronic Journal*. doi: 10.2139/ssrn.2885223

Charles, M., Harr, B., Cech, E., & Hendley, A. (2014). Who likes math where? Gender differences in eighth-graders' attitudes around the world. *International Studies in Sociology of Education*, *24*, 85–112. doi:10.1080/09620214.2014.895140

Duncan, O. D., & Duncan, B. (1955). A methodological analysis of segregation indexes. *American Sociological Review*, *20*, 210–217. doi:10.2307/2088328

Else-Quest, N. M., & Hamilton, V. (2018). Measurement and analysis of nation-level gender equity in the psychology of women. In C. B Travis, J. W. White, A. Rutherford, W. S. Williams, S. L. Cook, & K. F. Wyche (Eds.), *APA handbooks in psychology series. APA handbook of the psychology of women: Perspectives on women's private and public lives* (pp. 545–563). Washington, DC: American Psychological Association. doi: 10.1037/0000060-029

Else-Quest, N. M., Hyde, J. S., & Linn, M. C. (2010). Cross-national patterns of gender differences in mathematics: A meta-analysis. *Psychological Bulletin*, *136*, 103–127. doi:10.1037/a0018053

Fahle, E. M., & Reardon, S. F. (2018). Education. In Stanford Center on Poverty and Inequality (Series Ed.), *State of the Union: The poverty and inequality report* (pp. 9–12). Palo Alto, CA: Stanford Center on Poverty and Inequality. Retrieved from https://inequality.stanford.edu/sites/default/files/Pathways_SOTU_2018.pdf

Fan, X., Chen, M., Matsumoto, A. R., & Fan, X. (1997). Gender differences in mathematics achievement: Findings from the "National Education Longitudinal Study of 1988." *Journal of Experimental Education*, *65*, 229–242.

Feingold, A. (1992). Sex differences in variability in intellectual abilities: A new look at an old controversy. *Review of Educational Research*, *62*, 61–84. doi:10.2307/1170716

Fryer, R. G., & Levitt, S. D. (2010). An empirical analysis of the gender gap in Mathematics. *American Economic Journal: Applied Economics*, *2*, 210–240.

Guimaraes, J., & Sampaio, B. (2008). Mind the gap: Evidence from gender differences in scores in Brazil. *Anais Do XXXVI Encontro Nacional de Economia*. Retrieved from https://ideas.repec.org/p/anp/en2008/200807211527140.html

Guiso, L., Monte, F., Sapienza, P., & Zingales, L. (2008). Culture, gender, and math. *Science*, *320*, 1164–1165.

Han, S. W. (2016). National education systems and gender gaps in STEM occupational expectations. *International Journal of Educational Development, 49*, 175–187. doi:10.1016/j.ijedudev.2016.03.004

Hedges, L. V., & Friedman, L. (1993). Gender differences in variability in intellectual abilities: A reanalysis of Feingold's results. *Review of Educational Research, 63*, 94–105. doi:10.2307/1170561

Hedges, L. V., & Nowell, A. (1995). Sex differences in mental test scores, variability, and numbers of high-scoring individuals. *Science, 269*, 41–45.

Hyde, J. S., Lindberg, S. M., Linn, M. C., Ellis, A. B., & Williams, C. C. (2008). Gender similarities characterize math performance. *Science, 321*, 494–495. doi:10.1126/science.1160364

Hyde, J. S., & Mertz, J. E. (2009). Gender, culture, and mathematics performance. *Proceedings of the National Academy of Sciences of the U S A, 106*, 8801–8807. doi:10.1073/pnas.0901265106

Inglehart, R., & Norris, P. (2003). *Rising tide: Gender equality and cultural change around the world*. Cambridge, England: Cambridge University Press.

Janosz, M., LeBlanc, M., Boulerice, B., & Tremblay, R. E. (1997). Disentangling the weight of school dropout predictors: A test on two longitudinal samples. *Journal of Youth and Adolescence, 26*, 733–762. doi:10.1023/A:1022300826371

Jimerson, S., Egeland, B., Sroufe, L. A., & Carlson, B. (2000). A prospective longitudinal study of high school dropouts examining multiple predictors across development. *Journal of School Psychology, 38*, 525–549. doi:10.1016/S0022-4405(00)00051-0

Lindberg, S. M., Hyde, J. S., Petersen, J. L., & Linn, M. C. (2010). New trends in gender and mathematics performance: A meta-analysis. *Psychological Bulletin, 136*, 1123–1135. doi:10.1037/a0021276

Marteleto, L., Gelber, D., Hubert, C., & Salinas, V. (2012). Educational inequalities among Latin American adolescents: Continuities and changes over the 1980s, 1990s and 2000s. *Research in Social Stratification and Mobility, 30*, 352–375.

Mizala, A., & Torche, F. (2012). Bringing the schools back in: the stratification of educational achievement in the Chilean voucher system. *International Journal of Educational Development, 32*, 132–144.

OECD. (2015). *The ABC of gender equality in education: Aptitude, behavior, confidence*. Retrieved from https://www.oecd.org/pisa/keyfindings/pisa-2012-results-gender-eng.pdf

Park, H. (2008). The varied educational effects of parent-child communication: A comparative study of fourteen countries. *Comparative Education Review, 52*, 219–243. doi:10.1086/528763

Pekkarinen, T. (2008). Gender differences in educational attainment: Evidence on the role of tracking from a Finnish quasi-experiment. *Scandinavian Journal of Economics, 110*, 807–825. doi:10.1111/j.1467-9442.2008.00562.x

Penner, A. M. (2008). Gender differences in extreme mathematical achievement: An international perspective on biological and social factors. *American Journal of Sociology, 114*(Suppl.), S138–S170.

Reardon, S. F., Fahle, E. M., Kalogrides, D., Podolsky, A., & Zárate, R. C. (2019). Gender achievement gaps in U.S. school districts. *American Educational Research Journal, forthcoming*. Advance online publication. doi:10.3102/0002831219843824

Riegle-Crumb, C. (2005). The cross-national context of the gender gap in math and science. In L. V. Hedges & B. Schneider (Eds.), *The social organization of schooling* (pp. 227–243). New York, NY: Russell Sage Foundation.

Sikora, J., & Pokropek, A. (2012). Gender segregation of adolescent science career plans in 50 countries. *Science Education, 96,* 234–264. doi:10.1002/sce.20479

Torche, F. (2005). Privatization reform and inequality of educational opportunity: The case of Chile. *Sociology of Education, 78,* 316–343. doi:10.1177/003804070507800403

Torche, F. (2010). Economic crisis and inequality of educational opportunity in Latin America. *Sociology of Education, 83,* 85–110.

Tsui, M. (2007). Gender and mathematics achievement in China and the United States. *Gender Issues, 24*(3), 1–11. doi:10.1007/s12147-007-9044-2

UNESCO. (2014). *Latin America and the Caribbean Education for All 2015 regional review.* Retrieved from https://unesdoc.unesco.org/ark:/48223/pf0000232701

UNESCO. (2015). *Executive summary TERCE Third Regional Comparative and Explanatory Study-Learning achievements.* Retrieved from https://unesdoc.unesco.org/ark:/48223/pf0000243983_eng

UNESCO. (2016). *Gender inequality in learning achievement in primary education: What can TERCE tell us?* Retrieved from https://unesdoc.unesco.org/ark:/48223/pf0000244349

UNESCO. (2018). *Achieving gender equality in education: Don't forget the boys.* Retrieved from https://en.unesco.org/gem-report/node/2426

Valverde, G., & Näslund-Hadley, E. (2011). *La condición de la educación en matemáticas y ciencias naturales en América Latina y el Caribe* [The condition of education in mathematics and natural sciences in Latin America and the Caribbean]. Retrieved from https://publications.iadb.org/handle/11319/2757

Van de Werfhorst, H. G., & Mijs, J. J. B. (2010). Achievement inequality and the institutional structure of educational systems: A comparative perspective. *Annual Review of Sociology, 36,* 407–428. doi:10.1146/annurev.soc.012809.102538

Williams, R. (2005). *GOLOGIT2: Stata module to estimate generalized logistic regression models for ordinal dependent variables* (Statistical Software Components S453401). Chestnut, MA: Boston College, Department of Economics.

The World Bank. (2017). *The World Bank databank.* Retrieved from http://databank.worldbank:data/home.aspx

The World Economic Forum. (2013). *The Global Gender Gap Report 2013.* Geneva, Switzerland: Author.

Xie, Y., & Shauman, K. A. (2003). *Women in science: Career processes and outcomes.* Cambridge, MA: Harvard University Press.

Manuscript received July 6, 2018
Final revision received July 29, 2019
Accepted August 2, 2019

#  The Moderating Effect of Neighborhood Poverty on Preschool Effectiveness: Evidence From the Tennessee Voluntary Prekindergarten Experiment

Francis A. Pearman II
*Stanford University*

*This study drew data from a randomized trial of a statewide prekindergarten program in Tennessee and presents new evidence on the impacts of preK on third-grade achievement using administrative data on children's neighborhood environments. Results indicate that preK had no measurable impact on children's third-grade math achievement regardless of children's neighborhood conditions. However, preK significantly improved third-grade reading achievement for children living in high-poverty neighborhoods. The treatment effects on reading achievement were substantial: Among children living in high-poverty neighborhoods, those who took up an experimental assignment to attend preK scored over half a standard deviation higher on average than the control group in third grade. In contrast, preK enrollment had, if anything, a negative effect on third-grade reading achievement among children living in low-poverty neighborhoods. These differential effects were partially explained by alternative childcare options and contextual risk factors.*

KEYWORDS: academic achievement, moderation analysis, neighborhood poverty, preK, randomized controlled trial

## Introduction

Experimental evidence for the positive long-term effects of preschool comes from a small set of interventions implemented in confined geographical areas. For instance, the Perry Preschool Project, commonly cited as a model of effective preK, served low-income children living in the attendance zone of a single high-poverty elementary school in Ypsilanti, MI

---

FRANCIS A. PEARMAN II is an assistant professor of education at Stanford University, CERAS 524, 485 Lasuen Mall, Stanford, CA 94305, USA; e-mail: *apearman@stanford.edu*. His research focuses on social and educational inequality.

(Berruate-Clement, 1984). In contrast, states looking to expand access to prekindergarten for large segments of their child population will need to accommodate children living in neighborhoods that differ markedly from those served by the Perry Preschool Project. Despite an extensive neighborhood effects literature that has documented qualitative differences in the experience of poverty across places (Small, 2008; Votruba-Drzal, Miller, & Coley, 2016), researchers have largely assumed away what these contextual differences might mean for preschool effects.

This oversight is notable because it is reasonable to suspect that variation in neighborhood conditions may have implications for what children experience in the absence of preK—what is commonly referred to as the *counterfactual condition* (Bloom & Weiland, 2015). For instance, if a low-income child grows up in a high-poverty neighborhood relative to a more affluent or socioeconomically mixed neighborhood, existing evidence suggests that this child will likely experience heightened stress (Brooks-Gunn, Johnson, & Leventhal, 2010), increased exposure to violence and victimization (Sharkey, Tirado-Strayer, Papachristos, & Raver, 2012), greater likelihood of parental disengagement (Cohen, 2017), as well as fewer quality day care and out-of-school learning opportunities (Jacob & Wilder, 2010). Should there exist systematic variation in what children experience in the absence of preK that is tied to the structure and function of residential contexts, then even a standardized preK program implemented with fidelity across a broad range of locations may still produce variation in treatment effects across neighborhood contexts. Despite the potential importance of understanding this brand of impact heterogeneity in an era of preK expansion (Morris et al., 2017)—in particular, for informing decisions about how best to target preK interventions—little is known as to whether residential environments matter for how low-income children engage and respond to preschool.

This article revisits data from the Tennessee Voluntary Prekindergarten Program (TN-VPK), the first randomized trial of a scaled-up, statewide prekindergarten initiative in the United States. Data for this study were drawn from the full sample of eligible, randomly assigned children, as well as from a smaller subsample for whom a broader set of baseline measures was available. This study leverages the fact that there existed large variation in the level of poverty in children's residential neighborhoods at baseline, even though the sample itself was composed exclusively of economically disadvantaged children. This variation made it possible to explore whether the effect of preschool on later achievement varied based on where children lived. In particular, the following research question guides this study: Do the effects of assignment to or enrollment in preschool on children's third-grade achievement vary across levels of poverty in children's residential neighborhoods at baseline? Additionally, this study attempts to shed light on factors that might be responsible for potential impact heterogeneity across neighborhood contexts, including differential access to alternative childcare options, differential exposure to contextual risk factors, and differences in family socioeconomic status (SES) across neighborhood contexts.

This study advances research on preK education by providing the first estimates, to the author's knowledge, of whether preK effects on later academic achievement differ based on the characteristics of children's residential neighborhoods. This article begins with a brief review of why neighborhood poverty matters for children's academic achievement. Next, this paper describes relevant early childhood literature examining impact heterogeneity across community characteristics. After providing a brief overview of the Tennessee Voluntary PreK program, the paper turns attention to the data, randomization method, and the empirical strategy used to answer the question of whether preK effects differ across levels of neighborhood poverty. Finally, results are presented and possible explanations are explored before discussing implications for policy and research.

## Background

A considerable body of research has underscored the notion that neighborhoods play a prominent role in influencing children's early development (Chetty, Hendren, & Katz, 2015; Wodtke, Elwert, & Harding, 2016). However, surprisingly little attention has been devoted to the role that neighborhood environments play in shaping who benefits from early childhood education. As noted previously, this literature gap is notable because the opportunities and risk factors embedded in neighborhood environments help shape the nature of the counterfactual condition to which preschool exposure is compared.

### Neighborhoods, Preschool, and Academic Achievement

There are at least three reasons why neighborhoods may influence the extent to which children benefit from a statewide preschool program. First, neighborhoods differ in alternative childcare options (Swenson, 2008). Children in high-poverty neighborhoods are less likely than their peers in more affluent neighborhoods to have access to nursery schools, private or informal day care facilities, or other enrichment programs (Green, 2015; Jocson & Thorne-Wallington, 2013; Tate, 2008). Moreover, there is evidence that alternative childcare options in high-poverty neighborhoods are of lower quality than those in more affluent neighborhoods (McCoy, Connors, Morris, Yoshikawa, & Friedman-Krauss, 2015; Valentino, 2018). Therefore, it is conceivable that the effects of a statewide preK program could differ depending on whether children have limited access to community-based alternatives or if their access is circumscribed by lower quality services, with preK effects potentially being most pronounced if so.

Second, neighborhoods can differ in terms of contextual risk factors, including exposure to violence, housing instability, unemployment, and weakened family units (Brooks-Gunn et al., 2010; Sharkey et al., 2012). If preK effects operate, in part, by safeguarding children from developmental risk factors during periods of elevated developmental vulnerability (Dodge, Greenberg, & Malone, 2008), then one would expect preK to be most beneficial for those in high-risk communities where exposure to preK can presumably do the most safeguarding. Finally, prior research has indicated that preK effects are largest for children from economically disadvantaged families (Cooper & Lanza, 2014; Gormley, Phillips, & Gayer, 2008; Yoshikawa et al., 2013). Therefore, given that disadvantaged families often live in disadvantaged neighborhoods (Reardon & Bischoff, 2011), any independent, moderating effect of family SES on preschool effects would be observed to some extent at the neighborhood level.

The few studies that have directly tested whether preK effects hinge on neighborhood characteristics have used broad distinctions between urban and rural communities and have operationalized community characteristics based on the location of the preK center itself. Fitzpatrick (2008) drew observational data from the National Assessment of Educational Progress and found, based on a regression–discontinuity design, that disadvantaged children attending preK in small towns and rural areas of Georgia experienced the largest gains in math and reading scores after the expansion of universal preK. By contrast, McCoy, Morris, Connors, Gomez, and Yoshikawa (2016) found that Head Start was more effective at improving receptive vocabulary scores in urban areas but was more effective at improving oral comprehension scores in rural areas. Although these studies provide important insights into preK effect moderation by community characteristics in a general sense, their precise insights into how neighborhood conditions relate to preK effects are limited for two reasons.

First, although the urban-rural divide may have implications for the types of risk and protective factors embedded in each community type (Votruba-Drzal et al., 2016), this binary conceals important heterogeneity that exists within each category—heterogeneity that captures more directly the level of disadvantage that children encounter in their residential areas. After all, the term "urban" spans not only central city poverty but also *suburban* wealth (Milner, 2015), while the term "rural" similarly encompasses a broad range of communities that differ in socioeconomic composition (Flora, Flora, & Gasteyer, 2015). Second, prior preschool research on impact heterogeneity by community characteristics has focused on the location of the preK center itself rather children's own residential neighborhoods. This is notable because measuring disadvantage at the level of children's residential environment is likely a better approximation of the contextual determinants of children's counterfactual condition than the area in which the preK center is located, which may or may not be near children's homes.

In fact, the neighborhood environment of the preK center could differ markedly from children's home neighborhood and thus obscure what is understood about the counterfactual condition.

For instance, as shown in the Supplementary Appendix Table A.1 in the online version of the journal, approximately 44% of children in the full analytic sample who lived in high-poverty neighborhoods attempted to enroll (and were subsequently randomized) in preK centers that were located in low-poverty neighborhoods. (A full description of the data and randomization method is described below.) Likewise, 7% of children living in low-poverty neighborhoods were randomized at preK centers located in high-poverty neighborhoods. These crossovers were permissible because enrollment in the TN-VPK program, unlike many K–12 schools in Tennessee, was not based on catchment area assignment. Consequently, many children attempted to enroll in preK centers that were not necessarily local, reiterating the importance of capturing counterfactual variation in terms of residential contexts.

To assess variation in neighborhood conditions and examine whether this variation moderates the effect of preK, this study revisits data from a randomized trial of a statewide prekindergarten initiative, the TN-VPK program. Recent research on TN-VPK found that initial gains from preK faded by third grade (Lipsey, Farran, & Durkin, 2017), a finding echoed in several recent experimental studies of preschool effects (Puma, Bell, Cook, & Heid, 2010; Puma et al., 2012). The current study reexamines TN-VPK findings with an eye toward whether residential contexts mattered in these achievement patterns. In particular, this study asks whether neighborhood poverty moderates the effect of preK on children's achievement in third grade.

## Tennessee Voluntary PreK Program

The Tennessee Voluntary Prekindergarten (TN-VPK) program, a statewide early childhood initiative housed in the Tennessee Department of Education (TNDOE), enrolls over 18,000 4-year-olds in school districts across Tennessee. The structure of TN-VPK is similar to other statewide prekindergarten programs implemented within state departments of education (Karch, 2010). In particular, local school districts submit applications to TNDOE for funding to establish preK classrooms. TNDOE awards local school districts a funding amount based on the state's *Basic Education Program* formula, which accounts for a number of factors including a district's financial need. Districts are then expected to provide matching funds in order to cover the full cost of establishing the prekindergarten program. However, districts are permitted to allocate additional dollars if they wish to provide supplementary services such as transportation that are not required by TNDOE.

TN-VPK serves 4-year-old children eligible for kindergarten the following year whose families meet a number of eligibility requirements. Children eligible for the free-and-reduced-price lunch program are given top priority for admission. As space permits, children are also eligible if they are English language learners, in state custody, deemed unserved or underserved by a state advisory council, or have a disability. Per state mandates, TN-VPK requires 5 days of teaching with a minimum of 5.5 hours of instruction per day. Maximum class size is 20 students, with one teacher required for every 10 students. All VPK classrooms are required to be staffed by at least one teacher who is state-licensed and endorsed for Early Child Education by the state board of education. Additionally, preference is for at least one assistant per classroom to hold a child development certificate or an associate degree in early childhood.

Each classroom is required to operate a comprehensive curriculum that is approved by the Tennessee Department of Education's Office of Early Learning. These curricula must incorporate instruction in a broad range of developmental domains, including linguistic, cognitive, social-personal, and physical, and must be delivered through a mix of direct and individualized instruction, group activities, and center-based activities. The list of approved curricula during the 2014-2015 school year included 22 "comprehensive" curricula and 12 supplementary curricula. To ensure quality implementation of the curricula, TN-VPK drew from standards set forth by the National Institute of Early Education Research and are described in detail in the TNDOE's Scope of Services (TNDOE, 2017). TN-VPK was determined to have met 9 of the 10 current standards advocated by the National Institute of Early Education Research. These standards and curricular mandates ensure a minimum level of quality across TN-VPK classrooms.[1]

## Method

This study is situated in a larger evaluation of the TN-VPK program. The evaluation began during the 2009–2010 school year and leveraged the fact that many preschools across the state were oversubscribed. This oversubscription meant that many eligible children were denied admission. At the start of the 2009–2010 school year and again at the start of the subsequent school year, a group of sites identified as oversubscribed by TNDOE agreed to randomly assign applicants to a treatment condition that was granted admission and a control condition that was denied admission. This randomization produced across two cohorts of children during the 2009–2010 and 2010–2011 school years a total of 111 randomized applicant lists in 79 schools in 29 school districts.

Of the 3,131 eligible children in these randomized applicant lists, 2,990 were included in the state education database for at least 1 school year after prekindergarten. There is no evidence that the remaining children attended public schools in Tennessee; however, the parents of 11 of these children who were either home-schooled or enrolled in private school provided consent to be followed in the Intensive Substudy (ISS) sample (described in more detail below).

These 2,990 children composed the *full randomized sample* (hereafter referred to as "the RCT sample"). The randomized controlled trial (RCT) sample includes all children who were eligible for preK enrollment at the beginning of their respective prekindergarten year and who were randomly assigned to attend or not to attend VPK. In all, 1,852 children received offers of admission, and the remaining 1,138 did not. These two groups made up the intent-to-treat (ITT) treatment and control conditions for the RCT analysis. In addition, state databases were used to identify which of these students actually enrolled in VPK, regardless of their assignment status. Of the 2,990 children in the sample, 1,997 attended VPK for at least 1 day; the remaining 993 children had no VPK attendance during their respective prekindergarten year.

**Intensive Substudy Sample**

The ISS sample is a subset of the full sample for whom parental consent was obtained to collect annual assessment data through third grade as well as a broader set of baseline characteristics, including achievement scores and indicators of family SES. Attempts were made to contact parents of every child on eligible randomization lists at the start of their respective prekindergarten year, and few parents explicitly refused. However, logistical challenges limited the consent rate for the first cohort of children. In particular, for the 2009–2010 school year, consent could only be obtained by way of mail-in responses that were mailed to parents directly from TNDOE before the start of the school year. As a result, the consent rate for the first cohort was 24.4%.

For the second cohort, arrangements were made to obtain parental consent during the VPK application process, which was in-person and did not require any additional steps. This modification led to a higher consent rate of 67.9% for the second cohort. These procedures together yielded 1,331 consented children across both cohorts. However, additional restrictions including limiting the sample to age- and income-eligible children and restricting children to those on randomization lists that contained at least one consented child in the treatment and control condition yielded a final analytic sample of 1,076 children, who came to be known as the "Intensive Substudy" (ISS) sample. Of these children, 697 were randomly assigned to VPK, and the remaining 379 to the control group. (Supplementary Appendix Table E.1 shows that compliance rates for the RCT and ISS samples did not differ across levels of neighborhood poverty.)

As described by Lipsey, Hofer, Dong, Farran, and Bilbrey (2013), consent rates differed by treatment condition across both cohorts. In particular, parents of children randomly assigned to VPK were more likely to consent

*Pearman*

into the ISS sample than parents of children who were not assigned to VPK. (Consent rates for Cohorts 1 and 2 differed by 14 and 6 percentage points between treatment conditions, respectively.) The differential consent rates across treatment conditions for both cohorts introduced potential bias into the ISS sample with respect to the treatment-control contrast, despite that this subsample was drawn from a randomized applicant list. As described below, the analytic models used in the current study control for a host of baseline characteristics in an effort to mitigate potential bias within this subsample. (Exploratory analysis, in the Supplementary Appendix Table H.1, revealed no evidence that treatment-control differences in consent rates differed across levels of neighborhood poverty.) Moreover, to improve generalizability of the ISS sample with regard to the full RCT sample, consent weights were computed that effectively upweighted children least likely to consent into the ISS sample. These weights were computed from a logistic regression in which a binary indicator for consenting into the ISS sample was regressed on a full set of baseline characteristics. (Complete regression results are provided in the Supplementary Appendix Table F.1.) This regression was used to compute predicted probabilities of consent for each child. The inverse of these predicted probabilities became the consent weights, which were used in subsequent analyses for the ISS sample.

### Geolocating Residential Addresses

On children's initial application to TN-VPK, parents were required to indicate a home address, which was subsequently geocoded and matched to block-level data. However, not every parent provided a valid street address that could be geocoded. Of the 2,990 children in the RCT sample, 86% could be geocoded; of the 1,076 children in the ISS sample, 85% could be geocoded. (Supplementary Appendix Table G.1 shows no evidence that the likelihood of having a valid street address differed by treatment condition.) Children without valid street addresses were excluded from the analyses. Importantly, as described below, this study adjusts for differential selection into the respective analytic samples based on the probability of having a valid home address. And as described in more detail in the Results section, results were robust to the exclusion of these sampling weights, indicating that bias due to having a valid street address was unlikely.

## Measures

### Baseline Characteristics

Data on baseline descriptive characteristics for the full RCT sample were gathered from the state administrative database. These baseline descriptive characteristics include *date of birth, a binary indicator for female, race/ethnicity, age*, and *English as a primary language*. In addition to baseline data gathered from the state, children in the ISS sample were administered a series of assessments and questionnaires during the fall of their prekindergarten year that was used to create a more robust set of baseline controls for the ISS sample. The baseline parent questionnaire provided indicators of *mother's education, number of working parents, number of household magazine and newspaper subscriptions,* and *the frequency with which families used the library*. Baseline achievement was also assessed with a selection of scales from the Woodcock Johnson III Achievement Battery (Woodcock, McGre, & Mather, 2001). These scales included one measure of reading (Passage Comprehension), three measures of math (Applied Problem Solving, Quantitative Concepts, and Calculation), two measures of language (Oral Comprehension and Picture Vocabulary), and two measures of literacy (Letter-Word Identification and Spelling). For all baseline covariates (excluding neighborhood poverty), multiple imputation was used to fill in missing values due to item-specific nonresponse. Missing data on baseline covariates were rare for both the RCT and ISS samples, with rates ranging from 0.0% to 1.2% in the RCT sample and from 0.0% to 5.5% in the ISS sample. (A complete list of missing data patterns is provided in the Supplementary Appendix Table G.1.) Subsequent analyses were based on 25 multiply imputed datasets computed separately for each analytic sample that were combined based on Rubin's (1987) rules.

**Neighborhood Poverty**

Although neighborhood disadvantage can be measured using a variety of indicators, this study focuses on neighborhood poverty because prior research suggests that neighborhood poverty is closely associated with underlying social processes believed to be responsible for neighborhood effects (Wilson, 2012). Moreover, neighborhood poverty, unlike multidimensional scales of neighborhood disadvantage, has a straightforward, policy-relevant interpretation that is not reliant on distributional considerations (Wodtke, 2013). In particular, prior neighborhood effect research has noted the existence of thresholds, or "tipping points," after which neighborhoods begin to matter in shaping children's developmental outcomes (Galster, 2012).

This study uses as its primary measure of neighborhood poverty a three-level ordinal measure (coded 1, 2, and 3), to indicate whether a child lived at baseline in a low-poverty neighborhood (below 10% poor), moderate-poverty neighborhood (10% to 30% poor), or high-poverty neighborhood (at least 30% poor). Although thresholds of neighborhood poverty used in prior literature are somewhat varied (see D. J. Harding, 2003; Jargowsky, 1997; Wodtke, 2013), exploratory analysis indicated that measures based on the 10% and 30% thresholds best captured the relation between neighborhood poverty and preschool effects.

Measures of neighborhood poverty were based on block-group-level data gathered from the 2008–2012 American Community Survey. Block groups are standard geographical units made available by the U.S. Census Bureau that are composed of contiguous clusters of residential blocks that contain between 600 and 3,000 people. The boundaries of block groups are mostly defined by local participants in the Census Bureau's Participant Statistical Areas Program, suggesting that boundaries are oftentimes locally meaningful and reflective of residential divisions (Bureau of the Census, 2008). (As in the Results section, Supplementary Appendix Table B examines the robustness of estimates to alternative thresholds of neighborhood poverty, various measures of neighborhood disadvantage including a multidimensional composite index, as well as different conceptions of neighborhood boundaries.)

**Outcome**

The Tennessee Comprehensive Assessment Program (TCAP) requires students attending public school in Tennessee to be tested annually in core subject areas from third grade through eighth grades. The outcome variables in the present study were children's *reading* and *math* performance, respectively, on these statewide achievement tests in third grade. Like baseline data for the full sample, these achievement data were gathered from the state educational database. It is important to note that some children in the second cohort were retained and had not yet reached third grade by the time follow-up data collection occurred. However, as indicated in the Supplementary Appendix Table D.1, there were no statistically significant differences in retention across treatment conditions in any level of neighborhood poverty in either the RCT or ISS sample. Therefore, retention differences should not bias treatment contrasts among children who did take the test in third grade.

# Weighting

As noted previously, not every child in the RCT or ISS samples provided a valid street address at baseline that could be geolocated. In addition, not every child took the TCAP in third grade. To correct for potential nonrandom "attrition," this study used a poststratification weighting technique to adjust for differential probabilities of being included in the analytic samples. In effect, children more likely to misreport a valid street address or to have a missing TCAP score in third grade were up-weighted relative to their peers. These sampling weights were estimated from two logistic regressions in which a binary indicator of whether a child had a valid street address at baseline or whether a child took the TCAP in third grade, respectively, was regressed on the full set of sample-specific baseline characteristics as well as a set of subsequent child-level variables gathered from the state administrative database. Results from these regressions (and all included covariates) are provided in the Supplementary Appendix Table F.1. The inverse of these two predicted probabilities became the poststratification address and TCAP weights, respectively.

Effect estimates reported below for the full RCT sample were weighted by the product of the address weight and the TCAP weight. Effect estimates for the ISS sample were weighted similarly, except for the addition of the consent weight described earlier. That is, effect estimates for the ISS sample were weighted by the product of all three weights (see Wodtke, Harding, & Elwert, 2011 for similar weighting technique). As described in the Results section, results were robust to the exclusion of each of these weights, indicating that bias due to missing a valid street address or missing outcome data—or, for the ISS sample, bias due to differential consent rates—was unlikely.

## Balance Tests and Summary Statistics

Although randomization into treatment conditions eliminates observed and unobserved differences across treatment and control groups in expectation, randomization does not necessarily ensure randomization within subgroups defined by baseline characteristics (VanderWeele & Knol, 2011). Moreover, it is unclear whether the extent to which any nonrandom "attrition" based on misreporting of baseline addresses or incomplete outcome data may have affected randomization. Therefore, this section reports a series of balance tests across the RCT and ISS samples, respectively, to determine the extent of subgroup balance on baseline covariates. Unadjusted differences were assessed using ordinary least squares regressions with ITT condition as the only predictor. Adjusted differences were based on regressions that included the other baseline characteristics germane to each analytic sample as covariates. Adjusted models included indicators for preK center. All models used sampling weights.

Tables 1 and 2 present unadjusted and adjusted comparisons of baseline characteristics between treatment and control groups within each level of neighborhood poverty for the RCT and ISS samples, respectively. As shown in Columns (3), (6), and (9) of each table, the few covariate imbalances that were present in unadjusted comparisons were considerably reduced in the fully adjusted comparisons. In the RCT sample (Table 1), 1 of the 24 comparisons was significant with $p < .10$, and 1 of the 24 comparisons was significant with $p < .05$ based on $t$ tests that did not adjust for multiple comparisons; in the ISS sample (Table 2), 3 of the 57 comparisons were significant with $p < .10$, and 2 of the 57 comparisons were significant with $p < .05$ based on analogous $t$ tests that did not adjust for multiple comparisons. These patterns are generally consistent with what one would expect under random assignment (i.e., simply by chance, one would expect 1 out of 20 and 1 out of 10 comparisons to show up as significant at the $p < .05$ and $p < .10$ levels, respectively).

Table 1
Summary Statistics and Balance Tests for Children in TN-VPK-Neighborhood Data Linked Sample Across Levels of Neighborhood Poverty, Full Sample

| | Low Poverty (<10% Poor) | | | Moderate Poverty (10% to 30% Poor) | | | High Poverty (>30% Poor) | | |
|---|---|---|---|---|---|---|---|---|---|
| | Control Group, M (1) | Experimental Versus Control (2) | Adjusted Difference (3) | Control Group, M (4) | Experimental Versus Control (5) | Adjusted Difference (6) | Control Group, M (7) | Experimental Versus Control (8) | Adjusted Difference (9) |
| Available test score | 0.84 | 0.01 (0.03) | −0.01 (0.03) | 0.85 | 0.03 (0.02) | 0.02 (0.03) | 0.86 | −0.02 (0.04) | −0.02 (0.04) |
| Female | 0.51 | −0.02 (0.04) | 0.02 (0.04) | 0.49 | 0.02 (0.03) | 0.06 (0.04) | 0.48 | 0.07 (0.06) | 0.04 (0.07) |
| Black | 0.45 | −0.08 (0.06) | 0.01 (0.03) | 0.61 | −0.15* (0.07) | −0.02 (0.03) | 0.75 | 0.02 (0.11) | 0.01 (0.03) |
| Hispanic | 0.20 | −0.06* (0.03) | −0.01 (0.01) | 0.27 | −0.07* (0.03) | 0.00 (0.01) | 0.33 | −0.11* (0.05) | 0.01 (0.02) |
| White | 0.54 | 0.10* (0.04) | −0.01 (0.02) | 0.39 | 0.14*** (0.04) | −0.02 (0.01) | 0.28 | 0.09 (0.05) | −0.00 (0.02) |
| English language learner | 0.76 | 0.07* (0.03) | −0.04* (0.02) | 0.68 | 0.13*** (0.03) | 0.03 (0.02) | 0.64 | 0.14** (0.05) | 0.04† (0.02) |
| Age | 4.90 | 0.22*** (0.04) | 0.03 (0.02) | 4.97 | 0.09* (0.04) | −0.02 (0.03) | 4.97 | 0.17** (0.06) | −0.03 (0.03) |
| Neighborhood poverty | 0.05 | 0.00 (0.00) | −0.00 (0.00) | 0.18 | 0.01 (0.01) | −0.00 (0.00) | 0.42 | −0.01 (0.02) | −0.01 (0.01) |
| N | 395 | 513 | | 452 | 513 | | 169 | 198 | |

Note. TN-VPK = Tennessee Voluntary Prekindergarten Program. This table presents summary statistics and balance tests of equivalency for a subset of variables collected prior to randomization. The first row indicates the share of children living in each level of neighborhood poverty that took the state-wide achievement assessment in third grade. The remaining rows refer to the sample of children who had valid achievement data in third grade and who provided valid street addresses at baseline. Columns (1), (2), and (3) include children living in low-poverty neighborhoods; Columns (4), (5), and (6) include children living in moderate-poverty neighborhoods; Columns (7), (8), and (9) refer to children living in high-poverty neighborhoods. Columns (1), (4), and (7) show the control group mean for each variable across quartiles of neighborhood disadvantage. Columns (2), (5), and (8) report unadjusted differences between the experimental VPK group and the control group, which is estimated using a weighted ordinary least squares regression of each variable on a binary indicator for random assignment into the treatment condition. The adjusted comparisons in Columns (3), (6), (9), and (12) were based on regressions that included as covariates the complete set of baseline characteristics. Adjusted comparisons also included randomization pool fixed effects. All comparisons were weighted to adjust for children's differential probability of (1) having taken the state-wide achievement test in third grade and (2) having reported a valid street address at baseline. Standard errors are reported in parenthesis and are clustered at the neighborhood level.
†p < .10. *p < .05. **p < .01. ***p < .001, for two-tailed tests of significance.

Table 2

**Summary Statistics and Balance Tests for Children in TN-VPK-Neighborhood Data Linked Sample Across Levels of Neighborhood Poverty, ISS Sample**

| | Levels of Neighborhood Poverty ||||||||||
|---|---|---|---|---|---|---|---|---|---|
| | Low Poverty (<10% Poor) ||| Moderate Poverty (10% to 30% Poor) ||| High Poverty (>30% Poor) |||
| | Control Group, M (1) | Experimental Versus Control (2) | Adjusted Difference (3) | Control Group, M (4) | Experimental Versus Control (5) | Adjusted Difference (6) | Control Group, M (7) | Experimental Versus Control (8) | Adjusted Difference (9) |
|---|---|---|---|---|---|---|---|---|---|
| Available test score | 0.78 | −0.01 (0.05) | 0.00 (0.00) | 0.84 | 0.05 (0.04) | 0.00 (0.00) | 0.74 | 0.13 (0.09) | 0.00 (0.00) |
| Female | 0.49 | 0.03 (0.08) | 0.12 (0.12) | 0.43 | 0.06 (0.07) | 0.08 (0.08) | 0.54 | 0.07 (0.15) | −0.19 (0.30) |
| Black | 0.39 | −0.02 (0.13) | −0.07 (0.06) | 0.61 | −0.11 (0.15) | −0.00 (0.11) | 0.80 | −0.01 (0.25) | −0.11 (0.15) |
| Hispanic | 0.23 | −0.09 (0.07) | −0.04 (0.03) | 0.27 | −0.10† (0.06) | 0.06 (0.04) | 0.36 | −0.02 (0.11) | 0.20* (0.10) |
| White | 0.55 | 0.02 (0.08) | −0.16† (0.09) | 0.39 | 0.14† (0.07) | −0.06 (0.07) | 0.14 | 0.06 (0.08) | −0.04 (0.06) |
| English language learner | 0.72 | 0.11 (0.07) | −0.07 (0.06) | 0.60 | 0.25*** (0.07) | 0.13** (0.04) | 0.58 | 0.08 (0.12) | 0.08 (0.12) |
| Age | 4.96 | 0.12† (0.07) | −0.05 (0.06) | 5.02 | 0.03 (0.08) | 0.04 (0.04) | 4.90 | 0.13 (0.16) | −0.03 (0.08) |
| Letter-word | 318.79 | 0.61 (4.28) | 1.76 (4.15) | 315.11 | 5.89 (3.84) | 3.30 (3.17) | 317.86 | −4.91 (7.01) | 6.71 (6.07) |
| Spelling | 351.99 | −1.51 (3.94) | −2.17 (4.95) | 354.09 | −1.85 (4.03) | −0.24 (4.17) | 354.03 | −8.76 (5.51) | −0.60 (13.80) |
| Oral comprehension | 440.96 | 3.29 (2.36) | 0.10 (1.82) | 440.71 | 3.37 (2.63) | −2.71 (2.36) | 438.47 | 3.06 (4.18) | 0.37 (4.79) |
| Picture vocabulary | 451.77 | 5.81 (3.78) | 0.49 (2.76) | 448.64 | 8.36* (4.04) | −1.32 (2.07) | 443.20 | 5.41 (8.88) | 2.27 (5.61) |
| Applied problem solving | 387.14 | 6.63 (4.03) | 4.11 (3.29) | 386.20 | 7.44* (3.42) | 1.42 (3.25) | 385.41 | −1.32 (7.73) | 7.16 (4.78) |
| Quantitative concept | 406.48 | 0.39 (1.80) | −0.04 (2.17) | 408.00 | −0.11 (2.06) | −2.35 (1.79) | 408.45 | −4.56 (3.96) | −8.47† (4.85) |
| Mother high school noncompleter | 0.09 | 0.05 (0.04) | −0.14 (0.12) | 0.19 | −0.03 (0.05) | 0.18 (0.13) | 0.22 | 0.00 (0.13) | −0.14 (0.32) |
| % with two working parents | 0.35 | 0.02 (0.07) | −0.10 (0.15) | 0.35 | −0.03 (0.06) | −0.04 (0.12) | 0.24 | 0.00 (0.10) | 0.33† (0.19) |
| Library use | 0.45 | 0.01 (0.07) | 0.11 (0.18) | 0.42 | −0.08 (0.07) | −0.01 (0.12) | 0.60 | −0.09 (0.15) | 0.35 (0.23) |

*(continued)*

## Table 2 (continued)

### Levels of Neighborhood Poverty

|  | Low Poverty (<10% Poor) ||| Moderate Poverty (10% to 30% Poor) ||| High Poverty (>30% Poor) |||
| --- | --- | --- | --- | --- | --- | --- | --- | --- | --- |
|  | Control Group, M (1) | Experimental Versus Control (2) | Adjusted Difference (3) | Control Group, M (4) | Experimental Versus Control (5) | Adjusted Difference (6) | Control Group, M (7) | Experimental Versus Control (8) | Adjusted Difference (9) |
| Newspaper subscriptions | 0.86 | −0.02 (0.05) | −0.03 (0.11) | 0.80 | 0.03 (0.06) | −0.03 (0.15) | 0.85 | −0.00 (0.10) | −0.14 (0.30) |
| Magazine subscriptions | 0.76 | −0.02 (0.05) | 0.00 (0.09) | 0.75 | −0.02 (0.06) | −0.16 (0.10) | 0.85 | −0.12 (0.11) | −0.14 (0.22) |
| Neighborhood poverty | 0.04 | 0.00 (0.01) | 0.00 (0.00) | 0.18 | 0.01 (0.01) | 0.01 (0.01) | 0.43 | −0.03 (0.03) | −0.02 (0.03) |
| N | 110 | 169 |  | 156 | 205 |  | 37 | 55 |  |

*Note.* TN-VPK = Tennessee Voluntary Prekindergarten Program. ISS = Intensive Substudy. This table presents summary statistics and balance tests of equivalency for a subset of variables collected prior to randomization. The first row indicates the share of children living in each level of neighborhood poverty that took the state-wide achievement assessment in third grade. The remaining rows refer to the sample of children who had valid achievement data in third grade and who provided valid street addresses at baseline. Columns (1), (2), and (3) include children living in low-poverty neighborhoods; Columns (4), (5), and (6) include children living in moderate-poverty neighborhoods; Columns (7), (8), and (9) refer to children living in high-poverty neighborhoods. Columns (1), (4), and (7) show the control group mean for each variable across quartiles of neighborhood disadvantage. Columns (2), (5), and (8) report unadjusted differences between the experimental VPK group and the control group, which is estimated using a weighted ordinary least squares regression of each variable on a binary indicator for random assignment into the treatment condition. The adjusted comparisons in Columns (3), (6), (9), and (12) were based on regressions that included as covariates the complete set of baseline characteristics. Adjusted comparisons also included randomization pool fixed effects. All comparisons were weighted to adjust for children's differential probability of (1) having taken the state-wide achievement test in third grade and (2) having reported a valid street address at baseline. Standard errors are reported in parenthesis and are clustered at the neighborhood level.
†*p* < .10. **p* < .05. ***p* < .01. ****p* < .001, for two-tailed tests of significance.

*Differential Effects of PreK by Neighborhood*

Although covariate adjustment yielded statistically similar treatment conditions across nearly all the baseline characteristics in the three subgroups of neighborhood poverty, this approach may still be inadequate to account for bias due to unobserved confounding, which remains the primary threat to the internal validity of the present study (Rosenbaum & Rubin, 1983). In subsequent analytic models, each of the specified baseline characteristics was included as a covariate to improve precision and to adjust for any initial biases that may have been introduced by the compromised randomization inherent to a post hoc subgroup analysis with obvious attrition due to misreporting of baseline addresses and incomplete outcome data.

The bottom rows in Tables 1 and 2 report the distribution of the analytic samples across levels of neighborhood poverty. With regard to the RCT sample, 16.3% of children lived at baseline in high-poverty neighborhoods. The largest share of children in the RCT sample lived in moderate-poverty neighborhoods (43.1%), while similar percentages lived in low-poverty neighborhoods (40.5%). These patterns are similar with respect to the ISS sample, with the distinction being that a slightly larger share of children (49.1%) lived in the moderate-poverty neighborhoods, 38.6% in low-poverty neighborhoods, and 12.3% in high-poverty neighborhoods.

## Analysis

This study estimates moderated ITT effects of the VPK treatment across levels of neighborhood poverty. These estimates are essentially covariate-adjusted comparisons of control and treatment group means across levels of neighborhood poverty. Following prior research on variation in treatment effects (e.g., Schochet & Deke, 2014), this study estimates moderated ITT effects of preK on children's third-grade reading and math achievement ($y$), respectively, using ordinary least squares regression specification of the following form:

$$y_i = \alpha + VPK_i \left( \theta + \sum_{l=1}^{3} Z_i NP_{li} \right) + \sum_{j=1}^{K} X_{ji} \gamma_j + s_i \delta + \varepsilon_i \tag{1}$$

where *VPK* is an indicator variable for being randomly assigned to VPK, *NP* is a categorical measure of neighborhood poverty coded 1 through 3 that records whether children resided at baseline in a low-poverty neighborhood (below 10% poor), moderate-poverty neighborhood (at least 10% poor but less than 30% poor), or high-poverty neighborhood (at least 30% poor). Mechanically, *NP* is arrayed as series of binary indicators for each level of neighborhood poverty with low-poverty neighborhoods serving as the referent category.[2] *X* represents a vector of sample-specific baseline covariates included to improve precision and account for chance differences between groups in the distribution of baseline characteristics. Finally, *s* is a set of indicators for randomization site that ensure that third-grade achievement patterns were not associated with any unobserved characteristics that systematically varied across preK classrooms. All regressions were weighted to adjust for differential probabilities of being included in the sample as described previously. Standard errors were clustered at the neighborhood level.

Two assumptions are required for moderated effects to be interpreted causally. Although neighborhoods were not randomized in the study, a causal interpretation can be made if it is assumed that (a) randomization yielded subgroup randomization within levels of neighborhood poverty and (b) compliance rates did not vary across these subgroups. As discussed previously, Tables 1 and 2 provided no evidence of covariate imbalance across levels of neighborhood poverty. Moreover, Supplementary Appendix Table E.1 provides no evidence that compliance rates varied across levels of neighborhood poverty. In short, there is no evidence to discourage a causal interpretation of moderated effects. Thus, the estimate of $\theta$ in Equation (1) can be interpreted as identifying the causal impact of being offered admission into VPK for children in low-poverty neighborhoods, while estimates of $Z_2$ and $Z_3$ in Equation (1) identify the increment (or decrement) to the causal impact of being offered VPK admission for those in moderate- and high-poverty neighborhoods, respectively. It is also important to note that the inclusion of fixed effects for randomization site ($s$) controls for unobserved differences across preK classrooms and means that results are interpreted as within-randomization-list estimates, that is, as the potential differential effect of preK across neighborhood contexts for children within the same randomization list. To facilitate interpretation of results, this study also reports results in standard deviation differences, which are readily interpreted as effect sizes. Effect sizes were computed by dividing the coefficient of interest by a weighted average of the pooled standard deviation across levels of neighborhood poverty.

Because of the potential of crossovers, that is, some children offered admission into VPK did not enroll in VPK and some children assigned to the control group wound up attending VPK anyway, these ITT estimates are likely different from the causal effect of actually enrolling in VPK. (Supplementary Appendix Table E.1 shows compliance rates across categories of neighborhood poverty.) This study follows Chetty et al. (2015) and estimates the moderated impact of VPK enrollment—the moderated local average treatment effect (LATE)—by leveraging a two-stage least squares approach and instrumenting for VPK enrollment and the interaction between VPK enrollment and the three-level neighborhood poverty variable with an indicator for treatment assignment and interactions between treatment assignment and the three-level neighborhood poverty variable. Formally, this specification takes the following form:

## Differential Effects of PreK by Neighborhood

$$y_i = \alpha + EnrollVPK_i\left(\theta + \sum_{l=1}^{3} Z_i NP_{li}\right) + \sum_{j=1}^{K} X_{ji}\gamma_j + s_i\delta + \varepsilon_i \qquad (2)$$

where *EnrollVPK* and the interactions between *EnrollVPK* and *NP* are the instrumented indicators for enrollment in VPK and its interaction with neighborhood poverty. The remainder of components, clustering, and weighting strategy are identical to those in Equation (1). Under the assumptions that VPK assignment only affects outcomes through VPK enrollment and that subgroup randomization holds within levels of neighborhood poverty, $\theta$ can be interpreted as the causal effect of VPK enrollment for children living in low-poverty neighborhoods, while $Z_2$ and $Z_3$ can be interpreted as the change in the causal effect of VPK enrollment for children living in moderate- and high-poverty neighborhoods relative to those living in low-poverty neighborhoods.

In addition to examining moderated ITT and LATE effects of VPK across levels of neighborhood poverty, this study is also interested in potential explanations for why VPK effects might differ across levels of neighborhood poverty. This study sheds light on this issue by examining the extent to which factors correlated with neighborhood poverty—*rival moderators*—might account for why neighborhood poverty matters for differential VPK effects. This study focuses on four such rival moderators described in the background section of this article: differential access to alternative childcare options, contextual risk factors, differential sorting of families by SES, and urbanicity.[3] In particular, this *rival moderator analysis* is based on models that take the following form:

$$y_i = \alpha + VPK_i\left(\theta + \sum_{l=1}^{3} Z_i NP_{li}\right) + \delta(VPK_i \times AltMod_i) + \sum_{j=1}^{K} X_{ji}\gamma_j + s_i\delta + \varepsilon_i \qquad (3)$$

$$y_i = \alpha + EnrollVPK_i\left(\theta + \sum_{l=1}^{3} Z_i NP_{li}\right) + \delta(EnrollVPK_i \times AltMod_i) \\ + \sum_{j=1}^{K} X_{ji}\gamma_j + s_i\delta + \varepsilon_i. \qquad (4)$$

Equations (3) and (4) are identical to Equations (1) and (2) except that Equations (3) and (4) add an interaction between VPK and the rival moderator(s) of interest (*AltMod$_i$*). [The components, weighting, and clustering in Equations (3) and (4) are identical to those described for Equations (1) and (2).] Notably, for the LATE estimates described in Equation (4), *EnrollVPK* and the interactions between *EnrollVPK* and the moderators of interest were instrumented with an indicator for treatment assignment and interactions between treatment assignment and moderators of interest. Of interest in Equations (3) and (4) is the change in the coefficient for the interaction between preK and neighborhood poverty after the inclusion of each moderator(s) of interest compared with the coefficient for the interaction between preK and neighborhood poverty relative to Equations (1) and (2). It is

1339

*Pearman*

important to underscore that this rival moderator analysis is purely exploratory and should not be interpreted in causal terms because there are likely a number of unobserved neighborhood- and family-level characteristics that correlate with the rival moderators of interest.

The following section reports estimates of Equations (1) and (2) for children's third-grade reading and math achievement, respectively, before turning to the rival moderator analysis (Equations 3 and 4). For reference, the following section also reports main effect estimates of ITT and LATE that exclude interaction terms from Equations (1) and (2). The coefficients of interest in these main effect equations are interpreted as the average effect of VPK assignment and enrollment, respectively, across the entire sample of participants, controlling for neighborhood poverty. Also, given that this study is probing interactions that the original TN-VPK experiment was not set up to answer, the analyses have limited power to detect differential effects. Thus, a more flexible significance level is used throughout the Results section ($p < .10$) than is typical in preK literature.

## Results

This section begins by describing results for the full sample and concludes with results for the ISS sample. Overall, this study finds considerable evidence that the estimated effect of VPK on reading achievement depended on the residential environments in which children lived at baseline, with significant positive effects being observed for those children living in high-poverty neighborhoods but negative effects for children living in low-poverty neighborhoods. Table 3 reports full sample estimates of the effect of VPK on children's reading and math achievement in third grade, respectively. Table 4 reports analogous estimates for the ISS sample. Columns (1) through (4) refer to reading achievement; Columns (5) through (8) refer to math achievement. As noted above, the primary focus of the discussion concerns effect sizes because of their ease of interpretation: Effect sizes are readily interpreted as a standard deviation difference. Effect sizes are provided in brackets below the coefficient and standard errors in Tables 3 and 4.

### Full Sample: Reading Achievement

Column (1) of Table 3, which reports ITT estimates for the full sample, provides no evidence that, on average, random assignment into VPK classrooms had an effect on third-grade reading achievement. However, Column (2) in Table 3 shows that the estimated effect of VPK on third-grade reading achievement varied across levels of neighborhood poverty. In particular, the point estimate for the main effect of VPK in Column (2) indicates that, among children living in low-poverty neighborhoods, children assigned to VPK scored 0.13 standard deviations (*SD*s) lower than those not assigned to VPK in reading achievement in third grade ($\beta = -4.44$, $p = .067$). In

Table 3
**Moderated Effect of Tennessee Voluntary PreK on Children's Third-Grade Achievement Across Levels of Neighborhood Poverty at Baseline, Full Sample Results**

|  | Reading Achievement (n = 2,240) ||||  Math Achievement (n = 2,239) ||||
|---|---|---|---|---|---|---|---|---|
|  | ITT Estimates || TOT Estimates || ITT Estimates || TOT Estimates ||
|  | (1) | (2) | (3) | (4) | (5) | (6) | (7) | (8) |
| VPK | −1.51 | −4.44† | −3.84 | −10.02† | −3.32† | −3.91 | −8.43† | −9.59† |
|  | (1.82) | (2.42) | (4.51) | (5.33) | (1.75) | (2.65) | (4.40) | (5.80) |
|  | [−0.04] | [−0.13] | [−0.11] | [−0.29] | [−0.10] | [−0.11] | [−0.25] | [−0.28] |
| Moderate poverty |  | −2.31 |  | −3.07 |  | −2.15 |  | −0.16 |
|  |  | (2.58) |  | (4.91) |  | (2.57) |  | (4.99) |
|  |  | [−0.07] |  | [−0.09] |  | [−0.06] |  | [−0.00] |
| High poverty |  | −7.72* |  | −18.96** |  | −4.76 |  | −10.11† |
|  |  | (3.24) |  | (6.68) |  | (3.18) |  | (5.73) |
|  |  | [−0.22] |  | [−0.55] |  | [−0.14] |  | [−0.29] |
| VPK × Moderate Poverty |  | 1.61 |  | 3.02 |  | −1.33 |  | −3.37 |
|  |  | (3.34) |  | (6.82) |  | (3.48) |  | (7.15) |
|  |  | [0.05] |  | [0.09] |  | [−0.04] |  | [−0.10] |
| VPK × High Poverty |  | 13.53** |  | 28.68** |  | 6.94† |  | 14.59† |
|  |  | (4.70) |  | (9.90) |  | (4.21) |  | (8.65) |
|  |  | [0.39] |  | [0.83] |  | [0.20] |  | [0.42] |
| $R^2$ | 0.11 | 0.12 | 0.11 | 0.10 | 0.11 | 0.11 | 0.10 | 0.09 |

*Note.* ITT = intent-to-treat; TOT = treatment-on-the-treated; VPK = Voluntary Prekindergarten Program. Columns (1) through (4) refer to children's reading achievement in third grade. Columns (5) through (8) refer to children's math achievement in third grade. Columns (1) and (5) report ITT estimates from an ordinary least squares regression of children's achievement on an indicator for random assignment into the experimental preK group. Columns (2) and (6) interact the indicator of random assignment with a measure of neighborhood poverty at baseline. Columns (3) and (7) report LATE estimates using a two-stage least squares specification, instrumenting for preK enrollment with the experimental indicator of preK assignment. Columns (4) and (8) instrument for preK enrollment and the interaction of preK enrollment and levels of neighborhood poverty with the experimental indicator of preK assignment and the interaction of preK assignment and levels of neighborhood poverty. Moderate-poverty neighborhoods had poverty rates that met or exceeded 10% but were below 30%. High poverty neighborhoods had poverty rates that met or exceeded 30%. The referent category in the table is low-poverty neighborhoods, which had poverty rates below 10%. All models included randomization pool fixed effects and were weighted to adjust for differences in children's likelihood of having a nonmissing third-grade achievement score and a valid street address at baseline. All models controlled for baseline characteristics at the individual and neighborhood level. Individual-level controls included age, race, gender, primary language, and cohort. Neighborhood poverty rates were measured at the block-group level and were gathered from the 2008–2012 American Community Survey. Standard errors are in parenthesis and are clustered at the neighborhood level. Effect sizes are in brackets. Effect sizes were computed by dividing the coefficient of interest by a weighted average of the pooled standard deviation across levels of neighborhood poverty.
†$p < .10$. *$p < .05$. **$p < .01$. ***$p < .001$, for two-tailed tests of significance.

Table 4
**Moderated Effect of Tennessee Voluntary PreK on Children's Third Grade Achievement Across Neighborhood Poverty Levels at Baseline, Intensive Substudy Sample Results**

|  | Reading Achievement (n = 732) |  |  |  | Math Achievement (n = 732) |  |  |  |
|---|---|---|---|---|---|---|---|---|
|  | ITT Estimates |  | TOT Estimates |  | ITT Estimates |  | TOT Estimates |  |
|  | (1) | (2) | (3) | (4) | (5) | (6) | (7) | (8) |
| VPK | −1.67 | −4.00 | −3.30 | −6.13 | −5.82 | −6.76 | −11.50† | −10.90 |
|  | (3.35) | (4.82) | (6.14) | (6.89) | (3.58) | (6.83) | (6.67) | (9.71) |
|  | [−0.05] | [−0.12] | [−0.10] | [−0.18] | [−0.16] | [−0.19] | [−0.32] | [−0.30] |
| Moderate poverty |  | 0.82 |  | 4.89 |  | −1.97 |  | 3.67 |
|  |  | (5.28) |  | (8.82) |  | (7.68) |  | (11.84) |
|  |  | [0.02] |  | [0.15] |  | [−0.05] |  | [0.10] |
| High poverty |  | −12.77† |  | −38.67† |  | −13.47 |  | −25.29 |
|  |  | (7.63) |  | (21.13) |  | (9.02) |  | (15.88) |
|  |  | [−0.38] |  | [−1.16] |  | [−0.37] |  | [−0.70] |
| VPK × Moderate Poverty |  | −2.05 |  | −6.24 |  | −2.48 |  | −8.54 |
|  |  | (6.03) |  | (10.15) |  | (8.51) |  | (13.40) |
|  |  | [−0.06] |  | [−0.19] |  | [−0.07] |  | [−0.24] |
| VPK × High Poverty |  | 22.31* |  | 51.45* |  | 14.19 |  | 27.26 |
|  |  | (9.25) |  | (25.03) |  | (9.50) |  | (18.08) |
|  |  | [0.67] |  | [1.55] |  | [0.39] |  | [0.75] |
| $R^2$ | 0.32 | 0.33 | 0.32 | 0.28 | 0.30 | 0.30 | 0.29 | 0.29 |

*Note.* ITT = intent-to-treat; TOT = treatment-on-the-treated; VPK = Voluntary Prekindergarten Program. Columns (1) through (4) refer to children's reading achievement in third grade. Columns (5) through (8) refer to children's math achievement in third grade. Columns (1) and (5) report ITT estimates from an ordinary least squares regression of children's achievement on an indicator for random assignment into the experimental preK group. Columns (2) and (6) interact the indicator of random assignment with a measure of neighborhood disadvantage at baseline. Columns (3) and (7) report local average treatment effect estimates using a two-stage least squares specification, instrumenting for preK enrollment with the experimental indicator of preK assignment. Columns (4) and (8) instrument for preK enrollment and the interaction of preK enrollment and neighborhood disadvantage with the experimental indicator of preK assignment and the interaction of preK assignment and neighborhood disadvantage. All models included randomization pool fixed effects and were weighted to adjust for differences in children's likelihood of having a nonmissing third-grade achievement score and a valid street address at baseline. All models controlled for baseline characteristics at the individual and neighborhood level. Individual-level controls included age, race, gender, primary language, cohort, mother's education, number of working parents, number of household magazine and newspaper subscriptions, frequency of library usage, and children's baseline achievement on six cognitive assessments: Letter Word, Applied Problem Solving, Quantitative Concepts, Story Recall, Picture Vocabulary, and Oral Comprehension. Neighborhood poverty rates were measured at the block-group level and were gathered from the 2008-2012 American Community Survey. Moderate-poverty neighborhoods had poverty rates that met or exceeded 10% but were below 30%. High-poverty neighborhoods had poverty rates that met or exceeded 30%. The referent category in the table is low-poverty neighborhoods, which had poverty rates below 10%. Standard errors are in parenthesis and are clustered at the neighborhood level. Effect sizes are in brackets. Effect sizes were computed by dividing the coefficient of interest by a weighted average of the pooled standard deviation across levels of neighborhood poverty.
†$p < .10$. *$p < .05$. **$p < .01$. ***$p < .001$, for two-tailed tests of significance.

*Differential Effects of PreK by Neighborhood*

contrast, the point estimate for the interaction in Column 2 between VPK assignment and high-poverty neighborhood indicates that the effect of VPK assignment was 0.39 *SD*s larger for children living at baseline in high- compared with low-poverty neighborhoods ($\beta$ = 13.53, $p$ = .004).

This general pattern is echoed in Columns (3) and (4) of Table 3, which report LATE estimates of the main and moderated effect of VPK on third-grade reading achievement, respectively. Similar to the ITT estimates, Column (3) provides no evidence that VPK *enrollment* had an effect, on average, on third-grade reading achievement. However, Column (4) shows heterogeneity across levels of neighborhood poverty. The significant point estimate for VPK in Column (4) shows that children living at baseline in low-poverty neighborhoods who enrolled in VPK scored −0.29 *SD*s lower in reading achievement than their respective control group in third grade ($\beta$ = −10.02, $p$ = .060). However, the significant interaction term in Column 4 between VPK enrollment and high-poverty neighborhood indicates that the estimated difference in achievement between treatment and control group children was 0.83 *SD*s larger (more positive) for children living at baseline in high- compared with low-poverty neighborhoods ($\beta$ = 28.68, $p$ = .004).

Although the significant interaction terms provide evidence that the effect of VPK differs for children living in high- compared with low-poverty neighborhoods, what remains unclear is whether children living in high-poverty neighborhoods actually experienced a positive effect from VPK. Figures 1 and 2 plot the average marginal effect of VPK on third-grade reading achievement ($y$-axis) across levels of neighborhood poverty ($x$-axis). These average marginal effects are equivalent to the adjusted difference in means between treatment conditions within each level of neighborhood poverty. The panel on the left of each figure refers to the RCT sample; the panel on the right refers to the ISS sample, discussed below. The points in each figure specify the average marginal effect of VPK within each level of neighborhood poverty. The bars correspond to 90% confidence intervals.

Overall, Figures 1 and 2 show that the effect of VPK on third-grade reading achievement for the full RCT sample was positive and statistically meaningful for children in high-poverty neighborhoods at baseline. Among this subset, the average marginal effect of VPK *assignment* on third-grade reading achievement was 0.27 *SD*s ($p$ = .029); the average marginal effect of VPK *enrollment* on third-grade reading achievement was 0.54 *SD*s ($p$ = .041). In contrast, point estimates in the far left of the panel in each figure, illustrating average marginal effects of VPK for children in low-poverty neighborhoods, reiterate the earlier noted finding that both assignment to and enrollment in VPK exerted a negative effect on third-grade reading achievement in the full sample ($p$ = .067 and $p$ = .060, respectively). No evidence supports either VPK assignment or enrollment having an effect on third-grade achievement for children living in moderate-poverty neighborhoods at baseline.

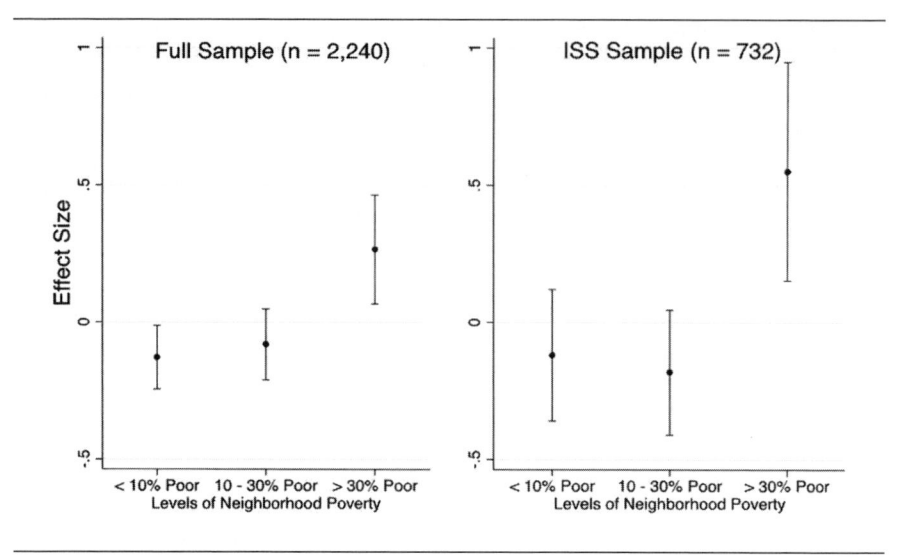

*Figure 1.* **ITT estimates of the effect of VPK on third-grade reading achievement across levels of neighborhood poverty.**

*Note.* ITT = intent to treat; VPK = Voluntary Prekindergarten program; ISS = Intensive Substudy.

## Full Sample: Math Achievement

Columns (5) to (8) in Table 3 report results for third-grade math achievement in the RCT sample. In contrast to the null main effects observed for reading achievement, this study finds some evidence that the control group, on average, outperformed the treatment group on math achievement in third grade. Column 5 in Table 3 provides evidence that, on average, children in the *control group* (i.e., those randomly assigned not to attend VPK) outperformed children randomly assigned to attend VPK by 0.10 *SD*s ($\beta = -3.32$, $p = .055$). Similarly, those *enrolled* in VPK scored 0.25 *SD*s lower on third-grade math achievement, on average, than the control group ($\beta = -8.43$, $p = .055$), as shown in Column (7). Similar to that reported for reading achievement, Columns (6) and (8) show that the effects of VPK varied across levels of neighborhood poverty. In particular, Column (6) shows that the effect of VPK assignment on children's third-grade math achievement was 0.20 *SD*s larger for children in high- compared with low-poverty neighborhoods ($\beta = 6.94$, $p = .098$). Similarly, Column (8) shows that the effect of VPK enrollment on children's third-grade math achievement was 0.42 *SD*s larger for children in high- compared with low-poverty neighborhoods ($\beta = 14.59$, $p = .090$).

Despite observing that VPK effects differed across levels of neighborhood poverty, the left panels of Figures 3 and 4 provide no evidence that

*Differential Effects of PreK by Neighborhood*

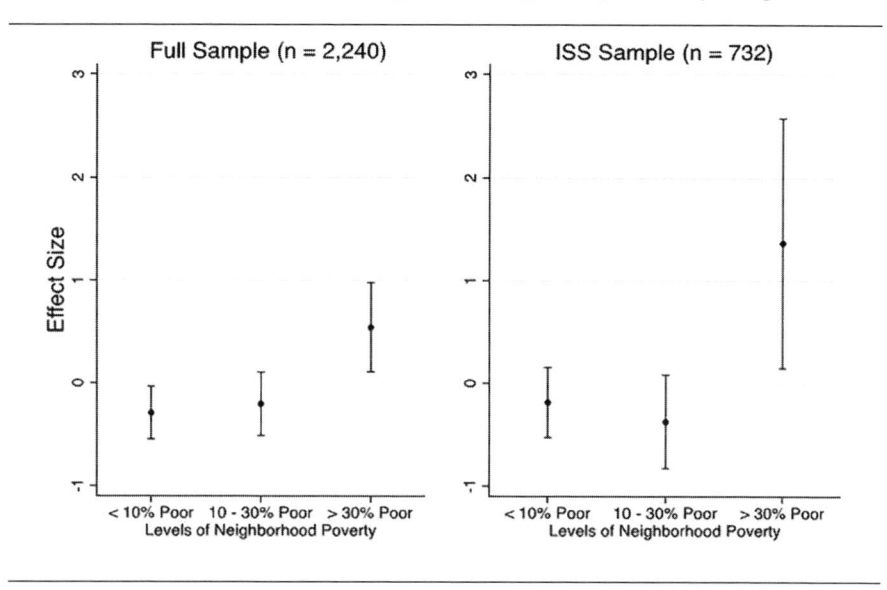

*Figure 2.* **LATE estimates of the effect of VPK on third-grade reading achievement across levels of neighborhood poverty.**

*Note.* LATE = local average treatment effect; VPK = Voluntary Prekindergarten program; ISS = Intensive Substudy.

VPK affected third-grade math achievement for children living in high-poverty neighborhoods. Said otherwise, there is no evidence that the effect of VPK for children living in high-poverty neighborhoods was statistically different from zero despite being significantly larger than the observed effect for children living in low-poverty neighborhoods. However, Figures 3 and 4 do provide evidence that among children living in moderate-poverty neighborhoods, those assigned to the control group outperformed their peers assigned to the treatment group on third-grade math achievement ($p$ = .048), while children who did not enroll in VPK outperformed their peers who did enroll in VPK on third-grade math achievement in low- ($p$ = .098) and moderate-poverty neighborhoods, respectively ($p$ = .042).

### Intensive Substudy Sample

Results for the ISS sample in Table 4 are arranged similarly to Table 3: Columns (1) to (4) report results for reading achievement; Columns (5) to (8) report results for math achievement; effect sizes are in brackets. Substantive conclusions for the ISS sample are generally consistent with those reported in the previous section. Based on a subset of the full sample for whom baseline achievement and socioeconomic variables were available,

1345

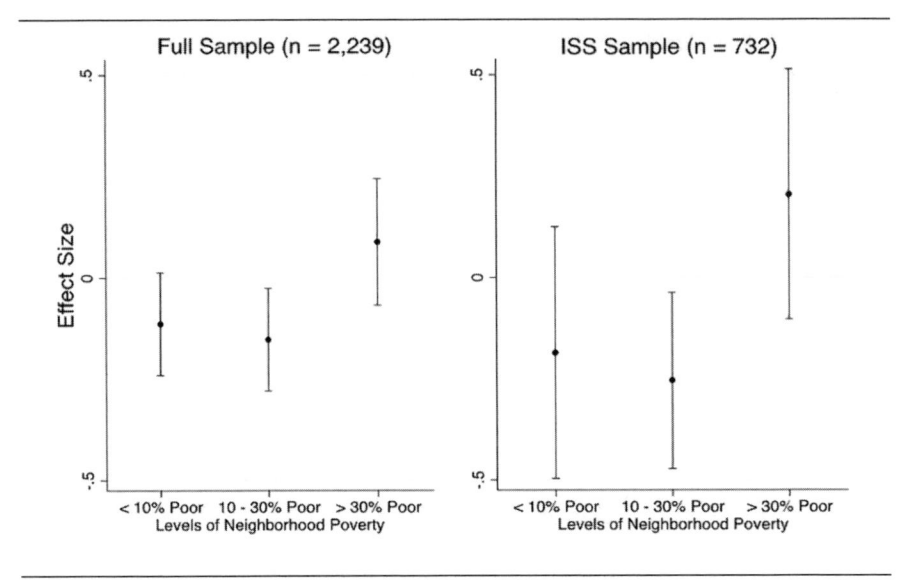

*Figure 3.* **ITT estimates of the effect of VPK on third-grade math achievement across neighborhood poverty.**
*Note.* ITT = intent to treat; VPK = Voluntary Prekindergarten program; ISS = Intensive Substudy.

this study finds that the effect of VPK differed across levels of neighborhood poverty, with the largest effects of VPK enrollment on third-grade reading achievement observed for children living in high-poverty neighborhoods. In contrast to what was observed for the full sample, however, there is no evidence that VPK enrollment or assignment affected the third-grade achievement of children living in low-poverty neighborhoods.

### Intensive Substudy Sample: Reading Achievement

Columns (1) and (3) in Table 4 provide no evidence of a significant main effect of either VPK assignment or enrollment on third-grade reading achievement in the ISS sample. However, there exists considerable impact heterogeneity across levels of neighborhood poverty. As shown in Column (2) of Table 4, the estimated effect of VPK *assignment* on third-grade reading achievement was 0.67 *SD*s larger for children living at baseline in high- compared with low-poverty neighborhoods ($\beta$ = 22.31, $p$ = .016). Similarly, Column (4) shows that the estimated effect of VPK *enrollment* on third-grade reading achievement was 1.55 *SD*s larger for children living in high- compared with low-poverty neighborhoods ($\beta$ = 51.45, $p$ = .040).

As before, to help provide some intuition for the magnitude and practical significance of these interactions, the panels on the right side of Figures 1

*Differential Effects of PreK by Neighborhood*

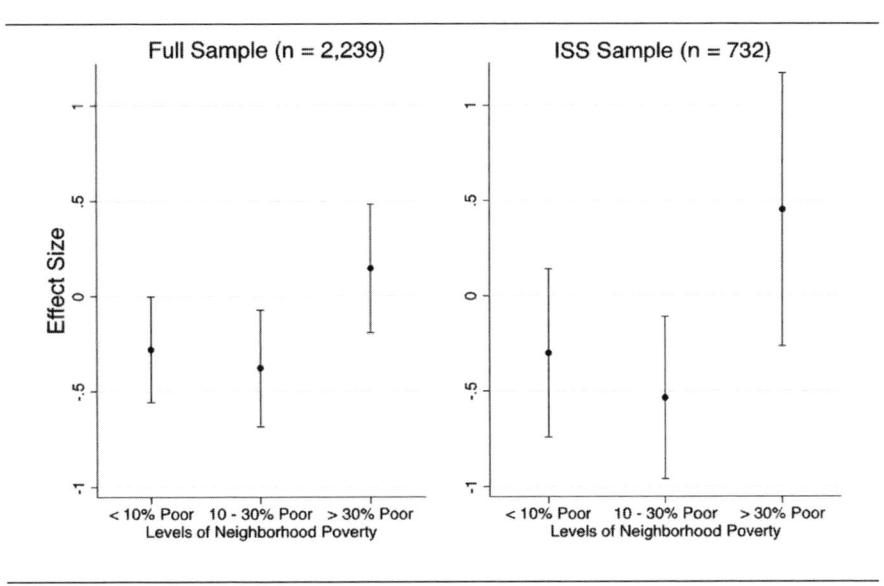

*Figure 4.* **LATE estimates of the effect of VPK on third-grade math achievement across neighborhood poverty.**
*Note.* LATE = local average treatment effect; VPK = Voluntary Prekindergarten program; ISS = Intensive Substudy.

and 2 show, for the ISS sample, the estimated effect of VPK assignment and enrollment, respectively, on third-grade reading achievement within each level of neighborhood poverty. These figures show that, among children living in high-poverty neighborhoods, VPK *assignment* was associated with a 0.55 *SD* gain in reading achievement in third grade ($p$ = .023), and VPK *enrollment* was associated with a 1.36 *SD* gain ($p$ = .065). As noted in the previous paragraph, no evidence was found that VPK assignment or enrollment affected the reading achievement of children in low-poverty neighborhoods. Neither was evidence found that VPK enrollment affected reading achievement of children in moderate-poverty neighborhoods.

### Intensive Substudy Sample: Math Achievement

Columns (5) through (8) in Table 4 report ITT and LATE estimates of the main and moderated effect of VPK on third-grade math achievement in the ISS sample. Despite that the magnitude of the point estimates for the main effects and interaction terms were largely consistent with those observed for the full sample (i.e., negative main effects but potential heterogeneity across neighborhood types), only the main effect of VPK *enrollment* in the ISS sample was significant ($\beta$ = −11.50, $p$ = .085, effect size = −0.32 *SDs*);

the remainder of estimates were imprecisely estimated. Consequently, this study cannot rule out the possibility that, for the ISS sample, the overall effect of VPK *assignment* on math achievement is indistinguishable from zero and that the null effect of VPK assignment—as well as the effect of VPK enrollment—are invariant to the level of disadvantage in children's residential neighborhood at baseline.

**Robustness Checks**

Part B of the online supplement provides a series of additional analyses that show that substantive conclusions are robust to different thresholds of neighborhood poverty, alternative weighting strategies that account differential selection into the analytic samples, and the inclusion in the analytic sample of children who were unable to be geolocated to a valid street address at baseline but who could be geolocated to the centroid of the zip code in which they lived. Interestingly, no evidence is found that VPK effects varied across levels of neighborhood poverty when neighborhood poverty was measured at the census-tract- or zip-code-level as opposed to the block-group level. This finding is likely attributable to census tracts and zip codes covering larger geographic areas than block groups. For instance, the median size of block groups in Tennessee is roughly two square miles, whereas the median sizes of census tracts and zip codes in Tennessee are roughly 8 and 70 square miles, respectively. In other words, when conceptualizing neighborhoods at a broader geographic scale than the surrounding block groups in which children lived, this study found no evidence that neighborhood poverty moderates preK effects. That is, neighborhoods mattered at a granular level (Chetty, Friedman, Hendren, Jones, & Porter, 2018) such that incorporating characteristics of communities only miles away diminishes the moderating effects of community-level poverty rates. This finding indicates that whatever mechanisms might be driving the observed impact heterogeneity documented in the preceding pages likely operated at an acutely local scale.

## Rival Moderator Analysis

Supplementary Appendix Tables I.1 to I.4 report results from the rival moderator analysis. This analysis examined the extent to which differential effects of VPK across levels of neighborhood poverty were explained by alternative childcare options, contextual risk factors, family SES, and urbanicity.[4] Overall, no evidence was found that the moderating role of neighborhood poverty on VPK effects was explained by lower-SES families being more likely to live in high-poverty neighborhoods. In fact, allowing the effect of VPK to vary across family SES amplified effect heterogeneity across levels of neighborhood poverty for both math and reading achievement. That is, impact heterogeneity across neighborhood poverty was simply not detecting differences in family SES, suggesting that there were likely "neighborhood effects" at play (Sharkey & Faber, 2014). For reading achievement, neighborhood effects operated to some extent through the distribution of alternative childcare options, which explained around 3% of the differential VPK effect across levels of neighborhood poverty, and contextual risk factors, which explained 6% to 7% of this variation. For math achievement, the relatively smaller differential effects described in the previous section were explained, in part, by contextual risk factors, which explained 1% to 3% of the differential VPK effects across levels of neighborhood poverty. Finally, urbanicity accounted for just 1% of the variation in VPK effects across levels of neighborhood poverty for reading achievement and 2% to 5% for math.

## Discussion

The promise of preK finds its origins in evaluations of several small-scale early education demonstration programs, designed and run by researchers during the 1970s and 80s, that served low-income children living in disadvantaged areas of their respective cities. However, the expansion of prekindergarten in recent years to communities both within and outside metropolitan areas, to disinvested inner cities and affluent suburbs alike, raises the natural question of whether it is reasonable to expect similar effects for children living in such varied environments, environments that existing evidence suggests likely influence the counterfactual condition to which preK effects are compared (e.g., Wodtke & Harding, 2012). This study builds on these observations and leverages data from a recently conducted randomized controlled trial of VPK in Tennessee and demonstrates that VPK effects were contingent on the neighborhoods in which children lived.

This study estimates that the effect of VPK *assignment* on third-grade reading achievement for the full RCT sample was 0.39 *SD*s larger for children living in the high- compared with low-poverty neighborhoods in Tennessee, and that the average effect of VPK assignment for those living in high-poverty neighborhoods was positive and statistically significant. Similarly, this study estimates that the effect of VPK *enrollment* on third-grade reading achievement in the full RCT sample was fully 0.83 *SD*s larger for children in high- compared with low-poverty neighborhoods; likewise, the marginal effect of VPK enrollment on reading for those in high-poverty neighborhoods was significant and substantively meaningful. Evidence for impact heterogeneity with regard to math achievement was also observed, but this heterogeneity did not correspond with statistically meaningful effects for children living in high-poverty neighborhoods as it did for reading achievement. These conclusions were largely corroborated in the ISS sample, which allowed for a broader set of baseline controls including indicators of family SES and baseline levels of academic achievement.

Moreover, this study found no evidence that this impact heterogeneity across neighborhoods was simply detecting differences in family-level SES.

However, this study found evidence that the differential VPK effect across neighborhood contexts may have been attributable, in part, to the distribution of alternative childcare options and the types of risk factors children were exposed to in their neighborhood contexts. In particular, larger VPK effects were observed for children living in higher poverty neighborhoods, in part, because these children had fewer alternative childcare options and because they were overexposed to contextual risk factors relative to their peers living in low-poverty neighborhoods.

Last, despite finding compelling evidence that VPK effects differed across neighborhood contexts for reading achievement, less evidence of impact heterogeneity was observed for math achievement. Moreover, despite finding strong evidence of positive VPK effects on the reading achievement of children in high-poverty neighborhoods, no such evidence was found with regard to children's math achievement. This differential pattern of effects for reading versus math achievement could be the result of several factors. One speculative explanation for the persistence of effects for reading but not for math may be related to the practical emphasis of VPK curriculum. Despite curricular mandates established by the Tennessee Department of Education that required preschool teachers to cover both math and literacy skills, there is often variation in how preschool teachers implement curricular requirements (Hamre et al., 2010), with some evidence that instructional time for literacy can exceed that for math (Farran, Meador, Christopher, Nesbitt, & Bilbrey, 2017). If preschools in the analytic sample placed heavier emphasis on the development of early literacy skills than numeracy skills, then it is reasonable that effects on literacy, relative to those on math, would be more likely to make up for any literacy-related disadvantages at home and in communities.

A related explanation for why neighborhood poverty had a strong moderating effect on VPK with regard to reading achievement but less so for math achievement is that neighborhood poverty may not be as consequential for children's math relative to their reading achievement. In fact, Tables 3 and 4, which report moderated ITT and LATE estimates for the full and ISS samples, respectively, reveal that children living in a high-poverty neighborhood at baseline in the absence of VPK consistently performed worse, on average, than children living in lower-poverty neighborhoods on third-grade reading achievement across the full and ISS samples and with respect to ITT and LATE estimates. In contrast, the adjusted difference in third-grade math achievement between children living in high- versus low-poverty neighborhoods in the absence of VPK were of a smaller magnitude and were measured less precisely than those for third-grade reading achievement. Therefore, the differential moderating effects of neighborhood poverty on third-grade reading versus math achievement may simply be due to the fact that neighborhood poverty was more consequential for children's reading than math achievement.

These findings highlight several key considerations for theory, policy, and research for early childhood education. First, this study provides considerable evidence that understanding the distribution and prevalence of neighborhood poverty may be important for understanding nonschooling determinants of preK effect persistence (Bailey, Duncan, Odgers, & Yu, 2017). Moreover, that VPK effects persisted through third grade for children living in high-poverty neighborhoods is particularly noteworthy in light of several recent experimental and quasi-experimental studies that documented within their respective study populations near-universal fadeout of preK effects by third grade (Hill, Gormley, & Adelstein, 2015; Lipsey et al., 2017; Puma, Bell, Cook, & Heid, 2010; Puma et al., 2012). Second, that positive VPK impacts accrued for children in high-poverty neighborhoods is consistent with recent studies that document the importance of early childhood interventions for safeguarding children from periods of elevated developmental vulnerability due, in this case, to disadvantaged neighborhood contexts (Dodge et al., 2008).

Third, that effects hinged in either direction depending on children's level of neighborhood poverty suggest that a responsive policy agenda that aims to maximize a population-level preschool effect would likely need to include provisions that aim to increase the share of children from high-poverty neighborhoods in preschool classrooms. This inclusion could be accomplished in at least two ways. First, site-selection policies could prioritize establishing VPK centers in high-poverty areas, which would presumably expand VPK access to the children most likely to benefit. This is especially relevant because there is evidence in the current study that more children attended VPK in low-poverty neighborhoods than there were VPK children residing in low-poverty neighborhoods (see Supplementary Appendix Table A.1). This pattern of results suggests that there may have been inequities in VPK access based on neighborhood poverty. However, it is possible that suboptimal effects may be observed if increasing the number of preK classrooms in high-poverty neighborhoods concentrates large numbers of disadvantaged children in the same preK classroom as opposed to more economically integrated classrooms (Miller, Votruba-Drzal, McQuiggan, & Shaw, 2017; van Ewijk & Sleegers, 2010).

A second, alternative approach could focus on mitigating potential barriers, such as transportation and information gaps, that prevent children living in high-poverty neighborhoods from attending VPK centers in low-poverty neighborhoods (Greenberg, Michie, & Adams, 2018; J. F. Harding & Paulsell, 2018). Moreover, a policy agenda that endeavors to provide such access—that is, access that might increase the number of low-income children attending VPK classrooms in lower-poverty neighborhoods—would likely promote racial and socioeconomic integration of VPK classrooms as well given the strong correlation between neighborhood poverty, family SES, and racial composition (Reardon & Bischoff, 2011).

*Pearman*

## Limitations

Although this study extends prior early childhood research by highlighting the importance of residential contexts for understanding who did and did not benefit from VPK, this study is not without several limitations. The analytic samples suffered from nonnegligible attrition due to misreporting of baseline addresses and incomplete outcome data.[5] Moreover, the analytic samples were composed exclusively of economically disadvantaged children, whom TN-VPK targets for enrollment. Therefore, it is unclear whether the variation in treatment effects across neighborhood environments that was documented in the preceding pages would hold in a more socioeconomically diverse sample of children.[6] Finally, this study concerned academic outcomes only during third grade. Future research will need to identify whether the results described in this study pertain to preschool programs in other states, whether the moderated effects also pertain to behavioral or noncognitive outcomes (e.g., disciplinary outcomes, grade retention, attendance, executive function), and whether differential effects on outcomes of interest pertain to earlier grades or persist beyond third grade. Finally, in light of the limited explanatory power of the rival moderators examined in this study, future research will need to examine further the underlying mechanisms linking neighborhood poverty to children's engagement and response to early childhood programs.

## Conclusion

Limitations notwithstanding, the current study provides important new evidence about variation in preK effects. Based on data from an RCT of a statewide preK program in Tennessee, this study finds that preK effects increase with level of poverty in children's residential neighborhoods, with positive effects accruing to low-income children in high-poverty neighborhoods but negative effects accruing to low-income children in low-poverty neighborhoods. This pattern was partially attributable to the distribution of alternative childcare options and the extent to which children were exposed to contextual risk factors in their neighborhoods. The analyses presented here should encourage future research into how children's neighborhoods relate to the implementation and efficacy of preschool interventions, especially those implemented at scale. In particular, if this study is to serve as any guide, recent efforts to expand preK to broader populations of children should double down on efforts at identifying *where* effects might be maximized.

## Notes

The author would like to thank Dale Farran, Mark Lipsey, Kelley Durkin, Caroline Christopher, Georgine Pion, Mark Lachowicz, Matt Springer, and Walker Swain for their invaluable support and critical feedback as this paper developed. The author is also greatly indebted to the irreplaceable team of research assistants and project coordinators associated with the Tennessee Voluntary PreKindergarten Study, especially Janie Hughart, Richard Feldser, and Ilknur Sekmen, as well as to the Tennessee Department of Education and the Tennessee Education Research Alliance. The author would also like to thank Greg Duncan and participants in the 2017 Society for Research on Educational Effectiveness Annual Conference for a constructive discussion about this study. Additionally, the author wishes to thank Christy McGuire, Elizabeth Votruba-Drzal, Portia Miller, Tom Dee, Sean Reardon, and several anonymous reviewers who provided helpful comments about various aspects of this study. Financial support from the National Institutes of Health (NIH) is gratefully acknowledged (5R01HD079461-02).

Supplemental material for this article is available in the online version of the journal.

[1] Despite the TN-VPK Program being a statewide program with curricular and staffing mandates, the quality of VPK programs may vary across the state (Valentino, 2018). For instance, McCoy et al. (2015) found neighborhood poverty to be negatively correlated with the quality of Head Start classrooms. Therefore, children living in disadvantaged neighborhoods may also be more likely to attend a lower quality VPK program. Thus, if one was interested in the effect of statewide VPK controlling for VPK quality, then one would expect the estimated effects of VPK on the achievement of children in high-poverty neighborhoods to be plausible lower bounds.

[2] Using moderate-poverty neighborhoods as the referent group does not change substantive conclusions. In particular, virtually all instances in which preK effects differed between low- and high-poverty neighborhoods also varied between moderate- and high-poverty neighborhoods. The key implication is that positive preK effects were restricted to children in high-poverty neighborhoods, and these effects differed from those for children in lower-poverty neighborhoods.

[3] Differential access to alternative childcare options was measured by the density of childcare establishments per capita in the surrounding community. The number of childcare establishments was gathered from the 2010 Zip Business Patterns database and rescaled as a per capita rate based on the number of children younger than 5 years in the zip code. (Note that these estimates were only available at the zip code level.) Differential exposure to contextual risk factors was approximated with two measures: the share of single-parent families in children's neighborhood and the per capita crime rate in the surrounding community. The share of single-parent families was measured at the block-group level and gathered from the 2008–2012 American Community Survey. Per capita crime rate was gathered from the Tennessee Bureau of Investigation and measured as the number of violent and property crime arrests per capita in year 2010. (Similar to data about alternative childcare options, crime data were only available at the zip code level.) Family SES was measured by the mother's level of education. As described above, data about children's family-level SES were only available for the ISS sample. Finally, urbanicity was measured as a binary indicator of whether a child lived in an urbanized area or urban cluster, as defined by the U.S. Census Bureau. Urban areas form the cores of metropolitan statistical areas and contain 50,000 or more people, whereas urban clusters form the cores of micropolitan statistical areas and contain 2,500 to 50,000 people.

[4] Supplementary Appendix C (Figure C.1, Table C.1) shows that each rival moderator is correlated with neighborhood poverty, with children living in high-poverty neighborhoods having (1) less access to alternative childcare options, (2) overexposure to crime and violence, (3) overexposure to single-parent households in their communities, (4) mothers with less education, and (5) increased likelihood of living in an urban area. These correlations suggest that a rival moderator analysis—one that examines changes in the moderating capacity of neighborhood poverty after accounting for its correlation with each factor—is warranted.

[5] This study showed that results were robust to the exclusion of sampling weights that accounted for students' differential probability of being included in the analytic samples based on observed characteristics (see Supplementary Tables C.8 to C.12). Moreover, results were robust to the inclusion in the analytic sample of a subset of children who could not be geolocated to a street address but could be geolocated to the centroid of their home zip code. Thus, there is little reason to suspect that analytic samples were substantively different from their full RCT and ISS samples.

*Pearman*

[6]Prior research has shown that children in nonpoor families are less vulnerable to adverse neighborhood conditions (Wodtke & Harding, 2012), suggesting that the low-income children included in the sample may have been especially susceptible to the type of impact heterogeneity observed. Consequently, findings should not be generalized to universal preK programs that "target" a broader range of children.

## ORCID iD

Francis A. Pearman II https://orcid.org/0000-0002-9516-8302

## References

Bailey, D., Duncan, G. J., Odgers, C. L., & Yu, W. (2017). Persistence and fadeout in the impacts of child and adolescent interventions. *Journal of Research on Educational Effectiveness, 10*, 1934–5747. doi:10.1080/19345747.2016.1232459

Berruate-Clement, J. R. (1984). *Changed lives: The effects of the Perry Preschool Program on youths through age 19* [Monographs of the High/Scope Educational Research Foundation, Number Eight]. Retrieved from http://files.eric.ed.gov/fulltext/ED313128.pdf

Bloom, H. S., & Weiland, C. (2015). *Quantifying variation in Head Start effects on young children's cognitive and socio-emotional skills using data from the National Head Start Impact Study*. Retrieved from https://www.mdrc.org/sites/default/files/quantifying_variation_in_head_start.pdf

Brooks-Gunn, J., Johnson, A. D., & Leventhal, T. (2010). Disorder, turbulence, and resources in children's homes and neighborhoods. In *Chaos and its influence on children's development: An ecological perspective* (pp. 155–170). Washington, DC: American Psychological Association. doi:10.1037/12057-000

Bureau of the Census. (2008). Census Block Group Program for the 2010 Decennial Census—Final criteria. *Federal Register, 73*(51). Retrieved from https://www2.census.gov/geo/pdfs/reference/fedreg/bg_criteria.pdf

Chetty, R., Friedman, J. N., Hendren, N., Jones, M. R., & Porter, S. R. (2018, October). *The Opportunity Atlas: Mapping the childhood roots of social mobility* (NBER Working Paper No. 25147). Cambridge, MA: National Bureau of Economic Research. doi:10.3386/w25147

Chetty, R., Hendren, N., & Katz, L. F. (2015). The effects of exposure to better neighborhoods on children: New evidence from the Moving to Opportunity experiment. *American Economic Review, 106*(4), 1–88. doi:10.1257/aer.20150572

Cohen, S. (2017). *3 principles to improve outcomes for children and families*. Cambridge, MA: Harvard University, Center on the Developing Child. Retrieved from https://46y5eh11fhgw3ve3ytpwxt9r-wpengine.netdna-ssl.com/wp-content/uploads/2017/10/HCDC_3PrinciplesPolicyPractice.pdf

Cooper, B. R., & Lanza, S. T. (2014). Who benefits most from Head Start? Using latent class moderation to examine differential treatment effects. *Child Development, 85*, 2317–2338. doi:10.1111/cdev.12278

Dodge, K. A., Greenberg, M. T., & Malone, P. S. (2008). Testing an idealized dynamic cascade model of the development of serious violence in adolescence. *Child Development, 79*, 1907–1927. doi:10.1111/j.1467-8624.2008.01233.x

Farran, D. C., Meador, D., Christopher, C., Nesbitt, K. T., & Bilbrey, L. E. (2017). Data-driven improvement in prekindergarten classrooms: Report from a partnership in an urban district. *Child Development, 88*, 1466–1479. doi:10.1111/cdev.12906

Fitzpatrick, M. D. (2008). Starting school at four: The effect of universal pre-kindergarten on children's academic achievement. *B. E. Journal of Economic Analysis & Policy, 8*(1), 1–38. doi:10.2202/1935-1682.1897

Flora, C. B., Flora, J. L., & Gasteyer, S. P. (2015). *Rural communities: Legacy+ change.* Boulder, CO: Westview Press.

Galster, G. (2012). The mechanism(s) of neighbourhood effects: Theory, evidence, and policy implication. In M. van Ham, D. Manley, N. Bailey, L. Simpson, & D. Maclennan (Eds.), *Neighbourhood effects research: New perspectives* (pp. 23–56). New York, NY: Springer.

Gormley, W. T., Phillips, D., & Gayer, T. (2008). Preschool programs can boost school readiness. *Science, 320,* 1723–1724. doi:10.1126/science.1156019

Green, T. (2015). Places of inequality, places of possibility: Mapping "opportunity in geography" across urban school-communities. *Urban Review, 47,* 717–741. doi:10.1007/s11256-015-0331-z

Greenberg, E., Michie, M., & Adams, G. (2018). *Expanding preschool access for children of immigrants.* Washington, DC: Urban Institute. Retrieved from https://www.urban.org/sites/default/files/publication/96546/expanding_preschool_access_for_children_of_immigrants_0.pdf

Hamre, B. K., Justice, L. M., Pianta, R. C., Kilday, C., Sweeney, B., Downer, J. T., & Leach, A. (2010). Implementation fidelity of MyTeachingPartner literacy and language activities: Association with preschoolers' language and literacy growth. *Early Childhood Research Quarterly, 25,* 329–347. doi:10.1016/J.ECRESQ.2009.07.002

Harding, D. J. (2003). Counterfactual models of neighborhood effects: The effect of neighborhood poverty on dropping out and teenage pregnancy. *American Journal of Sociology, 109,* 676–719. doi:10.1086/379217

Harding, J. F., & Paulsell, D. (2018). *Improving access to early care and education: An equity-focused policy research agenda.* Princeton, NJ: Mathematica.

Hill, C. J., Gormley, W. T., & Adelstein, S. (2015). Do the short-term effects of a high-quality preschool program persist? *Early Childhood Research Quarterly, 32,* 60–79. doi:10.1016/j.ecresq.2014.12.005

Jacob, B. A., & Wilder, T. (2010). *Educational expectations and attainment* (NBER Working Paper No. 15683). Cambridge, MA: National Bureau of Economic Research. Retrieved from https://www.nber.org/papers/w15683

Jargowsky, P. (1997). *Poverty and place: Ghettos, barrios, and the American city.* New York, NY: Russell Sage Foundation.

Jocson, K. M., & Thorne-Wallington, E. (2013). Mapping literacy-rich environments: Geospatial perspectives on literacy and education. *Teachers College Record, 115*(June), 1–24.

Karch, A. (2010). Policy feedback and preschool funding in the American states. *Policy Studies Journal, 38,* 217–234. doi:10.1111/j.1541-0072.2010.00359.x

Lipsey, M. W., Farran, D., & Durkin, K. (2017). Effects of the Tennessee Prekindergarten Program on children's achievement and behavior through third grade. *Early Childhood Research Quarterly, 45*(4th quarter 2018), 155–176. doi:10.1016/j.ecresq.2018.03.005

Lipsey, M. W., Hofer, K. G., Dong, N., Farran, D. C., & Bilbrey, C. (2013, August). *Evaluation of the Tennessee Voluntary Prekindergarten Program: Kindergarten and first grade follow-up results from the randomized control design.* Nashville, TN: Peabody Research Institute. Retrieved from https://my.vanderbilt.edu/tnprekevaluation/files/2013/10/August2013_PRI_Kand1st Followup_TN-VPK_RCT_ProjectResults_FullReport1.pdf

McCoy, D. C., Connors, M. C., Morris, P. A., Yoshikawa, H., & Friedman-Krauss, A. H. (2015). Neighborhood economic disadvantage and children's cognitive and social-emotional development: Exploring Head Start classroom quality as a mediating mechanism. *Early Childhood Research Quarterly, 32*, 150–159. doi: 10.1016/j.ecresq.2015.04.003

McCoy, D. C., Morris, P. A., Connors, M. C., Gomez, C. J., & Yoshikawa, H. (2016). Differential effectiveness of Head Start in urban and rural communities. *Journal of Applied Developmental Psychology, 43*, 29–42. doi:10.1016/j.appdev.2015.12.007

Miller, P., Votruba-Drzal, E., McQuiggan, M., & Shaw, A. (2017). Pre-K classroom-economic composition and children's early academic development. *Journal of Educational Psychology, 109*, 149–165. doi:10.1037/edu0000137

Milner, H. R. (2015). *Rac(e)ing to class: Confronting poverty and race in schools and classrooms.* Cambridge, MA: Harvard Education Press.

Morris, P. A., Connors, M., Friedman-Krauss, A., McCoy, D., Weiland, C., Feller, A., & Page, L. (2017). New findings on impact variation from the Head Start Impact Study: Informing the scale-up of early childhood programs. *AERA Open, 4*(2), 1–16. doi:10.1177/2332858418769287

Puma, M., Bell, S., Cook, R., & Heid, C. (2010). *Head Start impact study: Final report.* Washington, DC: U.S. Department of Health and Human Services. Retrieved from http://files.eric.ed.gov/fulltext/ED507845.pdf

Puma, M., Bell, S., Cook, R., Heid, C., Broene, P., Jenkins, F., . . . Downer, J. (2012). *Third grade follow-up to the Head Start Impact Study: Final report.* Washington, DC: U.S. Department of Health and Human Services. Retrieved from http://files.eric.ed.gov/fulltext/ED539264.pdf

Reardon, S. F., & Bischoff, K. (2011). Income inequality and income segregation. *American Journal of Sociology, 116*, 1092–1153. doi:10.1086/657114

Rosenbaum, P. R., & Rubin, D. B. (1983). The central role of the propensity score in observational studies for causal effects. *Biometrika, 70*, 41–55.

Rubin, D. B. (1987). *Multiple imputation for nonresponse in surveys.* New York, NY: Wiley.

Schochet, P. Z., & Deke, J. (2014). *Understanding variation in treatment effects in education impact evaluations: An overview of quantitative methods.* Washington, DC: U.S. Department of Education. Retrieved from https://ies.ed.gov/ncee/pubs/20144017/pdf/20144017.pdf

Sharkey, P., & Faber, J. W. (2014). Where, when, why, and for whom do residential contexts matter? Moving away from the dichotomous understanding of neighborhood effects. *Annual Review of Sociology, 40*, 559–579. doi:10.1146/annurev-soc-071913-043350

Sharkey, P., Tirado-Strayer, N., Papachristos, A. V., & Raver, C. C. (2012). The effect of local violence on children's attention and impulse control. *American Journal of Public Health, 102*, 2287–2293. doi:10.2105/ajph.2012.300789

Small, M. L. (2008). Four reasons to abandon the idea of the ghetto. *City and Community, 7*, 389–398. doi:10.1111/j.1540-6040.2008.00271_8.x

Swenson, K. (2008). *Child care arrangements in urban and rural areas.* Washington, DC: Office of the Assistant Secretary for Planning and Evaluation.

Tate, W. F. (2008). "Geography of opportunity": Poverty, place, and educational outcomes. *Educational Researcher, 37*, 397–411. doi:10.3102/0013189X08326409

Tennessee Department of Education. (2017). *Scope of services: Voluntary Pre-K 2017-2018.* Retrieved from https://www.ecschools.net/docs/district/welc/vpk%20scope%20of%20services%2017-18.pdf?id=661

Valentino, R. (2018). Will public pre-K really close achievement gaps? Gaps in prekindergarten quality between students and across states. *American Educational Research Journal, 55*, 79–116. doi:10.3102/0002831217732000

van Ewijk, R., & Sleegers, P. J. C. (2010). The effect of peer socioeconomic status on student achievement: A meta-analysis. *Educational Research Review, 5*, 134–150.

VanderWeele, T. J., & Knol, M. J. (2011). Interpretation of subgroup analyses in randomized trials: heterogeneity versus secondary interventions. *Annals of Internal Medicine, 154*, 680–683. doi:10.7326/0003-4819-154-10-201105170-00008

Votruba-Drzal, E., Miller, P., & Coley, R. L. (2016). Poverty, urbanicity, and children's development of early academic skills. *Child Development Perspectives, 10*, 3–9. doi:10.1111/cdep.12152

Wilson, W. J. (2012). *The truly disadvantaged: The inner city, the underclass, and public policy.* Chicago, IL: University of Chicago Press.

Wodtke, G. T. (2013). Duration and timing of exposure to neighborhood poverty and the risk of adolescent parenthood. *Demography, 50*, 1765–1788. doi:10.1007/s13524-013-0219-z

Wodtke, G. T., Elwert, F., & Harding, D. J. (2016). Neighborhood effect heterogeneity by family income and developmental period. *American Journal of Sociology, 121*, 1168–1222. doi:10.1086/684137

Wodtke, G. T., & Harding, D. J. (2012). *Poor families, poor neighborhoods: How family poverty intensifies the impact of* concentrated disadvantage (Population Studies Center Research Report No. 12-776). Retrieved from https://www.psc.isr.umich.edu/pubs/abs/7739

Wodtke, G. T., Harding, D. J., & Elwert, F. (2011). Neighborhood effects in temporal perspective: The impact of long-term exposure to concentrated disadvantage on high school graduation. *American Sociological Review, 76*, 713–736. doi:10.1177/0003122411420816

Yoshikawa, H., Weiland, C., Brooks-Gunn, J., Burchinal, M., Gormley, W., Ludwig, J., & Zaslow, M. J. (2013). *Investing in our future: The evidence base on preschool education.* New York, NY: Foundation for Child Development.

Manuscript received November 9, 2017
Final revision received July 16, 2019
Accepted August 2, 2019

# Giving Community College Students Choice: The Impact of Self-Placement in Math Courses

Holly Kosiewicz
*University of Southern California*
Federick Ngo
*University of Nevada, Las Vegas*

*This study examines the impact of a "natural experiment" that gave students the choice to place into or out of developmental math because of an unintended mistake made by a community college. During self-placement, more students chose to enroll in gateway college- and transfer-level math courses, however, greater proportions of female, Black, and Hispanic students enrolled in the lowest levels of math relative to test-placed counterparts. Difference-in-difference estimates show that self-placement led to positive outcomes, but mostly for White, Asian, and male students. This evidence suggests areas of concern and potential for improvement for self-placement policies. Self-determination theory, behavioral decision theory, and stereotype vulnerability provide possible explanations for the observed changes.*

KEYWORDS: behavioral decision theory, developmental education, self-determination theory, self-placement, stereotype vulnerability

Developmental education is an intervention explicitly designed and principally used by community colleges to develop or reinforce English, math, and reading skills for incoming students with weak academic preparation. While estimates indicate that over 60% of entering community college students are placed into at least one developmental education course (Chen, 2016), less than half actually finish their developmental education sequence, and far fewer pass a transfer-level college course within 3 years of initial

---

HOLLY KOSIEWICZ, received her PhD in education policy from the University of Southern California. Her contact address is 5214 Grover Avenue, Austin, TX 78789; e-mail: *hkosiewicz@gmail.com*. She is interested in evaluating higher education policy using quantitative and qualitative methods to improve outcomes for less advantaged students. Since the time of the research, her affiliation has changed to the Texas Higher Education Coordinating Board.

FEDERICK NGO, PhD, is an assistant professor at the University of Nevada, Las Vegas. His research examines how higher education policy and practice affects college access and success, with a particular focus on community college students.

enrollment (Bailey, Jeong, & Cho, 2010). Because of these low success-rates, some policymakers and researchers argue that developmental education has failed to fulfill its promise to help students succeed in and complete college (Bailey & Jaggars, 2016; Bettinger, Boatman, & Long, 2013; Burdman, 2012; MDRC, 2013; Smith, 2015).

An aspect of developmental education that policymakers and researchers have recognized as in need of significant overhaul is placement testing, which is commonly used to identify readiness for college-level coursework (Burdman, 2012). Although most community colleges have traditionally used placement tests to assist in assigning students to math and English courses (Hughes & Scott-Clayton, 2011), evidence suggests that these placement tests may actually misdiagnose a student's need for developmental education. Researchers have estimated that as many as one quarter of community college students who take a math placement test are incorrectly assigned to developmental education and that thousands of remediated students would have received an A or B in college-level courses based on their high school performance or other measures (Scott-Clayton, Crosta, & Belfield, 2014). Accordingly, a number of states are moving away from reliance on placement testing and seeking alternatives to solve problems associated with these tests (Rutschow & Mayer, 2018).

This evidence raises an important question: What would happen if we were to let community college students decide for themselves where to start their academic trajectory? For some time, reformers have been advocating for giving students greater latitude over decisions that determine the courses they take in college (Royer & Gilles, 2003). The strategy of self-placement, when students choose their own coursework, or directed self-placement (DSP), when students choose their own coursework with some guidance from an advisor or faculty, are seen as alternatives to test-based placement, which disregard student input in placement decisions (Royer & Gilles, 1998). Advocates of self-placement theorize that a placement strategy that involves students increases placement accuracy by factoring in information that placement tests miss, such as the student's motivation to succeed, and facilitates a student's intrinsic motivation, which is crucial for their short- and long-term academic success.

Self-placement strategies for the purposes of assigning students to developmental or gateway college-level courses have gained traction in recent years among decisionmakers seeking to improve community college student outcomes. For example, both Florida and Connecticut passed legislation in 2013 that embedded elements of self-placement into state-wide placement policies for developmental education (Fain, 2013). More specifically, in Florida, students who entered ninth grade in a Florida public school in 2013–2014 or after and earned a standard Florida high school diploma, or who are active duty service members, have discretion over whether to enroll in developmental education or directly in gateway college-level courses (SB

1720). Similarly, in Connecticut, community colleges are required to give students the option to enroll in an intensive college-readiness program instead of enrolling directly into a developmental education course (Public Act 12-140).

As self-placement strategies increasingly emerge as a way to improve course placement and student success in college, a small but growing body of research has examined whether such strategies actually benefit students. An evaluation of Florida's SB 1720 found that its version of self-placement was associated with an increase in the percentage of students passing developmental education courses and enrolling in gateway college-level courses; however, the study also found a decrease in the percentage of students passing gateway courses (Hu et al., 2016). In particular, Black students and female students were more likely to take developmental education courses relative to other groups following the policy change (Park, Woods, Hu, Jones, & Tandberg, 2018). Although these studies provide an assessment of Florida's policy change, the evaluation designs are descriptive and do not provide causal estimates of self-placement. Because of these limitations, it remains unclear whether giving students greater latitude over these placement decisions, particularly in the context of community colleges, has the potential to generate an academic payoff.

This study fills this knowledge gap by focusing on one community college's instituted change from a test-based placement strategy to a self-placement strategy to assign students into developmental education. In the spring of 2008, administrators unintentionally forgot to renew the college's license to use ACT's COMPASS, a computerized placement test, to assign students to developmental math. As a result, students in the summer and fall of 2008 bypassed previously mandatory testing and were instead allowed to choose their initial math course after receiving guidance from a college advisor. This meant that for two terms, the college allowed students who had not been tested for math to self-place into developmental or college-level math courses of their choice. This change in policy offers a unique natural experiment to estimate the effects of DSP on initial course fit and academic success.

Because the change to a self-placement policy happened so abruptly, and because other colleges in the same district did not make any contemporaneous placement policy changes, we use a difference-in-difference identification strategy to isolate the causal effects of DSP on short- and long-term academic success. A unique feature of the study is that we measure the impact of self-placement on course fit by comparing students' probability of withdrawal from and failure of initial math courses under both test-based placement and self-placement. We also evaluate short- and long-term success by examining the impact of DSP on completing a college-level math course (e.g., elementary algebra) or a transferable math course (e.g., precalculus), and the credits required to earn an associate's degree. We disaggregate our results by race, sex, and initial placement level to understand how DSP may affect different student populations.

*Giving Community College Students Choice*

Aggregate-level findings show that students who had the opportunity to self-place versus those who were test-placed into math experienced improved self-perceived course fit (i.e., reduced rates of withdrawal) and improved academic outcomes (i.e., meeting the math requirement for an associate degree). However, when we disaggregated our results by race, sex, and initial math level, we find that the benefits conferred by DSP were far greater for White, Asian, and male students and students who initially chose higher-level math courses. We found no or substantially reduced benefits from self-placement on most outcomes for female, Black, and Hispanic students. These discrepancies may stem from the fact that female, Black, and Hispanic students were more likely to choose or counseled into lower-level math courses than their test-placed counterparts. We draw on insights from self-determination theory (SDT), behavioral decision theory, and scholarship exploring stereotype vulnerability to explain these results. Overall, this evidence points to the nuanced effectiveness of education policy, and the danger of scaling policies that might have negative or unequal repercussions for certain student populations. Specifically, it should behoove decision makers considering mandating guided self-placement to offer evidence-based guidance to practitioners on how they can address potential inequities that may result from this policy change.

## Assessment and Placement in College Mathematics

Of the nearly 60% of community college students placed into developmental education, only one third complete developmental education sequences, and very few actually earn a postsecondary credential (Bailey et al., 2010; Chen, 2016). These poor academic outcomes have motivated inquiry into the effectiveness of strategies that assess college readiness (Burdman, 2012), such as placement tests, which have traditionally been the most common instruments used to assign students to developmental or college-level coursework (Hughes & Scott-Clayton, 2011). Sole reliance on these tests is decreasing as states and institutions pass placement reforms (Rutschow & Mayer, 2018), but they are still one of the first experiences many students have when they enroll in community colleges (Hughes & Scott-Clayton, 2011).[1]

Multiple reasons explain why community colleges use tests to help make developmental education assignments. First, commercially available instruments have been tested for validity and reliability, which can relieve colleges from needing to invest significant resources to conduct those analyses (Brown & Niemi, 2007). Second, commercially available placement tests—notably the ACCUPLACER and the now discontinued COMPASS—also allow community college administrators to test students efficiently since they can refer students to multiple subtests and produce a final placement within a single testing session (Melguizo, Kosiewicz, Prather, & Bos, 2014).

Despite these benefits, recent research raises questions about whether placement tests should be used as the sole means of assigning students to developmental math. For instance, several studies suggest that these tests have low predictive validity stemming from weak correlations between assessment scores and success rates in developmental and college-level courses (Scott-Clayton et al., 2014). Other studies show that students are oftentimes unaware of the high-stakes nature of placement tests. Consequently, many do not prepare for the test and receive results that do not reflect their true academic abilities (Safran & Visher, 2010; Venezia, Bracco, & Nodine, 2010).

To address the limitations of test-based placement policies, a growing number of states, including Texas, Connecticut, California, Florida, and North Carolina, have passed reforms to change how students are assigned to developmental education (Education Commission of the States, n.d.). These efforts have largely centered on incorporating "holistic advising practices" or "multiple measures," which are cognitive and noncognitive factors beyond test scores into placement policies. This revised placement approach is based on research demonstrating that integrating holistic factors or multiple measures (e.g., prior coursework, GPA, test scores, motivation, self-efficacy) into placement decisions can improve placement accuracy (Ngo, Chi, & Park, 2018; Ngo & Kwon, 2015; Scott-Clayton et al., 2014). An updated institutional survey found that more than half of community colleges now use multiple measures to make placement decisions (Rutschow & Mayer, 2018).

While multiple measures policies hold promise for improving placement accuracy, their incorporation nevertheless presents a significant logistical challenge for many institutions. The resources needed to gather and verify high school transcripts and batch process them to inform course placement may be administratively burdensome for already resource-strapped community colleges (Melguizo et al., 2014). Furthermore, although researchers from the RAND Corporation, American Institutes for Research, and the Center for the Analysis of Postsecondary Readiness are conducting studies to test the impact of alternative placement approaches on placement accuracy, institutions still lack evidence-based guidance to determine which factors actually improve placement for students with different characteristics.[2] Although high school GPA has been shown to be the best predictor of success in both college math and English courses (Bahr et al., 2019), research finds that there are few examples of high-quality data collaboration between K–12 and postsecondary education institutions (Grady, 2016). It is thus logistically difficult that community colleges would transition to a GPA placement policy to assign students to developmental education. For these reasons, policymakers may be seeking a simpler approach to reform the way that institutions assign students to developmental education. Allowing students to determine the extent that they are ready for college-level coursework is one such approach.

## Understanding How Directed Self-Placement May Affect Student Success

A self-placement strategy to assign students to developmental education may be a solution to the problems associated with test placement and the logistical difficulty of incorporating multiple measures into placement decisions. At its core, self-placement strategies involve students in decisions that determine whether an intervention is needed to ensure their success in school. DSP, a version of a self-placement strategy, is the placement process by which a student is allowed to choose, with consultation from faculty or an advisor, the developmental or entry-level college course they deem aligns closest with their scholastic background, academic and career goals, and constraints (Royer & Gilles, 1998). To guide our understanding of the ways in which DSP may improve over test-based placement policies in terms of increasing student success and improving placement accuracy, we describe two theories discussed in psychological and economic literatures and apply these theories within the context of this study.

### Self-Determination Theory

Self-determination theory (SDT), with roots in psychology, has made a consequential impact on education and is considered the theory advocates employ to support the implementation of DSP (Royer & Gilles, 1998, 2003). Developed in the 1970s, SDT is a macro-level framework used primarily to understand individual intrinsic motivation and how social contexts may promote or undermine an individual's inherent disposition for growth and development (Ryan & Deci, 2000). For decades, education researchers have drawn on SDT to explain differences in academic performance (Flink, Boggiano, & Barrett, 1990; Grolnick, Ryan, & Deci, 1991; Sheldon & Krieger, 2007), disposition toward self-regulated learning (Zimmerman, 1989), and depth of learning (Benware & Deci, 1984; Grolnick & Ryan, 1987). Practitioners have also employed SDT to reform pedagogical practice to improve student learning and achievement (Niemiec & Ryan, 2009).

In its essence, SDT proposes that students are more self-motivated to perform well in school when their educational environments grant them some degree of latitude over decisions that impact their education. Self-determination theorists contend that academic achievement is enhanced when a student's educational environment satisfies three specific psychological needs—autonomy, competence, and relatedness (Niemiec & Ryan, 2009). Defined, autonomy refers to acting in the interest of the self. For example, a student experiences autonomy when they are given the ability to self-direct how they solve a problem set. Competence refers to the idea that an individual feels effective in carrying out the task at hand. For example, a student may feel competent if they sense they can adequately respond to an assignment. Educational environments can support competence when they offer a student constructive feedback and that student in return feels more capable of completing a task. Finally, relatedness refers to feeling supported in one's environment and a belonging to one's community. While SDT posits that autonomy and competence set the foundation for supporting intrinsic motivation to perform, relatedness for some is also a necessary pillar (Ryan & Deci, 2000). The absence of any of these foundational pillars, according to SDT, can lead to alienation, a lack of initiative, and diminished functioning, factors that decrease academic achievement, persistence, and learning (Schunk, 2011).

Through the lens of SDT, the implementation of DSP theoretically improves success in college over a test-placement strategy primarily by providing an educational environment that fosters a student's psychological disposition toward growth and development. First, it fundamentally shifts the balance of power and puts the decision about whether to take developmental education in the hands of the student. While students receive direction from advisors and faculty, they ultimately determine if, and the level at which, they need developmental education to succeed in college. In this sense, DSP satisfies a student's basic psychological need to feel autonomous in their decision making. This may generate improvements compared with students placed by a test, which, in contrast, strips the student of any input in the placement decision process. Second, because DSP guides a student's decision making with support from a counselor or faculty, DSP may increase a student's sense of competency since the process provides feedback but nevertheless still leaves the decision in the hands of the student. Finally, DSP may signal to students that their input matters in decisions, which may increase the student's sense of belonging in higher education. We examine the validity of SDT within the context of this study by examining how self-placed students fare relative to test-placed students on both short- and long-term academic outcomes.

**Behavioral Decision Theory**

The notion that DSP can improve student achievement relative to a test-based placement policy also rests on the assumption that students are capable of selecting a course that will lead them to greater academic success. A growing body of behavior decision research within the field of education suggests that students often make less than ideal decisions because they have difficulty navigating complex tasks, such as how to apply for college or financial aid (Castleman, 2017), casting doubt that students will make better decisions under a self-placement strategy.

Much of that research draws on behavioral decision theory (BDT), an overarching framework that explains how goal-oriented individuals make decisions under constrained circumstances (Simon, 1955). Rather than viewing individuals as "omniscient calculators" capable of adequately and rationally evaluating the costs and benefits of each alternative (Lupia, McCubbins, & Popkin, 2000), behavioral theorists argue that the decisions individuals make are highly sensitive to individual differences (e.g., memory retrieval, prior experiences, demographic and psychological characteristics), the task's complexity (e.g., choosing between multiple alternatives), and the task's display (e.g., the type of and the way information and informational supports are presented) (Häubl & Trifts, 2000, Payne, Bettman, & Johnson, 1992), among other factors. Empirical studies framed by BDT have consistently found that individuals make suboptimal decisions because they have imperfect information, over-rely on recent experiences or irrelevant features, or misattribute probabilities to certain outcomes (Payne et al., 1992).

Because of these limitations, individuals frequently employ strategies that lead them to ignore alternatives or draw on irrelevant information (Croskerry, 2003), effectively increasing their risk of making an ill-informed decision (Payne et al., 1992). Thus, even if two individuals are presented with the same task and are supported by the same information, they will likely reach different decisions based on their own experiences and how they approach the decision-making process. In this sense, decisions are not calculated with accuracy, but are constructed.

Students faced with the decision to enroll in developmental education, and furthermore in a specific course, are forced into complex situations that make it difficult to determine which course will ultimately lead to success. Under a DSP regime, a student must consider multiple course alternatives; gather, use, and appraise different pieces of information; and weigh the potential benefits against the potential costs of each alternative to determine which course is best suited for them. At each point of this decision-making process, students can reach different determinations on which courses to consider or exclude based on how they read the information and bring their own experience to bear. Given community college students are diverse, not only in terms of their academic experiences and goals but also in terms of the ways in which they interpret and value information and have been socialized to think and feel about math (Cox, 2009), it is likely that they engage with the math DSP process differently, and as a result, make different placement decisions. Students of color and females, in particular, may be more susceptible to inaccurate perceptions of ability because of a vulnerability toward internalizing negative stereotypes about their academic aptitude (Aronson & Good, 2002; Aronson & Inzlicht, 2004; Inzlicht, McKay, & Aronson, 2006; Oyserman, Gant, & Ager, 1995). For these reasons, we examine the placement decisions that self-placed students relative to test-placed students make and disaggregate these results by student demographic characteristics and initial math level.

## Prior Research on Self-Placement

There is relatively little prior empirical research examining the effectiveness of self-placement for improving student success in college. The majority of research is focused on the use of DSP to assign students to developmental or college-level writing courses within the context of 4-year institutions.[3] As a result, research examining the impacts of DSP on success in developmental math in community colleges is almost nonexistent. The few studies that have examined the success of DSP students are descriptive and do not systematically compare the outcomes of students exposed to DSP to those placed under other assignment strategies. This is problematic because students who self-place compared to those who do not may be different along a number of observable and unobservable factors. Without considering these differences, estimates measuring the impact of a self-placement strategy are likely to be biased.

What little research has been done casts DSP in a positive light. For example, Blakesley, Harvey, and Reynolds (2003) descriptively found that DSP was associated with increased pass rates in a basic writing course and a college writing course. In another study, Royer and Gilles (1998) found that students who self-placed in English were more likely to earn credit in a freshman composition class than students who placed into the writing sequence based on their institution's old placement method of examining ACT scores and a writing sample.

There is emerging research investigating the unique version of self-placement being implemented in Florida under SB 1720, which allows students to "opt-out" of remediation. Findings thus far indicate that after the policy shift in 2014, significantly more students took the gateway math courses required for college graduation but pass rates in these courses declined (Hu et al., 2016; Park et al., 2018). But researchers also found that a greater number of students ultimately completed the math required for college graduation. Because these analyses are descriptive, it is difficult to know whether these changes are attributed to the policy shift.

Another challenge facing this analysis is that it does not disentangle the effects of changing placement policy from the effects of reforming advising and instruction. In addition to changing how institutions assign students to developmental education, SB1720 also requires colleges to offer more robust academic advising and to use new instructional methods to teach developmental education courses (Woods et al., 2017).

Although researchers and policymakers are building a better understanding of the overall effects of self-placement on student outcomes, they still lack basic information about how it affects different types of students, an important consideration in developmental education reforms (Kosiewicz, Ngo, & Fong, 2016). Some empirical research has examined the differential effects of DSP, but only with regard to the student's sex. Blakesley et al. (2003) found that males were close to 4 percentage points more likely to choose a lower-level

*Giving Community College Students Choice*

English class than female students (Blakesley et al., 2003). Since research shows that female students and students of color are particularly anxious about math and lean toward believing ability-based stereotypes, they tend to underestimate their math abilities (Maloney & Beilock, 2012; Shapiro & Williams, 2012; Spencer, Steele, & Quinn, 1999; Walton & Spencer, 2009). In the same way that stereotypes might affect performance on tests, students might unintentionally draw on these biased perceptions to self-place in math. Thus, DSP may actually generate worse outcomes than placement testing if it encourages certain student groups to select a less rigorous course than they would have otherwise qualified for, which risks unnecessarily lengthening the amount of time needed to earn a degree. Indeed, Park et al. (2018) noted equity concerns with Florida's self-placement policy, since Black students and female students were less likely to enroll in gateway math courses than White and male students, respectively. Again, whether this is attributable to the policy itself or merely to demographic changes in the college student population is unknown.

The goal of the present study is to provide causal evidence on the effects of DSP on course fit and short- and long-term academic outcomes and to lend theoretical insights into the relationship between granting students greater latitude over education decisions and postsecondary success. We draw on SDT to examine the extent to which DSP affects a student's probability of reaching critical academic milestones. We also draw on insights from BDT to better understand the choices made under DSP, the extent to which these decisions improve self-perceived course fit, and how they differ by key student demographics and initial math level.

## Setting

The study is made possible due to variation in placement policies across community colleges in a large urban community college district (LUCCD). The LUCCD is composed of nine colleges serving over 200,000 students each year. Each college enrolls a diverse group of students, with the majority identifying as students of color.

Importantly, the LUCCD colleges have similar developmental math sequences. The sequence of courses starts with arithmetic (four levels below transfer-level math) and is followed by pre-algebra (three levels below), elementary algebra (two levels below), and intermediate algebra (one level below).[4] Before 2009, students were required to pass at least elementary algebra to receive an associate's degree.

### A Natural Experiment in Self-Placement

In the spring of 2008, one of the LUCCD colleges (henceforth "College X") unintentionally failed to renew their COMPASS testing license with ACT Inc. in time, and thus was forced to allow students entering in the 2008 summer and fall semesters to choose math courses without taking a placement test. During these semesters, College X implemented DSP and told students that they would be allowed to determine which math course best matched their abilities and skill set. Interviews conducted with College X's research staff and a review of 2008 course catalogs indicate students were required to meet with a counselor before making a self-placement decision. In addition, students interested in enrolling in elementary algebra or a more rigorous math course were required to submit high school transcripts for counselor review. Institutional researchers who we were able to interview had only peripheral knowledge about the placement policy change, but they noted that some students faced long wait times to meet with a counselor. Thus, we do not know whether all self-placing students received guidance from a counselor or received additional support to make a placement decision. However, to the best of our knowledge, students at College X were not affected by any other changes in policy. Because of this abrupt change in placement regimes, this event can be considered a natural experiment and capable of generating causal claims on the benefits of employing DSP relative to test-placement on self-perceived course fit and scholastic achievement (Angrist & Pischke, 2009).

Before the placement policy change, College X employed a mostly test-based approach to assign students to math courses. To place into the math sequence, students entered the assessment office, answered a student background questionnaire, and took the COMPASS placement test. The student's placement test score, in addition to multiple measure points awarded to the student based on their survey answers, were used to assign the student to a math course. With their testing license renewed in spring 2009, College X reverted to using this test-based plus multiple measures placement approach to assign students to developmental math.

Since the remaining eight LUCCD colleges did not use DSP or experience any changes in their placement policies during the 2005–2008 study period, we consider these colleges as our control campuses, and College X as our treatment campus.

## Data

This study employed a rich set of longitudinal data obtained from the LUCCD. We drew on roughly 20,000 administrative records for students who placed *and* enrolled in math in any one of the nine LUCCD community colleges. These data link each students' demographic information, course placement, and enrollment and performance histories. Because of the way the data are structured, we were able to track a student's enrollment and performance patterns across time for cohorts assessed for and enrolled in math for each semester between the summer of 2005 and the fall of 2008. Enrollment and performance data extend from the summer of 2005 to the summer semester of 2012, which means there are at least 4 years of outcome data for each entering cohort. To reduce the influence of seasonal variation in student enrollment, and because DSP affected only the summer and fall 2008 cohorts, we removed the winter and spring enrollment cohorts from our data.

College X serves approximately 10,000 students each year, with about 1,200 first-time students enrolling each fall. Student records indicate that about 50% of all College X students identify as Hispanic, 16% as Asian/Pacific Islander, 17% as White, and 11% as Black.[5] Although the majority of students are high school graduates, on average over 80% of them place into elementary algebra or lower under test-based placement.[6]

Given the placement policy change, we would ideally compare all students in College X who had the opportunity to self-place with all students who took placement tests in the other colleges. Unfortunately, College X did not collect data on the total number of students who were given the opportunity to self-place into the math sequence; we can only observe students who ultimately enrolled in math for the first time during these two terms. In other words, we cannot identify students who began the DSP process but did not enroll in math. Our analysis therefore focuses on all first-time community college students, excluding concurrent high school students, who enrolled in a math course. Enrollment in a math course can be considered as an indicator of degree-aspiration, since math is required for an associate's degree and also for transfer to a 4-year college.

The sample of interest in College X consists of 385 students who self-placed and enrolled in the math sequence in the summer or fall of 2008. A total of 19,711 community college students test-placed and enrolled into math between the summer of 2005 and the fall of 2008, and these students serve as the counterfactuals to DSP students in College X. Table 1 provides sample sizes and variable means across treated and control institutions in each placement policy period. To minimize bias in causal estimates, the sample is restricted to students who assessed and enrolled in math the same semester. Raw first differences in covariates and outcomes for College X and the control colleges are calculated in columns (5) and (6). A naïve difference-in-difference is presented in column (7), and the formal regression-based estimate is presented next.

## Method

We employed a difference-in-difference (DID) estimation strategy to identify the average treatment effect of self-placement in College X relative to test-based placement in control colleges on outcomes for students who placed and enrolled in math. The strategy relies on the exogenous "natural experiment" of DSP in College X and thereby controls for observable and unobservable factors that jointly correlate with the opportunity to self-place in the math sequence and with academic outcomes.

## Table 1
### Sample Means Before and After Directed Self-Placement

| | (1) | | (2) | | (3) | | (4) | | (5) | (6) | (7) |
|---|---|---|---|---|---|---|---|---|---|---|---|
| | \multicolumn{4}{c}{College X} | \multicolumn{4}{c}{Control} | Difference (College X) | Difference (Control) | Difference-in-Difference |
| | Pre | | Post | | Pre | | Post | | (2)−(1) | (4)−(3) | (5)−(6) |
| | M | SD | M | SD | M | SD | M | SD | | | |
| **Demographics** | | | | | | | | | | | |
| Age | 21.06 | 6.20 | 21.19 | 6.96 | 22.07 | 7.09 | 22.14 | 7.25 | 0.12 | 0.07 | 0.05 |
| Female | 0.56 | 0.50 | 0.55 | 0.50 | 0.55 | 0.50 | 0.54 | 0.50 | −0.01 | −0.01 | 0.00 |
| Asian | 0.16 | 0.37 | 0.15 | 0.35 | 0.13 | 0.34 | 0.12 | 0.33 | −0.02 | −0.01 | −0.01 |
| Black | 0.11 | 0.32 | 0.11 | 0.31 | 0.16 | 0.36 | 0.16 | 0.36 | −0.01 | 0.00 | −0.01 |
| Hispanic | 0.50 | 0.50 | 0.53 | 0.50 | 0.53 | 0.50 | 0.54 | 0.50 | 0.03 | 0.01 | 0.02 |
| White | 0.17 | 0.37 | 0.14 | 0.34 | 0.13 | 0.33 | 0.11 | 0.32 | −0.03 | −0.01 | −0.02 |
| Other | 0.05 | 0.22 | 0.08 | 0.27 | 0.06 | 0.23 | 0.07 | 0.25 | 0.02 | 0.01 | 0.01 |
| English is primary language | 0.78 | 0.42 | 0.87 | 0.34 | 0.66 | 0.47 | 0.72 | 0.45 | 0.09 | 0.06 | 0.04 |
| **Outcomes** | | | | | | | | | | | |
| Withdrew from first math course | 0.26 | 0.44 | 0.18 | 0.39 | 0.19 | 0.39 | 0.18 | 0.38 | −0.07 | −0.01 | −0.06 |
| Failed first math course | 0.25 | 0.43 | 0.24 | 0.43 | 0.25 | 0.44 | 0.23 | 0.42 | −0.01 | −0.02 | 0.01 |
| Passed college-level math within 1 year | 0.25 | 0.43 | 0.29 | 0.45 | 0.36 | 0.48 | 0.38 | 0.49 | 0.04 | 0.03 | 0.01 |
| Passed college-level math within 2 years | 0.31 | 0.46 | 0.35 | 0.48 | 0.41 | 0.49 | 0.44 | 0.50 | 0.04 | 0.02 | 0.01 |
| Passed college-level math within 4 years | 0.36 | 0.48 | 0.37 | 0.48 | 0.46 | 0.50 | 0.48 | 0.50 | 0.01 | 0.02 | −0.01 |
| Passed transfer-level math within 1 year | 0.06 | 0.24 | 0.15 | 0.36 | 0.10 | 0.30 | 0.11 | 0.31 | 0.09 | 0.01 | 0.08 |
| Passed transfer-level math within 2 years | 0.11 | 0.31 | 0.20 | 0.40 | 0.14 | 0.35 | 0.15 | 0.36 | 0.09 | 0.01 | 0.08 |
| Passed transfer-level math within 4 years | 0.16 | 0.37 | 0.24 | 0.43 | 0.19 | 0.39 | 0.20 | 0.40 | 0.07 | 0.01 | 0.06 |
| Completed 30 units | 0.45 | 0.50 | 0.46 | 0.50 | 0.41 | 0.49 | 0.42 | 0.49 | 0.02 | 0.01 | 0.00 |
| N | 1,107 | | 385 | | 12,097 | | 6,507 | | | | |

*Note.* All outcomes are dichotomous except age. College-level math is elementary algebra. Transfer-level math is any course for which intermediate algebra is a pre-requisite (e.g., precalculus).

Model 1 is the DID equation estimating this treatment effect using a linear probability model:

$$y_{itc} = \gamma_0 + \lambda * CollegeX_c + \psi * Post2008_t + \delta * (CollegeX_c * Post2008_t) \\ + \vartheta_t + \vartheta_c + \theta * X_{itc} + \varepsilon_{itc} \quad (1)$$

Here, $y_{itc}$ is the academic outcome of interest for individual $i$ in college $c$ in semester-term $t$. The coefficient $\lambda$ captures the difference in outcomes for College X students; $\psi$ captures the difference between cohorts entering before and after summer 2008, when the DSP policy was enacted; the parameter of interest is $\delta$, which measures the effect of DSP relative to test-based placement on self-perceived course fit and academic success. The model also includes semester-cohort fixed effects $\vartheta_t$, institution-level fixed-effects $\vartheta_c$, and $X_{itc}$, a vector of student baseline characteristics, which control for observed and unobserved characteristics of community colleges and student cohorts that likely correlate with our outcomes. Student baseline characteristics include a students' age, sex, race, and primary language (dichotomized as English vs. other). To test the robustness of our results and protect against false positives, we estimate various specifications of our main model, including cluster-robust standard errors by campus and semester-cohort (Bertrand, Duflo, & Mullainathan, 2004).

For our analysis, we examine outcomes considered consequential for understanding placement in relation to community college student persistence and success (Melguizo, Bos, & Prather, 2011). With regard to placement and course fit, we examined whether a student (a) withdrew from their first math course on their first attempt (after the no-penalty drop deadline) and (b) failed their first math course on their first attempt. These outcomes may be particularly salient for understanding the impact of placement policies because a concern of test-based placement is that students may be placed incorrectly (Scott-Clayton et al., 2014), thus increasing their likelihood of withdrawal or failure. DSP, in contrast, may decrease withdrawal or failure if students choose courses that they think are a best-fit for themselves and feel confident in their decision—hence, self-perceived course fit. With regard to persistence and completion outcomes, we focused on outcomes that are important milestones toward degree completion and/or transfer but are also directly related to initial math placement. These include whether a student (c) successfully passed the college-level math (CLM) required for an associate's degree at the time (i.e., elementary algebra or higher); and (d) successfully passed a transfer-level math (TLM) course (e.g., precalculus or higher). A key concern of postsecondary math remediation is that it may add significant amounts of time to students' academic journeys (Melguizo, Bos, Ngo, Mills, & Prather, 2016; Ngo & Kosiewicz, 2017). We therefore examine math course completion (i.e., outcomes c and d) in a 1-, 2-, and 4-year window in order to ascertain whether DSP accelerated progress toward completion of these courses relative to students who were test-placed. Finally, since DSP essentially removed remediation requirements, we investigated whether students were more likely to make progress toward degree completion. We therefore examined whether a student (e) completed 30 degree-applicable credit hours (i.e., half of the credit hour requirement to complete an associate's degree). Since our goal is to understand the impact of DSP as a placement policy relative to test-based placement, we imputed zeros in cases where an outcome was not observed. This ensures all students were treated consistently.[7] In these instances, students who attempted and failed a subsequent course were treated equally as students who never attempted a subsequent course.

## Assumptions for Causal Inference

### Parallel Trends

To produce valid causal estimates of DSP on student success, trends in our identified outcomes between College X and the remaining eight colleges must have been probabilistically equivalent prior to the switch in placement regimes. Else, differences before the implementation of DSP will likely lead to biased impact estimates (Murnane & Willett, 2010). A visual inspection of pretreatment fall-to-fall trends (with summer and fall combined) suggests that there are no significant differences in trends between College X and other colleges (see Figure 1). However, there are significant divergences after DSP (dotted lines), a preview of the DID results. The figures suggest a decrease in the probability of withdrawal from the initial math course and an increase in the probability of passing TLM. To improve on this visual inspection, we also empirically test the parallel trends assumption by regressing each outcome on an interaction between each pre-treatment term and the College X dummy variable. These results are presented in the Appendix (Table A1) and show no evidence that pretreatment trends across all outcomes were significantly different between College X and the control colleges. These results suggest that the other eight LUCCD colleges together serve as a valid counterfactual for College X if it had continued using a test to assign students to developmental education.

### Covariate Analysis

One additional threat that could jeopardize causal inference is a change in the types of students enrolling in math following the implementation of DSP. For example, the opportunity to self-place may have attracted different sorts of students to College X since it provided them with a way out of taking developmental math courses. If demographic changes coincided with the implementation of DSP, the internal validity of the DSP DID estimates would be threatened. To test for the existence of this threat, we reran Model 1 each

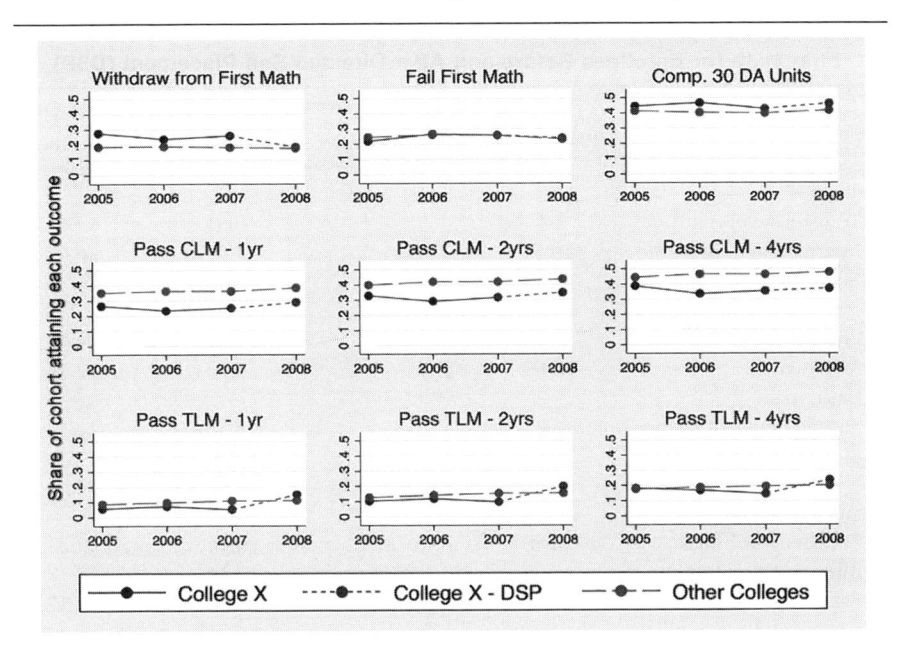

*Figure 1.* **Trends in cohort outcomes in College X and control colleges, before (solid) and after (dotted) implementation of directed self-placement (DSP). Summer/fall cohorts each year are shown.**

*Note.* CLM = college-level math; TLM = transfer-level math.

time with a different student baseline characteristic (e.g., age, sex, race, and language) as the dependent variable. The results, which are also presented in the Appendix (Table A2), suggest no significant changes across student baseline characteristics after the implementation of DSP at College X, leading us to conclude that our causal estimates are internally valid.

## Results

### Student Placement Decisions Under DSP Versus Test-Based Placement

Because DSP leaves decision making in the hands of the student, we first examine whether students made different placement decisions under DSP than under test-based placement. Table 2 shows test-based math placement and enrollment patterns in all colleges before and after 2008, and math enrollment patterns under DSP in College X.

We find that when colleges used a test-based placement policy, most students placed in the middle and lower levels of the developmental math

## Table 2
### First Math for Enrollees Before and After Directed Self Placement (DSP)

|  | (1) Pre-DSP (Placement) | (2) Post-DSP (Placement) | (3) Pre-DSP (Enrollment) | (4) Post-DSP (Enrollment) |
|---|---|---|---|---|
| **College X** | | | | |
| Transfer-level math | — | — | 0.81 | 14.55 |
| Intermediate algebra | 19.51 | — | 17.07 | 16.62 |
| Elementary algebra (college-level math) | 46.7 | — | 44.08 | 21.30 |
| Pre-algebra | 33.69 | — | 24.21 | 24.42 |
| Arithmetic | 0.09 | — | 6.32 | 12.73 |
| Basic math[a] | — | — | — | — |
| Supplemental/tutoring | — | — | 7.50 | 10.39 |
| N | 1,107 | | 1,107 | 385 |
| | | | $\chi^2 = 180.089$*** | |
| **Control colleges** | | | | |
| Transfer-level math | 6.58 | 7.01 | 6.94 | 7.25 |
| Intermediate algebra | 18.05 | 19.12 | 17.14 | 18.21 |
| Elementary algebra (college-level math) | 25.57 | 25.53 | 24.86 | 24.37 |
| Pre-algebra | 21.09 | 28.63 | 21.01 | 28.34 |
| Arithmetic | 26.74 | 17.87 | 24.46 | 16.17 |
| Basic math[a] | 1.97 | 1.84 | 0.92 | 0.66 |
| Supplemental/tutoring | | | 4.65 | 4.98 |
| N | 12,097 | 6,507 | 12,097 | 6,507 |
| | $\chi^2 = 246.333$*** | | $\chi^2 = 239.349$*** | |

*Note.* Each column shows the percent of students enrolling in each level of math during test-placement and during summer/fall 2008, when directed self-placement (DSP) was implemented in College X. The chi-square test-statistic for differences in each pre/post distribution is also shown. College-level math (CLM) is elementary algebra. Transfer-level math (TLM) is any course for which intermediate algebra is a prerequisite (e.g., precalculus).
[a] One college in the district offered a course below arithmetic focused on basic math content.
*$p < .05$. **$p < .01$. ***$p < .001$.

sequence, in either pre-algebra or elementary algebra (Columns 1 and 3). This means that the majority of test-placed students were required to take at a minimum two developmental math courses before reaching math courses counting toward upward transfer. Only about 7% of students assessed into TLM in the control colleges during this period, and no students in College X assessed at this math level. Table 2 also shows that during test-placement, the distribution of assigned and enrolled math courses closely matched. This can be explained by the district's strict policy requiring students to comply with their placement results.[8]

After DSP took effect, chi-square tests reveal statistically significant changes in enrollment distributions across the math sequence. Results show that self-placing students enrolled more evenly across the sequence of available math courses (Column 4). Table 2 shows that nearly 15% of self-placing students started their math trajectory by taking a TLM course compared to just 7% of test-placed math enrollees in the control colleges. Interestingly, approximately 13% of self-placers also enrolled in the lowest level of the math sequence, compared to just 6% of math enrollees under test-based placement at College X.

Drawing on insights from BDT, we also examined math enrollment decisions under DSP by sex and race, under the assumption that different placement decisions might be made along these lines. Table 3 shows math enrollment patterns by these characteristics before and after DSP implementation in College X. While the proportion of students enrolling in TLM increased overall, males, White, and Asian students experienced the largest gains. Conversely, a greater proportion of female, Black, and Hispanic students chose to enroll in arithmetic, the lowest level in the developmental math sequence. Chi-square tests indicate that these distributions significantly differed before and after DSP.

### Estimated Effects of DSP on Course Fit and Academic Success

We next measure the estimated impact of DSP relative to test-based placement on our outcomes. Table 4 shows different DID specifications for two course fit outcomes: (1) withdrawing from the first math course and (2) failing the first math course. Model 1 is a baseline DID specification with no covariates. Model 2 adds student-level demographic covariates to increase precision of the DID estimator, along with campus and semester-cohort dummies to account for differences across campuses and cohorts of entering students. Model 3 includes standard errors clustered at the campus level, and Model 4 includes standard errors clustered by semester-cohorts to assess sensitivity to model specification.

Across all specifications, results show that there is a consistent negative effect of DSP on the probability of withdrawal. Students given the opportunity to choose their own math courses were about 6 percentage points less likely to withdraw from their math course than students who were assigned to their math courses based on the results of a placement test. In turn, we find that DSP did not statistically affect a student's chances of failing a course. Taken together, these results suggest that DSP may have improved course fit, since students were more likely to stay enrolled in their chosen courses and no more likely to fail them compared to test-placed students. However, as described earlier, female, Black, and Hispanic students were more likely to

Table 3
**First Math Enrollment (%) by Student Subgroup, College X, 2005–2008**

| | Female | | Male | | White or Asian | | Hispanic Only | | Black Only | |
|---|---|---|---|---|---|---|---|---|---|---|
| | Test-Placement | DSP | Test-Placement | DSP | Test-Placement | DSP | Test-Placement | DSP | Test-Placement | DSP |
| Transfer-level math | 0.48 | 9.86 | 1.23 | 20.35 | 1.19 | 23.91 | 0.72 | 8.78 | 0.00 | 11.90 |
| Intermediate algebra | 14.98 | 15.96 | 19.75 | 17.44 | 23.28 | 15.94 | 15.03 | 19.02 | 5.51 | 7.14 |
| Elementary algebra (CLM) | 42.51 | 21.60 | 46.09 | 20.93 | 47.74 | 21.74 | 43.29 | 22.44 | 35.43 | 14.29 |
| Pre-algebra | 27.86 | 25.35 | 19.55 | 23.26 | 16.39 | 21.74 | 27.91 | 25.37 | 33.86 | 28.57 |
| Arithmetic | 5.64 | 15.49 | 7.20 | 9.30 | 5.46 | 4.35 | 5.01 | 15.61 | 14.96 | 26.19 |
| Supplemental/tutoring | 8.53 | 11.74 | 6.17 | 8.72 | 5.94 | 12.32 | 8.05 | 8.78 | 10.24 | 11.90 |
| N | 621 | 213 | 486 | 172 | 421 | 138 | 559 | 205 | 127 | 42 |
| Chi-square ($\chi^2$) | 88.223*** | | 97.978*** | | 106.132*** | | 74.587*** | | 22.534*** | |

*Note.* Each column shows the percent of students in each subgroup enrolling in each level of math during test-placement and directed self-placement (DSP). The test-statistic for differences in each pre/post distribution is also shown. College-level math is elementary algebra. Transfer-level math is any course for which intermediate algebra is a prerequisite (e.g., precalculus).
*$p < .05$. **$p < .01$. ***$p < .001$.

## Table 4
### Impact of Directed Self-Placement on Math Course Withdrawal and Failure

| | Withdrawal From First Enrolled Math | | | | Failed First Enrolled Math | | | |
|---|---|---|---|---|---|---|---|---|
| | (1) | (2) | (3) | (4) | (1) | (2) | (3) | (4) |
| College X * post-2008 | −0.062** (0.024) | −0.064** (0.024) | −0.064** (0.015) | −0.064* (0.020) | 0.010 (0.026) | 0.007 (0.026) | 0.007 (0.006) | 0.007 (0.007) |
| Post-2008 | −0.009 (0.006) | 0.008 (0.025) | 0.008 (0.022) | 0.008 (0.009) | −0.022** (0.007) | 0.061* (0.027) | 0.061*** (0.012) | 0.061*** (0.006) |
| College X | 0.070*** (0.012) | −0.008 (0.018) | −0.008 (0.008) | −0.008 (0.011) | −0.004 (0.014) | 0.025 (0.019) | 0.025 (0.011) | 0.025 (0.013) |
| Covariates | | x | X | X | | X | x | x |
| College FE | | x | X | X | | X | x | x |
| Cohort FE | | x | X | X | | X | x | x |
| Standard errors | | | Cluster (College) | Cluster (Cohort) | | | Cluster (College) | Cluster (Cohort) |
| Constant | 0.186*** (0.004) | 0.249*** (0.031) | 0.249*** (0.023) | 0.249*** (0.025) | 0.254*** (0.004) | 0.151*** (0.034) | 0.151*** (0.045) | 0.151*** (0.018) |
| N | 20,096 | 20,096 | 20,096 | 20,096 | 20,096 | 20,096 | 20,096 | 20,096 |

*Note.* Covariates include age, sex, race, and language. Some models include college and/or cohort fixed effects (FE).
*$p < .05$. **$p < .01$. ***$p < .001$.

enroll in lower-level math courses under DSP. Thus, decreased withdrawal may be the result of increased enrollment in lower-level courses. Since the cluster-robust semester-cohort standard errors are larger than standard errors clustered by college (Model 4), we employ this model for subsequent analyses to reduce the likelihood of finding significant results when in fact there are none (Bertrand et al., 2004).

Turning to short- and long-term academic outcomes in Table 5, we observe that DSP increased the probability of completing CLM within one year by about 2 percentage points. This effect disappears if we track students out for 2 or 4 years, which suggests that DSP enabled students to meet the math requirement more quickly than test-placed students. We also find that self-placers were 8 percentage points more likely to complete a math course required to transfer to a 4-year institution after 1 year. Notably, the positive effect remains even after following students for 2 and 4 years. This finding suggests that, overall, DSP increased the probability that entering students completed transferable math courses, most likely because more students directly enrolled in these courses under DSP. Since we did not find any significant effects of DSP on completion of 30 degree-applicable credits, the primary conclusion that emerges is that DSP may have set more students on the path to transfer but did not significantly change the likelihood of degree completion.

## Heterogeneous Effects of DSP by Subgroup

The BDT literature characterizes choices as informed by environmental contexts, personal experiences, and beliefs. Therefore, to determine whether DSP had differential effects across student groups, we interacted the main treatment coefficient with a set of demographic dummy variables.[9] We calculated the total treatment effect for each subgroup and report these in Table 6 to facilitate interpretation of the results. We compared the DID effects between male and females, Black versus White/Asian, Hispanic versus White/Asian, and high math versus low math (those who chose high/advanced math at intermediate algebra and above vs. those who chose lower-level math courses) who self-placed into the math sequence.

Findings from this analysis suggest that some students experienced a significantly higher payoff from DSP than others. While self-placing male students saw a decrease in the probability of failing the first enrolled math course, females experienced an increase of 4 percentage points. Most concerningly, male students in College X were over 8 percentage points more likely to complete CLM in 2 years and over 13 percentage points more likely to complete TLM over the time frame, but self-placing female students experienced no change in the probability of these outcomes. DSP also increased the probability that males earned 30 degree-applicable units after 4 years of math enrollment by 4 percentage points but did not do so for females.

## Table 5
### Impact of Directed Self-Placement in Mathematics on Short- and Long-Term Outcomes

| | Withdrawal From First Enrolled Math | Failed First Enrolled Math | Pass CLM in 1 Year | Pass CLM in 2 Years | Pass CLM in 4 Years | Pass TLM in 1 Year | Pass TLM in 2 Years | Pass TLM in 4 Years | Completed 30 Units |
|---|---|---|---|---|---|---|---|---|---|
| College X * post-2008 | −0.064* | 0.007 | 0.021* | 0.020 | 0.000 | 0.082*** | 0.087*** | 0.066** | 0.008 |
| | (0.020) | (0.007) | (0.008) | (0.012) | (0.014) | (0.011) | (0.013) | (0.018) | (0.011) |
| Post-2008 | 0.008 | 0.061*** | −0.005 | −0.013* | −0.024** | 0.028*** | 0.024*** | −0.007 | 0.060*** |
| | (0.009) | (0.006) | (0.003) | (0.004) | (0.005) | (0.002) | (0.003) | (0.005) | (0.008) |
| College X | −0.008 | 0.025 | −0.042 | −0.03 | −0.0z32 | −0.006 | −0.006 | 0.021 | 0.095** |
| | (0.011) | (0.013) | (0.023) | (0.021) | (0.024) | (0.009) | (0.015) | (0.016) | (0.019) |
| N | 20,096 | 20,096 | 20,096 | 20,096 | 20,096 | 20,096 | 20,096 | 20,096 | 20,096 |

*Note.* Results from estimation using Model 4 (see Table 4) are shown, which include all covariates (age, sex, race, and language), college and cohort dummies, and standard errors clustered by cohort. College-level math (CLM) is elementary algebra. Transfer-level math (TLM) is any course for which intermediate algebra is a requisite (e.g., precalculus).
\*$p < .05$. \*\*$p < .01$. \*\*\*$p < .001$.

Table 6
Impact of Directed Self-Placement in Mathematics by Subgroup

| | Withdrawal From First Enrolled Math | Failed First Enrolled Math | Pass CLM in 1 Year | Pass CLM in 2 Years | Pass CLM in 4 Years | Pass TLM in 1 Year | Pass TLM in 2 Years | Pass TLM in 4 Years | Completed 30 Units |
|---|---|---|---|---|---|---|---|---|---|
| Gender | | | | | | | | | |
| Treatment effect for female students | −0.064** (0.013) | 0.043** (0.009) | −0.031 (0.019) | −0.032 (0.026) | −0.054 (0.027) | 0.040 (0.019) | 0.036 (0.018) | 0.015 (0.020) | −0.019 (0.015) |
| Treatment effect for male students | −0.063 (0.040) | −0.038* (0.012) | 0.087** (0.019) | 0.086* (0.028) | 0.069 (0.029) | 0.133*** (0.013) | 0.151*** (0.023) | 0.130** (0.030) | 0.042* (0.014) |
| Race/ethnicity | | | | | | | | | |
| Treatment effect for Black students | −0.144* (0.060) | 0.057 (0.065) | 0.041 (0.023) | 0.020 (0.041) | −0.051 (0.051) | 0.062** (0.015) | 0.070** (0.014) | 0.036 (0.035) | −0.100* (0.035) |
| Treatment effect for Hispanic students | −0.033 (0.026) | −0.016 (0.013) | −0.001 (0.018) | −0.018 (0.023) | −0.018 (0.028) | 0.063*** (0.006) | 0.057*** (0.006) | 0.035 (0.015) | 0.018 (0.029) |
| Treatment effect for White/Asian students | −0.103*** (0.014) | 0.015 (0.026) | 0.058*** (0.006) | 0.092*** (0.008) | 0.057* (0.020) | 0.128*** (0.021) | 0.144** (0.030) | 0.139** (0.031) | 0.040 (0.052) |
| Math level | | | | | | | | | |
| Treatment effect for students choosing lower-level math | −0.141** (0.033) | −0.020 (0.014) | −0.031* (0.010) | −0.034* (0.014) | −0.060** (0.015) | 0.000 (0.006) | 0.007 (0.006) | −0.016 (0.009) | −0.022 (0.031) |
| Treatment effect for students choosing higher-level math | 0.007 (0.024) | 0.021** (0.006) | 0.115*** (0.020) | 0.118*** (0.019) | 0.103*** (0.020) | 0.171*** (0.016) | 0.182*** (0.018) | 0.165*** (0.018) | 0.065** (0.018) |

*Note.* The full regression results from which these treatment effects are calculated are available from the authors on request. The subgroup categories are drawn from the enrollment application. Math level is a dichotomous variable splitting the sample by initial math level (those who chose high/advanced math at intermediate algebra and above vs. those who chose lower-level math courses). College-level math (CLM) is elementary algebra. Transfer-level math (TLM) is any course for which intermediate algebra is a requisite (e.g., precalculus).
*p < .05. **p < .01. ***p < .001.

We also find evidence of differential effects of DSP for Black, Hispanic, as well as White and Asian students. Across most outcomes, White and Asian students benefitted from DSP relative to test-placement. For example, White and Asian students were more likely to complete CLM and TLM within various time frames and were 10 percentage points less likely to withdraw from their first math course under DSP. However, for Black and Hispanic students, the benefits of DSP were not as extensive. Results show that while DSP decreased the probability that Black self-placing students withdrew from their first math course by 14 percentage points and increased their likelihood of completing TLM within 2 years by 7 percentage points, they were also 10 percentage points less likely to complete 30 degree-applicable units. Hispanic self-placing students, relative to test-placing counterparts, increased their likelihood of completing TLM by 6 percentage points, but received no other benefit. Comparing the effects of DSP across these subgroups, we see that Whites and Asian students benefitted twice as much as Black and Hispanic students in terms of completing TLM. We believe these results can be explained by the fact that larger proportions of these subgroups enrolled in lower levels of math when DSP was policy. Overall, the findings suggest that although DSP may lead to positive outcomes, the effects are concentrated among male, White, and Asian students and may thereby have the potential to widen already existing racial and gender completion gaps.

Somewhat to be expected, students who chose to enroll in higher-level math during DSP benefitted the most from DSP. Although the probability of withdrawal decreased for those students who choose to enroll in lower-level math, they were also 3 percentage points less likely after the policy's implementation to complete CLM. This setback grew to 6 percentage points in size after 4 years, suggesting a longer-term penalty for making this choice. In contrast, the group of students who self-placed into higher-level math courses reaped significant benefits. Although the likelihood of failing their first math course increased by 2 percentage points, they were about 12 percentage points more likely to subsequently complete CLM and 17 percentage points more likely to complete TLM than test-placing students. In sum, these results reveal that the benefits of DSP are not uniform. In fact, as these results demonstrate, DSP may hurt rather than help students, particularly those who may be underestimating their abilities and enrolling in lower-level math courses of their own volition.

**Falsification Exercise**

We conducted a falsification exercise to assess whether the estimates we obtained were related to the "natural experiment" of DSP in 2008 and not to fluctuations in student outcomes over time. This involved specifying a placebo policy change at College X one year prior (summer/fall 2007) to DSP implementation and reestimating the main DID model, dropping observations from summer and fall 2008. A significant treatment effect estimate with the placebo policy change would indicate that the main findings may not be attributable to DSP but rather to other contemporaneous changes. The results from this test, shown in Table 7, indicate that the placebo estimates converge to zero in terms of practical and statistical significance. This, along with the covariate analysis described earlier, suggest the observed main results are indeed the consequence of DSP implementation in 2008 in College X.

## Limitations

The primary limitation of the study is that it focuses on a small sample of students in one college, thus limiting the generalizability of the estimates. Given we only can track DSP students for 4 years, we also were not able to assess the impact of DSP on associate's degree completion, upward transfer, or bachelor's degree completion; few students achieve these goals within the 4-year time frame. Nevertheless, the DID design results are a robust and internally valid treatment estimate of a unique policy prescription and examines important early and intermediary milestones in the path toward college completion.

Another limitation of the study is related to understanding how this version of DSP was implemented in practice. We learned of DSP several years after College X had reverted back to test-based placement from self-placement. Consequently, faculty members and administrators closely involved in the implementation of DSP had left College X, requiring us to gather secondhand information from college catalogs and administrators who had peripheral knowledge of the intervention. Thus, we have a limited understanding of the role and influence of college advisors or faculty on student self-placement decisions. The fact that we found female students and students of color were more likely to choose lower courses under a self-placement regime relative to test placement may suggest two problems. First, it may be the case that implicit bias related to race, gender, and ability influences the guidance that college advisors offer students in the process of making a placement decision; this has been documented among community college counselors (Maldonado, 2019). Second, it may also suggest that particular student groups prone to underestimating their academic abilities may ignore the guidance that they receive and choose a course that is lower than what they were recommended. Investigating how self-placement decisions are made would help us to contextualize our findings and identify where the implementation of DSP may be faltering.

Finally, whether College X changed their teaching practices in response to DSP students is also unknown and an area for further investigation. Since faculty may be used to certain types of students enrolling in their courses

Table 7

**Falsification Test With Placebo Treatment Term (Summer/Fall 2007)**

| | Withdrawal From First Enrolled Math | Failed First Enrolled Math | Pass CLM in 1 Year | Pass CLM in 2 Years | Pass CLM in 4 Years | Pass TLM in 1 Year | Pass TLM in 2 Years | Pass TLM in 4 Years | Completed 30 Units |
|---|---|---|---|---|---|---|---|---|---|
| Placebo policy (2007) | | | | | | | | | |
| College X * post-2007 | 0.001 (0.025) | 0.007 (0.028) | −0.001 (0.029) | −0.002 (0.030) | −0.011 (0.031) | −0.035 (0.018) | −0.037 (0.021) | −0.046 (0.024) | −0.021 (0.031) |
| N | 13,204 | 13,204 | 13,204 | 13,204 | 13,204 | 13,204 | 13,204 | 13,204 | 13,204 |

*Note.* The falsification exercise specifies a placebo policy change at College X one year prior (summer/fall 2007) to actual directed self-placement (DSP) implementation and re-estimates the main difference-in-difference (DID) model, dropping observations from summer and fall 2008. College-level math (CLM) is elementary algebra. Transfer-level math (TLM) is any course for which intermediate algebra is a requisite (e.g., precalculus).

*$p < .05$. **$p < .01$. ***$p < .001$.

based on placement test results, it is possible than an influx of different students would lead faculty to adjust teaching approaches.

## Discussion

For years, state and local policymakers have struggled to design and implement a placement approach that accurately identifies who is not academically ready for college. With evidence building that some students are inappropriately being assigned to developmental education and are floundering in college as a result, a real impetus exists to rethink how colleges are diagnosing college-readiness. Though several states and institutions are tweaking test-placement approaches to include other measures of academic preparation (e.g., GPA) and nonacademic indicators, some are implementing reforms that completely do away with historical test-based placement approaches. This study adds new evidence revealing the extent to which approaches that significantly involve students in placement decisions add value over policies that employ tests to assign students to developmental education.

Our findings show that relative to test placement, DSP may lead to improved outcomes for some students, but not for *all* students. On average, self-placed students who were identified as White, Asian, and male were less likely to withdraw from their first math course and were no more at risk of failing their first math course than test-placed students. They were also more likely to meet the math requirements needed for an associate degree or upward transfer than their counterparts. However, we found no or substantially reduced benefits from self-placement on most outcomes for female, Black, and Hispanic students. For example, under self-placement, women were more likely to fail their initial math course, and Black and Hispanic students were no more likely to pass CLM. Although there were positive effects on completion of transferable math courses for Black and Hispanic students, the benefits were half as large as they were for White and Asian students. We hypothesize the fact that a higher percentage of female, Black, and Hispanic self-placing students chose a lower course than their counterparts played a role in the gender and racial disparities we observed. These findings may be explained by other research demonstrating that female students and students of color are prone to underestimating their abilities in the STEM (science, technology, engineering, and mathematics) fields if they endorse math and science stereotypes (Aronson & Good, 2002).

But why may DSP have different effects for different types of students? As BDT would suggest, it might be that students affected by DSP incorporate different types of information or engage in different decision-making processes that lead them to make different placement decisions, some more optimal than others. For example, students who have been socialized to think that they are bad at math may believe in these ill-informed impressions and incorporate them into their placement decisions. As a result, these students introduce error into their decision making, discount appropriate alternatives, and choose a course for which they are overqualified, consequently increasing their chances of failure. While SDT implies that leaving students to make educational decisions leads to better academic outcomes, it appears, based on the results of this study, that this theory only holds when the decision was sound in the first place. In other words, though giving students latitude over determining the extent to which they are ready for college may confer some benefits, the extent to which these benefits translate into improving course fit and achievement rests on whether the self-placement decision was better than the decisions that would have been determined by a placement test. Our study shows that race/ethnicity and gender matters in these choices and may need to be explicitly addressed as community colleges move away from test-based placement and toward self-placement and other reforms.

So, what should policy makers consider when contemplating the practicality of DSP as an alternative to test-based placement approach to determining college readiness? One is to examine the kinds of guidance college counselors or faculty give students as they make their placement decisions. Because we were not able to gather firsthand information about how College X implemented DSP in practice, it is unclear the extent to which students received proper advising, if any advising, from counselors. We suspect, based on our results, that self-placement advising may have derailed some students from making better placement decisions. Indeed, research on community college counselors by Maldonado (2019) suggests one possibility—that counselors' views of student potential are linked to racialized and gendered beliefs. In a context of guided self-placement, counselors may have advised self-placing students differently, and partly responsible for the disparate outcomes we observed. Expanding and differentiating approaches to advising, increasing resources to decrease the counselor-to-student ratio, and promoting professional development training focused on equity-mindedness may help self-placement policies work for more students. These strategies may serve as safeguards against counselor bias, as well as math anxieties and apprehension that may lead students to choose low-level courses or the problem of overconfidence that may lead some to pick courses for which they are academically underprepared.

This study reveals curious patterns in how students may respond to being given additional latitude over the placement decisions that determine assignment to development or CLM courses. Giving community college students a voice in the matter suggests that some of these students will aim higher and be more likely to achieve their academic goals, but not all may do so.

Table A1

Pretreatment Differences Between College X and Control Colleges

| | Withdrawal From First Enrolled Math | Failed First Enrolled Math | Pass CLM in 1 Year | Pass CLM in 2 Years | Pass CLM in 4 Years | Pass TLM in 1 Year | Pass TLM in 2 Years | Pass TLM in 4 Years | Completed 30 Units |
|---|---|---|---|---|---|---|---|---|---|
| College X * summer/fall 2006 | 0.018 (0.035) | −0.026 (0.034) | 0.014 (0.035) | 0.022 (0.037) | 0.039 (0.038) | 0.022 (0.018) | 0.026 (0.024) | 0.047 (0.029) | −0.004 (0.039) |
| College X * summer/fall 2007 | −0.024 (0.032) | 0.004 (0.032) | −0.018 (0.032) | −0.026 (0.034) | −0.021 (0.035) | 0.032 (0.018) | 0.032 (0.023) | 0.028 (0.027) | 0.034 (0.037) |
| Summer/fall 2006 | −0.001 (0.009) | −0.013 (0.010) | 0.001 (0.010) | −0.009 (0.011) | −0.005 (0.011) | −0.017* (0.006) | −0.020** (0.008) | −0.009 (0.008) | 0.022* (0.011) |
| Summer/fall 2007 | 0.003 (0.009) | 0.003 (0.010) | 0.002 (0.010) | 0.001 (0.011) | 0.003 (0.011) | −0.011 (0.007) | −0.011 (0.008) | −0.005 (0.009) | 0.006 (0.011) |
| College X | −0.016 (0.029) | −0.011 (0.029) | 0.001 (0.028) | 0.030 (0.030) | 0.035 (0.031) | −0.010 (0.015) | −0.005 (0.019) | 0.030 (0.022) | 0.137*** (0.031) |
| College A | −0.026 (0.022) | 0.008 (0.022) | 0.056* (0.022) | 0.065** (0.023) | 0.057* (0.024) | 0.028* (0.012) | 0.024 (0.015) | 0.040* (0.016) | 0.053* (0.023) |
| College B | −0.164*** (0.020) | −0.020 (0.020) | 0.156*** (0.020) | 0.187*** (0.021) | 0.204*** (0.022) | 0.061*** (0.011) | 0.075*** (0.014) | 0.128*** (0.016) | 0.148*** (0.021) |
| College C | −0.079** (0.024) | −0.099*** (0.024) | 0.091*** (0.026) | 0.121*** (0.027) | 0.120*** (0.028) | 0.018 (0.014) | 0.040* (0.018) | 0.082*** (0.020) | 0.055* (0.027) |
| College D | −0.118*** (0.021) | −0.099*** (0.020) | 0.286*** (0.022) | 0.307*** (0.022) | 0.314*** (0.023) | 0.137*** (0.013) | 0.156*** (0.016) | 0.191*** (0.017) | 0.214*** (0.022) |
| College E | −0.001 (0.024) | 0.046 (0.025) | −0.055* (0.023) | −0.057* (0.024) | −0.064** (0.025) | −0.026* (0.011) | −0.047*** (0.014) | −0.041** (0.016) | −0.079** (0.024) |
| College F | −0.079*** (0.022) | 0.024 (0.023) | −0.068** (0.021) | −0.052* (0.023) | −0.048* (0.024) | −0.034*** (0.010) | −0.047*** (0.014) | −0.034* (0.015) | 0.027 (0.024) |
| College G | −0.083*** (0.020) | 0.029 (0.020) | 0.117*** (0.020) | 0.147*** (0.021) | 0.171*** (0.022) | 0.050*** (0.011) | 0.046*** (0.014) | 0.064*** (0.015) | 0.138*** (0.021) |
| Constant | 0.275*** (0.019) | 0.269*** (0.019) | 0.250*** (0.019) | 0.287*** (0.020) | 0.320*** (0.020) | 0.060*** (0.010) | 0.102*** (0.013) | 0.114*** (0.014) | 0.291*** (0.020) |
| N | 13,204 | 13,204 | 13,204 | 13,204 | 13,204 | 13,204 | 13,204 | 13,204 | 13,204 |

*Note.* We regressed each outcome on an interaction between each pretreatment term and the College X dummy variable. Observations from summer and fall 2008 (when DSP [directed self-placement] was implemented) are omitted. College-level math (CLM) is elementary algebra. Transfer-level math (TLM) is any course for which intermediate algebra is a requisite (e.g., precalculus).
* $p < .05$. ** $p < .01$. *** $p < .001$.

## Table A2
### The Impact of DSP on Student Background Characteristics (Difference-In-Difference)

| | Age | Female | Asian | Hispanic | Black | White | Other | English |
|---|---|---|---|---|---|---|---|---|
| College X * post-2008 | 0.028 | 0.004 | −0.012 | 0.026 | 0.002 | −0.027 | 0.010 | 0.042 |
| | (0.265) | (0.023) | (0.006) | (0.030) | (0.011) | (0.027) | (0.005) | (0.021) |
| Post-2008 | −1.464*** | 0.012 | 0.106*** | 0.184** | −0.356*** | 0.083** | −0.016 | −0.027 |
| | (0.201) | (0.028) | (0.010) | (0.040) | (0.022) | (0.022) | (0.011) | (0.021) |
| College X | −2.660*** | 0.054*** | 0.001 | 0.001 | −0.007** | −0.003 | 0.008* | 0.013*** |
| | (0.091) | (0.003) | (0.007) | (0.006) | (0.001) | (0.002) | (0.003) | (0.002) |
| College A | 0.559 | −0.052* | 0.160*** | 0.162*** | −0.357*** | 0.059*** | −0.024 | −0.261*** |
| | (0.250) | (0.016) | (0.022) | (0.029) | (0.018) | (0.008) | (0.013) | (0.012) |
| College B | −1.441** | 0.026* | 0.128*** | 0.451*** | −0.458*** | −0.070*** | −0.051* | −0.238*** |
| | (0.336) | (0.008) | (0.019) | (0.029) | (0.017) | (0.008) | (0.015) | (0.010) |
| College C | −1.605** | −0.001 | −0.006 | 0.483*** | −0.435*** | −0.015* | −0.028* | −0.087*** |
| | (0.333) | (0.016) | (0.006) | (0.019) | (0.018) | (0.005) | (0.008) | (0.011) |
| College D | −1.750** | −0.047** | 0.136*** | 0.042 | −0.411*** | 0.199*** | 0.034* | −0.086*** |
| | (0.376) | (0.013) | (0.007) | (0.021) | (0.015) | (0.012) | (0.013) | (0.009) |
| College E | 2.005*** | 0.143*** | −0.042** | −0.140*** | 0.302*** | −0.078*** | −0.042* | 0.092** |
| | (0.213) | (0.012) | (0.010) | (0.021) | (0.018) | (0.007) | (0.015) | (0.021) |
| College F | 1.803*** | 0.028 | −0.003 | 0.193*** | −0.085*** | −0.064*** | −0.040* | −0.115*** |
| | (0.305) | (0.012) | (0.008) | (0.025) | (0.015) | (0.007) | (0.013) | (0.017) |
| College G | −1.400*** | −0.022 | 0.044** | 0.208*** | −0.410*** | 0.135*** | 0.023** | −0.144*** |
| | (0.181) | (0.010) | (0.011) | (0.011) | (0.014) | (0.007) | (0.006) | (0.007) |
| Fall 2005 | −2.295*** | 0.086*** | 0.007 | −0.008 | 0.002 | 0.008** | −0.009* | −0.064*** |
| | (0.088) | (0.005) | (0.008) | (0.006) | (0.001) | (0.002) | (0.003) | (0.003) |
| Summer 2006 | −0.435*** | 0.058*** | 0.038*** | −0.059*** | −0.027*** | 0.025*** | 0.023*** | −0.051*** |
| | (0.041) | (0.003) | (0.003) | (0.005) | (0.003) | (0.001) | (0.001) | (0.005) |
| Fall 2006 | −2.626*** | 0.063*** | 0.009 | 0.002 | 0.002 | −0.001 | −0.012* | −0.049*** |
| | (0.098) | (0.005) | (0.007) | (0.006) | (0.002) | (0.002) | (0.004) | (0.003) |
| Summer 2007 | −0.152*** | 0.040*** | 0.021*** | −0.047*** | 0.029*** | −0.002 | −0.002 | 0.021*** |
| | (0.019) | (0.001) | (0.002) | (0.002) | (0.001) | (0.001) | (0.001) | (0.001) |
| Fall 2007 | −2.500*** | 0.049*** | 0.014 | −0.011 | −0.004 | 0.000 | 0.001 | −0.014** |
| | (0.093) | (0.005) | (0.006) | (0.006) | (0.002) | (0.002) | (0.003) | (0.003) |
| Summer 2008 | 3.053*** | −0.012** | 0.057*** | −0.067*** | −0.010*** | 0.024*** | −0.005* | −0.016*** |
| | (0.052) | (0.002) | (0.003) | (0.003) | (0.002) | (0.001) | (0.002) | (0.002) |
| Constant | 25.014*** | 0.485*** | 0.045** | 0.327*** | 0.471*** | 0.081*** | 0.076*** | 0.844*** |
| | (0.306) | (0.007) | (0.011) | (0.022) | (0.015) | (0.008) | (0.013) | (0.006) |
| N | 20,096 | 20,096 | 20,096 | 20,096 | 20,096 | 20,096 | 20,096 | 20,096 |

*Note.* Each column shows the results of Model 1 (see Table 4) estimated with each of the student baseline characteristics (e.g. age, sex, race, and language) as the dependent variable.

\*$p < .05$. \*\*$p < .01$. \*\*\*$p < .001$.

## Notes

This research was supported by the National Academy of Education/Spencer Foundation Dissertation Fellowship Program and the University of Southern California Graduate School Final Year Fellowship Program. Opinions do not necessarily reflect those of either granting institution, or the authors' current affiliated institutions.

[1] Some community colleges use diagnostic tests or state standardized tests to place students Hughes & Scott-Clayton, 2011; Ngo & Melguizo, 2016), but these instruments are used far less prevalently.

[2] Continuous improvement research to support the implementation of a statewide reform to postsecondary developmental education—A RAND-THECB Research Partnership, IES Award No. R305H150069; Principal Investigator—Trey Miller.

[3] One notable exception is Tompkins (2003), who investigates the implementation of DSP for writing placement in John Tyler Community College (2003).

[4] In one college, a course focused on basic numeracy precedes arithmetic and is located five levels below TLM. Examples of courses counting toward TLM include statistics, math for liberal arts majors, and precalculus or trigonometry.

[5] We employ racial and ethnic identifiers that appear in data used for this study.

[6] In the LUCCD, not all first-time students are required to assess into developmental math; a minority meet exemption requirements that allow them to bypass this requirement, which include meeting other academic thresholds or taking courses that do not lead to a degree.

[7] Consider the case where a student never reaches a course that counts toward TLM credit. In this instance, we do not know whether that student would have passed that course had they had the opportunity. However, since the goal of this study is to examine whether self- placement encourages greater success in these outcomes, attributing zero to students who never reached, and therefore, never obtained this outcome is justified.

[8] Typically, students must pass an ad hoc test administered by a faculty member to enroll in a higher course than where they assessed. Students can enroll in a lower course without taking any prior course of action.

[9] The full set of regression results are available on request.

## References

Angrist, J. D., & Pischke, J. S. (2009). *Mostly harmless econometrics: An empiricist's companion*. Princeton, NJ: Princeton University Press.

Aronson, J., & Good, C. (2002). The development and consequences of stereotype vulnerability in adolescents. In F. Pajares & T. Urdan (Eds.), *Academic motivation of adolescents* (pp. 299–330). Charlotte, NC: Information Age.

Aronson, J., & Inzlicht, M. (2004). The ups and downs of attributional ambiguity: Stereotype vulnerability and the academic self-knowledge of African American college students. *Psychological Science, 15*, 829–836.

Bahr, P. R., Fagioli, L. P., Hetts, J., Hayward, C., Willett, T., Lamoree, D., . . . Baker, R. B. (2019). Improving placement accuracy in California's community colleges using multiple measures of high school achievement. *Community College Review, 47*, 178–211.

Bailey, T., & Jaggars, S. S. (2016). *When college students start behind* (College Completion Series: Part Five). New York, NY: The Century Foundation. Retrieved from https://tcf.org/content/report/college-students-start-behind/

Bailey, T., Jeong, D. W., & Cho, S. W. (2010). Referral, enrollment, and completion in developmental education sequences in community colleges. *Economics of Education Review, 29*, 255–270.

Benware, C. A., & Deci, E. L. (1984). Quality of learning with an active versus passive motivational set. *American Educational Research Journal, 21*, 755–765.

Bertrand, M., Duflo, E., & Mullainathan, S. (2004). How much should we trust differences-in-differences estimates? *Quarterly Journal of Economics, 119,* 249–275.

Bettinger, E., Boatman, A., & Long, B. (2013). Student supports: Developmental education and other academic programs. *Future of Children, 23,* 93–115.

Blakesley, D., Harvey, E. J., & Reynolds, E. J. (2003). Southern Illinois University Carbondale as an institutional model: The English 100/101 stretch and directed self-placement program. In D. J. Royer & R. Gilles (Eds.), *Directed self-placement: Principles and practices* (pp. 207–241). Cresskill, NJ: Hampton Press.

Brown, R. S., & Niemi, D. N. (2007). *Investigating the alignment of high school and community college assessments in California* (National Center Report No. 07-3). San Jose, CA: National Center for Public Policy and Higher Education.

Burdman, P. (2012). *Where to begin? The evolving role of placement exams for students starting college.* Washington, DC: Jobs for the Future.

Castleman, B. (2017). *Behavioral insights for federal higher education policy.* Washington, DC: Urban Institute. Retrieved from https://files.eric.ed.gov/fulltext/ED578890.pdf

Chen, X. (2016). *Remedial coursetaking at U.S. public 2- and 4-year institutions: Scope, experiences, and outcomes* (NCES 2016-405). Washington, DC: National Center for Education Statistics. Retrieved from http://nces.ed.gov/pubsearch

Cox, R. D. (2009). "I would have rather paid for a class I wanted to take": Utilitarian approaches at a community college. *Review of Higher Education, 32,* 353–382.

Croskerry, P. (2003). The importance of cognitive errors in diagnosis and strategies to minimize them. *Academic Medicine, 78,* 775–780.

Education Commission of the States. (n.d.). *State legislation.* Retrieved from http://www.ecs.org/state-legislation-by-state/

Fain, P. (2013, June 5). Remediation if you want it. *Inside Higher Ed.* Retrieved from http://www.insidehighered.com/news/2013/06/05/florida-law-gives-students-and-colleges-flexibility-remediation

Flink, C., Boggiano, A. K., & Barrett, M. (1990). Controlling teaching strategies: Undermining children's self-determination and performance. *Journal of Personality and Social Psychology, 59,* 916–924.

Grady, M. (2016). *How high schools and colleges can team to use data and increase student success.* Boston, MA: Jobs for the Future. Retrieved from https://files.eric.ed.gov/fulltext/ED567871.pdf

Grolnick, W. S., & Ryan, R. M. (1987). Autonomy in children's learning: An experimental and individual difference investigation. *Journal of Personality and Social Psychology, 52,* 890–898.

Grolnick, W. S., Ryan, R. M., & Deci, E. L. (1991). The inner resources for school performance: Motivational mediators of children's perceptions of their parents. *Journal of Educational Psychology, 83,* 508–517.

Häubl, G., & Trifts, V. (2000). Consumer decision making in online shopping environments: The effects of interactive decision aids. *Marketing Science, 19*(1), 4–21.

Hu, S., Park, T. J., Woods, C. S., Tandberg, D. A., Richard, K., & Hankerson, D. (2016). *Investigating developmental and college-level course enrollment and passing before and after Florida's developmental education reform* (REL 2017-203). Retrieved from https://files.eric.ed.gov/fulltext/ED569942.pdf

Hughes, K. L., & Scott-Clayton, J. (2011). *Assessing developmental assessment in community colleges: A review of the literature* (CCRC Working Paper 19). New York, NY: Community College Research Center, Teachers College, Columbia University.

Inzlicht, M., McKay, L., & Aronson, J. (2006). Stigma as ego depletion: How being the target of prejudice affects self-control. *Psychological Science, 17,* 262–269.

Kosiewicz, H., Ngo, F., & Fong, K. (2016). Alternative models to deliver developmental math issues of use and student access. *Community College Review, 43*, 205–231.

Lupia, A., McCubbins, M. D., & Popkin, S. L. (Eds.). (2000). *Elements of reason: Cognition, choice, and the bounds of rationality.* Cambridge, England: Cambridge University Press.

MDRC. (2013). *Developmental education: A barrier to a postsecondary credential for millions of Americans.* New York, NY. Retrieved from http://www.mdrc.org/sites/default/files/Dev%20Ed_020113.pdf

Maldonado, C. (2019). "Where your ethnic kids go?": How counselors as first responders legitimate proper course placements for community college students. *Community College Journal of Research and Practice, 43*, 280–294.

Maloney, E. A., & Beilock, S. L. (2012). Math anxiety: Who has it, why it develops, and how to guard against it. *Trends in Cognitive Sciences, 16*, 404–406.

Melguizo, T., Bos, J. M., Ngo, F., Mills, N., & Prather, G. (2016). Using a regression discontinuity design to estimate the impact of placement decisions in developmental math. *Research in Higher Education, 57*, 123–151.

Melguizo, T., Bos, J., & Prather, G. (2011). Is developmental education helping community college students persist? A critical review of the literature. *American Behavioral Scientist, 55*, 173–184.

Melguizo, T., Kosiewicz, H., Prather, G., & Bos, J. (2014). How are community college students assessed and placed in developmental math? Grounding our understanding in reality. *Journal of Higher Education, 85*, 691–722.

Murnane, R. J., & Willett, J. B. (2010). *Methods matter: Improving causal inference in educational and social science research.* Oxford, England: Oxford University Press.

Niemiec, C. P., & Ryan, R. M. (2009). Autonomy, competence, and relatedness in the classroom: Applying self-determination theory to educational practice. *School Field, 7*, 133–144.

Ngo, F., Chi, W. E., & Park, E. S. Y. (2018). Mathematics course placement using holistic measures: Possibilities for community college students. *Teachers College Record, 120*(2). Retrieved from https://eric.ed.gov/?id=EJ1162841

Ngo, F., & Kosiewicz, H. (2017). How extending time in developmental math impacts student persistence and success: Evidence from a regression discontinuity in community colleges. *Review of Higher Education, 40*, 267–306.

Ngo, F., & Kwon, W. (2015). Using multiple measures to make math placement decisions: Implications for access and success in community colleges. *Research in Higher Education, 56*, 442–470.

Ngo, F., & Melguizo, T. (2016). How can placement policy improve math remediation outcomes? Evidence from experimentation in community colleges. *Educational Evaluation and Policy Analysis, 38*, 171–196.

Oyserman, D., Gant, L., & Ager, J. (1995). A socially contextualized model of African American identity: Possible selves and school persistence. *Journal of Personality and Social Psychology, 69*, 1216–1232.

Park, T., Woods, C. S., Hu, S., Jones, B. T., & Tandberg, D. (2018). What happens to underprepared first-time-in-college students when developmental education is optional? The case of developmental math and Intermediate Algebra in the first semester. *Journal of Higher Education, 89*, 318–340.

Payne, J. W., Bettman, J. R., & Johnson, E. J. (1992). Behavioral decision research: A constructive processing perspective. *Annual Review of Psychology, 43*, 87–131.

Royer, D. J., & Gilles, R. (1998). Directed self-placement: An attitude of orientation. *College Composition and Communication, 50*(1), 54–70.

Royer, D. J., & Gilles, R. (2003). *Directed self-placement: Principles and practices*. Cresskill, NJ: Hampton Press.

Rutschow, E. Z., & Mayer, A. K. (2018). *Early Findings from a National Survey of Developmental Education Practices* (Research brief). Retrieved from https://postsecondaryreadiness.org/wp-content/uploads/2018/02/early-findings-national-survey-developmental-education.pdf

Ryan, R. M., & Deci, E. L. (2000). Self-determination theory and the facilitation of intrinsic motivation, social development, and well-being. *American Psychologist, 55*, 68–78.

Safran, S., & Visher, M. G. (2010). *Case studies of three community colleges: The policy and practice of assessing and placing students in developmental education courses*. New York, NY: MDRC.

Scott-Clayton, J., Crosta, P., & Belfield, C. (2014). Improving the targeting of treatment: Evidence from college remediation. *Educational Evaluation and Policy Analysis, 36*, 371–393

Schunk, D. (2011). *Learning theories: An educational perspective* (6th ed.). Boston, MA: Pearson Education.

Shapiro, J. R., & Williams, A. M. (2012). The role of stereotype threats in undermining girls' and women's performance and interest in STEM fields. *Sex Roles, 66*, 175–183.

Sheldon, K. M., & Krieger, L. S. (2007). Understanding the negative effects of legal education on law students: A longitudinal test of self-determination theory. *Personality and Social Psychology Bulletin, 33*, 883–897.

Simon, H. A. (1955). A behavioral model of rational choice. *Quarterly Journal of Economics, 69*, 99–118.

Smith, A. A. (2015, May 8). *Legislative fixes for remediation*. Inside HigherEd. Retrieved from https://www.insidehighered.com/news/2015/05/08/states-and-colleges-increasingly-seek-alter-remedial-classes

Spencer, S. J., Steele, C. M., & Quinn, D. M. (1999). Stereotype threat and women's math performance. *Journal of Experimental Social Psychology, 35*(1), 4–28.

Steele, C. M. (1997). A threat in the air: How stereotypes shape intellectual identity and performance. *American Psychologist, 52*, 613–629.

Tompkins, P. (2003). Directed self-placement in a community college context. In D. Royer & R. Gilles (Eds.), *Directed self-placement: Principles and practices* (pp. 193–206). Cresskill, NJ: Hampton Press.

Venezia, A., Bracco, K. R., & Nodine, T. (2010). *One shot deal? Students' perceptions of assessment and course placement in California's community colleges*. San Francisco, CA: WestEd.

Walton, G. M., & Spencer, S. J. (2009). Latent ability grades and test scores systematically underestimate the intellectual ability of negatively stereotyped students. *Psychological Science, 20*, 1132–1139.

Woods, C. S., Richard, K., Park, T., Tandberg, D., Hu, S., & Jones, T. B. (2017). Academic advising, remedial courses, and legislative mandates: An exploration of academic advising in Florida community colleges with optional developmental education. *Innovative Higher Education, 42*, 289–303.

Zimmerman, B. J. (1989). A social cognitive view of self-regulated academic learning. *Journal of Educational Psychology, 81*, 329–339

Manuscript received May 24, 2018<br>
Final revision received July 18, 2019<br>
Accepted August 2, 2019

# Will Mentoring a Student Teacher Harm My Evaluation Scores? Effects of Serving as a Cooperating Teacher on Evaluation Metrics

Matthew Ronfeldt
Emanuele Bardelli
Stacey L. Brockman
Hannah Mullman
*University of Michigan*

*Growing evidence suggests that preservice candidates receive better coaching and are more instructionally effective when they are mentored by more instructionally effective cooperating teachers (CTs). Yet teacher education program leaders indicate it can be difficult to recruit instructionally effective teachers to serve as CTs, in part because teachers worry that serving may*

---

MATTHEW RONFELDT is an associate professor of educational studies at the University of Michigan School of Education, 610 East University Avenue, Ann Arbor, MI 48109; e-mail: *ronfeldt@umich.edu*. His scholarship focuses on identifying preservice and in-service factors that improve teaching quality and other teacher outcomes, particularly among teachers working with marginalized student populations, in order to inform policy and practice.

EMANUELE BARDELLI is a doctoral candidate in educational studies and a fellow in the causal inference in education policy research predoctoral training program at the University of Michigan School of Education. His research interests include teacher professional development, teacher learning, and instructional practices in mathematics education.

STACEY L. BROCKMAN is a doctoral candidate in educational studies and a fellow in the causal inference in education policy research predoctoral training program at the University of Michigan School of Education. A former high school history teacher and intervention specialist, her scholarship seeks to identify educational policies and practices that support at-risk secondary students' academic and social-emotional growth. She is also interested in how teacher education can promote teaching quality and student learning.

HANNAH MULLMAN is a doctoral student in educational studies and a fellow in the causal inference in education policy research predoctoral training program at the University of Michigan School of Education. Her research interests include pre- and in-service teacher learning, particularly as they relate to developing practices that promote justice and equity.

*negatively impact district evaluation scores. Using a unique data set on over 4,500 CTs, we compare evaluation scores during years these teachers served as CTs with years they did not. In years they served as CTs, teachers had significantly better observation ratings and somewhat better achievement gains, though not always at significant levels. These results suggest that concerns over lowered evaluations should not prevent teachers from serving as CTs.*

KEYWORDS: cooperating teacher, mentor teacher, clinical preparation, teacher evaluation, teacher education

## Introduction

A growing body of evidence suggests that certain characteristics of teachers' preservice training, including aspects of student teaching, are related to better workforce outcomes (Boyd, Grossman, Lankford, Loeb, & Wyckoff, 2009; Krieg, Theobald, & Goldhaber, 2016; Ronfeldt, 2012, 2015; Ronfeldt, Schwartz, & Jacob, 2014). Of particular relevance to the present study, three new studies have found recent graduates to be more instructionally effective when they learned to teach with more instructionally effective cooperating teachers (CTs) during their preservice training (Goldhaber, Krieg, & Theobald, 2018a; Ronfeldt, Brockman, & Campbell, 2018; Ronfeldt, Matsko, Greene Nolan, & Reininger, 2018).

Yet many teacher education program (TEP) leaders and state policy makers suggest that, despite their best efforts, teacher candidates are often placed with CTs who are not instructionally effective (Greenberg, Pomerance, & Walsh, 2011). As we describe in more detail below, there are a number of possible reasons why this might be the case. At least one possible explanation, commonly cited in Tennessee, where our study takes place, is that instructionally effective teachers fear mentoring teacher candidates will negatively impact their teacher evaluations. Given substantial evidence that new teachers are far less effective than more experienced teachers, allowing a candidate to take over the classroom for part of the year may indeed affect student achievement scores. However, early empirical evidence suggests these fears may be unwarranted. Though there are no existing studies published in peer-reviewed journals, a working paper in Washington (Goldhaber, Krieg, & Theobald, 2018b) finds that there are no average effects of supervising candidates on their student achievement gains, though lower performing teachers have worse achievement gains in math.

More studies in different labor markets and policy environments are needed—like the present study in Tennessee—in order to test whether these findings are specific to the Washington context. Additionally, in Tennessee, student achievement gains are only one aspect of the teacher evaluation system. This study also tests whether teachers' observation ratings, which receive equal weight in state evaluations, are impacted by mentoring a candidate. We also contribute to the existing empirical base by testing whether serving as a CT affects teacher evaluations in years after mentoring a candidate. We investigate this, in part, because some existing literature suggests that mentoring can function like a form of professional development for the CT (Spencer, 2007). Finally, we test whether the effects of serving as a CT are concentrated among teachers who are more or less instructionally effective or among teachers who work at specific school levels (elementary, middle, secondary).

Results from this study suggest that, compared with other years, teachers receive better observation ratings and similar achievement gains in years that they serve as CTs. We find positive effects on observation ratings for teachers across quartiles of instructional effectiveness, though effects are the most positive for teachers in the bottom quartile. When considering achievement gains, we detect small, positive effects for top-quartile teachers and small, negative (but nonsignificant) effects for bottom-quartile teachers; this is somewhat inconsistent with Goldhaber et al. (2018b), who found negative effects across quartiles and significantly negative effects in the bottom quartile. We also find the positive effects of serving as a CT on observation ratings to be concentrated among elementary teachers and the effects on student achievement gains to be similar across school level. Finally, in years after serving as a CT, teachers perform similarly on observation ratings and slightly worse on student achievement gains, though the latter results may be explained in part by student achievement gains' regression to the mean (Atteberry, Loeb, & Wyckoff, 2015).

The results of this study suggest that concerns that serving as a CT will harm teacher evaluations seem unwarranted; in fact, mentoring a candidate may increase evaluations. As TEPs and policy makers strive to recruit instructionally effective teachers to serve as CTs, this study suggests that these teachers should consider serving as CTs because, beyond benefiting the next generation of teachers, doing so may also improve their own evaluations.

## Literature Review

The vast majority of existing literature on CTs focuses on the effects of CTs on teacher candidates; however, this study investigates the effects of supervising a candidate on CTs themselves. At the present moment, we know of no published articles about the latter. In order to motivate this study, we begin by focusing on growing evidence that CTs who are instructionally effective teachers significantly impact candidate learning and performance. We then review literature about who serves as a CT and how placements are made in order to illustrate why candidates are not always placed with CTs who are highly effective teachers. We conclude with

*Effects of Serving as a Cooperating Teacher*

a review of a working paper and an unpublished report that are, to our knowledge, the only existing evidence for the impacts of supervising a candidate on student achievement.

### The Impact of CTs on Candidates' Instructional Effectiveness

Recent evidence suggests that new teachers are more instructionally effective in their first year if, during their preservice preparation, they received mentoring from more instructionally effective CTs. In a study evaluating statewide data from Tennessee, Ronfeldt, Brockman, and Campbell (2018) found that candidates who completed their student teaching or residency in a classroom with CTs who received observation ratings of 5.0 (significantly above expectations—the highest score on Tennessee's ratings scale) performed as if they had an additional year of teaching experience when they began teaching as compared with peers whose CTs received ratings of 3.0 (at expectations). They also found the student achievement gains of candidates and their CTs to be significantly and positively correlated. Likewise, in a study using data from Chicago Public Schools, Ronfeldt, Matsko, et al. (2018) found that an increase of one point in CTs' observational ratings (on a scale of 1–4) was associated with a 0.16 point gain for their preservice candidates' ratings in their first year, an amount comparable to the average difference on observation ratings between teachers in their first year and teachers with between 2 and 5 years of experience in Chicago (Jiang & Sporte, 2016). More recently in Washington, Goldhaber et al. (2018b) also found strong, positive associations between the math student achievement gains of mentees and mentors and more modest, but still positive, associations in English Language Arts (ELA).

### CT Recruitment and Selection

The literature reviewed thus far suggests that being assigned to an instructionally effective CT predicts teacher candidates becoming more instructionally effective themselves. Yet both existing qualitative literature and anecdotal evidence indicate that teacher candidates are often assigned to CTs who are not the most instructionally effective teachers in their schools or districts (Greenberg et al., 2011). There are many possible explanations for this. First, there is evidence that some TEPs privilege recruiting CTs who are known to provide good or supportive coaching to preservice candidates over recruiting the most instructionally effective teachers of P–12 students (Mullman & Ronfeldt, 2019). Additionally, recent research conducted on student teaching placements suggests that proximity to the program or the preservice candidate's home might be the most influential factor in selection of CTs, rather than instructional quality (Krieg et al., 2016; Maier & Youngs, 2009). Prior research also suggests that different stakeholders—including program staff, district leader, school administrators, and candidates themselves—in different programs take primary responsibility for making placement decisions (Grossman, Hammerness, McDonald, & Ronfeldt, 2008; Matsko et al., 2018), and these stakeholders likely differ in terms of how much they prioritize CT instructional effectiveness as a selection criterion. Finally, district and school leaders are sometimes hesitant to select their most instructionally effective teachers to serve as CTs because this could mean rookie teachers take over instruction for their best teachers, which they fear may have negative short-term effects on student learning and achievement, especially given the rise of high-stakes testing (St. John, Goldhaber, Krieg, & Theobald, 2018). In fact, some TEP leaders indicate that principals occasionally want to put candidates in the classrooms of struggling teachers so that they can help out and serve as "an extra set of hands" (Mullman & Ronfeldt, 2019). Similarly, St. John et al. (2018) find that principals sometimes make these matches "with the hope of either supporting or motivating a [CT's] practice" (p. 14).

Most relevant to this study, though, are reports by TEP leaders, and the district and school leaders with whom they collaborate, that teachers can be hesitant to serve as CTs for concerns that their annual evaluation scores may suffer. We initially learned about these concerns anecdotally, during conversations with TDOE policy makers and TEP leaders. These concerns were subsequently confirmed during interviews—as part of a research study on the variation in clinical preparation—by TEP leaders responsible for designing and implementing clinical experiences across Tennessee (Mullman & Ronfeldt, 2019). When asked about the basis for these concerns, some mentioned an unpublished report by the SAS Institute (2014) from a pilot study in Tennessee that concluded,

> For most grades and subjects, supervising student teachers had no significant difference in terms of teacher effectiveness, particularly for teachers who are considered average or high performing. However, the initial findings do suggest that low performing teachers might have a small negative impact in their effectiveness in Mathematics and Science when supervising student teachers as compared to not supervising. This finding has potential implications for the assignment of student-teachers to licensed teachers. (p. 2)

This potential harm to evaluation scores might worry teachers of all levels of effectiveness, given the climate of high-stakes testing.

### The Impact of Mentoring on CTs' Instructional Effectiveness

Despite these concerns, one might hypothesize that instructional quality would improve in classrooms with a teacher candidate/student teacher, given the higher student-to-teacher ratio, opportunities for collaborative teaching, and the introduction of potentially new knowledge/pedagogy by the teacher candidate.

We are aware, though, of only one recent working paper that has directly tested the impact of supervising a candidate on CTs' instructional performance. In Washington, Dan Goldhaber et al. (2018b) tested whether mentoring a candidate affected student achievement gains, and whether effects were heterogenous across levels of CT instructional effectiveness, as measured by teachers' value-added scores. Using data from 14 TEPs in Washington state, they found that there was no concurrent effect of mentoring a candidate on average student math or ELA achievement gains. However, there were differential effects by quartile of prior performance, namely, mentoring a candidate had a large and negative effect on students' math achievement for CTs in the lowest value-added quartile. The authors suggest that more effective CTs are able to "mitigate" the impact of letting an inexperienced candidate take over instruction in the classroom. Conversely, they also found modest, positive impacts on student math and reading performance in subsequent years of serving as a CT. In the next section, we consider more extensively different mechanisms by which mentoring a candidate might impact a teacher's concurrent and future performance.

## Contributions of the Present Study

In keeping with recent calls for more replication studies in educational research (Makel & Plucker, 2014), the present study replicates and extends the Goldhaber et al. study in a different teacher labor market and state context. Like Goldhaber et al. (2018b), we are interested in the effect that mentoring a candidate has on teachers' evaluation metrics. The present study, though, also extends prior research by incorporating both value-added measures and observational ratings as our outcomes of interest. While Goldhaber and colleagues only considered value-added measures, in many states (including Tennessee) observation ratings carry equal, and sometimes greater, weight in final evaluations. Especially given prior evidence that observation ratings may be prone to rater tendencies, biases, and subjectivities (Campbell, 2014; Campbell & Ronfeldt, 2018; White, 2018), it may be that the effects of supervising a candidate on observation ratings differ from the effects on value-added measures. For example, the elevated status of being a CT may cause raters to inflate scores of teachers supervising candidates. We further extend the literature through our use of an analytic sample from Tennessee state administrative data, a state with a labor market and cultural context that differs from Washington.

Similar to Goldhaber et al. (2018b), we consider heterogeneity of effects by prior performance level as well as heterogeneity by school level. We add the latter focus because elementary teachers typically have self-contained classrooms and teach all subjects to the same group of students, whereas secondary teachers typically work with different students (classes/preps) across the day and usually specialize in terms of subject matter. These different arrangements require different approaches and decisions about how to integrate candidates into classrooms and lead teaching responsibilities, and thus, may have different implications for impacts on a CT's own performance. Below, we elaborate on different school-level considerations regarding student teaching arrangements and potential mechanisms by which these arrangements may impact a CT's performance.

We also consider the impacts of mentoring a candidate on the CT's own learning and professional growth and interrogate whether serving as a CT changes future performance. The literature suggests that effective professional development include long-term, active learning (Desimone, 2009), and it is possible that mentoring a novice teacher meets these requirements. In fact, in a recent survey of CTs in Chicago, almost one fifth of CTs indicated that their primary reason for serving as a CT was because it helped them to improve as a teacher (Matsko, Ronfeldt, & Greene Nolan, 2019). This is further supported by a review of mentoring programs for novice teachers in the United Kingdom, where Shanks (2017) finds that mentors, in coaching novices, sometimes engage in the same kinds of critical inquiry and reflection as mentees, creating opportunities for learning for both parties. Thus, we test for lagged effects of serving as a CT on teachers' instructional performance in years after they mentored candidates.

## Logic Model

While most existing research has focused on the effects of CTs on the performance of those candidates working with them, this study investigates the effects on the performance of CTs themselves. The perception among some teachers, teacher educators, and policy makers in Tennessee and elsewhere—a central motivation for this study—is that serving as a CT can harm teachers' evaluation scores[1] in the year that they serve. How might this occur? To our knowledge, there is no existing research on how serving as a CT might impact one's own evaluation scores; thus, we can only speculate. In this section, we begin by considering a number of possible mechanisms by which mentoring a candidate might impact, positively and negatively, teachers' evaluation scores in the year that they serve; after, we consider how serving as a CT might affect their future performance, during postservice years.

Our logic model (see Figure 1) begins with an assumption that observation ratings and value-added measures reflect the quality of underlying teaching skills/competencies, an assumption that is supported by a number of studies demonstrating their validity and reliability (Cantrell & Kane, 2013; Gitomer & Bell, 2013; Hill, Kapitula, & Umland, 2011; Kane & Staiger, 2012). We also acknowledge, though, that these evaluation measures are unlikely to capture all dimensions of teaching quality and are known to measure other aspects of classrooms beyond teaching quality, so are prone to

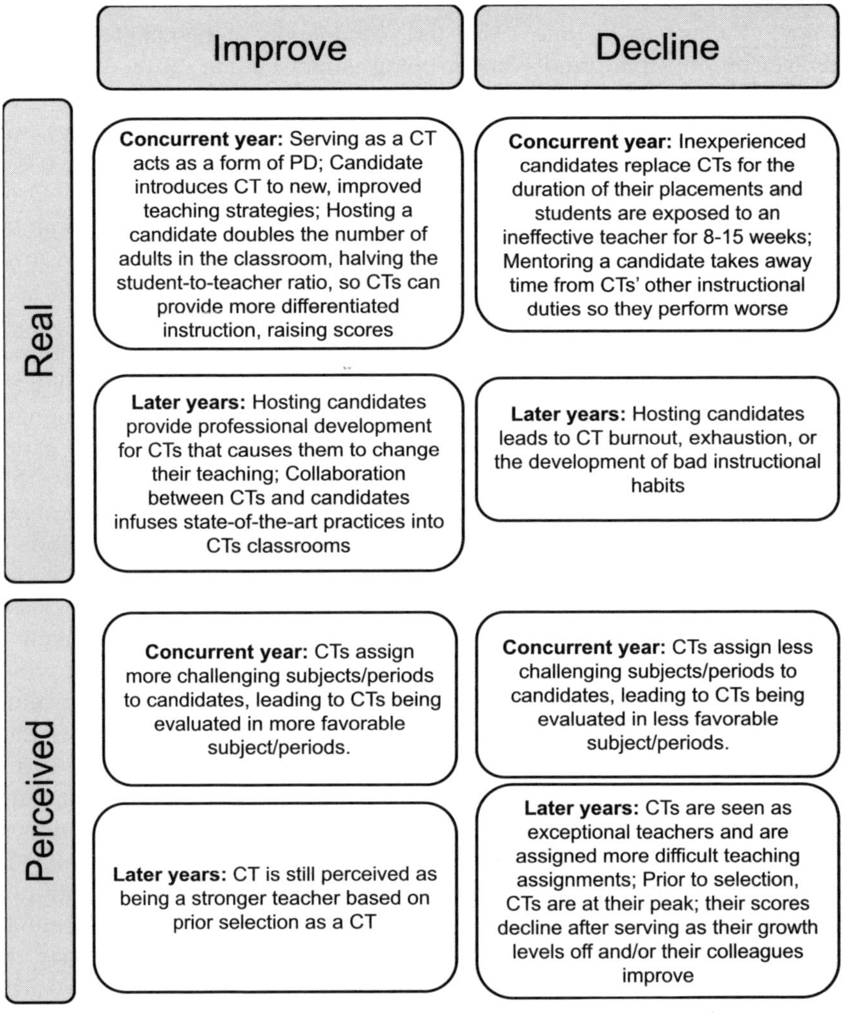

*Effects of Serving as a Cooperating Teacher*

|  | **Improve** | **Decline** |
|---|---|---|
| **Real** | **Concurrent year:** Serving as a CT acts as a form of PD; Candidate introduces CT to new, improved teaching strategies; Hosting a candidate doubles the number of adults in the classroom, halving the student-to-teacher ratio, so CTs can provide more differentiated instruction, raising scores | **Concurrent year:** Inexperienced candidates replace CTs for the duration of their placements and students are exposed to an ineffective teacher for 8-15 weeks; Mentoring a candidate takes away time from CTs' other instructional duties so they perform worse |
|  | **Later years:** Hosting candidates provide professional development for CTs that causes them to change their teaching; Collaboration between CTs and candidates infuses state-of-the-art practices into CTs classrooms | **Later years:** Hosting candidates leads to CT burnout, exhaustion, or the development of bad instructional habits |
| **Perceived** | **Concurrent year:** CTs assign more challenging subjects/periods to candidates, leading to CTs being evaluated in more favorable subject/periods. | **Concurrent year:** CTs assign less challenging subjects/periods to candidates, leading to CTs being evaluated in less favorable subject/periods. |
|  | **Later years:** CT is still perceived as being a stronger teacher based on prior selection as a CT | **Later years:** CTs are seen as exceptional teachers and are assigned more difficult teaching assignments; Prior to selection, CTs are at their peak; their scores decline after serving as their growth levels off and/or their colleagues improve |

*Figure 1.* **Logic model for effects of serving as a cooperating teacher on evaluation scores.**

*Note.* CT = cooperating teacher; PD = professional development.

manipulation, error, and bias, as previously stated. We know, for example, that observation ratings tend to be lower in classrooms with students who are lower achieving, Black and Latinx, receiving special education services, and secondary, and in classrooms of teachers who are male and, in some cases, Black (Campbell & Ronfeldt, 2018; Harris, Ingle, & Rutledge, 2014;

Jiang & Sporte, 2016; Steinberg & Garrett, 2016). We also know that observation ratings may vary by time of day and year, identity of the rater (e.g., master rater versus principal), and content being taught (White, 2018; Whitehurst, Chingos, & Lindquist, 2014). Similarly, value-added measures (VAMs) are somewhat unstable across years and can vary by student characteristics, subject area, and prior achievement (Loeb & Candelaria, 2012). Thus, as we consider how ratings change in response to serving as a CT, we consider not only how serving as a CT may influence the quality of teaching in a classroom but also how evaluations may change as a result of manipulation or bias, even where the quality of teaching remains constant.

As we consider mechanisms by which CTs' performance may be affected by mentoring a candidate, it is also important to consider different ways that CTs might hand over lead teaching responsibilities to candidates. In some cases, CTs only have one classroom or section (e.g., elementary teachers in self-contained classrooms). In these cases, CTs can allow candidates to take over lead teaching responsibilities across subject areas; they can also hand over lead teaching responsibilities in some subjects (e.g., reading, science) but not others. In other cases, CTs teach multiple periods or classes (e.g., secondary science teachers with Biology and AP Biology classes). Here, CTs may hand over lead teaching responsibilities in some classes but retain them in others. Given that school principals/leaders are likely to observe CTs when they are personally teaching, the decision about which classes to hand over to a candidate can have implications for CTs' evaluations. For example, if a CT hands over lead teaching responsibilities in their most challenging classes or subjects, then they are likely to be observed and evaluated in less challenging contexts which could boost their evaluations. Thus, as we discuss mechanisms by which mentoring may impact a teacher's own evaluations, we consider ways in which the quality of teaching may be impacted versus ways that evaluated performance may change without necessarily impacting underlying teaching quality; in Figure 1 we differentiate mechanisms by which "real" and "perceived" performance may be impacted. Here we intend to differentiate ways in which teachers may actually alter their teaching (quality) in response to service as a CT versus ways in which evaluations of their practice might be altered without necessarily changing their teaching at all. Our study is not designed to test which, if any, of these postulated mechanisms is at work; rather, we include this section to give the reader an orientation to plausible ways that serving as a CT might affect a teacher's evaluations. We encourage future research to investigate which of these mechanisms explain the results we observe.

**Possible Mechanisms for Impacting Performance During Service Year**

We begin by considering mechanisms by which CTs' performance may be harmed in years that they mentor candidates. Perhaps the most obvious is that candidates are inexperienced, and there is a great deal of literature demonstrating that rookie teachers tend to be less effective teachers, which could lead to lower student achievement scores. That is, when candidates take over some/all lead teaching responsibilities, students in their classrooms likely encounter less effective teaching, on average. If, as a result, students perform worse on state tests, this would be reflected in CTs' achievement gains, given that they are still the teacher of record. This is an example of a real change in teaching quality resulting from serving as a CT.

How could mentoring a candidate also negatively affect CTs' observation ratings? In Tennessee, a teacher must still be observed by a principal or school leader even when mentoring a candidate. In cases where CTs teach a self-contained class, the evaluator presumably observes the CT teaching the same students as those taught by their candidate. As a result of inexperienced candidates taking responsibility for some of the prior classroom activities, it is possible that the classroom culture will be worse and/or the students will be less prepared and, as a result, the CTs may struggle more when being observed and evaluated. In cases where CTs have multiple preps/classrooms and hand over lead teaching responsibility in only one/some classes, then it is possible that candidates will take over in classes with stronger classroom cultures (e.g., to make learning to teach somewhat easier). This would then leave CTs to be evaluated in classrooms/preps where they may be more likely to struggle and, hence, receive lower observation ratings. This scenario describes a case in which perceived teaching quality changes as a result of CT service. Yet another possibility is that mentoring a candidate takes CTs' time and effort away from improving their own teaching with P–12 students.

On the other hand, there are potential mechanisms that could lead to an improvement in CT performance. One possible mechanism is that mentoring a candidate effectively doubles the numbers of teachers in a classroom, thus decreasing the student-to-teacher ratio and raising the likelihood that students will receive more individualized attention. Moreover, mentoring a candidate allows for teacher collaboration, which can increase teaching quality and student performance; for example, candidates might share new curriculum or pedagogy perhaps from their TEPs. It is also possible that when mentoring a candidate, CTs may be more motivated to model exceptional teaching, thus putting extra time and effort into teaching. Finally, if they choose to place their candidates in their most challenging preps/classes/subjects, CTs will be more likely to be observed in settings that are favorable to their performance.

### Possible Mechanisms for Impacting Performance in Future Years

It is also possible that a teacher's performance after serving as a CT may be affected. Hosting a candidate could be a form of professional development for CTs (Spencer, 2007). By observing and providing feedback to a new teacher, or in engaging in planning and reflective conferences with them, CTs might sharpen their practice and become more effective teachers themselves. Student teachers also may bring innovative teaching practices from their methods courses that they implement during their clinical placements. CTs could learn from these practices and add them to their teaching repertoire, which could manifest in improved performance in subsequent years.

Of course, a CT's performance could also decline in postservice years. Hosting a candidate could be taxing for a CT, either personally or professionally. This could lead to CT burnout and exhaustion. Having to fulfill the always-increasing demands of teaching on top of mentoring a new teacher could lead to CTs not having enough time to properly rest, which could result in a decline in performance in the years following mentoring a candidate. It is also possible that teachers develop poor habits when mentoring a candidate and then carry these habits into future years (e.g., adopt ineffective practices used by the candidate).

School leaders might also believe that teachers who serve as CTs are exceptional, and, in subsequent years, assign these teachers more difficult preps/classes, leading to lower evaluations in subsequent years. It is also possible that school leaders inflate evaluations of teachers in years they serve as CTs (e.g., leaders may be more lenient given that being a CT is a form of service that effectively increases workload); consequently, postservice evaluations might subsequently decline mechanically even where the quality of teaching performance is consistent.

## Research Questions

Drawing upon this logic model, in this article we ask broadly, "What effects might serving as a CT have on teachers' concurrent and subsequent performance?" More specifically, the following research questions guided our analysis:

*Research Question 1:* Do teachers perform differently in years that they serve as CTs?
*Research Question 2:* Are the effects different for different groups of CTs?
*Research Question 3:* Do teachers perform differently in years after they serve as CTs?

## Data

Data for this article come from a unique data set of CTs collected by the Tennessee Department of Education. This data set includes information from 17 TEPs[2] in the state and identifies the teachers who served as CTs for these programs between the 2010–2011 and the 2013–2014 school years. We merge these data onto Tennessee's teacher and school databases. The teacher database includes information about teachers' work experience, licensing status, and evaluation scores. School-level data come from Tennessee school universe files which include information about student body characteristics, average attendance, and school improvement status.

**Descriptive Statistics**

Our analytic data set includes all teachers in Tennessee from the 2010–2011 school year through the 2016–2017 school year. Our sample includes 458,717 teacher-by-year observations. Table 1 presents descriptive information about types of evaluation data we have for teachers, including observation ratings and their value-added to student achievement measures (Teacher Value-Added Assessment System, or TVAAS; see Vosters, Guranio, and Wooldridge, 2018, for more information on how these scores are calculated). Similar to other states' use of value-added measures, TVAAS is calculated using state test data and intends to capture an individual teacher's effect on student achievement; teachers receive scores for specific tested subjects as well as composite scores. TVAAS was piloted in the 2010–2011 school year and fully implemented the following year, so we report value-added measures starting in 2011. TVAAS scores are available only for about half of the teachers in our sample because of variation in testing requirements across grade levels and school settings. Observation ratings are available starting from the 2011–2012 school year. We have a total of 4,522 teacher-by-year observations for teachers who served as CTs between the 2010–2011 and 2013–2014 school years.[3] Teachers in Tennessee are assessed multiple times per year using the Tennessee educator acceleration model, a rubric that includes four domains and multiple indicators within each domain.[4] The four domains are instruction, environment, planning, and professionalism. Professionalism is only assessed one time, at the end of the school year. For the other three, multiple domains and indicators are scored simultaneously, during the same observation, and teachers receive scores on a scale from 1 (*significantly below expectations*) to 5 (*significantly above expectations*). For this article, domain and overall ratings are an average of indicator scores and domain scores, respectively.

Table 2 presents summary statistics comparing those teachers who served as CTs with those who did not. Reading the table from left to right, we present the average statistics for our entire analytic sample of teachers, CTs, all other teachers, and the difference between CTs and other teachers. Teachers who served as CTs are, on average, statistically different from other teachers when it comes to their observation ratings, TVAAS scores, teacher covariates, and school covariates. On average, we find that CTs are more likely to be White (7.6 percentage point difference) and female (3.6 percentage point difference), have 1.88 years more experience, are more likely to hold an advanced degree, and work in schools with a greater proportion

Table 1
Number of Teachers With Valid Evaluation Data

| | 2011 | 2012 | 2013 | 2014 | 2015 | 2016 | 2017 | Total |
|---|---|---|---|---|---|---|---|---|
| All teachers | 21,708 | 74,512 | 71,756 | 73,906 | 71,966 | 72,807 | 72,062 | 458,717 |
| Observation ratings | 0 | 70,616 | 65,894 | 57,637 | 60,873 | 70,523 | 69,681 | 395,224 |
| TVAAS scores | 21,291 | 21,843 | 30,551 | 31,714 | 25,523 | 8,879 | 21,158 | 160,959 |
| Did serve as cooperating teacher | 417 | 1,163 | 1,561 | 1,381 | *Cooperating teacher data unavailable* | | | 4,522 |
| Observation ratings | 0 | 1,151 | 1,507 | 1,291 | | | | 3,949 |
| TVAAS scores | 417 | 472 | 983 | 786 | | | | 2,658 |
| Did not serve as cooperating teacher | 21,291 | 73,349 | 70,195 | 72,525 | | | | 237,360 |
| Observation ratings | 0 | 69,465 | 64,387 | 56,346 | | | | 190,198 |
| TVAAS scores | 20,874 | 21,371 | 29,568 | 30,928 | | | | 102,741 |

*Note.* TVAAS = Teacher Value-Added Assessment System. The Cooperating Teacher database is available for a subset of teacher education programs in the state for school years 2010–2011 through 2013–2014. The Tennessee Teacher Value-Added Assessment System was piloted during the 2010–2011 school year and fully implemented in the 2011–2012 school year. Observation ratings are available starting from the 2011–2012 school year.

*Effects of Serving as a Cooperating Teacher*

Table 2
**Comparing Observable Characteristics of All Teachers With Those of Cooperating Teachers**

|  | All Teachers | Cooperating Teachers | Other Teachers | Difference | p Value |
|---|---|---|---|---|---|
| Outcomes of interest | | | | | |
| Observation ratings | 3.888 | 4.042 | 3.879 | 0.163 | *** |
| TVAAS—All subjects | 0.042 | 0.098 | 0.038 | 0.061 | *** |
| TVAAS—Mathematics | 0.083 | 0.158 | 0.078 | 0.081 | *** |
| TVAAS—ELA | 0.022 | 0.054 | 0.019 | 0.035 | *** |
| Teacher covariates | | | | | |
| Percent female | 0.799 | 0.833 | 0.797 | 0.036 | *** |
| Percent White | 0.870 | 0.941 | 0.866 | 0.076 | *** |
| Percent Black | 0.122 | 0.055 | 0.125 | −0.070 | *** |
| Percent other | 0.005 | 0.004 | 0.005 | −0.002 | *** |
| Percent bachelor's degree | 0.408 | 0.326 | 0.414 | −0.088 | *** |
| Percent master's degree | 0.503 | 0.547 | 0.500 | 0.047 | *** |
| Percent PhD | 0.009 | 0.012 | 0.009 | 0.002 | * |
| Age | 42.55 | 42.95 | 42.53 | 0.42 | *** |
| Years of teaching experience | 11.96 | 13.74 | 11.86 | 1.88 | *** |
| School assignment | | | | | |
| Elementary school | 0.433 | 0.498 | 0.429 | 0.068 | *** |
| Middle school | 0.185 | 0.193 | 0.185 | 0.009 | ** |
| High school | 0.278 | 0.230 | 0.280 | −0.051 | *** |
| School covariates | | | | | |
| Percent White | 0.678 | 0.759 | 0.673 | 0.086 | *** |
| Percent Black | 0.216 | 0.140 | 0.221 | −0.081 | *** |
| Percent Hispanic | 0.075 | 0.074 | 0.075 | −0.001 | |
| Percent FRPL | 0.587 | 0.567 | 0.589 | −0.022 | *** |
| Percent proficient | 0.514 | 0.537 | 0.513 | 0.024 | *** |
| N | 241,882 | 4,522 | 237,360 | | |

*Note.* ELA = English Language Arts; FRPL = free or reduced-priced lunch; TVAAS = Teacher Value-Added Assessment System. The Cooperating Teacher database is available for a subset of teacher education programs in the state for school years 2010–2011 through 2013–2014. The Tennessee Teacher Value-Added Assessment System was piloted during the 2010–2011 school year and fully implemented in the 2011–2012 school year. Observation ratings are available starting from the 2011–2012 school year.
+$p$ < .10. *$p$ < .05. **$p$ < .01. ***$p$ < .001.

of students who are White and meet proficiency levels on state exams and with a smaller proportion of students who qualify for free or reduced-priced lunch. CTs also tend to have higher observation ratings and TVAAS scores. In our sample, the average observation rating for a CT was 4.04, compared with 3.88 for teachers who did not serve. The average TVAAS score for CTs was

*Ronfeldt et al.*

0.061 student standard deviation units higher than other teachers. These findings are consistent with other prior research which has found CTs to have stronger evaluation scores, on average, than non-CTs (Goldhaber et al., 2018b; Matsko et al., 2018; Ronfeldt, Brockman, & Campbell, 2018).

## CT Blocks

In order to conduct a more appropriate comparison of those who serve as CTs with those who do not, we construct blocks of all eligible teachers for a student teaching placement in a given year. We identify teachers who served each year and then group them with all other teachers in their districts with the same teaching endorsement (e.g., secondary math, elementary, secondary ELA, etc.). This allows us to create a hypothetical pool of all teachers who could have potentially served as CT for a particular candidate.[5] We merge Tennessee's Personnel Information Reporting System and teacher assignment data onto our analytic sample and then compare the courses they taught that year and assigned them an endorsement. If for example, a seventh-grade social studies teacher in district $D$ serves as a CT, we create a block with all secondary social studies teachers in district $D$ in order to build a sample of all possible CTs for that year.

Table 3 presents descriptive differences between blocks with and without eligible CTs, as well as differences between CTs and the rest of the teachers in their block. Compared with blocks without CTs, on average, blocks with CTs have lower observation ratings and years of experience but higher TVAAS scores. Blocks with CTs have a higher share of elementary and middle school teachers and lower share of secondary teachers. They also tend to have more female and Black teachers but fewer White teachers.

When we look within blocks, we find that CTs outperform non-CTs. In our sample, CTs have average observation ratings of 4.03, approximately 0.19 higher than non-CTs in the same blocks. CTs also have higher TVAAS scores than non-CTs (a 0.08 standard deviation difference for teachers not in blocks and a difference of 0.06 for those in blocks). CTs were more likely to be female, White, and hold a graduate degree, but were less likely to be Black. When compared with the rest of their block, CTs were also more likely, on average, to teach in schools with higher proportions of White and higher achieving students.

## Method

### Research Question 1

We use a generalized differences-in-differences method with teacher fixed-effects model to investigate the effects of serving as a CT on evaluation metrics. This modeling strategy allows us the estimate the within-teacher changes in years during which they serve as a CT as compared with the other

## Effects of Serving as a Cooperating Teacher

### Table 3
### Descriptive Statistics by Block

|  | Blocks Without CTs | Blocks With CTs All | Non-CTs | CTs |
|---|---|---|---|---|
| Outcomes of interest |  |  |  |  |
| Observation ratings | 3.911 | 3.855 | 3.838 | 4.026 |
| TVAAS | 0.033 | 0.051 | 0.045 | 0.109 |
| TVAAS—Mathematics | 0.063 | 0.100 | 0.093 | 0.171 |
| TVAAS—ELA | 0.019 | 0.024 | 0.020 | 0.058 |
| Teacher covariates |  |  |  |  |
| Percent female | 0.781 | 0.824 | 0.822 | 0.845 |
| Percent White | 0.878 | 0.857 | 0.849 | 0.940 |
| Percent Black | 0.110 | 0.137 | 0.145 | 0.056 |
| Percent other | 0.005 | 0.005 | 0.005 | 0.004 |
| Percent bachelor's degree | 0.401 | 0.419 | 0.429 | 0.330 |
| Percent master's degree | 0.509 | 0.493 | 0.487 | 0.548 |
| Percent PhD | 0.010 | 0.008 | 0.008 | 0.011 |
| Age | 42.87 | 42.10 | 42.05 | 42.69 |
| Years of teaching experience | 12.13 | 11.72 | 11.57 | 13.41 |
| School assignment |  |  |  |  |
| Elementary school | 0.39 | 0.50 | 0.50 | 0.53 |
| Middle school | 0.17 | 0.21 | 0.21 | 0.19 |
| High school | 0.33 | 0.20 | 0.20 | 0.21 |
| School covariates |  |  |  |  |
| Percent White | 0.693 | 0.657 | 0.648 | 0.752 |
| Percent Black | 0.207 | 0.230 | 0.238 | 0.145 |
| Percent Hispanic | 0.072 | 0.079 | 0.080 | 0.074 |
| Percent FRPL | 0.588 | 0.586 | 0.589 | 0.564 |
| Percent proficient or above | 0.517 | 0.509 | 0.507 | 0.537 |
| N | 102,560 | 118,562 | 114,037 | 4,222 |

*Note.* CT = cooperating teacher; ELA = English Language Arts; FRPL = free or reduced-priced lunch; TVAAS = Teacher Value-Added Assessment System. Blocks were calculated according to whether a CT served in particular district in a given year. We group them with all other teachers in that district with the same teaching endorsement, so blocks represent all eligible CTs for a preservice teacher that year.

years during which they did not mentor a teacher candidate while accounting for differences between teachers who are selected and teachers who are not selected to serve as a CT.

Our preferred model is

$$Y_{it} = \beta_{0i} + \beta_1 CT_{it} + \delta Exp_{it} + \lambda_t + \varepsilon_{it}, \tag{1}$$

where $Y_{it}$ is the outcome of interest. $\beta_{0i}$ is the individual-level fixed effect. $CT_{it}$ is an indicator variable taking the value of 1 for all years $t$ during which teacher $i$ is reported as serving as a CT. $Exp_{it}$ is a set of indicators for years of work experience that we add to the model to increase efficiency and account for the timing of being selected to be a CT. $\lambda_t$ is the year fixed effect. We use these fixed effects to account for any secular variation in evaluation scores. $\varepsilon_{it}$ is the stochastic error term adjusted for clustering of teachers at the school level.

Our coefficient of interest is $\beta_1$. This term captures the causal effect of serving as a CT on evaluation scores and teacher value-added estimates. Our causal claim rests on two identifying assumptions. First, any individual-level characteristics that lead to selection into serving as a CT are constant over time and can be accounted for by an individual-level fixed effect. Second, these characteristics have a linear and additive functional form to the model's intercept (Angrist & Pischke, 2008).

### Research Question 2

*Heterogeneity by Quartile*

Goldhaber et al. (2018b) suggested that the effects of serving as a CT vary for teachers in different effectiveness quartiles, with a negative effect among the lowest quartile of teachers.[6] We calculate effectiveness quartiles using a two-step approach. First, we estimate the teacher fixed effect from this model:

$$Y_{it} = \tau_i + \pi_1 CT_{it} + \varepsilon_i, \qquad (2)$$

where $\tau_i$ is the teacher fixed effect for teacher $i$. This captures the evaluation score averages for teacher $i$ over all observation years, controlling for effects of serving as a CT on evaluation scores. We use these teacher fixed effects to calculate the quartile of effectiveness for each teacher or, more formally, $Q_i|\tau_i$. We use these quartiles to estimate the effect of serving as a CT for teachers across the quality distribution using the model

$$Y_{it} = \beta_{0i} + \beta_1 CT_{it} \times Q_i + \delta Exp_{it} + \lambda_t + \varepsilon_{it}, \qquad (3)$$

where $CT_i \times Q_i$ is the interaction term between the CT indicator and the quartile of effectiveness for each $Y$. $\beta_1$ is a vector of four estimates, one for each quartile of effectiveness, that allow us to test whether the effects of serving as a CT vary at different points of the teacher performance continuum.

*Heterogeneity by School Level*

A major difference between elementary and secondary teachers is that the former are typically with the same group of students throughout the day while the latter tend to work with different students during the school day. As we discuss in our logic model, differences in school level could have differential impacts on CTs' evaluation scores. It is, therefore, important to investigate whether or not the effects of serving as a CT vary by school level.

Thus, we divide schools into four categories, elementary (Grades K–5), middle (Grades 6–8), high (Grades 9–12), and other (e.g., K–8 schools). We estimate a model similar to Equation (4) where we interact the CT indicator with indicators for school level, allowing us to test whether the effects of serving as a CT vary by instructional setting.

**Research Question 3**

We modify our preferred model described in Equation (1) to answer the third research question:

$$Y_{it} = \beta_{0i} + \beta_1 CT_{it} + \beta_2 CT_{after_{it}} + \delta Exp_{it} + \lambda_t + \varepsilon_{it}, \qquad (4)$$

where we divide the counterfactual for serving as a CT in Equation (1) into two parts using the $CT_{after_{it}}$ indicator. This indicator takes the value of 1 for all teachers who were reported as being a CT for at least 1 year and for all years following serving as a CT. This allows us to separately estimate the effects of serving as a CT on evaluation metrics for the years during which a teacher serves as a CT and for the years following serving as a CT.[7] The coefficient of interest for these analyses is $\beta_2$. This captures the effects of serving as a CT in the period following this experience as compared with evaluation scores during the period preceding serving as a CT.

**Robustness Checks**

We run several robustness checks to test whether our results are sensitive to our model specification, to the sample of teachers that we use, and to our estimation strategy. We find that the results from our preferred model are robust against all of these checks.

First, we test whether our results are sensitive to the inclusion of teacher experience. Papay and Kraft (2015) argued that experience coefficients could be biased when used in a fixed-effects model that includes year terms. We address this concern by estimating our preferred model without the experience terms and by adjusting the experience coefficients using the technique described by Papay and Kraft (2015).

Second, we include school-level covariates to control for possible unobserved differences among workplaces that could confound selection to be a CT and evaluation scores. For example, researchers have found a relationship between teachers' evaluation ratings and the characteristics of their students (Campbell & Ronfeldt, 2018; Jiang & Sporte, 2016; Steinberg & Garrett, 2016).

Third, we estimate our preferred model on progressively more restrictive samples of teachers in order to account for various forms of likely

*Ronfeldt et al.*

selection. We restrict our sample to teachers who teach in the same school district and subject area as the teachers who we observe serving as a CT, to teachers who teach in the same school and subject, and to teachers who are reported as being CTs at least once.

Last, we use traditional difference-in-differences and matched-sample model specifications to check whether our results are sensitive to model specification. In this specification, we use the equation:

$$Y_{it} = \beta_{0i} + \beta_1 CT_{ever\,it} + \beta_2 CT_{it} + \delta Exp_{it} + \pi_{it} + \lambda_t + \varepsilon_{it},$$

where $CT_{ever}$ is an indicator variable that takes the value of 1 for any teacher who served as a CT at least once, $CT$ is the indicator variable taking the value of 1 during all years $t$ for which teacher $i$ is reported as serving as a CT, $\pi_{it}$ is a school fixed effect, and $\lambda_t$ is a year fixed-effect term. Conceptually, this difference-in-differences model compares teachers who serve as CTs with teachers who did not serve within the same school. The first difference is between teachers who ever serve as CT ("ever CTs") and teachers who never serve as a CT. This difference accounts for average unobserved differences on evaluation scores between the group of teachers that is ever selected to serve as CTs and those who were never chosen to serve. The second difference is within the group of "ever CTs" and compares the evaluation scores for the years during which these teachers serve as CT and years during which they do not. This second difference estimates the effect of serving as a CT on evaluation scores.

This difference-in-differences specification relies on more permissive assumptions than our preferred model: that the evaluation scores of teachers who were ever selected to be CTs and the ones for teachers who were not have parallel trends before CT selection. Evaluation and CT data availability make it difficult to formally assess the degree to which evaluation scores for CTs and non-CTs followed parallel trends before serving as CT. In particular, when teachers served as CTs early in our observation window, we sometimes have no data on pretrends or only scores for a single year. Thus, we limit our analysis to the 2013 and 2014 CT cohorts where we have at least 2 years of pretrend data, and we present an event study using this cohort (see Appendix Figure A1). While we observe parallel trends between CTs and non-CTs in terms of TVAAS, we observe that observation ratings (OR) seem to increase the year prior to being selected to serve as a CT.

These results could hint at CTs being selected to serve based on prior year observation score data. Given the mixed evidence during the pre-CT period and the data limitations (especially the shifting cohorts of teachers across years), it is difficult to be certain that the parallel trends assumption has been met; therefore, we only include these difference-in-differences results as a robustness check to our preferred model. We recommend caution when interpreting the results from these difference-in-differences models, as relaxing the assumptions of our preferred model could introduce bias when the parallel trend assumptions are not met. In fact, we find that the difference-in-differences estimates have greater magnitude than our fixed-effects estimates. Two possible sources of bias can explain these results. First, if the difference-in-differences models do not meet the parallel trend assumption, the estimates will be biased upward. Second, unobserved variables could lead to an increase in observation ratings that is unrelated to serving as a CT. For example, a teacher might take on a leadership role at the school at the same time as mentoring a candidate. We could expect that taking on that leadership role could lead to higher observation ratings and that this increase is unrelated to serving as a CT.

In part because of possible concerns that, prior to serving, teachers who become CTs may be increasing on observation ratings at relatively greater rates than other teachers, we restrict the estimate sample to a matched sample of teachers. This matched-sample model allows us to construct a comparison group that is similar on observed characteristics, including pretrends on evaluation data, to teachers who serve as a CT. We do this in a two-step process. First, we identify a sample of teachers who have observed characteristics similar to our CT sample. Second, we use this matched group to calculate the effect of serving as a CT on evaluation scores. Specifically, we match CTs and non-CTs using a nearest neighbor matching algorithm that uses an exact match on teacher demographic characteristics (i.e., race/ethnicity and gender), highest level of education completed (i.e., bachelor's, postbachelor's, or master's degree), school level (i.e., elementary, middle, or high school), and CT block. We fuzzy match using Mahalanobis distance on up to two prior years of evaluation data and years of experience at the time of serving as a CT. We remove two CTs from these analyses who did not match with other teachers in the state on background characteristics. Appendix Figure A2 reports the density distributions for the fuzzy matched variables pre- and postmatching. We note that the matching procedure was able to identify similar teachers across the demographic variables for all four outcomes of interest, suggesting that these models appear to meet the common support assumption. We also note that we were not able to have quality matches on TVAAS mathematics scores 2 years prior to serving as a CT, which could introduce some bias in the matched estimates for this particular measure.

This matched-sample specification relies on the assumption that we match teachers who are reported as being CTs to teachers similar to them in all observed characteristics included in the model except for being selected to be a CT. However, these estimates could be biased if selection to be a CT is driven by unobserved teacher characteristics.

The estimates from these two alternative model specifications have generally the same sign and are larger in magnitude than the estimates from our preferred model. The results confirm that our preferred model provides the most conservative estimates of our outcomes of interest.

## Results

### Research Question 1: Do Teachers Perform Differently in Years That They Serve as CTs?

We present the main results for the four outcomes of interest—observation ratings, average TVAAS, mathematics TVAAS, and ELA TVAAS—in Table 4. We begin by summarizing results from our preferred models with teacher fixed effects. Across the first row, we notice that the effects of serving as a CT on evaluation metrics is either small and positive, in the case of observation ratings, or not significantly different from zero, in case of all three TVAAS estimates. Regarding observation ratings, estimates suggest that teachers' observation scores increase by 0.04 points in years that they serve as CTs as compared with other years; this is roughly equivalent to about one fifth of the expected growth in observation ratings for a first-year teacher (Ronfeldt, Brockman, & Campbell, 2018). It is worth noting that CTs have, on average, almost 14 years of experience, a point in teachers' careers when their observation ratings tend not to increase substantially (i.e., after the 10-year mark, see Papay & Kraft, 2015, for an in-depth analysis).

The even-numbered columns in Table 4 display the estimates from the difference-in-differences models. We note that the point estimates for our coefficient of interest tend to be greater in magnitude in these models than in the teacher fixed-effects ones.[8] In fact, the estimate on models for TVAAS (all subjects), is now positive and statistically significant at the 5% level, suggesting that teachers have greater achievement gains in years that they serve as CTs. These models also allow us to estimate the difference in evaluation scores (across years) between teachers who are reported as serving as CTs at least once in our data set (see row "Ever cooperating teacher") and teachers who are not reported as serving as CTs during our observation period. We interpret this coefficient as the baseline difference in evaluation scores that might have led specific teachers to be selected as CTs. Across all four outcomes, we note that teachers who serve as CTs at least once have significantly and meaningfully higher evaluation scores than their peers. In other words, teachers who serve as CTs are, on average, higher performing teachers and seem to be positively selected on their evaluation scores.

### Sensitivity to Sample Selection

As discussed in the "Data" section (see Table 2), we find that teachers in the same districts and subject areas as our CT sample seem to differ from teachers in other districts/subjects. While our teacher fixed-effects models effectively compare a teacher's performance in years in which they serve as CTs with performance in years in which they do not, in our preferred specification we use the full sample of teachers—including those teachers

## Table 4
### Effects of Serving as a Cooperating Teacher on Evaluation Metrics

|  | Observation Ratings || TVAAS—All Subjects || TVAAS—Math || TVAAS—ELA ||
|---|---|---|---|---|---|---|---|---|
|  | (1) Fixed Effects | (2) Diff-in-Diff | (3) Fixed Effects | (4) Diff-in-Diff | (5) Fixed Effects | (6) Diff-in-Diff | (7) Fixed Effects | (8) Diff-in-Diff |
| Cooperating teacher | 0.040*** | 0.053*** | 0.008 | 0.014* | −0.001 | 0.005 | −0.003 | 0.005 |
|  | (0.007) | (0.007) | (0.006) | (0.006) | (0.012) | (0.013) | (0.007) | (0.007) |
| Ever cooperating teacher |  | 0.108*** |  | 0.040*** |  | 0.075*** |  | 0.028*** |
|  |  | (0.007) |  | (0.006) |  | (0.011) |  | (0.005) |
| Mean outcome | 3.885 | 3.885 | 0.063 | 0.040 | 0.133 | 0.081 | 0.037 | 0.021 |
| Standard deviation | 0.572 | 0.582 | 0.358 | 0.379 | 0.495 | 0.515 | 0.238 | 0.252 |
| Year fixed effects | Yes | Yes | Yes | Yes | Yes | Yes | Yes | Yes |
| Teacher fixed effects | Yes | No | Yes | No | Yes | No | Yes | No |
| Experience fixed effects | Yes | Yes | Yes | Yes | Yes | Yes | Yes | Yes |
| School fixed effects | No | Yes | No | Yes | No | Yes | No | Yes |
| N | 174,214 | 242,339 | 91,726 | 127,669 | 40,943 | 61,943 | 46,771 | 69,642 |
| $R^2$ | .771 | .292 | .689 | .125 | .702 | .194 | .619 | .131 |
| Adjusted $R^2$ | .639 | .286 | .541 | .110 | .537 | .167 | .412 | .105 |

*Note.* TVAAS = Teacher Value-Added Assessment System; ELA = English Language Arts. Robust standard error clustered by teacher in parentheses. Cooperating teacher is a time-varying indicator taking the value of 1 during the school year in which a teacher is reported as serving as a cooperating teacher. Experience is included as single indicators for Years 0 to 30 and as a pooled indicator for experience >30 years. We drop singleton observations from models with teacher fixed effects. This leads to different sample sizes between the fixed-effects and difference-in-differences (diff-in-diff) models.

+$p$ < .10. *$p$ < .05. **$p$ < .01. ***$p$ < .001.

in non-CT blocks—to estimate coefficients for teaching experience and for the intercept term in a generalized difference-in-differences model. Thus, we wondered whether our estimates could be sensitive to our choice of analytic sample. To test this, we constrain our analyses to successively more restricted samples: (1) to teachers who teach in the same districts, same subject areas, and years as CTs in our sample; (2) to teachers who teach in the same school, subject areas, and years as CTs in our sample; and (3) only to teachers who served as CTs at least once. For all outcomes, the estimates for our preferred models have qualitatively similar estimates over the different estimation samples (see Appendix Table A1). However, the estimates for observation ratings decrease by about a quarter when we restrict the sample to teachers in the same blocks or same schools. This might be in line with our prior finding indicating that the blocks and schools where we observe CTs are different on baseline characteristics than other blocks and schools in the state. This will lead to a mechanical change in the coefficients for the covariates that we include in the model which could in part explain the difference in point estimates across the different sample specifications.[9]

Alternatively, this might indicate the presence of positive selection bias that is not fully accounted for in our preferred models but is accounted for in the models that restrict the sample to teachers in the same block or school. As an additional robustness check for our sample choice, we use a nearest neighbor matching algorithm to construct a sample of teachers who have observed characteristics similar to CTs but that were not picked to serve as CTs. We report these estimates in Table 5, alongside the estimates for the teacher fixed-effects and difference-in-differences models. Overall, we observe that the estimates for observation ratings have the same sign and magnitude across the different estimation models. Our matching algorithm matches on up to 2 years of prior evaluation scores; thus, we are matching CTs with non-CTs that have similar patterns of returns to experience preceding the CT years. While differences in pretrends could explain the positive effects on observation ratings in CT service years for our difference-in-differences specifications, these pretrends are unlikely to explain observed effects in our matched-sample models. Results for TVAAS appear to be significant and greater in magnitude for the matched-sample model. This might indicate that the matched-sample models, and to an extent the difference-in-differences models, could fail to account for self-selection bias. Said another way, teachers who were selected to be CTs could be different in unobserved ways from teachers who were not selected (e.g., stronger motivation). The teacher fixed-effects models account for these unobserved differences by leveraging the within-teacher variation in evaluation scores for CTs, assuming that these unobserved differences are constant within a teacher during our observation period. Failing to account for these unobserved differences could lead to estimates that are biased upward.

### Table 5
### Coefficient Sensitivity to Estimation Method

|  | (1) Observation Ratings | (2) TVAAS All Subjects | (3) TVAAS Math | (4) TVAAS ELA |
|---|---|---|---|---|
| Fixed-effects model | 0.040*** | 0.008 | −0.001 | −0.003 |
| Difference-in-differences model | 0.053*** | 0.014* | 0.005 | 0.005 |
| Matched sample | 0.059*** | 0.039** | 0.092** | 0.007 |

*Note.* ELA = English Language Arts; TVAAS = Teacher Value-Added Assessment System. This table reports the sensitivity of the cooperating teacher (CT) coefficient to various model specifications. The fixed-effects models include controls for years of experience, year and teacher fixed-effects. The difference-in-differences models include controls for years of experience, year and school fixed effects. The matched sample models report the Average Treatment Effects on the Treated (ATET) on teachers who serve as CTs. We fuzzy match using Mahalanobis distance on up to two prior years of evaluation data and years of experience at time of serving as a CT. We exact match on teacher background characteristics. We remove two CTs who do not match with other teachers in the state on background characteristics.
+$p < .10$. *$p < .05$. **$p < .01$. ***$p < .001$.

**Research Question 2: Are the Effects Different for Different Groups of CTs?**

In this section, we investigate whether the effects of serving as a CT differ for different groups of teachers. We begin by examining heterogeneity for different quartiles of effectiveness. We then consider differences in estimates for teachers who work in different school levels.

*Heterogeneity by Effectiveness Quartile*

Table 6 reports the estimates that include an interaction term between the CT indicator and quartile indicator. We interpret the estimate for these interaction terms as the effect of serving as a CT for teachers in the various quartiles of effectiveness. We find positive effects of serving as a CT for teachers in all four quartiles of observation ratings. Moreover, we find that teachers in the lowest quartile benefit the most from serving as a CT compared with teachers in the other quartiles. A possible explanation for this pattern of results is that the ceiling effect built into the observation score rubric negatively biases the effects of serving as CT for teachers in the upper quartiles of effectiveness. In this case, the observation scores of more effective teachers do not have as much room for improvement as the scores of less effective teachers.[10] It is also possible that any contemporaneous professional development benefits of serving as a CT might affect lower performing CTs most.

Table 6
**Heterogeneity by Quartile for Research Question 1**

|  | (1) Observation Ratings | (2) TVAAS All Subjects | (3) TVAAS Math | (4) TVAAS ELA |
|---|---|---|---|---|
| Cooperating Teacher # Quartile 1 | 0.108*** (0.023) | 0.001 (0.013) | −0.049+ (0.029) | −0.031 (0.024) |
| Cooperating Teacher # Quartile 2 | 0.026* (0.013) | −0.003 (0.009) | −0.028 (0.019) | −0.020+ (0.011) |
| Cooperating Teacher # Quartile 3 | 0.030** (0.011) | 0.008 (0.008) | −0.004 (0.019) | 0.004 (0.010) |
| Cooperating Teacher # Quartile 4 | 0.023** (0.009) | 0.031* (0.014) | 0.069* (0.027) | 0.027+ (0.015) |
| Mean outcome | 3.885 | 0.040 | 0.081 | 0.021 |
| Standard deviation | 0.582 | 0.379 | 0.515 | 0.252 |
| Year fixed effects | Yes | Yes | Yes | Yes |
| Teacher fixed effects | Yes | Yes | Yes | Yes |
| Experience fixed effects | Yes | Yes | Yes | Yes |
| $N$ | 233,149 | 117,034 | 52,933 | 60,082 |
| $R^2$ | .748 | .662 | .671 | .585 |
| Adjusted $R^2$ | .643 | .527 | .521 | .398 |

*Note.* ELA = English Language Arts; TVAAS = Teacher Value-Added Assessment System. Robust standard error clustered by teacher in parentheses. Cooperating teacher is a time-varying indicator taking the value of 1 during the school year in which a teacher is reported as serving as a cooperating teacher. Experience is included as single indicators for Years 0 to 30 and as a pooled indicator for experience above 30 years. We drop singleton observations from models with teacher fixed effects. Quartiles are calculated for each outcome using the teacher fixed effect from a regression that includes an indicator for being a cooperating teacher, time-varying school characteristics, teacher fixed effects, and year fixed effects.

+$p < .10$. *$p < .05$. **$p < .01$. ***$p < .001$.

## Effects of Serving as a Cooperating Teacher

Results for TVAAS tell a different story. We find that the effects of serving as a CT increase along the instructional effectiveness continuum. We observe positive effects on TVAAS scores only for teachers in the fourth quartile of effectiveness and possibly negative, but imprecisely estimated and nonsignificant, effects for teachers in the first and second quartiles of effectiveness. Finding effects to be more negative for lower performing teachers is consistent with what Goldhaber et al. (2018b) found in their sample of teachers from Washington state, where they hypothesize that more effective teachers are better able to buffer any potential negative effects of their mentees. However, while they found significant, negative effects overall (across quartiles), with the most negative effects concentrated in the lowest quartile, we find no significant effect overall (across quartiles) and instead small positive effects among teachers in the top quartile of instructional effectiveness.

One possibility is that our results are entirely driven by regression to the mean in evaluation scores (see Goldhaber et al., 2018b). In this case, we might conflate random year-to-year variation in evaluation scores with effects of serving as a CT. Specifically, if teachers' service (as CT) years coincide with years in which they also happen to randomly be at their peak performance, then they will tend to naturally regress to their average performance in post-CT years; this could lead to estimates like the ones that we observe for observation ratings in Table 6.[11] We check whether our results are sensitive to the way that we calculated the quartile of effectiveness by conducting a Monte Carlo simulation of the effects of serving as a CT on a placebo sample of CTs. Using the CT blocks described earlier, we randomly select 1,000 cohorts of teachers who were not actually selected as CTs during our observation period; these cohorts serve as a placebo for serving as a CT. We then calculate the effects of serving as a placebo CT for these 1,000 cohorts in order for us to test the extent to which our results are sensitive to regression to the mean. Appendix Table A3 shows the results from this Monte Carlo simulation. If regression to the mean were at play, then we would expect that the effects of regression to the mean on evaluation scores as a result of serving as a CT for the placebo group to have estimates similar to what we found for our CT sample. Instead, we find all the point estimates for the placebo CT sample are all close to zero and their 95% confidence intervals are centered at zero. More simply, we find no placebo effect of serving as a CT on evaluation scores for teachers along the effectiveness continuum. This suggests that our estimates for the effects of serving as a CT for each quartile of effectiveness are robust against teachers' evaluation scores regressing to the mean.

*Heterogeneity by School Level*

Table 7 displays the results for the effects of serving as a CT by school level. We find that the positive results on observation ratings are driven by

Table 7
**Heterogeneity by School Level for Research Question 1**

| | (1) Observation Ratings | (2) TVAAS All Subjects | (3) TVAAS Math | (4) TVAAS ELA |
|---|---|---|---|---|
| Cooperating Teacher # Elementary School | 0.052*** (0.009) | 0.002 (0.008) | −0.017 (0.014) | −0.002 (0.011) |
| Cooperating Teacher # Middle School | 0.032* (0.016) | 0.002 (0.009) | −0.006 (0.022) | −0.004 (0.011) |
| Cooperating Teacher # High School | 0.022 (0.014) | 0.021 (0.016) | 0.079 (0.058) | −0.016 (0.015) |
| Cooperating Teacher # Other School | 0.031 (0.025) | 0.033 (0.023) | 0.009 (0.045) | 0.016 (0.035) |
| Mean outcome | 3.882 | 0.065 | 0.135 | 0.037 |
| Standard deviation | 0.570 | 0.356 | 0.494 | 0.238 |
| Year fixed effects | Yes | Yes | Yes | Yes |
| Teacher fixed effects | Yes | Yes | Yes | Yes |
| Experience fixed effects | Yes | Yes | Yes | Yes |
| $N$ | 170,464 | 87,080 | 38,666 | 4,4243 |
| $R^2$ | .771 | .689 | .700 | .618 |
| Adjusted $R^2$ | .639 | .534 | .527 | .402 |

*Note.* ELA = English Language Arts; TVAAS = Teacher Value-Added Assessment System. Robust standard error clustered by teacher in parentheses. Cooperating teacher is a time-varying indicator taking the value of 1 during the school year in which a teacher is reported as serving as a cooperating teacher. Experience is included as single indicators for Years 0 to 30 and as a pooled indicator for experience >30 years. We drop singleton observations from models with teacher fixed effects. This leads to different sample sizes between the fixed effects and diff-in-diff models. Quartiles are calculated for each outcome using the teacher fixed effect from a regression that includes an indicator for being a cooperating teacher, time-varying school characteristics, teacher fixed effects, and year fixed effects.
+$p$ < .10. *$p$ < .05. **$p$ < .01. ***$p$ < .001.

CTs who teach in elementary and middle schools and that the evaluation scores of teachers in high schools or other schools do not change when serving as a CT. Though we are not entirely sure why we observe these differences by school level, we explore possible explanations in the Discussion and Conclusion section. We also find that estimates for serving as a CT on TVAAS scores are mostly similar across the different school settings. However, the results seem to suggest that high school mathematics teachers' scores increase the year they serve as CTs but that these point estimates are imprecisely estimated.

## Research Question 3: Do Teachers Perform Differently in Years After They Serve as CTs?

Based on our most conservative (i.e., teacher fixed effects) estimates, we find small and positive effects on observation ratings and null effects on TVAAS scores for years in which teachers serve as CTs. One possibility, though, is that the effects of serving as a CT are not immediate, but instead are observed in subsequent years. As discussed in our logic model, for example, if serving as a CT functions as a form of professional development, then we might not expect to observe increases in performance during the year a teacher serves, but perhaps in following years. In the next section, we turn to Research Question 3, where we estimate different effects for the years during which teachers serve as CTs and for years following that experience.

Table 8 reports the results of the teacher fixed-effects and difference-in-differences estimates of the effects of serving as a CT during postservice years. Specifically, we compare performance while serving as a CT and after serving as a CT with the evaluation scores during the time before serving as a CT. For observation ratings, CTs' evaluations do not increase, on average, in years following serving as a CT (see Columns 1 and 2). We note that the point estimates from the difference-in-differences model change sign but remain nonsignificant. That is, both specifications indicate that serving as a CT does not have a lasting impact on observation ratings beyond the years during which teachers serve as CTs.

On the other hand, results for TVAAS scores show that CTs' scores decline in the period after serving as a CT. TVAAS scores for the years in which teachers serve as CTs are similar to their scores for years before serving as a CT. However, scores in years after serving are lower than scores in years prior to serving. These results might highlight unobserved differences between CTs and non-CTs that we are not able to control in our main models. In fact, the negative effects for TVAAS scores disappear once we restrict our analyses to only teachers who ever serve as a CT (see Appendix Table A4, Column 3). This could suggest that CTs have differential returns to experience on TVAAS scores (this is supported by Atteberry et al., 2015, who find differential returns to experience by quartile of performance). Specifically,

Table 8
**Effects of Serving as a Cooperating Teacher on Growth of Evaluation Metrics**

|  | Observation Ratings |  | TVAAS—All Subjects |  | TVAAS—Math |  | TVAAS—ELA |  |
|---|---|---|---|---|---|---|---|---|
|  | (1) Teacher FE | (2) Diff-in-Diff | (3) Teacher FE | (4) Diff-in-Diff | (5) Teacher FE | (6) Diff-in-Diff | (7) Teacher FE | (8) Diff-in-Diff |
| Cooperating teacher | 0.046*** | 0.049*** | −0.002 | −0.004 | −0.010 | −0.013 | −0.011 | 0.001 |
|  | (0.009) | (0.009) | (0.007) | (0.008) | (0.015) | (0.015) | (0.008) | (0.008) |
| After cooperating teacher | 0.015 | −0.005 | −0.020* | −0.030*** | −0.016 | −0.024 | −0.017* | −0.007 |
|  | (0.009) | (0.010) | (0.008) | (0.009) | (0.017) | (0.016) | (0.008) | (0.008) |
| Ever cooperating teachers |  | 0.112*** |  | 0.058*** |  | 0.091*** |  | 0.032*** |
|  |  | (0.009) |  | (0.008) |  | (0.014) |  | (0.006) |
| Mean outcome | 3.885 | 3.885 | 0.040 | 0.040 | 0.081 | 0.063 | 0.021 | 0.021 |
| Standard deviation | 0.582 | 0.582 | 0.379 | 0.379 | 0.515 | 0.502 | 0.252 | 0.252 |
| Year FE | Yes | Yes | Yes | Yes | Yes | Yes | Yes | Yes |
| Teacher FE | Yes | No | Yes | No | Yes | No | Yes | No |
| Experience FE | Yes | Yes | Yes | Yes | Yes | Yes | Yes | Yes |
| School FE | No | Yes | No | Yes | No | Yes | No | Yes |
| $N$ | 233,149 | 242,339 | 117,034 | 127,669 | 52,933 | 61,943 | 60,082 | 69,642 |
| $R^2$ | .748 | .292 | .662 | .125 | .671 | .194 | .585 | .131 |
| Adjusted $R^2$ | .643 | .286 | .527 | .111 | .521 | .167 | .398 | .105 |

*Note.* ELA = English Language Arts; TVAAS = Teacher Value-Added Assessment System; FE = fixed effects. Robust standard error clustered by teacher in parentheses. Cooperating teacher is a time-varying indicator taking the value of 1 during the school year in which a teacher is reported as serving as a cooperating teacher. Experience is included as single indicators for Years 0 to 30 and as a pooled indicator for experience >30 years. We drop singleton observations from models with teacher fixed effects. This leads to different sample sizes between the fixed effects and (difference-in-differences) diff-in-diff models.
+$p < .10$. *$p < .05$. **$p < .01$. ***$p < .001$.

CTs experience relatively higher growth on TVAAS in years leading up to service years. This performance may increase the likelihood that teachers are tapped to serve as CTs; given a bump in performance during years leading up to serving. A postserving decline may be expected if non-CTs close the TVAAS gap during the post-CT period.

Similar to Research Question 1, we explore whether our results are sensitive to sample selection. Appendix Table A4 presents the results for the teacher fixed-effects models on restricted samples of teachers. The estimate directions and magnitudes are similar across the estimation samples for the main effect of serving as a CT. The estimates for the period following serving as a CT appear to change somewhat depending on the sample that we use. For observation ratings, we find that the positive but insignificant estimate for the years following serving as a CT appears to move toward a null but imprecise estimate. For TVAAS scores, we find that the negative effect on the years following serving as a CT appears to move toward zero.[12]

## Discussion and Conclusion

There is growing evidence that recruiting more instructionally effective teachers to serve as CTs is a promising approach to improving the preparation that teacher candidates receive and, subsequently, the instructional effectiveness of the incoming supply of new teachers. So why are program leaders reporting that it can be difficult to get our most instructionally effective teachers to serve as CTs? The challenge appears to be multifaceted, and this study investigates one factor: that teachers are hesitant to mentor a teacher candidate for fear that they may receive lower evaluations. Our results suggest that any concerns over declining evaluations are not warranted. Rather, we find observation ratings may even increase while student achievement gains are unaffected. The implications are that instructionally effective teachers who are considering becoming CTs should not let fears over evaluation scores deter them from serving. Moreover, program and district leaders charged with recruiting these teachers to serve can assure potential CTs that such fears are likely unjustified.

Since this article is the first to examine effects on observation ratings, a unique contribution is finding that teachers' concurrent observation ratings may actually benefit by mentoring a candidate. Regarding student achievement gains, our main results are similar to those of Goldhaber et al. (2018b). Both studies found no effects of serving as a CT on average student achievement gains, though our study found small, positive effects in alternate model specifications.

Goldhaber et al. (2018b) also found that the effects of serving as a CT on math achievement were negative among lowest performing teachers, but effectively zero for other quartiles. They point out that this finding is somewhat surprising in light of prior evidence that principals sometimes place candidates in classrooms of less effective teachers to help boost performance in those classrooms; in fact, the authors find the reverse to be true—placing candidates in these classrooms appears to harm performance. They conclude that more instructionally effective teachers are likely better able to buffer against the negative effects of mentoring a teacher candidate. In comparison, we find that the coefficients for serving as a CT decrease as CT effectiveness decreases. However, we find positive and significant effects for teachers in the top quartile, and negatively trending but nonsignificant effects for teachers in lower quartiles. These results do not seem to be consistent with an explanation that higher performing teachers are mitigating the negative effects of mentoring a candidate; rather, our results seem to suggest that higher performing teachers may actually benefit from mentoring a candidate.

In terms of observation ratings, we find that all teachers, across all quartiles of prior performance, receive significantly higher ratings in years that they serve as CTs. However, we find that lowest quartile CTs tend to benefit most.[13] One might be tempted to conclude, based on this latter finding, that program and school leaders should place candidates with lower performing teachers. However, such a conclusion, we believe, would likely be premature in light of other evidence. First, there is strong evidence that graduates have better early career performance when they learn to teach with more instructionally effective CTs (Goldhaber et al., 2018b; Ronfeldt, Brockman & Campbell, 2018; Ronfeldt, Matsko, et al., 2018). Second, results from our study and Goldhaber et al. (2018b) indicate that student achievement gains among lower performing teachers tend to decline in years they mentor candidates. Finally, as we describe above, we do not actually know if the overall teaching and learning quality are benefitting in classrooms of lowest performing teachers who mentor candidates or if teachers' new roles as CTs somehow change their evaluation procedures in ways that benefit their ratings without necessarily benefitting their teaching (see Figure 1 Perceived Changes). We encourage future work to examine the specific mechanisms by which teachers' evaluations change as a result of mentoring candidates, including why there appear to be differences by the level of instructional effectiveness of CTs.

Prior research has established that recruiting the most instructionally effective teachers to serve as CTs is likely to benefit the new supply of prospective teachers and those schools and districts that hire them. Our present study suggests that this strategy is likely also to benefit those teachers who serve as CTs, or at least cause the least harm. Though there are some subtle differences between our results and those of Goldhaber et al. (2018b) and SAS Institute (2014), a common conclusion from all three studies is that the student achievement gains of the least instructionally effective teachers are likely to suffer when they mentor candidates, whereas the achievement gains and observation ratings of the most instructionally effective teachers

## Effects of Serving as a Cooperating Teacher

are likely to go unchanged and possibly even improve. Taken together, the existing research then tends to suggest that placing candidates with the most instructionally effective teachers is likely to result in the greatest overall benefit and least harm.

A limitation of our study is that we do not have comprehensive data identifying all teachers who served as CTs across the state and across the years included in our study. Rather, our CT data come from only those TEPs that kept and were willing to share these data and only for years that were included in their records. Thus, our coverage across TEPs and years is uneven, and our results are not necessarily generalizable across the state. It is possible that the CTs for the particular TEPs and years in our sample may respond differently to serving than the CTs we do not observe. Though unlikely, it is possible that teachers in our sample improved on observation ratings when they served but teachers outside our sample declined in performance. This might occur, for example, if CTs in unobserved programs may have had different motivations for serving than those in the programs we observed. The study by Goldhaber et al. (2018b) also did not have full coverage of programs in Washington state and so may be subject to similar limitations. We are currently in negotiations with the TDOE to see if we can access comprehensive data on CTs across all programs in Tennessee for future cohorts.

More research is needed to understand possible mechanisms by which teachers get a boost in observation ratings during the years in which they serve as a CT. As identified in our logic model, one possibility is that serving as a CT does indeed boost the quality of instruction. It might be, for example, that, in years they are serving as CTs, having an additional adult in the classroom helps with instruction by increasing the amount of individual instructional time each student has with a teacher. In years they serve as CTs, teachers also might invest more in instructional planning as a result of needing to onboard another teacher and ensure they are modeling good practice.

Another possibility is that teachers who serve as CTs must schedule evaluations on days or during sections/periods when their candidates are not lead teaching. This likely means that unscheduled observations (for evaluation) are less common in years that teachers mentor candidates. It also could mean that CTs are able to be more strategic about when they schedule observations/evaluations—for example, during easier periods/classes or during subjects in which they especially excel—thus, reducing the impact of unannounced observations. In these ways, teachers could effectively boost their evaluations, possibly explaining the bumps in performance we observe during years they serve as CTs. These explanations are consistent with finding that lowest quartile teachers benefit most on observation ratings when serving as CTs, as one might expect strategically planned evaluations to benefit less-effective teachers most. They are also consistent with finding little to no effects of serving as a CT on TVAAS scores, where student achievement, rather than scheduled observations by raters, dictate performance.

One additional consideration is that we find the positive effects of serving as a CT on observation ratings to be concentrated in elementary schools. There are many reasons why this might be true. It could be, for example, that elementary candidates are better able to form personal relationships with students because they are with them all day or because younger students are more willing to connect with new teachers in their classrooms; this, in turn, likely translates into a stronger instructional environment. If this were the case, though, we would expect teachers' TVAAS, and not just observation ratings, to increase in years they serve as CTs.

Alternatively, one of the main differences between elementary and secondary teachers is that the former are tasked with teaching all subjects, even ones in which they are less knowledgeable or effective. It is possible that elementary teachers who mentor candidates are more inclined to hand over lead teaching responsibilities in subjects in which they feel less proficient. If so, this could result in evaluators being more likely to evaluate CTs when teaching their stronger subjects and, thus, to rate them higher than in other years.[14] If this were the case, then we would expect mentoring a candidate to likely benefit lower performing elementary CTs the most, which is what we observe (see Appendix Table A6).

It is true that secondary teachers also often teach multiple courses. Secondary teachers could also then assign their teacher candidates to the subjects that are their weakest. However, we believe that secondary candidates are often more specialized in their subject matter/content focus and more likely to request a specific class that is a match. Compared with elementary teachers, this would likely place more constraints on secondary CTs in terms of which parts of the school day that they would be able to hand over lead teaching responsibilities to candidates, that is, secondary CTs may have somewhat less flexibility than elementary CTs in how they assign their candidates.

An important next step for future research is to interrogate the mechanisms by which observation ratings increase in years that teachers serve as CTs and, relatedly, whether the increases reflect improvements in actual teaching quality or instead changes in evaluation processes that result in better evaluations but not necessarily better teaching. If boosted performance among CTs is explained by having opportunities to game the evaluation system, whether intentional or unintentional, then we expect that some will argue that it seems inequitable for teachers who mentor a candidate to gain such an advantage. Though we understand this perspective, we also recognize that mentoring a candidate is a tremendous amount of additional work for classroom teachers and is often unrecognized, underappreciated, and not rewarded. One recent study found that CTs typically received only about $300 for mentoring a candidate, and that many were not compensated at all (Matsko et al., 2019). Especially given that CTs can have meaningful, positive impacts on the instructional effectiveness of the incoming supply of teachers, they may be deserving of advantages during the years in which they mentor. In fact, one consideration might be to relieve teachers of being evaluated in the years in which they mentor a candidate. Doing so would remove any concerns about CTs gaming the evaluation system and would offset fears that mentoring a candidate might harm evaluation scores—even though our results suggest that these fears may be unwarranted—while providing a low-cost incentive to serve.

Finally, our research extends prior work by testing whether or not serving as a CT has a longer term effect on evaluation during post-CT years. A positive effect in post-CT years could suggest that mentoring serves a professional development function. For observation ratings, we find post-CT performance to decline back to pre-CT levels. For TVAAS, we find post-CT performance to actually be somewhat worse than pre-CT levels. However, when we constrain models only to individuals who ever served as CTs, the postserving estimates are similar to pre-CT estimates; this may suggest that CTs are not actually doing worse in post-CT years, but that instead, non-CTs tend to have stronger relative returns. Either way, while we find some boost to evaluations during the years in which teachers serve as CTs, we find no evidence that serving as a CT makes individuals better teachers in post-CT years. Thus, serving as a CT does not appear to function as a form of long-term professional development. These results differ from Goldhaber et al. (2018b), who find student achievement gains to increase significantly in post-CT years and conclude that serving as a CT may serve as a form of professional development for teachers.

More research needs to investigate why these results in Washington differ from ours in Tennessee. One possibility is that the different labor markets, policy, and evaluation contexts afford and constrain different responses by teachers who serve as CTs. Another possibility is that the differences result from different ways that the two studies measured value-added to student achievement gains (VAMs) and modeled effects on achievement gains. In particular, in their construction of VAMs, Goldhaber and colleagues used student-level data and controlled for many student characteristics, whereas our study, due to our data sharing agreements, depends on teacher-level TVAAS measures which do not adjust for the same covariates. If such differences in VAM construction were responsible for differences in effects on post-CT outcomes, then we would have likely also expected differences in concurrent effects, but our results are similar.

While more research is clearly needed to understand the mechanisms by which serving as a CT impact evaluations—during concurrent and post-CT years—the main policy conclusion from this study is generally promising and consistent with conclusions from prior research: Serving as a CT does not appear to harm a teacher's concurrent or future performance evaluations and may even be of benefit. For teacher education program leaders, a related implication is that they should not be concerned about potential unintended consequences to teachers they recruit to serve as CTs. In fact, sharing that there may even be some evaluation benefits could assist in their recruitment efforts.

## Appendix

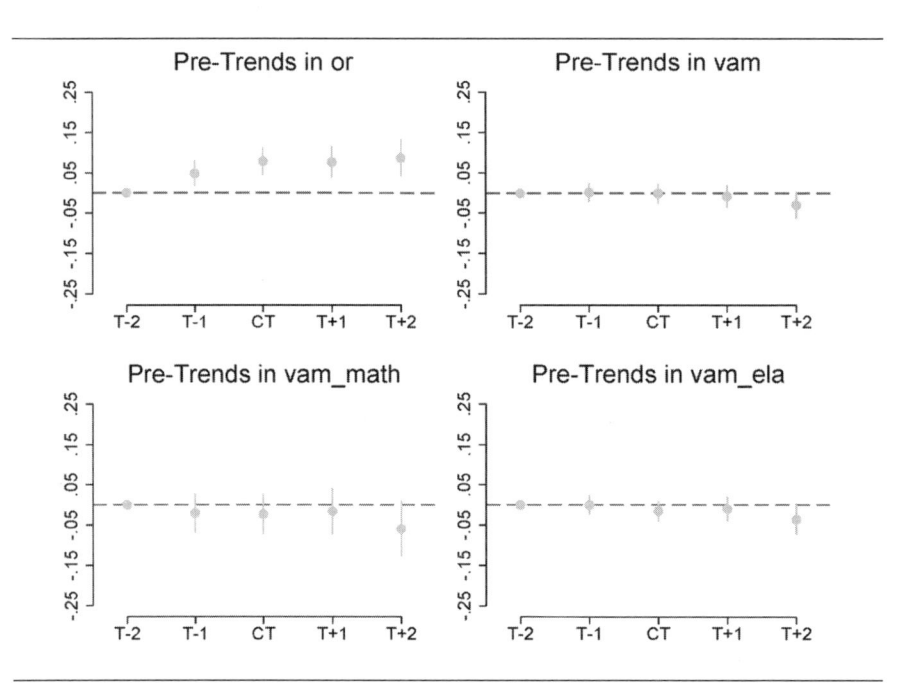

*Figure A1.* **Event study of serving as a cooperating teacher (CT) on outcomes of interest.**

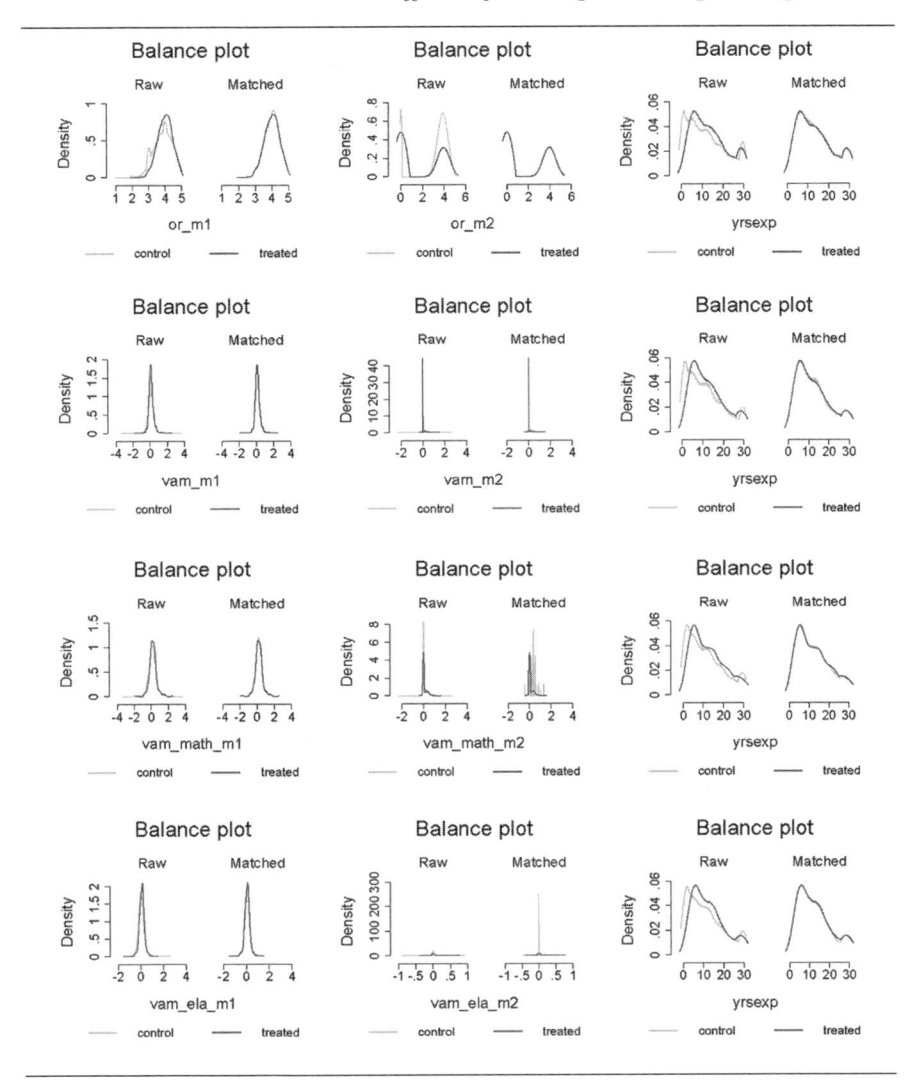

*Figure A2.* **Density distribution of fuzzy matched covariates.**

Table A1
Robustness Checks of Teacher Fixed-Effects Models on Different Subsamples of Teachers

|  | (1) Main Effects | (2) CT Blocks | (3) CT Schools | (4) CT Ever |
|---|---|---|---|---|
| *Panel A: Observation ratings* | | | | |
| CT | 0.040*** (0.007) | 0.028*** (0.007) | 0.030** (0.009) | 0.035*** (0.007) |
| Mean outcome | 3.870 | 3.868 | 3.878 | 4.022 |
| Standard deviation | 0.578 | 0.570 | 0.534 | 0.485 |
| Year fixed effects | Yes | Yes | Yes | Yes |
| Teacher fixed effects | Yes | Yes | Yes | Yes |
| Experience fixed effects | Yes | Yes | Yes | Yes |
| $N$ | 174214 | 81513 | 19880 | 10127 |
| $R^2$ | .771 | .784 | .816 | .741 |
| Adjusted $R^2$ | .639 | .638 | .671 | .603 |
| *Panel B: TVAAS—All subjects* | | | | |
| CT | 0.008 (0.006) | 0.008 (0.006) | 0.008 (0.008) | 0.007 (0.006) |
| Mean outcome | 0.063 | 0.068 | 0.077 | 0.113 |
| Standard deviation | 0.358 | 0.326 | 0.309 | 0.305 |
| Year fixed effects | Yes | Yes | Yes | Yes |
| Teacher fixed effects | Yes | Yes | Yes | Yes |
| Experience fixed effects | Yes | Yes | Yes | Yes |
| $N$ | 91,726 | 52,838 | 11,817 | 7,042 |
| $R^2$ | .689 | .709 | .728 | .668 |
| Adjusted $R^2$ | .541 | .537 | .522 | .517 |

*Note.* CT = cooperating teacher; TVAAS = Teacher Value-Added Assessment System. Robust standard error clustered by teacher in parentheses. CT is a time-varying indicator taking the value of 1 during the school year in which a teacher is reported as serving as a CT. Experience is included as single indicators for Years 0 to 30 and as a pooled indicator for experience above 30 years. Models (1) and (4) include singleton observations by teacher-year. This leads to different sample sizes between the fixed-effects and difference-in-differences (diff-in-diff) models.
+$p$ < .10. *$p$ < .05. **$p$ < .01. ***$p$ < .001.

## Table A2
**Nearest Neighbor Matching Quality**

|  | Mean | | Variance | |
|---|---|---|---|---|
|  | Raw | Matched Sample | Raw | Matched Sample |
| *Panel A*: Observation ratings (OR) | | | | |
| OR—2 years prior | −0.847 | 0.000 | 1.874 | 1.000 |
| OR—1 year prior | 0.271 | 0.002 | 0.613 | 1.004 |
| Years of experience | 0.175 | 0.016 | 0.849 | 1.017 |
| N | 45,220 | 4,608 | | |
| *Panel B*: TVAAS | | | | |
| TVAAS—2 years prior | −0.097 | 0.012 | 0.503 | 1.135 |
| TVAAS—1 year prior | 0.255 | 0.013 | 0.777 | 1.180 |
| Years of experience | 0.181 | 0.008 | 0.866 | 0.987 |
| N | 18,424 | 2,236 | | |
| *Panel C*: TVAAS Mathematics | | | | |
| TVAAS Math—2 years prior | −0.127 | 0.015 | 0.427 | 1.088 |
| TVAAS Math—1 year prior | 0.186 | 0.007 | 0.757 | 1.122 |
| Years of experience | 0.172 | 0.021 | 0.865 | 0.987 |
| N | 8,753 | 988 | | |
| *Panel D*: TVAAS ELA | | | | |
| TVAAS ELA—2 years prior | −0.036 | 0.009 | 0.704 | 1.081 |
| TVAAS ELA—1 year prior | 0.265 | 0.008 | 0.823 | 1.126 |
| Years of experience | 0.162 | 0.016 | 0.817 | 0.969 |
| N | 9,684 | 1,176 | | |

*Note.* ELA = English Language Arts; TVAAS = Teacher Value-Added Assessment System. This table reports the standardized difference between cooperating teachers (CTs) and non-CTs on the variables we used to construct the nearest neighbor matched sample. Values close to 0 for the matched sample means and close to 1 for the matched sample variance indicate that the matching procedure was able to identify a non-CT sample similar to the observed CT sample.

Table A3
Monte Carlo Simulation

| | (1) Observation Ratings | | (2) TVAAS—All Subjects | | (3) TVAAS—Math | | (4) TVAAS—ELA | |
|---|---|---|---|---|---|---|---|---|
| Placebo CT # Q1 | −0.003 | [−0.031, 0.027] | −0.006 | [−0.033, 0.024] | 0.004 | [−0.058, 0.063] | −0.019 | [−0.046, 0.009] |
| Placebo CT # Q2 | −0.008 | [−0.032, 0.016] | −0.006 | [−0.019, 0.009] | 0.001 | [−0.030, 0.030] | −0.012 | [−0.029, 0.004] |
| Placebo CT # Q3 | −0.004 | [−0.024, 0.017] | 0.002 | [−0.011, 0.015] | 0.010 | [−0.017, 0.039] | −0.004 | [−0.021, 0.012] |
| Placebo CT # Q4 | 0.002 | [−0.016, 0.020] | 0.006 | [−0.015, 0.028] | 0.007 | [−0.033, 0.047] | −0.005 | [−0.030, 0.020] |

*Note.* ELA = English Language Arts; TVAAS = Teacher Value-Added Assessment System. This table reports the results of a Monte Carlo simulation that draws 1,000 placebo CTs and calculates the placebo effect of serving as a cooperating teacher (CT) on evaluation scores. The model adjusts for year fixed effects. Quartiles are calculated using the method that we used to estimate the heterogeneity by quartile using evaluation years 2011–2015. 95% credible intervals are in brackets.
+$p < .10$. *$p < .05$. **$p < .01$. ***$p < .001$.

Table A4
Robustness Checks of Growth Models on Different Subsamples of Teachers

|  | (1) Main Effects | (2) CT Blocks | (3) CT Schools | (4) CT Ever |
|---|---|---|---|---|
| *Panel A: Observation ratings* | | | | |
| CT | 0.046*** (0.009) | 0.040*** (0.010) | 0.034* (0.013) | 0.032** (0.011) |
| Years following serving as a CT | 0.015 (0.009) | 0.027+ (0.014) | 0.007 (0.019) | −0.003 (0.016) |
| Mean outcome | 3.885 | 3.847 | 3.841 | 4.038 |
| Standard deviation | 0.582 | 0.582 | 0.561 | 0.497 |
| Year fixed effects | Yes | Yes | Yes | Yes |
| Teacher fixed effects | Yes | Yes | Yes | Yes |
| Experience fixed effects | Yes | Yes | Yes | Yes |
| $N$ | 233,149 | 81,807 | 19,880 | 13,157 |
| $R^2$ | .748 | .784 | .816 | .707 |
| Adjusted $R^2$ | .643 | .638 | .671 | .599 |
| *Panel B: TVAAS—All subjects* | | | | |
| CT | −0.002 (0.007) | −0.001 (0.008) | 0.009 (0.011) | 0.010 (0.009) |
| Years following serving as a CT | −0.020* (0.008) | −0.018+ (0.011) | 0.005 (0.016) | 0.008 (0.013) |
| Mean outcome | 0.040 | 0.048 | 0.046 | 0.093 |
| Standard deviation | 0.379 | 0.349 | 0.330 | 0.314 |
| Year fixed effects | Yes | Yes | Yes | Yes |
| Teacher fixed effects | Yes | Yes | Yes | Yes |
| Experience fixed effects | Yes | Yes | Yes | Yes |
| $N$ | 117,034 | 52,918 | 11,817 | 8,468 |
| $R^2$ | .662 | .709 | .728 | .645 |
| Adjusted $R^2$ | .527 | .537 | .522 | .515 |

*Note.* CT = cooperating teacher; TVAAS = Teacher Value-Added Assessment System. Robust standard error clustered by teacher in parentheses. CT is a time-varying indicator taking the value of 1 during the school year in which a teacher is reported as serving as a CT. Experience is included as single indicators for Years 0 to 30 and as a pooled indicator for experience above 30 years. In addition, experience variable is interacted with the years following indicator, allowing differential returns to experience after a teacher first serves as a CT. Models (1) and (4) include singleton observations by teacher-year.
+$p < .10$. *$p < .05$. **$p < .01$. ***$p < .001$.

## Table A5
### Two-Stage Estimation for Cooperating Teacher Growth Trajectories

|  | Observation Ratings || TVAAS All Subjects || TVAAS Math || TVAAS ELA ||
| --- | --- | --- | --- | --- | --- | --- | --- | --- |
|  | (1) Teacher FE | (2) Diff-in-Diff | (3) Teacher FE | (4) Diff-in-Diff | (5) Teacher FE | (6) Diff-in-Diff | (7) Teacher FE | (8) Diff-in-Diff |
| Cooperating teacher | .051 | .048 | −.002 | −.011 | −.015 | −.015 | −.012 | −.008 |
| After cooperating teacher | .023 | −.005 | −.021 | −.043 | −.025 | −.038 | −.020 | −.027 |
| Ever cooperating teachers |  | .112 |  | .066 |  | .098 |  | .043 |

*Note.* ELA = English Language Arts; TVAAS = Teacher Value-Added Assessment System; FE = fixed effects. Standard errors are not reported because they are not calculated for the two-stage Papay-Kraft estimation correction. Cooperating teacher is a time-varying indicator taking the value of 1 during the school year in which a teacher is reported as serving as a cooperating teacher. Experience is included as single indicators for Years 0 to 30 and as a pooled indicator for experience above 30 years. We drop singleton observations from models with teacher fixed effects. This leads to different sample sizes between the fixed-effects and difference-in-differences (diff-in-diff) models.

Table A6
Heterogeneity by School Type and Quartile

|  | Observation Ratings ||||  TVAAS—All Subjects ||||
|---|---|---|---|---|---|---|---|---|
|  | (1) Elementary School | (2) Middle School | (3) High School | (4) Other School | (5) Elementary School | (6) Middle School | (7) High School | (8) Other School |
| CT # Quartile 1 | 0.153*** (0.030) | 0.012 (0.018) | 0.051 (0.035) | 0.032 (0.025) | 0.003 (0.020) | −0.011 (0.012) | −0.007 (0.013) | −0.008 (0.027) |
| CT # Quartile 2 | 0.058 (0.051) | 0.025+ (0.014) | 0.041 (0.026) | 0.027 (0.025) | 0.081* (0.037) | 0.004 (0.012) | 0.002 (0.013) | 0.028 (0.021) |
| CT # Quartile 3 | 0.089* (0.037) | 0.035** (0.012) | 0.029 (0.024) | −0.005 (0.019) | 0.009 (0.038) | 0.010 (0.018) | 0.005 (0.024) | 0.088** (0.034) |
| CT # Quartile 4 | 0.006 (0.027) | 0.104* (0.048) | 0.042 (0.050) | −0.020 (0.109) | 0.036 (0.051) | 0.018 (0.018) | −0.039 (0.031) | 0.077 (0.052) |
| N | 228,417 |||| 112,476 ||||
| $R^2$ | .749 |||| .663 ||||
| Adjusted $R^2$ | .644 |||| .523 ||||

*Note.* TVAAS = Teacher Value-Added Assessment System. Robust standard error clustered by teacher in parentheses. Cooperating teacher is a time-varying indicator taking the value of 1 during the school year in which a teacher is reported as serving as a cooperating teacher. Experience is included as single indicators for Years 0 to 30 and as a pooled indicator for experience above 30 years. We drop singleton observations from models with teacher fixed effects. This leads to different sample sizes between the fixed-effects and difference-in-differences (diff-in-diff) models. Quartile are calculated for each outcome using the teacher fixed effect from a regression that includes an indicator for being a cooperating teacher, time-varying school characteristics, teacher fixed effects, and year fixed effects.
+$p$ < .10. *$p$ < .05. **$p$ < .01. ***$p$ < .001.

## Notes

We appreciate the generous financial support that was provided for this research by the Institute of Education Sciences, U.S. Department of Education through the Statewide, Longitudinal Data Systems Grant (PR/Award R372A150015). Stacey Brockman, Emanuele Bardelli, and Hannah Mullman also received predoctoral support from the Institute of Education Sciences, U.S. Department of Education (PR/Award R305B150012). We also appreciate comments from John Papay, Dan Goldhaber, James Cowan, Kevin Schaaf, reviewers at the Tennessee Education Research Alliance on earlier drafts of this article, and from seminar attendees at our Causal Inference in Education Research Seminar presentation at the University of Michigan and conference attendees at the 2019 American Education Finance and Policy conference in Kansas City, MO. This project would not have been possible without the partnership, support, and data provided by the Tennessee Department of Education. Please note that the views expressed are those of the authors and do not necessarily reflect those of this study's sponsors, the Tennessee Department of Education, or the institutions to which the authors are affiliated.

[1] In this article, we use measures from Tennessee's educator evaluation system as our outcomes of interest. While this evaluation system is designed to capture multiple aspects of teaching, we are also aware that these measures could leave out important dimensions of teaching practice that are not addressed on standardized tests or observation rubrics. We want to stress that we are measuring the impacts of mentoring a candidate on these evaluation scores rather than changes in CTs' skills while mentoring candidates.

[2] The Tennessee Department of Education asked all educator preparation programs in the state to share their placement data for this project, including cooperating teacher-teacher candidate match data that is not currently collected as part of the teacher licensing process. Seventeen programs agreed to share their data. These data cover about 40% of the teacher candidates prepared in Tennessee during our period of observation.

[3] In most of our models, we restrict the evaluation data to cover the same time span as the CT data set. As a robustness check, we use the full evaluation data. Our results are robust against the data set that we use to estimate the effects of serving as a CT on evaluation scores.

[4] About 20% of Tennessee teachers are assessed using other rubrics than the Tennessee educator acceleration model rubric. As part of the state-wide educator assessment system, the Tennessee department of education calculates equated scores among the Tennessee educator acceleration model rubric scores and these other observational rubrics. We use these equated scores in our analyses.

[5] It is common for TEPs to ask candidates for their preferences in terms of districts in which they are willing to complete their student teaching and to then select placements in the requested districts (Krieg et al., 2016; Maier & Youngs, 2009). One reason for this is that candidates often have geographic and travel constraints.

[6] Goldhaber et al. (2018b) also found evidence of regression to the mean in their sample. We test for this issue using a Monte Carlo simulation described below. We do not find evidence that evaluation scores regress to the mean in our sample. This fact could be due to differences in the way that we calculated the effectiveness quartile for teachers and the way in which Tennessee calculates teacher value-added scores. First, we calculate quartile of effectiveness using all evaluation data available for each teacher. This is because we do not have access to evaluation data for the period preceding serving as a cooperating teacher. Second, TVAAS models differ from traditional value-added models insofar that scores for each teacher are calculated separately for each student cohort and that teacher value-added are calculated using empirical Bayes's estimates (Vosters et al., 2018).

[7] We observe that 83% of CTs are reported to serve only once during our observation period, 14% of CTs serve twice, 3% serve three times or more.

[8] As we discussed in the "Method" section, these results rely on a different set of assumptions than the teacher fixed-effects estimates. Namely, we are assuming that the evaluation scores for CTs and non-CTs follow parallel trends during the pre-CT period. The difference in results between the two specifications could suggest that this assumption is not met, that is, CTs have different returns to experience than non-CTs. While our analysis of parallel trends is partial, we find some potential evidence that pretrends are not parallel for observation ratings (see the "Method" section and Appendix Figure

A2). However, the results that we report in Appendix Tables A1 and A3 seem to suggest that that the estimates from our preferred models are well specified.

[9]To test whether the change in covariate coefficients could explain our results, we adjust the year fixed effects using the method that Papay and Kraft (2015) describe. These models' results are qualitatively identical to our preferred model estimates.

[10]Observation scores averages are 3.36 ($SD$ = 0.36) for CTs in Quartile 1, 3.73 ($SD$ = 0.30) for teachers in Quartile 2, 4.01 ($SD$ = 0.28) for Quartile 3, and 4.39 ($SD$ = 0.38) for Quartile 4.

[11]In detail, regression to the mean happens when a variable is measured with error. Point estimates from a regression model will include both the true effect and the effect of random measurement error. Since the effect of the random measurement error changes year-to-year, the point estimate for the effect of interest will also vary around its true point estimate. In our case, regression to the mean could explain the improvement in observation scores that we observe for teachers in the lower quartiles of effectiveness by suggesting that our CT estimates are based on a "good evaluation" (i.e., positive measurement error) year and that these teachers' observation ratings regress back to their mean for years following serving as CT.

[12]A possible explanation for these unstable estimates could be collinearity between the CT and following CT indicators, the experience fixed effects, and the years fixed effects. This would lead to unstable point estimates that are sensitive to the estimation sample that we use to identify the main effects. To address this concern, we use the two-stage adjustment strategy for year fixed-effects described in Papay and Kraft (2015). We first estimate the year fixed effect using a model that does not include teacher fixed effects. We then use the year-specific coefficients estimated in Stage 1 in our preferred model. The results from these models are consistent with the estimates from our preferred models (see Appendix Table A5), confirming a null effect on observation ratings for years following serving as a CT and a possible small and negative effect on TVAAS scores.

[13]It is important to point out that the lowest performing teachers in our CT sample are actually still more instructionally effective than the average teacher in the state.

[14]It also may be more difficult for evaluators to do unplanned observations and evaluations of elementary CTs, since they are more likely than secondary CTs to be with a student teacher throughout the entire day. Since teachers can prepare for planned observations, this may effectively boost evaluations for elementary CTs.

## References

Angrist, J. D., & Pischke, J. S. (2008). *Mostly harmless econometrics: An empiricist's companion*. Princeton, NJ: Princeton University Press.

Atteberry, A., Loeb, S., & Wyckoff, J. (2015). Do first impressions matter? Predicting early career teacher effectiveness. *AERA Open*, 1(4). doi:10.1177/2332858415607834

Boyd, D. J., Grossman, P. L., Lankford, H., Loeb, S., & Wyckoff, J. (2009). Teacher preparation and student achievement. *Educational Evaluation and Policy Analysis*, *31*, 416–440.

Campbell, S. L. (2014). *Quality teachers wanted: An examination of standards-based evaluation systems and school staffing practices in North Carolina middle schools* (Order No. 3633946). Available from ProQuest Dissertations & Theses A&I (1612601875).

Campbell, S. L., & Ronfeldt, M. (2018). Observational evaluation of teachers: Measuring more than we bargained for? *American Educational Research Journal*. Advanced online publication. doi:10.3102/0002831218776216

Cantrell, S., & Kane, T. J. (2013). *Ensuring fair and reliable measures of effective teaching: Culminating findings from the MET project's three-year study* (MET Project Research Paper). Retrieved from https://files.eric.ed.gov/fulltext/ED540958.pdf

Desimone, L. M. (2009). Improving impact studies of teachers' professional development: Toward better conceptualizations and measures. *Educational Researcher*, *38*, 181–199.

Gitomer, D., & Bell, C. (2013). Evaluating teachers and teaching. In K. F. Geisinger, B. A. Bracken, J. F. Carlson, J. I. C. Hansen, N. R. Kuncel, S. P. Reise, & M. C. Rodriguez (Eds.), *APA handbook of testing and assessment in psychology. Vol. 3: Testing and assessment in school psychology and education* (pp. 415–444). Washington, DC: American Psychological Association.

Goldhaber, D., Krieg, J., & Theobald, R. (2018a). *Effective like me? Does having a more productive mentor improve the productivity of mentees?* (CALDER Working Paper No. 208-1118-1). Washington, DC: National Center for Analysis of Longitudinal Data in Education Research.

Goldhaber, D., Krieg, J., & Theobald, R. (2018b). *The costs of mentorship? Exploring student teaching placements and their impact on student achievement* (CALDER Working Paper). Washington, DC: National Center for Analysis of Longitudinal Data in Education Research.

Greenberg, J., Pomerance, L., & Walsh, K. (2011). *Student teaching in the United States*. Retrieved from the National Council on Teacher Quality https://www.nctq.org/dmsView/Student_Teaching_United_States_NCTQ_Report

Grossman, P., Hammerness, K. M., McDonald, M., & Ronfeldt, M. (2008). Constructing coherence: Structural predictors of perceptions of coherence in NYC teacher education programs. *Journal of Teacher Education, 59*, 273–287.

Harris, D. N., Ingle, W. K., & Rutledge, S. A. (2014). How teacher evaluation methods matter for accountability: A comparative analysis of teacher effectiveness ratings by principals and teacher value-added. *American Educational Research Journal, 51*, 73–112.

Hill, H. C., Kapitula, L., & Umland, K. (2011). A validity argument approach to evaluating teacher value-added scores. *American Educational Research Journal, 48*, 794–831.

Jiang, J. Y., & Sporte, S. (2016). *Teacher evaluation in Chicago: Differences in observation and value-added scores by teacher, student, and school characteristics*. Retrieved from https://consortium.uchicago.edu/sites/default/files/2018-10/Teacher%20Evaluation%20in%20Chicago-Jan2016-Consortium.pdf

Kane, T. J., & Staiger, D. O. (2012). *Gathering feedback for teaching: Combining high-quality observations with student surveys and achievement gains*. Seattle, WA: Bill and Melinda Gates Foundation. Retrieved from http://www.metproject.org/downloads/MET_Gathering_Feedback_Research_Paper.pdf

Krieg, J., Theobald, R., & Goldhaber, D. (2016). A foot in the door: Exploring the role of student teaching assignments in teachers' initial job placements. *Educational Evaluation and Policy Analysis, 38*, 364–388.

Loeb, S., & Candelaria, C. (2012). How stable are value-added estimates across years, subjects, and student groups? *Carnegie Knowledge Network*. Retrieved from https://cepa.stanford.edu/content/how-stable-are-value-added-estimates-across-years-subjects-and-student-groups

Maier, A., & Youngs, P. (2009). Teacher preparation programs and teacher labor markets: How social capital may help explain teachers' career choices. *Journal of Teacher Education, 60*, 393–407.

Makel, M. C., & Plucker, J. A. (2014). Facts are more important than novelty. *Educational Researcher, 43*, 304–316.

Matsko, K. K., Ronfeldt, M., & Greene Nolan, H. (2019). *How different are they? Comparing preparation offered by traditional, alternative, and residency pathways*. Manuscript submitted for publication.

Matsko, K. K., Ronfeldt, M., Green Nolan, H., Klugman, J., Reininger, M., & Brockman, S. L. (2018). Cooperating teacher as model and coach: What leads to candidates' perceptions of preparedness? *Journal of Teacher Education*. Advance online publication. doi:10.1177/0022487118791992

Mullman, H., & Ronfeldt, M. (2019). *The landscape of clinical preparation in Tennessee*. Manuscript submitted for publication.

Papay, J. P., & Kraft, M. A. (2015). Productivity returns to experience in the teacher labor market: Methodological challenges and new evidence on long-term career improvement. *Journal of Public Economics, 130*, 105–119.

Ronfeldt, M. (2012). Where should candidates learn to teach? Effects of field placement school characteristics on teacher retention and effectiveness. *Educational Evaluation and Policy Analysis, 34*, 3–26.

Ronfeldt, M. (2015). Field placement schools and instructional effectiveness. *Journal of Teacher Education, 66*, 304–320.

Ronfeldt, M., Brockman, S. L., & Campbell, S. L. (2018). Does cooperating teachers' instructional effectiveness improve preservice teachers' future performance? *Educational Researcher, 47*, 405–418. doi:10.31102/0013189X18782906

Ronfeldt, M., Goldhaber, D., Cowan, J., Bardelli, E., Johnson, J., & Tien, C. D. (2018). *Identifying promising clinical placement using administrative data: Preliminary results from ISTI placement initiative pilot* (CALDER Working Paper). Retrieved from https://caldercenter.org/sites/default/files/WP%20189.pdf

Ronfeldt, M., Matsko, K. K., Greene Nolan, H., & Reininger, M. (2018). *Who knows if our teachers are prepared? Three different perspectives on graduates' instructional readiness and the features of preservice preparation that predict them* (CEPA Working Paper No.18-01). Retrieved from Stanford Center for Education Policy Analysis: http://cepa.stanford.edu/wp18-01

Ronfeldt, M., Schwartz, N., & Jacob, B. (2014). Does pre-service preparation matter? Examining an old question in new ways. *Teachers College Record, 116*(10), 1–46.

SAS Institute. (2014). *Preliminary report: The impact of candidates on teacher value-added reporting*. Cary, NC: Author.

Shanks, R. (2017). Mentoring beginning teachers: Professional learning for mentees and mentors. *International Journal of Mentoring and Coaching in Education, 6*(3), 158–163.

Spencer, T. L. (2007). Cooperating teaching as a professional development activity. *Journal of Personnel Evaluation in Education, 20*, 211–226.

Steinberg, M. P., & Garrett, R. (2016). Classroom composition and measured teacher performance: What do teacher observation scores really measure? *Educational Evaluation and Policy Analysis, 38*, 293–317.

St. John, E., Goldhaber, D., Krieg, J., & Theobald, R. (2018). *How the match gets made: Exploring candidate placements across teacher education programs, districts, and schools* (CALDER Working Paper). Retrieved from https://caldercenter.org/sites/default/files/CALDER%20WP%20204-1018-1.pdf

Vosters, K. N., Guranio, C. M., & Wooldridge, J. M. (2018). Understanding and evaluating the SAS® EVAAS® univariate response model (URM) for measuring teacher effectiveness. *Economics of Education Review, 66*, 191-205. doi:10.1016/j.econedurev.2018.08.006

White, M. (2018). *Generalizability of observation instruments* (Unpublished doctoral dissertation). University of Michigan, Ann Arbor.

Whitehurst, G., Chingos, M. M., & Lindquist, K. M. (2014, May 13). *Evaluating teachers with classroom observations*. Washington, DC: Brown Center on Education Policy, Brookings Institute.

Manuscript received November 16, 2018
Final revision received August 7, 2019
Accepted August 8, 2019

# Standards for Educational and Psychological Testing
*Now Available in Spanish!*

The *Standards for Educational and Psychological Testing* represents the gold standard in guidance on testing in the United States and in many other countries. The *Standards* is a product of the American Educational Research Association, the American Psychological Association, and the National Council on Measurement in Education, and has been published collaboratively by the three organizations since 1966.

The current edition offers important updates in five major areas. It establishes fairness as a foundational concept in testing and highlights the need for accessibility for all examinees; incorporates a broader set of accountability issues for the use of tests in educational policy; represents more comprehensively the role of tests in the workplace; recognizes the expanding role of technology in testing; and features an improved structure for better communication of the standards. Each of the standards has been written for the professional and for the educated layperson and addresses professional and technical issues of test development and use in education, psychology, and employment. This book is a vitally important reference for professional test developers, researchers, sponsors, publishers, policymakers, employers, students, and other users in education and psychology.

ISBN 978-0-935302-35-6 (Paperback)
ISBN: 978-0935302-41-7 (eBook)

**Print-Only and eBook-Only Pricing**
Member Price: $49.95
Non-Member and Institutional* Price: $69.95

**Print + eBook Bundle Pricing**
Member Price: $59.95
Non-Member Price: $79.95

*eBooks not available for institutional use.

**AERA**
FOUNDED 1916
AMERICAN EDUCATIONAL RESEARCH ASSOCIATION

eBook available from ebooks.aera.net
Print + eBook Bundle available from www.aera.net

**ORDER YOUR COPY TODAY**

# Review of Research in Education

**AERA** — AMERICAN EDUCATIONAL RESEARCH ASSOCIATION — FOUNDED 1916

## Volume 44: Emergent Approaches for Education Research: What Counts as Innovative Educational Knowledge and What Education Research Counts?

**Edited by:**
**Jeanne M. Powers, Gustavo E. Fischman,** and **Margarita Pivovarova** all at *Arizona State University, USA*

The 2020 volume of *RRE*, "Emergent Approaches for Education Research," publishes original research syntheses that further our understanding of how innovative methodological approaches can provide more comprehensive explanations in response to enduring questions, challenge existing theoretical frameworks and paradigms, and address novel challenges in the field of education.

### AERA Members
AERA members can order *Review of Research in Education, Volume 44* as a benefit of membership or at the members-only price of $15.

www.aera.net

### Institutions and Individuals
**Subscriptions**
ISSN: 0091-732X
Individuals: $74.00
Volume 44
(Print volume + online access back to 1999)
Institutions: $461.00

**Single Print Copies**
ISBN: 9781071818862
Individuals: $74.00
Volume 44 (print only)
800-818-7243

### Authors
J. W. Hammond
Pamela A. Moss
Minh Q. Huynh
Carl Lagoze
Heela Goren
Miri Yemini
Claire Maxwell
Efrat Blumenfeld-Lieberthal
Alisha Butler
Kristin A. Sinclair
Casey D. Cobb
Christian Fischer
Zachary A. Pardos
Ryan Shaun Baker
Joseph Jay Williams
Padhraic Smyth
Renzhe Yu
Stefan Slater
Rachel Baker
Mark Warschauer
Judith L. Green
W. Douglas Baker
Monaliza Maximo Chian
Carmen Vanderhoof
LeeAnna Hooper
Gregory J. Kelly
Audra Skukauskaite
Melinda Z. Kalainoff
Thomas M. Philip
Ayush Gupta
Maithreyi Gopalan
Kelly Rosinger
Jee Bin Ahn
Dominik E. Froehlich
Sara Van Waes
Hannah Schäfer
Richard Miller
Katrina Liu
Arnetha F. Ball
Samantha Viano
Dominique J. Baker
Sebnem Cilesiz
Thomas Greckhamer
Rachel E. Schachter
Donald Freeman
Naivedya Parakkal
Maxwell M. Yurkofsky
Amelia J. Peterson
Jal D. Mehta
Rebecca Horwitz-Willis
Kim M. Frumin

**SAGE Publishing**

journals.sagepub.com/home/rre

# New from TC Press

**Excluded by Choice: Urban Students with Disabilities in the Education Marketplace**
978-0-8077-6400-8
Federico R. Waitoller

**Trauma Doesn't Stop at the School Door: Strategies and Solutions for Educators, PreK–COLLEGE**
978-0-8077-6410-7
Karen Gross

**The Art of Reflective Teaching: Practicing Presence**
978-0-8077-6364-3
Carol R. Rodgers

**Learning to Teach in an Era of Privatization: Global Trends in Teacher Preparation**
978-0-8077-6159-5
Christopher A. Lubienski
T. Jameson Brewer
EDITORS

**Holler If You Hear Me, Comic Edition**
978-0-8077-6325-4
Gregory Michie
Ryan Alexander-Tanner

**Transformative Ethnic Studies in Schools: Curriculum, Pedagogy, & Research**
978-0-8077-6345-2
Christine E. Sleeter
Miguel Zavala

**Letting Go of Literary Whiteness: Antiracist Literature Instruction for White Students**
978-0-8077-6305-6
Carlin Borsheim-Black
Sophia Tatiana Sarigianides

**Race, Justice, and Activism in Literacy Instruction**
978-0-8077-6321-6
Valerie Kinloch
Tanja Burkhard
Carlotta Penn
EDITORS

**Making School Integration Work: Lessons from Morris**
978-0-8077-6362-9
Paul Tractenberg
Allison Roda
Ryan Coughlan
Deirdre Dougherty

## Teachers College Press
TEACHERS COLLEGE | COLUMBIA UNIVERSITY